Crisis and Conflict in Agriculture

Crisis and Conflict in Agriculture

Edited by

Rami Zurayk

*Faculty of Agricultural and Food Sciences,
American University of Beirut, Lebanon*

Eckart Woertz

*CIDOB (Barcelona Centre for International Affairs), Spain
and
Scientific Advisor to the Kuwait Chair at Sciences Po, Paris, France*

Rachel Bahn

*Faculty of Agricultural and Food Sciences,
American University of Beirut, Lebanon*

CABI is a trading name of CAB International

CABI
Nosworthy Way
Wallingford
Oxfordshire OX10 8DE
UK

CABI
745 Atlantic Avenue
8th Floor
Boston, MA 02111
USA

Tel: +44 (0)1491 832111
Fax: +44 (0)1491 833508
E-mail: info@cabi.org
Website: www.cabi.org

Tel: +1 (617)682-9015
E-mail: cabi-nao@cabi.org

© CAB International 2018. All rights reserved. No part of this publication may be reproduced in any form or by any means, electronically, mechanically, by photocopying, recording or otherwise, without the prior permission of the copyright owners.

A catalogue record for this book is available from the British Library, London, UK.

Library of Congress Cataloging-in-Publication Data

Names: Zurayk, Rami, editor. | Woertz, Eckart, editor. | Bahn, Rachel (Rachel Anne), editor.
Title: Crisis and conflict in agriculture / edited by Rami Zurayk, Eckart Woertz, Rachel Bahn.
Description: Boston, MA : CABI, 2018. | Includes bibliographical references and index.
Identifiers: LCCN 2018015198| ISBN 9781786393647 (hardback) | ISBN 9781786393654 (epdf)
Subjects: LCSH: Agriculture and politics--Developing countries. | Rural development--Developing countries. | Crops and climate--Developing countries.
Classification: LCC HD1417 .C75 2018 | DDC 338.1/8091724--dc23 LC record available at https://lccn.loc.gov/2018015198

ISBN-13: 9781786393647 (hbk)
         9781786393654 (PDF)
         9781786393661 (ePub)

Commissioning editor: David Hemming
Editorial assistant: Alexandra Lainsbury
Production editor: James Bishop

Typeset by SPi, Pondicherry, India
Printed and bound by CPI Group (UK) Ltd, Croydon, CR0 4YY

Contents

List of Contributors	vii
Acknowledgements	xiii
Introduction to the Volume	xv

Part 1   Theoretical Exploration of and Methodological Approaches to Agriculture, Crisis and Conflict — 1

1. **Agriculture, Conflict and the Agrarian Question in the 21st Century** — 3
   Rachel Bahn and Rami Zurayk

2. **Geopolitics, Food and Agriculture** — 28
   Eckart Woertz

3. **Climate Change and Conflict: Agriculture, Migration and Institutions** — 40
   Martin Smidt and Ole Magnus Theisen

4. **Water, Agriculture and Conflict: Global, National and Local Analysis of Conflict in MENA, sub-Saharan Africa and the United States** — 53
   Martin Keulertz

5. **Illegal Drug Plant Cultivation and Armed Conflicts: Case Studies from Asia and Northern Africa** — 64
   Pierre-Arnaud Chouvy

6. **Remote Sensing and GIS-based Technologies for Assessing the Impact of Conflict on Agricultural Production** — 73
   Hadi Jaafar

Part 2   Case Studies on Agriculture, Crisis and Conflict — 89

Middle East and North Africa

7. **The 'Arab Spring' in North Africa: Egypt and Tunisia** — 91
   Ray Bush

8	**Degraded Capital Formation: the Achilles' Heel of Syria's Agriculture** *Linda Matar*	105
9	**Crisis and Agricultural Change in the Kurdistan Region of Iraq, 1980s–2010s: an Interdisciplinary Approach** *Lina Eklund and Katharina Lange*	118
10	**Yemen's Agricultural World: Crisis and Prospects** *Max Ajl*	131
11	**Farming for Freedom: the Shackled Palestinian Agricultural Sector** *Alaa Tartir*	144

Pastoral Belts in Africa and Central Asia

12	**Games without Frontiers: Development, Crisis and Conflict in the African Agro-Pastoral Belt** *Michele Nori and Edoardo Baldaro*	157
13	**Border Change and Conflict in Central Asia: the Case of Agro-Pastoral Communities in Cross-Border Areas of the Ferghana Valley** *Asel Murzakulova and Irène Mestre*	176

South and Southeast Asia

14	**Conflict and Resistance in Southern Punjab: a Political Ecology of the 2010 Floods in Pakistan** *Ali Nobil Ahmad*	190
15	**India: Rural Roots of Naxalite–Maoist Insurgency** *Archana Prasad*	203
16	**Agrarian Transition, Adaptation and Contained Conflict in Cambodia and Vietnam since the 1990s** *Christophe Gironde*	214

Latin America

17	**Beyond Displacement by Armed Conflict: the Relationship Between Environmental, Economic and Armed Displacement in Colombia** *Carolina Castro Osorio and Edinso Culma*	232
18	**Prior Consultation and the Defence of Indigenous Lands in Latin America** *Marcela Torres Wong*	246
19	**The Political Mediation of Indigenous Land Conflicts in Argentina** *Matthias vom Hau*	261

Europe

20	**The Role of Land Reform in Rural Development: Promoting Productivity or Democracy?** *Matthew Hoffman*	275

Index   287

List of Contributors

Ali Nobil Ahmad is a Fellow at the Leibniz-Zentrum Moderner Orient, Berlin. His research focuses on the politics of resources in Pakistan's peripheries. He was previously Visiting Professor of South Asian Studies at Brandeis University, Assistant Professor of History at the Lahore University of Management Sciences, and a recipient of the Scott Trust Bursary for journalists. In 2017, he curated 'anthropoSCENE', a festival of films and talks on climate justice sponsored by the Rosa Luxembourg Foundation, and has made two short documentaries on the political ecology of Southern Punjab. He received his PhD from the European University Institute in Florence.
Contact: Leibniz-Zentrum Moderner Orient, Kirchweg 33, 14129 Berlin, Germany. Email: alinobil@googlemail.com or ali.nobilahmad@zmo.de

Max Ajl is a doctoral student in the Department of Development Sociology at Cornell University. His dissertation is on colonial underdevelopment, the transition to independence, and post-colonial planning in Tunisia, especially in the rural sector. His writings have been published in *Review of African Political Economy*, *Middle East Report*, and *Journal of Peasant Studies*, amongst other academic and non-academic fora. He is an editor at *Jadaliyya* and *Viewpoint*.
Contact: Cornell University, Department of Development Sociology, 240 Warren Hall, 137 Reservoir Road, Ithaca, NY 14853, United States. Email: msa95@cornell.edu

Rachel Bahn is the coordinator of the Food Security Program and an instructor of agribusiness at the American University of Beirut. Her research interests include food security and food systems, and economic policies and programming in developing country contexts. Previously based in Washington, DC, she served as an economist with the United States Department of Treasury and the United States Agency for International Development. She holds degrees from the Johns Hopkins University School of Advanced International Studies (SAIS) and from Saint Joseph's University in Philadelphia, Pennsylvania.
Contact: Rachel Bahn, American University of Beirut, Faculty of Agricultural and Food Sciences, PO Box 11-0236, Riad El Solh 1107 2020, Beirut, Lebanon. Email: rb89@aub.edu.lb

Edoardo Baldaro is a research fellow in International Relations at the University of Naples 'L'Orientale' - Department of Human and Social Sciences. He holds a PhD in Political Sciences from the Scuola Normale Superiore - Institute of Human and Social Sciences (Florence). His main research interests are critical security studies, foreign policy analysis (strategies and decision-making), North–South relations, African politics, region-building processes, statehood and authority in Africa. He is currently affiliated to the research project 'STREETPOL: Participatory challenges from Tunisia to Oman' and the Scientific Independence of Young Researchers (SIR) program funded by the Italian Ministry for Higher Education and Research.

Contact: University of Naples 'L'Orientale', Department of Human and Social Sciences, Largo San Giovanni Maggiore 30, 80134 Naples, Italy. Email: edoardo.baldaro@sns.it

Ray Bush is Professor of African Studies and Development Politics at the University of Leeds. He works in and on Africa and the Near East with particular focus on the political economy of rural development, economic transformation and alternative development strategies. He is a member of the editorial working group of the *Review of African Political Economy* and edits the Critical Agrarian Studies page for www.roape.net.
Contact: School of Politics and International Studies (POLIS), University of Leeds. Leeds LS2 9JT, United Kingdom. Email: r.c.bush@leeds.ac.uk

Carolina Castro Osorio holds a Master's in sociology from the National University of Colombia. She was a researcher at the National Center for Historical Memory and the Office of the High Commissioner for Peace. Her work has focused on research in culture, both in the aesthetic and anthropological senses. She has worked as faculty chair of public policy on culture in the Universidad Externado de Colombia. She belongs to the research group of environmental history at National University of Colombia, and is currently a researcher at the Secretary of Culture in Bogotá.
Contact: Secretary of Culture, Carrera 8 No. 8-55, Bogotá, Colombia. Email: tantalia16@gmail.com

Pierre-Arnaud Chouvy, PhD, is a geographer and research fellow at CNRS-Prodig, Paris, France. He is the author of *Opium. Uncovering the Politics of the Poppy* (2009/2010, London/Cambridge: I.B. Tauris/Harvard University Press) and numerous articles on both illegal opium production and cannabis cultivation in the world. His work can be consulted at www.geopium.org.
Contact: Centre national de la recherche scientifique (CNRS), PRODIG Research team, 2, rue Valette, 75005 Paris, France. Email: pachouvy@geopium.org

Edinso Culma is a sociologist and holds a Master's in anthropology. His work and academic interests include research and intervention projects related to political violence; the internal Colombian armed conflict; drug-trafficking and the study of socio-economic and cultural issues in the Amazon region. He has worked for the IOM-ICBF by accompanying families and rural communities from the municipality of San Miguel (Putumayo) with the objective of seeking strategies that counter the illegal recruitment of children and youth to armed groups; with the Murui of Putumayo and Caquetá in the process of elaborating an Ethnic Safeguarding Plan; and is currently researcher for the National Center for Historical Memory.
Contact: Cra 31a #25a-44, Bogotá, Colombia. Email: edmoo75@gmail.com

Lina Eklund is a Physical Geography researcher working at the Center for Middle Eastern Studies (CMES) at Lund University and at the Department of Planning at Aalborg University. For her PhD she conducted research on the link between people and land in Iraqi Kurdistan. Her current research is placed within the fields of land system science and environmental security. With the use of satellite images and spatial methods she explores the potential connections between drought, migration, land use, and conflict in Syria and Iraq.
Contact: Lund University, Center for Middle Eastern Studies (CMES), Box 201, 221 00, Lund, Sweden. Email: lina.eklund@cme.lu.se or linae@plan.aau.dk

Christophe Gironde is a political economist, currently working as a senior lecturer at the Graduate Institute of International and Development Studies (IHEID) in Geneva. His main domains of interest are agrarian change and human development. He has extensive field research experience in rural Vietnam and Cambodia, and previously in the Republic of Congo and Burundi. Before joining IHEID in 2004, he worked four years as a researcher in Norway (Fafo, International Studies) and lived two years in Vietnam (1996–97) in the frame of his doctoral thesis. He is currently working on the process of land commercialization and their consequences on rural livelihoods in Cambodia and Ghana.
Contact: The Graduate Institute of International and Development Studies, Case Postale 1672, 1211 Genève 1, Switzerland. Email: christophe.gironde@graduateinstitute.ch

Matthew Hoffman is Assistant Professor in the Department of Sociology and Economics at the University of Southern Maine and a visiting scholar at RURALIS Institute for Rural and Regional Research in Trondheim, Norway. Prof. Hoffman received his MS and PhD in Rural Sociology from

Cornell University. His research focuses on property rights, natural resource governance, and institutions for collaborative decision making in landscapes fragmented by private property. At the University of Southern Maine he teaches courses on food systems and social justice in the Food Studies Program.

Contact: RURALIS – Institute for Rural and Regional Research, Universitetssenteret Dragvoll, Loholt Alle 81 (4. etg), N-7049 Trondheim, Norway. Email: mdh32@cornell.edu

Hadi Jaafar has international and Middle Eastern expertise in hydrology, irrigation engineering and water resources management; he has managed water rights adjudication projects for government agencies in the United States and worked on several strategic water resources and irrigation projects in many countries in the Middle East. He is currently an Assistant Professor of Irrigation and Water Management at the Department of Agriculture at the American University of Beirut, with research interests focusing on remote sensing and GIS applications in agriculture, water and food security.

Contact: American University of Beirut, Faculty of Agricultural and Food Sciences, PO Box 11-0236, Riad El Solh 1107 2020, Beirut, Lebanon. Email: hj01@aub.edu.lb

Martin Keulertz is Assistant Professor of food security at the American University of Beirut. He previously worked as a post-doctoral research fellow at Purdue University (USA) and Humboldt University Berlin (Germany). He obtained his PhD at King's College London (UK) in 2013 with a thesis on foreign direct investment in Sudanese agriculture by Jordan and Qatar. He earned an MSc in Middle East Politics at the School of Oriental and African Studies (University of London) and a BA in Political and Social Sciences at the University of Wales, Bangor. Martin's research interests centre around the water-food-energy nexus with a particular focus on the Arab world, North America and Sub-Sahara Africa. He has published on the global political economy of water and food. He is an associate editor of the journal *Food Security* (Springer Link) and editor of the *Handbook of Water, Food and Society* (OUP, 2017).

Contact: American University of Beirut, Faculty of Agricultural and Food Sciences, PO Box 11-0236, Riad El Solh 1107 2020, Beirut, Lebanon. Email: mk219@aub.edu.lb

Katharina Lange is a research fellow at the Leibniz-Zentrum Moderner Orient (ZMO) in Berlin where she heads an interdisciplinary group of researchers investigating *The Politics of Resources*. An anthropologist by training, she has conducted extensive fieldwork in Syria and, more recently, in the Kurdistan Region of Iraq. Her current research project '(Re)valuations of Land in the Kurdistan Region, Iraq' is an ethnographic exploration of shifting valuations of land in the Kurdistan Region, using biographical and family history interviews as well as participant observation.

Contact: Leibniz-Zentrum Moderner Orient, Kirchweg 33, 14129 Berlin, Germany. Email: katharina.lange@zmo.de

Linda Matar is a Senior Research Fellow at National University of Singapore's Middle East Institute. She is also a Lecturer at National University of Singapore's College of Alice and Peter Tan. Her research and teaching involve the political economy and economic development of the Arab Near East and Southeast Asia. She is the author of *The Political Economy of Investment in Syria* (Palgrave Macmillan, 2016). She obtained her PhD in Economics from the School of Oriental and African Studies (SOAS), University of London.

Contact: National University of Singapore (NUS), 29 Heng Mui Keng Terrace, Block B, unit 06-06, Singapore 119620. Email: linda@nus.edu.sg

Irène Mestre is a geographer from Jean Moulin-Lyon III University, Research Unit UMR 5600 Environnement-Ville-Société (Lyon, France). She is also affiliated to the French Institute of Research on Central Asia (Bishkek, Kyrgyzstan). Her main fields of expertise are pasture management and conflicts over natural resources.

Contact: Université Jean Moulin-Lyon 3, Research Unit UMR 5600 Environnement Ville Société, 18 rue Chevreul, 69007 Lyon, France. Email: irene_mestre@hotmail.com

Asel Murzakulova is a Senior Research Fellow with the Mountain Societies Research Institute of the University of Central Asia (Bishkek, Kyrgyzstan) and co-founder of the analytical club 'Mongu'. She is involved in research on the management of natural resources, conflicts and the impact of migration on agriculture.

Contact: University of Central Asia, Mountain Sciences Research Institute, 138 Toktogul Street, 720001 Bishkek, Kyrgyz Republic. Email: asel.murzakulova@ucentralasia.org

Michele Nori is a tropical agronomist with a specialization in rural sociology and specific expertise on the livelihood systems of pastoral communities. Through interfacing field practices, academic research and policy making levels, he has developed a horizontal career based on more than 20 years' collaborations in different regions of the globe with a number of organizations. His publications range from scientific papers to technical notes to advocacy documents on matters related to sustainable agro-pastoral livelihoods and rural development. He is currently involved with Ian Scoones of the Institute of Development Studies in elaborating and developing a project that links pastoralism to our society in a provocative way – *can we learn from pastoralists in managing risks, living with uncertainties and building resilience* (www.pastres.org)?
Contact: Robert Schuman Centre, European University Institute (EUI), Via Boccaccio 121. 50133 Firenze, Italy. Email: michele.nori@eui.eu

Archana Prasad is Professor at Centre for Informal Sector and Labour Studies, Jawaharlal Nehru University, Delhi. She is involved with several grassroots and working class movements and has served on several government committees concerning women's and adivasi issues. She has also published several books, scholarly and popular articles on a wide range of subjects related to adivasi and women's issues. Her latest book entitled *The Red Flag of the Warlis: History of An Ongoing Struggle* (Leftword, 2017) is based on an oral history of a grassroots movement.
Contact: Centre for Informal Sector and Labour Studies, Jawaharlal Nehru University
New Delhi 110067, India. Email: archie.prasad11@gmail.com or archanaprasad@mail.jnu.ac.in

Martin Smidt is a PhD-candidate at the Norwegian University of Science and Technology (NTNU). His thesis is on the topic of climate change, institutions and conflict. Working mainly with quantitative data, including GIS, he examines links between climatic shocks, vegetation health, food production and violent conflict.
Contact: Norwegian University of Science and Technology (NTNU), Department of Sociology and Political Science, Dragvoll, Building 9, Level 5, 7491 Trondheim, Norway. Email: martin.smidt@ntnu.no

Alaa Tartir is a research associate at the Centre on Conflict, Development and Peacebuilding, and a visiting fellow at the Department of Anthropology and Sociology, at the Graduate Institute of International and Development Studies in Geneva, Switzerland. He is also a programme advisor at Al-Shabaka: The Palestinian Policy Network, and a visiting professor at Paris School of International Affairs, Sciences Po. Amongst other positions, Tartir was a post-doctoral fellow at The Geneva Centre for Security Policy, a visiting scholar at Utrecht University, and a researcher in international development studies at the London School of Economics (LSE), where he earned his PhD.
Contact: The Graduate Institute of International and Development Studies (IHEID),
Centre on Conflict, Development and Peacebuilding (CCDP), Maison de la Paix, Chemin Eugène-Rigot 2A, Petal 2, 8th Floor, Case Postale 1672, 1211 Geneva 1, Switzerland.
Email: alaa.tartir@graduateinstitute.ch

Ole Magnus Theisen is Associate Professor of Political Science at the Norwegian University of Science and Technology (NTNU) and External Associate with the Peace Research Institute Oslo (PRIO). His current research focus is on the role of institutions in resolving disputes and containing violence as well as intermediate linkages between climate anomalies and violence, such as the individual support for violence subject to resource shocks and mobility patterns after resource shocks. He is currently involved in projects funded by the European Commission and the Research Council of Norway. He has published in journals such as *Climatic Change, Environmental Research Letters, International Security, Journal of Peace Research* and others.
Contact: Norwegian University of Science and Technology (NTNU), Department of Sociology and Political Science, Dragvoll, Building 9, Level 5, 7491 Trondheim, Norway. Email: ole.magnus.theisen@ntnu.no

Marcela Torres Wong has a PhD in Political Science by American University in Washington DC. Since 2017, she works as a fulltime professor and researcher in the Latin American Faculty of Social Science (FLACSO) in Mexico City. Her research interests include indigenous movements, extractive industries and socio-environmental conflicts. She has received grants from the Inter-American Foundation (IAF), the Tinker Foundation, the Pontificia Universidad Catolica del Peru and the Centre National de la Recherche Scientifique (CNRS).
Contact: Latin American Faculty of Social Sciences (FLACSO-México), Carr. Picacho-Ajusco 377, Héroes de Padierna, C.P. 14200, Ciudad de México, CDMX, Mexico. Email: marcela.torres@flacso.edu.mx or mt6112b@student.american.edu

Matthias vom Hau is an associate professor at the Institut Barcelona d'Estudis Internacionals (IBEI). His work is centrally concerned with the relationship between identity politics, institutions and development, with a regional focus on Latin America. He has published widely on how states construct a sense of national belonging, and how non-state actors negotiate official nationalisms. Another line of research challenges the fundamentally ahistorical approach that underpins the supposedly negative relationship between ethnic diversity and public goods provision. His most recent research investigates the consequences of indigenous movements in Latin America and beyond, with a particular focus on land conflict.
Contact: Institut Barcelona d'Estudis Internacionals (IBEI), Carrer de Ramon Trias Fargas, 25–27, 08005 Barcelona, Spain. Email: mvomhau@ibei.org

Eckart Woertz is senior research fellow at CIDOB, the Barcelona Centre for International Affairs, scientific advisor to the Kuwait Chair at Sciences Po in Paris and teaches at the Barcelona Institute of International Studies (IBEI). Formerly he was a visiting fellow at Princeton University, director of economic studies at the Gulf Research Center in Dubai and worked for banks in Germany and the United Arab Emirates. He is author of *Oil for Food* (Oxford University Press, 2013) and has published numerous journal articles on development issues in the Middle East. He is on the editorial boards of *Food Security* and the *Journal of Arabian Studies* and holds a PhD in economics from Friedrich-Alexander University, Erlangen-Nuremberg.
Contact: c/ Elisabets, 12 08001 Barcelona, Spain. Email: ewoertz@cidob.org

Rami Zurayk (DPhil, Oxon) is a professor at the Faculty of Agricultural and Food Sciences at the American University of Beirut. He is a member of the Steering Committee of the High-Level Panel of Experts on Food Security and Nutrition (HLPE) of the Committee of World Food Security. His work addresses the political ecology of Arab food security and its linkages with the agrarian question. His latest work on the subject includes *Control Food, Control People: The Struggle for Food Security in Gaza* (IPS, 2013) and *The Arab Uprisings Through an Agrarian Lens* (Palgrave, 2016).
Contact: American University of Beirut, Faculty of Agricultural and Food Sciences, PO Box 11-0236, Riad El Solh 1107 2020, Beirut, Lebanon. Email: rzurayk@aub.edu.lb

Acknowledgements

The editors wish to thank Kuwait Foundation for the Advancement of Sciences (KFAS) and the Kuwait Program at Sciences Po for their generous financial support of a conference conducted in March 2017, at which selected chapters from this volume were presented and discussed.

Introduction to the Volume

Rami Zurayk,[1] Eckart Woertz[2] and Rachel Bahn[3],*

[1]*Department of Landscape Design and Ecosystem Management, American University of Beirut, Beirut, Lebanon;* [2]*CIDOB (Barcelona Centre for International Affairs), Spain and Scientific Advisor to the Kuwait Chair at Sciences Po, Paris, France;* [3]*Department of Agriculture and Food Security Program, American University of Beirut, Beirut, Lebanon*

The interactions between agriculture, crisis and conflict may be of several varieties. Agriculture can be a cause of crisis or conflict, for example, by serving as a source of dispute. Agriculture can provide the funds that sustain crisis or conflict, by generating the commodity goods that are marketed to finance military operations. Agriculture may also be a victim of crisis or conflict. Rural areas, rural populations and agricultural systems may be the disproportionate focus of armed conflict, particularly civil conflicts. The effects of conflict may include physical damage; destruction of assets, incomes and livelihoods; and generalized food insecurity. The interactions of agriculture, crisis and conflict are hardly new, and date to the invention of agriculture. Nevertheless, our knowledge about these interactions is partial, and expanding this knowledge is the motivation for this volume.

Conflict has been with agriculture since its inception. Both have influenced each other in a dialectic relationship. Early city-states relied on slavery and other forms of coerced labour to produce their grain supplies, and that labour was often obtained through armed campaigns in surrounding areas that relied on pastoral lifestyles, hunting and shifting cultivation (Scott, 2017). This structural relationship with surrounding areas meant conflict and state interference in agriculture, yet irrigation management in these states was less centralized than Wittfogel's notion of 'hydraulic societies' suggests. It involved decentralized agency and knowledge on the part of cultivators. Whether it was the Egyptian pharaohs or later Ottoman sultans and Sudanese modernization regimes, top-down administration of the Nile waters, for example, was more chimera than reality, and was also accompanied by social and political bargains (Hassan, 1997; Verhoeven, 2011; Mikhail, 2014).

In more modern times agriculture has been at the heart of social conflict over land and resource allocation. The crisis of feudalism and the transition to increased commodification and market capitalism was closely linked to issues of agriculture and class-based conflict (Aston and Philpin, 1985). Marx described the early beginnings of capitalism as a process of 'primitive accumulation' that was driven by colonial plunder and fed on coercive forms of labour control – most notably plantation slavery in the Americas, which was imposed in the non-European periphery by European conquistadors from the 16th century onwards. Later, Karl Kautsky would put the agrarian question at the centre of his analysis of capitalist development in the 19th century, defining it as the capitalist transformation of agriculture, and analysing its impacts on society and its potential to trigger

---

* Email: rb89@aub.edu.lb

conflict among social classes (McLaughlin, 1998). Smallholder agriculture in Europe showed considerable staying power and was often integrated into modern economies as a producer of subsistence for commodified labour power working outside the agricultural sector. However, the transition process led to the relative decline of such smallholders at the expense of larger, mechanized farms. Agricultural labour displacement and the resulting rural migration flows tested the absorption capacities of cities.

The expansion of market-oriented agriculture also eyed colonies abroad. Land grabbing was not only essential for expanding markets and procuring raw materials. Indeed, the colonialist Cecil Rhodes regarded imperialism as a necessity to rid his mother country of its excess population and to prevent civil war at home (Weis, 2007, p. 51). The result of such views and policies was the destruction of socio-economic structures in colonies and a conflict-ridden process of land tenure change that were instrumental in the famines of the 'Late Victorian Holocausts' during the latter half of the 19th century (Davis, 2001). By the mid-20th century, concentration of land ownership in the global south was widely perceived as a hindrance to economic development and a major reason for social conflict in post-colonial struggles: between 1945 and 1950 nearly half of the global population lived in countries that were undertaking land reform (Hobsbawm, 1994, p. 354ff).

If the expansion of market-oriented agriculture entailed crisis and conflict, the track record of communist regimes that pursued state-led forms of commodity production and development was even worse. Forced collectivization, misguided development experiments and the deliberate elimination of political opponents via engineered famines cost the lives of millions in the Stalinist Soviet Union in the early 1930s and during Mao's Great Leap Forward in China (1958–1962) (Dikötter, 2010; Snyder, 2010).

The post-colonial era in the former colonies was marked both by land reform and by the pursuit of economic development through import-substituting industrialization. These development models were in crisis by the 1980s and the resulting structural adjustment measures led again to crisis and considerable conflict within agrarian relations, as Ray Bush points out in this volume. Echoes of the colonial past resurfaced in the global land grab debate, when foreign investors from Asia, the Gulf countries and western financial institutions engaged in land acquisitions in often food-insecure developing countries in the wake of the global food crisis of 2007/2008 (Cotula et al., 2009; Deininger et al., 2011; Woertz, 2013).

More recently, the environment and climate change have been associated with conflict through agriculture. The Syrian civil war and the crisis in Darfur have been attributed – at least partially – to increased drought and environmental stress factors, which significantly affect agricultural communities (Mazo, 2010; Kelley et al., 2015). Such neo-Malthusian tales have met with criticism, questioning their empirical evidence and explanatory power in light of political economy issues that contributed to such conflicts (Verhoeven, 2011; de Chatel, 2014; Bromwich, 2015; Fröhlich, 2016; Selby et al., 2017). Nevertheless, popular narratives about the path from climate change to conflict via agriculture persist.

Beyond state-building and conflicts over land tenure, agriculture and food supplies have also been crucial for warfare, whether during the time of the Roman Empire or the two World Wars of the 20th century (Erdkamp, 2005; Collingham, 2011). Revenues from agriculture have also played a role in financing violent non-state actors in a number of countries, such as warlords in Afghanistan, the Islamic State in Iraq and Syria (ISIS) or various factions in Myanmar's (Burma's) civil war (see Chouvy, Chapter 5, this volume; Jaafar and Woertz, 2016, Eklund et al., 2017).

Thus, agriculture and conflict have historically been connected in a myriad of ways, ranging from state-building to development, class-based conflicts, warfare and funding of violent non-state actors. The dynamics have played out in domestic and international settings alike, but what do we make out of these connections? Can they be treated in a unified theoretical approach in any meaningful way?

Efforts to understand conflict have attempted to identify causes, proximate and distant, for its onset, duration, intensity and outcome. Conflict in its various forms has been the focus of study and debate, ranging from local conflict to revolution to war between states. These efforts at understanding date back at least to the 19th century, while the field of peace and conflict studies emerged as an academic discipline in the mid-20th century. With the end of the Cold War, attention shifted from international conflict to the increasing incidence of civil conflicts in recent decades. The availability of quantitative modelling techniques and new datasets allowed for a number of studies exploring the linkages between conflict (particularly civil conflict) and factors such as: geography and proximity to conflict; levels of economic development, economic incentives including inequality and international economic relationships (Collier and Hoeffler, 1998; 2004; Collier et al., 2003; Martin et al., 2008); the presence or abundance of natural resources (Ross, 2004), their absence or degradation (Maxwell and Reuveny, 2000) or their management (Cramer, 2006); ethnic fragmentation and social relationships; a history of earlier conflict; and political regimes and governance structures (Hegre et al., 2001; Brückner and Ciccone, 2007). The contemporary debate has not decisively proven any single theory of conflict. Rather, there has been a general acknowledgement that conflict, which is diverse in its forms, has no single cause and therefore no simple explanation (World Bank, 2011; FAO, 2016).

A sub-set of the conflict scholarship has explored conflict and agriculture. As noted above, the agrarian question perceives the relationship between conflict and agriculture through the potential for transformation of the agricultural sector to contribute to class-based conflict. In recent studies of the causes of civil conflict, agriculture has been implicated, particularly when considering factors such as economic development, natural resources (or the lack thereof) (Midlarsky, 1988; Cotula, 2013; Breisinger et al., 2015) or environmental shocks in more agricultural societies (Brinkman and Hendrix, 2011; Harris et al., 2013; von Uexkull et al., 2016). The persistence of civil conflict into the 21th century within developing countries has led to calls for more robust exploration of the relationship between agriculture and conflict (Cramer and Richards, 2011), and of conflict outside the urban context (Kalyvas, 2004). Recent conflict experience has informed a deeper understanding of the effect of conflict in damaging or reshaping agriculture and in contributing to food insecurity, as well as the cyclical relationship between conflict and agriculture (FAO, 2010; 2016; FAO et al., 2017).

This volume sets out to explore the dialectic relating agriculture, crisis and conflict, and attempts to expand our knowledge of these interactions. Part 1 of the volume discusses thematic issues and methodological approaches to understanding the intersection of agriculture, crisis and conflict. Part 2 provides case studies that take a detailed approach to understanding agricultural contexts facing crisis and conflict, or the role played by agriculture within crisis and conflict.

Rachel Bahn and Rami Zurayk (Chapter 1) attempt to establish a unified construct, bringing together theories of conflict with theoretical approaches to agrarian transformation, as understood through the agrarian question. Through an overview of current knowledge on the intersection of agriculture and conflict, the authors first emphasize the complexity of interactions, which can move in multiple directions. In the absence of a singular theoretical framework linking agriculture and conflict – theories of conflict have been broader, not focused solely on agriculture, moreover no single theory of conflict appears to have gained widespread acceptance – they find the agrarian question helps to illustrate these interactions but does not act as a definitive or compelling theory of conflict per se: agrarian transformation and the social reorganization associated with it do not necessarily lead to widespread violent conflict, whose onset and persistence is in many cases at least partially determined by additional factors such as historical experience, social attitudes and local institutions.

Eckart Woertz (Chapter 2) looks beyond domestic development issues to the geopolitical dynamics of food supplies. He analyses the historic importance of agricultural production and trade for urbanization, state-building and warfare, looking at examples from ancient times and various food regimes since the 19th century. With the 2008 global food crisis and contemporary debate over land grabs in sub-Saharan Africa this geopolitical dimension of food and agriculture has gained renewed interest. Woertz predicts that the importance of food trade will increase in the 21st century, in response to continued urbanization and shifting agricultural production patterns.

Martin Smidt and Ole Magnus Theisen (Chapter 3) discuss suggested linkages between climate change and conflict, after exploring the social effects of climate change and offering an overview of the quantitative literature. The authors focus on the role of migration and of institutions as factors intermediating the relationship between climate and conflict. They note limitations to previous quantitative climate–conflict analyses and provide recommendations to integrate insights from qualitative studies into such analyses in the future.

Martin Keulertz (Chapter 4) explores conflict over water at three levels – global, national and sub-national. He demonstrates that conflict thus far is primarily experienced at the sub-national level between individual actors, for instance, farmers. In contrast, inter-state conflict over water has been often predicted, but has hardly ever materialized because of the ability of virtual water trade to distribute scarce water resources in a broadly acceptable manner. In light of water scarcity and climate change, Keulertz anticipates difficult decisions, namely whether to withdraw from water-intensive agriculture in selected areas of the global south or to experience greater sub-national conflict over water resources into the future.

Pierre-Arnaud Chouvy (Chapter 5) explores the special case of illicit crops in relation to armed conflict. He looks to examples from Asia and North Africa to understand the systemic relationships that exist between civil war economies and the economies built around illegal agricultural production and trade.

Turning to methodological approaches, **Hadi Jaafar** (Chapter 6) utilizes geographic information systems (GIS) and vegetation-related remotely sensed indices to detect the effects of conflict on agricultural production in rural areas. Using data from Syria, Jaafar then illustrates methods to derive crop yields from remote sensing data while correcting for natural variations in weather. His work provides one example of the direct impact of conflict in limiting production, including as a result of abandonment of agricultural lands.

Part 2: Case Studies on Agriculture, Crisis and Conflict brings a diversity of case studies across geographic regions. Studies are selected from areas that might be expected to feature in such a volume – the Middle East and North Africa, sub-Saharan Africa, and South and Southeast Asia – as well as less obvious regions where conflict within agriculture refers not to widespread violence or wars but rather latent or simmering crisis (Central Asia, Europe).

Many of the case studies presented in Part 2 build on theoretical frameworks or theories of conflict, and oftentimes offer an original contribution to the methodological approaches discussed within Part 1. As an example, **Christophe Gironde** (Chapter 16), **Matthias vom Hau** (Chapter 19) and **Matthew Hoffman** (Chapter 20) offer studies from Vietnam and Cambodia, Argentina and Scotland, respectively, that rely heavily on ethnographic and sociological approaches. Similarly, **Lina Eklund and Katharina Lange** (Chapter 9) offer a case study of the Kurdistan Region of Iraq (KRI) that illustrates a mixed methods approach drawing on GIS techniques highlighted by **Jaafar**, as well as ethnographic and sociological methods.

Overt conflict in countries of the Middle East and North Africa appear to be rooted in prolonged crisis within or affecting the agricultural sector and agrarian communities. For example, looking to the cases of Egypt and Tunisia, **Ray Bush** (Chapter 7) reviews agricultural and rural development policies in both countries and points to evidence of low-level resistance to central government policies over many decades. Agricultural crisis is linked to the civil uprisings in Egypt and Tunisia through a context of continued resistance even under new regimes, as opposed to a clear role for farmers and agrarian interests in the disquiet that culminated in revolutions.

Linda Matar (Chapter 8) considers the armed conflict in Syria, especially relevant in view of its profound geopolitical dimension. This chapter explores a continuous decline in investment in the country's agricultural sector in the decade before open conflict broke out. This decline was a product of explicit government policy and subsequent negligence of smallholders at the expense of commercial interests of well-connected urban clients of the Assad regime. As a result agriculture became susceptible to shocks such as drought, in which peasants were pushed out of the agricultural sector and off their lands. Matar argues that the prolonged crisis in the country's agricultural sector laid the foundations for the ongoing conflict that has affected Syria since 2011.

Lina Eklund and Katharina Lange (Chapter 9) look to another case of prolonged agricultural crisis from northern Iraq. They combine satellite image analysis and qualitative interviews to assess changes in agricultural production, finding expansion, retraction and shifting cultivation across different areas within the agricultural system. The authors link changes in agricultural production to differential control of land by small-scale producers and agricultural entrepreneurs, profitability of different crops and physical displacement of farmers. Eklund and Lange thus demonstrate the value of using integrated approaches combining both quantitative and qualitative methods to understand challenges for local food production systems, and note the benefits and some practical challenges of multidisciplinary collaboration.

Focusing on the case of Yemen, **Max Ajl** (Chapter 10) argues that the agricultural sector has experienced a protracted crisis due to multiple factors: the abandonment and physical erosion of terraces and highland production, foreign agricultural subsidies that undermine Yemeni agriculture, the dominance of water-intensive qat production and government policies that foster overextraction of water for irrigation. Protracted crisis has been made acute by civil war and the Saudi–US war on Yemen, which has led to serious reductions in food security and famine. Yemen's agricultural recovery requires first an end to the war and, thereafter, development policies including investment in traditional crops, terrace restoration and sustainable irrigation techniques.

Alaa Tartir (Chapter 11) looks to both domestic and geopolitical explanations to understand the current state of Palestinian agriculture. He describes the effects of occupation as well as faulty domestic policies on the Palestinian agricultural sector, and thereby demonstrates that the effects of conflict on agriculture may not be clearly untangled from those of other causes.

Using comparative case studies from Africa, **Michele Nori and Edoardo Baldaro** (Chapter 12) argue that pastoral areas overlap with marginal and peripheral areas that are governed by 'limited states' that do not exercise full control and are vulnerable to the entry of insurgent groups. In the Sahel such groups draw membership and support partially from pastoral communities. In the Horn of Africa, an extended period of state collapse has prompted pastoral communities to expand existing resilience networks stretching across international borders. The authors conclude that the current intersection of global, regional and local dynamics has reshaped pastoral areas and communities. They reject efforts to limit pastoral practices, and instead call for pastoral communities to be recognized as natural allies to secure vast and remote territories.

Asel Murzakulova and Irène Mestre (Chapter 13) also consider crisis within agro-pastoral communities, but offer a different perspective on the role of frontiers: in contrast to Nori and Baldaro's implication that fading frontiers stoke conflict, the case from the Ferghana Valley in Central Asia shows that the imposition of international borders has provoked crisis by exacerbating disputes over access to resources and within pastoral and farming communities.

Turning to south Asia, **Ali Nobil Ahmad** (Chapter 14) focuses on the 2010 floods in Pakistan's Southern Punjab province and their aftermath. Ahmad's central argument is that the devastating impact of the floods was partially manmade. They hit poorer farming communities especially hard, while connected commercial farming interests lobbied successfully to spare their operations from relief flooding along the canal and irrigation infrastructure. Poor water management practices rooted in the colonial era thus served as a major source of grievance for public protests and conflict against the central government. Ahmad views these within the ecological agrarian question and rejects the narrative that emphasizes only Islamic or ethno-nationalist dimensions in explaining post-flood conflicts.

Archana Prasad (Chapter 15) looks to the Naxalite–Maoist insurgency in India and similarly finds a problematic understanding of the conflict there. She first reviews the Naxal movement in the 1970s, ongoing agrarian transformation in Adivasi societies and the effects of Maoist insurgency in affected areas of India. Prasad then argues that a misreading of capitalism and the agrarian question – specifically, a belief that Indian agrarian society is feudal rather than capitalist – lies behind the failure of the contemporary Maoists to achieve social transformation in India.

Like Prasad, **Christophe Gironde** applies the agrarian lens to make sense of crisis and (the absence of) conflict in agrarian communities. In a comparative case study of Cambodia and Vietnam, Gironde (Chapter 16) explores the evolution of crisis and protest in agricultural and rural communities. He points to mitigating factors including overall satisfaction with material living conditions and socio-cultural considerations to conclude that Vietnam and Cambodia's failure to experience outright agrarian conflict reflects a shift in dialectic from crisis-conflict to transition-adaptation.

Moving to the case of Colombia, **Carolina Castro Osorio and Edinso Culma** (Chapter 17) use a typology of causes of displacement to consider the effects of one of the most protracted conflicts in Latin America on rural communities and farming, as well as the effects of oil production and palm oil cultivation in two different areas of the country. The authors note that displacement due to economic and environmental forces in fact happens within the context of armed conflict, and may occur in waves. Like the Palestine case study, this chapter highlights multiple causes of crisis within the agricultural sector.

Next, **Marcela Torres Wong** (Chapter 18) explores tension and conflicts surrounding natural resource extraction in indigenous communities in Bolivia, Peru and Mexico. She focuses on prior consultation rights, which when adopted should allow indigenous communities to be consulted by their government regarding all projects that may impact their territories. Torres then uses examples to contrast the pro- and anti-extractivist movements and outcomes, highlighting the implementation of prior consultation procedures and the agrarian dynamics at play in each case.

In another study focused on the rights of indigenous communities, **Matthias vom Hau** (Chapter 19) considers indigenous movements for communal land rights in Argentina. The study is placed against the backdrop of re-emerging extractivist development models in the wake of new mining codes (1990s) and a commodities boom (2000s). Vom Hau concludes that, while economic pressures are often at the root of demands for the right to land, patterns of land governance are the result of more than economic structures and commercial interests: the organizational capacity of actors and states and institutions play a critical role.

Finally, in a case study from northern Europe, **Matthew Hoffman** (Chapter 20) looks to Scotland to understand the role of land reform in rural development and the role of property rights. This reform effort was largely unrelated to agricultural production and has done little to change the size of land holdings, increase the rights of landholders or raise agricultural productivity. Rather, Hoffman argues that land reform in Scotland has been a vehicle for rural community development and greater democracy. Community development may be better achieved through reforms to support democracy than efforts targeting higher agricultural productivity. The case of post-agrarian Scotland may, however, have little to say about the path of land reform or options for peaceful resolution to agricultural crisis and conflict in more agrarian countries. Indeed, in a global context widespread migration is no longer an option to resolve tensions over scarce resources or to escape the effects of climate change. Agrarian populations are largely trapped in place and under the purported control of powers exercising weak or limited governance mechanisms. Hence, agriculture may continue to intersect with crisis and conflict into the 21st century.

With this volume, we set out to expand our knowledge of the interactions of conflict, crisis and agriculture at the level of theory, methodological approaches and illustrations from a range of contemporary contexts. We find that these interactions are non-linear, do not move in a single direction and do not have uniform effects. The complexity of interactions reflects the complexity found in socio-ecological systems, and the study of agriculture, crises and conflicts must therefore adopt ontological tools adapted to the study of complex systems. For instance, technologies such as GIS provide researchers the ability to make specific observations and collect detailed data in conflict settings, without being present on the ground – traditionally a limiting factor – and without reliance on institutions that may be party to the conflict. These remote technologies are, by definition, limited to the physical environment, but leave the human dimensions largely unexplored. These latter dimensions need to be captured through methods that are more rooted in the social sciences, such as ethnographies and anthropological studies. We therefore highlight the need to link approaches that are commonly used in distant disciplinary fields, and call for more transdisciplinary researchers, not just transdisciplinary teams.

References

Aston, T.H. and Philpin, C.H.E. (1985) *The Brenner Debate: Agrarian Class Structure and Economic Development in Pre-Industrial Europe*. Past and Present Publications. Cambridge University Press, Cambridge.

Breisinger, C., Ecker, O. and Trinh Tan, J.F. (2015) Conflict and food insecurity: how do we break the links? In: 2014–2015 Global Food Policy Report. International Food Policy Research Institute, Washington, DC. Available at: http://ebrary.ifpri.org/utils/getfile/collection/p15738coll2/id/129072/filename/129283.pdf (accessed 21 April 2017).

Brinkman, H.-J. and Hendrix, C. (2011) Food insecurity and violent conflict: causes, consequences, and addressing the challenges. Occasional Paper No. 24. World Food Programme, Rome. Available at: http://documents.wfp.org/stellent/groups/public/documents/newsroom/wfp238358.pdf?_ga=1.23246 6971.546340923.1487918749 (accessed 24 February 2017).

Bromwich, B. (2015) Nexus meets crisis: a review of conflict, natural resources and the humanitarian response in Darfur with reference to the water–energy–food nexus. *International Journal of Water Resources Development* 31(3), 375–392. DOI: 10.1080/07900627.2015.1030495.

Brückner, M. and Ciccone, A. (2007) Growth, democracy, and civil war. CEPR Discussion Paper No. 6568. Centre for Economic Policy Research, London. Available at: www.antoniociccone.eu/?p=240 (accessed 24 February 2017).

Collier, P. and Hoeffler, A. (1998) On economic causes of civil war. *Oxford Economic Papers* 50, 563–573.

Collier, P. and Hoeffler, A. (2004) Greed and grievance in civil war. *Oxford Economic Papers* 56, 563–595.

Collier, P., Elliott, V.L., Hegre, H., Hoefler, A., Reynal-Querol, M. and Sambanis, N. (2003) *Breaking the Conflict Trap: Civil War and Development Policy*. World Bank and Oxford University Press, Washington, DC.

Collingham, E.M. (2011) *The Taste of War: World War Two and the Battle for Food*. Allen Lane, London.

Cotula, L. (2013) *The Great African Land Grab? Agricultural Investments and the Global Food System*. Zed Books, London.

Cotula, L., Vemeulen, S., Leonard, R. and Keeley, J. (2009) Land grab or development opportunity? Agricultural investment and international land deals in Africa. IIED, FAO and IFAD, London.

Cramer, C. (2006) *Civil War is Not a Stupid Thing: Accounting for Violence in Developing Countries*. Hurst & Company, London.

Cramer, C. and Richards, P. (2011) Violence and war in agrarian perspective. *Journal of Agrarian Change* 11(3), 277–297.

Davis, M. (2001) *Late Victorian Holocausts: El Niño Famines and the Making of the Third World*. Verso, London.

de Chatel, F. (2014) The role of drought and climate change in the Syrian uprising: untangling the triggers of the revolution. *Middle Eastern Studies* 50(4), 521–535. DOI: http://dx.doi.org/10.1080/00263206.2 013.850076.

Deininger, K., Byerlee, D., Lindsay, J., Norton, A., Selod, H. and Stickler, M. (2011) *Rising Global Interest in Farmland. Can It Yield Sustainable and Equitable Benefits? In Agriculture and Rural Development*. World Bank, Washington, DC.

Dikötter, F. (2010) *Mao's Great Famine: The History of China's Most Devastating Catastrophe, 1958–1962*, 1st US edn. Walker & Co., New York.

Eklund, L., Degerald, M., Brandt, M., Prishchepov, A.V. and Pilesjö, P. (2017) How conflict affects land use: agricultural activity in areas seized by the Islamic State. *Environmental Research Letters* 12(5), 054004.

Erdkamp, P. (2005) *The Grain Market in the Roman Empire: A Social, Political and Economic Study*. Cambridge University Press, Cambridge, UK.

Food and Agriculture Organization (FAO) (2010) The state of food insecurity in the world 2010. Food and Agriculture Organization of the United Nations, Rome. Available at: www.fao.org/docrep/013/i1683e/i1683e03.pdf (accessed 12 January 2017).

Food and Agriculture Organization (FAO) (2016) Peace and food security: investing in resilience to sustain rural livelihoods amid conflict. Food and Agriculture Organization of the United Nations, Rome. Available at: www.fao.org/3/a-i5591e.pdf (accessed 12 January 2017).

FAO, IFAD, UNICEF and WHO (2017) The state of food security and nutrition in the world 2017: building resilience for peace and food security. Food and Agriculture Organization of the United Nations, Rome. Available at: www.fao.org/3/a-I7695e.pdf (accessed 25 October 2017).

Fröhlich, C.J. (2016) Climate migrants as protestors? Dispelling misconceptions about global environmental change in pre-revolutionary Syria. *Contemporary Levant* 1(1), 38–50. DOI: 10.1080/20581831.2016.1149355.

Harris, K., Keen, D. and Mitchell, T. (2013) When disasters and conflicts collide: improving links between disaster resilience and conflict prevention. Overseas Development Institute, London. Available at: https://www.odi.org/sites/odi.org.uk/files/odi-assets/publications-opinion-files/8228.pdf (accessed 30 October 2017).

Hassan, F. (1997) The dynamics of a riverine civilization. A geoarchaeological perspective on the Nile Valley, Egypt. *World Archaelogy* 29(1), 51–74.

Hegre, H., Ellingsen, T., Gates, S. and Gleditsch, N.P. (2001) Toward a democratic civil peace? Democracy, political change, and civil war, 1816–1992. *American Political Science Review* 95(1), 33–48.

Hobsbawm, E.J. (1994) *The Age of Extremes: A History of the World, 1914–1991*. 1st US edn. Pantheon Books, New York.

Jaafar, H. and Woertz, E. (2016) Agriculture as a funding source of ISIS: a GIS and remote sensing analysis. *Food Policy* 64, 14–25. DOI: http://dx.doi.org/10.1016/j.foodpol.2016.09.002.

Kalyvas, S. (2004) The urban bias in research on civil wars. *Security Studies* 13(3), 160–190. DOI: 10.1080/09636410490914022.

Kelley, C.P., Mohtadi, S., Cane, M.A., Seager, R. and Kushnir, Y. (2015) Climate change in the fertile crescent and implications of the recent Syrian drought. *Proceedings of the National Academy of Sciences of the USA* 112(11), 3241–3246. DOI: 10.1073/pnas.1421533112.

Martin, P., Mayer, T. and Thoenig, M. (2008) Civil wars and international trade. *Journal of the European Economic Association* 6(2–3), 541–550.

Maxwell, J. and Reuveny, R. (2000) Resource scarcity and conflict in developing countries. *Journal of Peace Research* 37(3), 301–322.

Mazo, J. (2010) *Climate Conflict: How Global Warming Threatens Security and What to Do About It*. Adelphi, Abingdon, UK; Routledge, for the International Institute for Strategic Studies.

McLaughlin, P. (1998) Rethinking the agrarian question: the limits of essentialism and the promise of evolutionism. *Human Ecology Review* 5(2), 25–39.

Midlarsky, M. (1988) Rulers and the ruled: patterned inequality and the onset of mass political violence. *American Political Science Review* 82(2), 491–509.

Mikhail, A. (2014) Oriental Democracy. *Global Environment* 7(2), 381–404.

Ross, M. (2004) What do we know about natural resources and civil war? *Journal of Peace Research* 41(3), 337–356.

Scott, J.C. (2017) *Against The Grain: A Deep History of the Earliest States*. Yale University Press, New Haven, Connecticut.

Selby, J., Dahi, O.S., Fröhlich, C. and Hulme, M. (2017) Climate change and the Syrian civil war revisited. *Political Geography* 60 (Supplement C), 232–244. DOI: https://doi.org/10.1016/j.polgeo.2017.05.007.

Snyder, T. (2010) *Bloodlands: Europe Between Hitler and Stalin*. Basic Books, New York.

Verhoeven, H. (2011) Climate change, conflict and development in Sudan: global neo-Malthusian narratives and local power struggles. *Development and Change* 42(3), 679–707.

von Uexkull, N., Croicu, M., Fjelde, H. and Buhaug, H. (2016) Civil conflict sensitivity to growing-season drought. *Proceedings of the National Academy of Sciences of the USA* 113(4), 12391–12396.

Weis, T. (2007) *The Global Food Economy: The Battle for the Future of Farming*. Zed Books, London.

Woertz, E. (2013) *Oil for Food: The Global Food Crisis and the Middle East*. Oxford University Press, Oxford.

World Bank (2011) World development report 2011: conflict, security, and development. World Bank Group, Washington, DC. Available at: http://siteresources.worldbank.org/INTWDRS/Resources/WDR2011_Full_Text.pdf (accessed 12 January 2017).

Part 1

Theoretical Exploration of and Methodological Approaches to Agriculture, Crisis and Conflict

1 Agriculture, Conflict and the Agrarian Question in the 21st Century

Rachel Bahn[1,]* and Rami Zurayk[2]

[1]Department of Agriculture and Food Security Program, American University of Beirut, Beirut, Lebanon; [2]Department of Landscape Design and Ecosystem Management, American University of Beirut, Beirut, Lebanon

Introduction

Understanding the causes of conflict can help to prevent conflict, inform resolutions to conflict and avoid its most harmful effects. There are indications that armed conflict has especially strong effects on agricultural systems (Baumann and Kuemmerle, 2016), while most civil conflicts affect rural areas and their populations (Kalyvas, 2004), whose economies and livelihoods still rely heavily on agriculture. Conflict damages agricultural sectors, disrupts food production systems, destroys assets and household incomes, and contributes to food insecurity and malnutrition (FAO, 2016b). Reports in early 2017 linked famine affecting 20 million individuals to conflict in primarily agricultural settings (Sengupta, 2017), an extreme but compelling argument as to the importance of the intersection of agriculture and conflict. Yet, the relationship between agriculture and conflict is not fully understood and no complete critical theoretical framework has been proposed to understand these linkages.

To contribute to the development of such a framework, this chapter explores the current state of knowledge with regard to the following questions: How does agriculture relate to conflict, and vice versa? Does agriculture feature within prominent theories of conflict? Is there a role for agrarian transformation in explaining conflict and crisis in the 21st century?

The chapter opens with a data-centric overview of the current state of global agriculture, the situation of conflict-affected countries and the role of agriculture therein. Next, we summarize current thinking on agriculture and conflict. We then consider the agrarian question, a prominent framework linking agricultural transformation with societal change, and its role in potentially explaining certain types of conflict. Selected cases of conflict are considered to identify common characteristics of agrarian transformation, and illustrate that the agrarian question may be useful in analysing and explaining some but not all conflicts in the agrarian world.

Agriculture, the Agrarian World and Conflict: What Do We Know?

Data reveal that, despite global trends, agriculture remains the dominant activity across much of the world's physical area and for much of the global population. Agriculture is even more important to the populations and economies of fragile and conflict-affected states, though less integrated into global trading networks. Recent conflict has affected a significant share of the

---

* Email: rb89@aub.edu.lb

world's population and agricultural production. Data therefore demonstrate the important, persistent inter-relation of agriculture, the agrarian world and conflict.

The global importance of agriculture

More than a third of the world's land area – 49 million km^2 or 37.7% of global land area – remained under agricultural production as of 2013 (World Bank, 2016b). The agricultural sector employed approximately 1 billion people globally in 2013. Sub-Saharan Africa, North Africa, Southeast Asia and Pacific, and the Middle East reported an increase in the number of agricultural workers over the period 2000–2013, the combined result of rapid population growth and the persistence of labour-intensive agricultural production systems. However, the share of labour within the agricultural sector declined in favour of services and industry across all regions over the period 2000–2013 (ILO, 2014). Similarly, the economic value added generated by the agricultural sector has increased in absolute terms but fallen in relative terms over recent decades. Globally, agricultural value added reached $2979 billion (constant US$2010) in 2014, extending a steady increase since the 1960s. Agricultural value added as a share of global gross domestic product (GDP) has declined since the 1990s, from more than 8% to 3.9% in 2014, as industry and services grew relatively faster (World Bank, 2016b).

Smallholder agriculture[1] is the dominant production system for significant portions of the globe, particularly developing countries. Small farms of less than 2 ha occupy approximately 12% of the world's agricultural land, but account for 84% of all farms by number (Lowder *et al.*, 2016) and are home to two-thirds of the developing world's rural population (Samberg *et al.*, 2016). In addition, smallholder agriculture plays an important role in feeding the globe, though there is disagreement as to what extent. Samberg *et al.* (2016) report that smallholder agriculture in Latin America, sub-Saharan Africa, and South and East Asia accounts for more than 70% of all food calories produced in these regions; 50% of all food calories produced globally; and the majority of production of selected food crops including rice. Herrero *et al.* (2017) estimate that small farms of less than 2 ha produce more than 25% of seven essential nutrients in sub-Saharan Africa, Southeast Asia, South Asia, China and East Asia Pacific. However, Lowder *et al.* (2016) challenge claims that small farms produce a large share of the world's food given that they occupy only a fraction of total agricultural land. Moreover, larger farms intensively producing high-yielding crops produce much of the tradeable surpluses of food and nutrients to feed import-dependent areas including urban centres (Herrero *et al.*, 2017).

Agriculture in fragile and conflict-affected states

The World Bank (2016a) identified 35 countries in fragile and conflict-affected situations (FCS)[2] in 2017, listed in Table 1.1.

The FCS countries collectively account for 10.6% of the world's agricultural land (2013

Table 1.1. Fragile and conflict-affected countries, 2017. (From World Bank, 2016a.)

Afghanistan	Iraq	Sierra Leone
Burundi	Kiribati	Solomon Islands
Central African Republic	Kosovo	Somalia
Chad	Lebanon	South Sudan
Comoros	Liberia	Sudan
Democratic Republic of Congo	Libya	Syrian Arab Republic
Côte d'Ivoire	Madagascar	Togo
Djibouti	Mali	Tuvalu
Eritrea	Marshall Islands	West Bank and Gaza
The Gambia	Federated States of Micronesia	Yemen
Guinea-Bissau	Myanmar (Burma)	Zimbabwe
Haiti	Papua New Guinea	

figures), 6.3% of all arable land (temporary cropland) (2013 figures) and 10.5% of all rural land (2010 figures) (FAO, 2016a; World Bank, 2016b). However, FCS countries have potential to expand the land area under irrigated agricultural production: they account for only 4.8% of the global land area equipped for irrigation (2013 figures) (FAO, 2016a), below their share of arable land.

The data presented above provide a country-level perspective to understand the intersection between conflict and agriculture, using a set of globally standardized indicators. Sub-national data on the incidence of conflict, which does not adhere to national borders and may particularly affect certain regions, provide a fuller understanding. A complementary dataset of Koren and Bagozzi (2016b) has been used to assess the prevalence of conflict and civil conflict in agricultural areas over the period 1991–2008.[3] This dataset reveals that conflict has affected approximately 15.1% of all global cropland[4] in recent decades, and civil conflict accounted for most of that impact (authors' calculations based on Koren and Bagozzi, 2016b).

The prevalence of smallholder farming differs significantly across FCS countries, with respect to the relative land area occupied. For example, Samberg et al. (2016) assess the structure of farming systems in selected countries to find that smallholder farms of less than 5 ha dominate agricultural production systems in Haiti and Liberia but are much less prominent in Mali, Sierra Leone, Sudan and south Sudan.[5] Average farm sizes are larger in those (FCS) countries where mixed livestock and cropping systems necessitate a more extensive agricultural model (L. Samberg, Minnesota, personal communication, 2017). A general lack of reporting of Gini coefficients for land across FCS countries unfortunately prevents any systematic review of patterns in land ownership and control, or comparison to global averages.[6]

FCS countries are home to nearly 500 million people, the majority of whom live in rural areas. Of a total population of 486 million, 286 million or 59.2% lived in rural areas (2015 figures) (World Bank, 2016b). FCS countries accounted for 6.6% of the global population but 8.5% of the world's rural population as of 2015 (World Bank, 2016b). These population figures may understate the number of affected individuals if conflict has provoked out-migration to urban areas or to countries not considered fragile or conflict-affected. Indeed, according to the Food and Agriculture Organization of the United Nations (FAO) (2016b), conflict and violence prompted an average of 42,500 people to seek safety outside of their homes, whether internally or in another country, each day in 2014 – approximately 15 million people that year. This caveat seems appropriate in light of the fact that FCS countries[7] reported significant populations of internally displaced persons (IDPs) – more than 23 million people – due to conflict and violence as of 31 December 2015 (IDMC, 2016).

Again, the dataset of Koren and Bagozzi (2016b) offers an alternative measure of the reach of conflict and civil conflict to the global population over the period 1991–2008. The granular data indicate that significantly more of the world's population, an average of approximately 20.3% over the period 1991–2008, has lived in direct proximity to conflict than reflected by the narrower list of countries put forth by the World Bank (6.6%) (authors' calculations based on Koren and Bagozzi, 2016b).

FCS countries account for very little of the global economy, according to national-level figures. These countries represented only 0.95% of global GDP as of 2015 (World Bank, 2016b). Granular data for the period 1991–2008, however, suggest that significantly more of the world's economy has been in proximity to conflict than indicated by the World Bank data. Approximately 6.2% of the global GDP was generated in areas in direct proximity to conflict over the period 1991–2008 (authors' calculations based on Koren and Bagozzi, 2016b).

FCS countries produced a relatively higher share of global agricultural GDP (2.7%) than total GDP in 2013 (latest data available) (FAO, 2016a). These countries similarly accounted for approximately 2.7% of the gross production value of food globally in 2013 (latest data available) (FAO, 2016a). These figures most likely understate the countries' potential contribution to global GDP and global agriculture, if in fact conflict reduces economic activity. However, FCS countries produce agriculture that is largely not entering global commodity markets. These countries accounted for only 0.7% of the total value of agricultural-based (food and fibre) exports[8] over the period 2013–2015 (latest available data)

(UN, 2017), significantly less than their share of global agriculture.⁹

Agriculture remains an important source of economic activity and livelihoods in FCS countries, more than for the world on average. For example, agriculture employs two-thirds of the total workforce in countries in protracted crisis, and accounts for one-third of all economic activity (FAO, 2016b). These figures are significantly higher than the global averages (19.8% and 3.9%, respectively) (World Bank, 2016b).

Cross-country comparisons reveal interesting but non-conclusive patterns about agricultural activity, agricultural populations, socio-economic status and the incidence of conflict. Full comparison of indicators is shown in the accompanying table for the world, FCS countries collectively and FCS countries individually (see Appendix). FCS countries collectively report more rural populations, greater dependence on agriculture for their economies and slightly higher unemployment rates than the world average. A selection of scatter plots using country-level data indicate that FCS countries (black points) are more likely to experience higher levels of rural poverty and less equal distribution of income (Gini coefficient of income), and to have more rural populations, than non-FCS countries (grey points) (see Figs 1.1, 1.2 and 1.3). Again, a lack of comprehensive data on access to or distribution of land (such as the Gini coefficient of land) precludes a systematic comparison among FCS and non-FCS countries. The scatter plots point to broad patterns rather than conclusive, linear relationships.

In summary, a review of available, contemporary data indicates that the intersection between conflict and agriculture remains significant in terms of people, land and economies. The large degree of overlap between conflict and agriculture is materially observable and raises questions as to the nature of the relationship, including causal linkages and multi-layered explanations.

Elucidating the Linkages Between Conflict and Agriculture

The academic literature has extensively explored the topic of conflict and its drivers, proposing many theories to understand its causes. Yet the understanding of the relationship between agriculture and conflict remains incomplete. Accordingly, there is a need to apply or develop a conceptual frame to understand the linkages, due to a shortage of critical readings of conflict and agriculture.

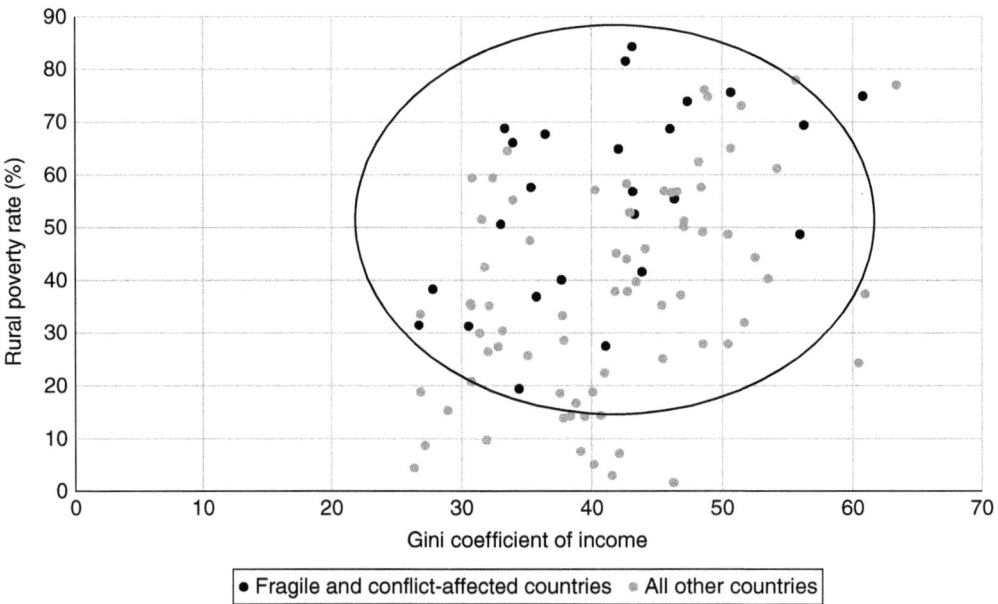

Fig. 1.1. Scatter plot – Gini coefficient of income and rural poverty. (From: World Bank, 2016b; UNDP, 2015.)

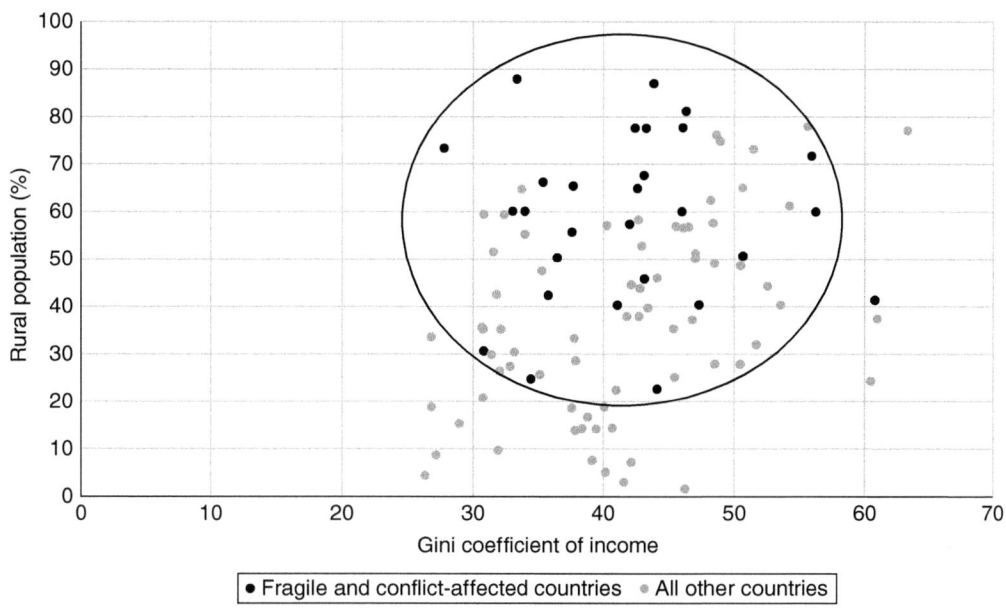

Fig. 1.2. Scatter plot – Gini coefficient of income and rural population. (From: World Bank, 2016b; UNDP, 2015.)

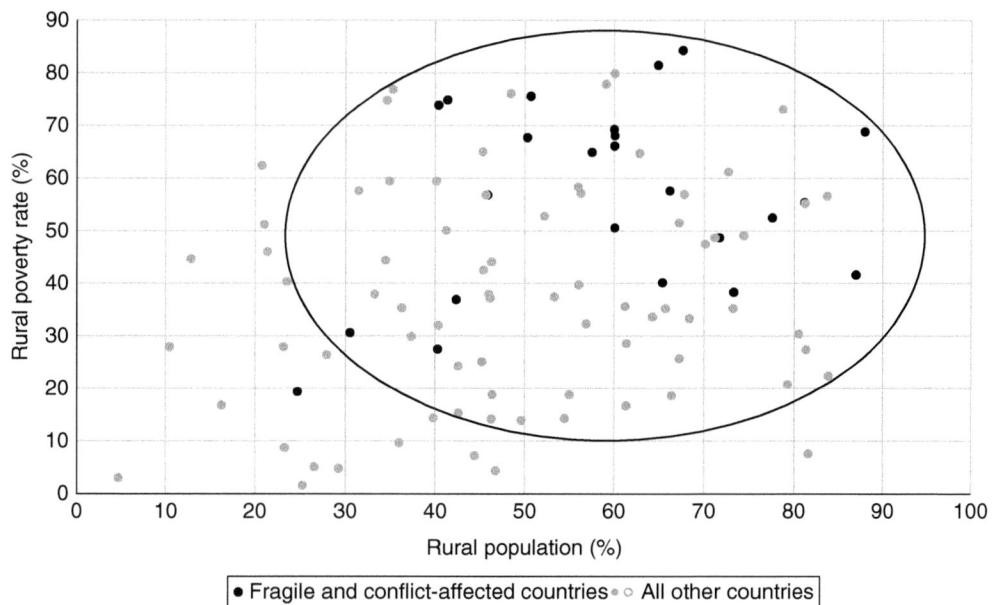

Fig. 1.3. Scatter plot – rural population and rural poverty. (From: World Bank, 2016b.)

Extent and evolution of conflict: conflict, crisis and violence in the 21st century

Significant changes in the types and incidence of conflict and violence have been observed between the 20th and the 21st centuries. The 20th century principally experienced inter-state conflict and episodes of civil war, executed by sovereign states or defined rebel movements. Such conflict was ultimately resolved either through victory and defeat or negotiated settlement, and followed by a brief post-conflict period before restoration of peace. In contrast, the 21st century has experienced fewer incidents of inter-state war, and the death toll of civil war has fallen to one-quarter of levels in the 1980s (World Bank, 2011).

Nevertheless, changes in the structure of violent conflict in the 21st century have not meant the elimination of conflict or fragility. Forty wars were reported in 2014, the highest figure since 1999 (Baumann and Kuemmerle, 2016), and the number of violent conflicts and conflict-related deaths has risen rapidly from a record low in 2005 (FAO et al., 2017). Patterns of violence have become protracted, with single countries exiting and re-entering conflict in cycles of repeated violence, weak governance and instability. Four explanatory factors are proposed for this. First, conflicts have become ongoing and repeated.[10] Second, contemporary forms of conflict and violence (including violent crime) undermine economic development. Third, different forms of violence are inter-linked. Fourth, in countries where political, economic, and social change fail to meet expectations, grievances may escalate into popular demand for change and violent conflict (World Bank, 2011).

It is these evolving patterns of conflict that theories have been developed to explain, and thereby, *inter alia*, reduce the incidence of conflict or bring about faster, less costly and more durable resolutions.

Theoretical models of conflict

Extensive research has generated a range of theoretical models to explain conflict, and the academic literature offers numerous studies offering evidence for and against various theories. The causal mechanisms identified by the literature may be more or less broadly classified; for example, in a wide-ranging empirical review of factors associated with the onset of armed conflict and civil war, Hegre and Sambanis (2006) group these according to 18 concepts: ethnic fragmentation, ethnic dominance or polarization, level of democracy, inconsistency of political institutions, political instability, political system, centralization, neighbourhood political economy, region, war in neighbouring countries, terrain/geography/population distribution, economic growth, economic policies, social welfare, natural resources, militarization, time and colonial war.[11] Thomson (2011) divides key literature exploring the causal mechanisms driving civil (as opposed to inter-state) conflict more generally, as follows:

- *Economic demand and opportunity*: Collier et al. (2003) assert that the failure of economic development – reflected in low, unequal and stagnant per capita incomes and dependence on primary commodity exports (including agricultural products) – is the key root cause of civil conflict. Initial efforts of Collier and Hoeffler (1998) to explain the incidence and duration of conflict by assessing the potential economic costs and opportunities of prospective combatants, set the framework for analysis of additional economic factors including initial incomes, economic inequality, and availability of financial resources to fund rebellion. Collier and Hoeffler (2004) determine that the greed motivation is a more powerful explanatory factor for civil conflict than grievance. Other studies have concluded that economic opportunity may influence the duration, intensity and character of conflict but does not serve as primary cause (Ballentine, 2003). Looking internationally, researchers contend that the expansion of international trade has reduced the incidence of internal conflict (Thomson, 2011, citing Griswold, 2007), while others posit that trade openness may deter severe civil war but heighten the risk of smaller-scale internal conflict (Martin et al., 2008).
- *Natural resources*: Evidence is mixed as to whether an abundance of or dependence upon natural resources ('resource curse') makes conflict more likely to begin, endure or cause death (Ross, 2004); whether a scarcity of natural resources or environmental degradation is linked to conflict (Maxwell and Reuveny, 2000); and the importance of management of resources as distinct from their abundance or scarcity in stimulating conflict (Cramer, 2006; Murzakulova and Mestre, Chapter 13, this volume).
- *Political structures*: Studies posit that political and governance structures may have a moderating influence on conflict. For example, lower economic growth is associated

with greater likelihood of civil conflict in non-democracies, but has no such impact in countries with democratic institutions (Brückner and Ciccone, 2007), while other studies conclude that both strong democracies and authoritarian states are less likely to experience internal conflict than intermediate regimes (Hegre et al., 2001).

Cramer and Richards (2011) observe that analysis of violent conflict in the post-Cold War period has considered a range of factors, but has disregarded considerations including the agrarian roots and dynamics of violent conflict, which may be relevant to violent conflict in the 21st century and have important contributions to conflict resolution and post-conflict reconstruction and peace-building. To expand current and future scholarship, they call for analytical and empirical research to shift from national-level to local exploration of rural areas, to consider social and spatial contexts on communities and individuals, and to widen the timeline of analysis to reach beyond the period immediately preceding the conflict (Cramer and Richards, 2011). Similarly, Kalyvas (2004) points to an urban bias within conflict studies generally and particularly with regard to civil wars, which may result in an underestimation of the importance of agricultural, agrarian or rural elements in studying the determinants of conflict as well as the impacts of conflict.

While research into the causes and drivers of conflict continues, there is simultaneous acknowledgement of the complexity of conflict, the interaction of different causal factors[12] and the failure of any single model to explain all instances. 'The causes of armed conflict have been the subject of lengthy and controversial debate. The conclusion of a synthesis of this literature is that there is no simple causal explanation for conflict. Conflict comes in many forms, and its causes are complex, non-linear, and mediated by a host of factors' (FAO, 2016b, p. 18). Moreover, the causes and effects of violent conflict are entwined, difficult to distinguish (World Bank, 2011), and may vary from conflict to conflict (Cramer, 2006).

Conflict and agriculture: causality or dialectic?

Studies of the relationship between conflict and agriculture have considered both the role of agriculture in causing conflict, the effects of conflict on agriculture, and more nuanced, bidirectional effects. We review those briefly here, paying particular attention to the role of agricultural land given its prominence in many case illustrations.

Agriculture as cause of conflict

The relationship between agriculture and conflict encompasses a variety of types of conflict, ranging from grain riots and peasant revolt to revolution and civil war. The relationship between agriculture and both peasant revolts and grain riots is more evident: Peasant revolts or rural uprisings by definition take place in rural settings, which tend to be agriculturally focused in developing countries. Grain riots are mobilizations to protest food shortages or high prices, typically linked to shortfalls in agricultural production or disrupted distribution (Goldstone, 2014). The relationship between agriculture and wider conflicts including civil conflict is less immediately obvious and is the focus of the following discussion (see also Table 1.2).

Evidence of the causal relationship between agriculture and conflict often focuses on access to and competition over resources needed for agriculture, including land and water.[13] Land is one apparent nexus point for agriculture and conflict (both internal and cross-border): Conflict may arise due to issues of land tenure, customary land rights and sustainable use of natural resources. For example, land issues played a significant role in 27 of 30 inter-state conflicts in Africa over the period 2000–2016 (FAO, 2016b).[14] Access to water may also serve as a source of conflict. For example, some contend that competition over water resources may make conflict more likely between states sharing a river basin, particularly in countries at lower levels of economic development (Brinkman and Hendrix, 2011), though this conclusion may not hold in all cases (de Waal, 2015 citing Islam and Susskind, 2013; Keulertz, Chapter 4, this volume). Competition between farmers and pastoralists for access to both land and water, or policies that favour farmers or disfavour pastoralists, may be another point of conflict within agrarian contexts. Violent conflict between farmers and pastoralists has been reported in Mali, in a context of water scarcity and shifting desert boundaries (Breisinger et al., 2015).

Table 1.2. Illustrative linkages of agriculture and conflict. (Adapted from: FAO, 2010; Brinkman and Hendrix, 2011; Baumann and Kuemmerle, 2016; FAO, 2016b.)

Phenomenon and causal mechanism

Unidirectional: agriculture as a causal factor of conflict
Agricultural resources
- Contest over agricultural land or water resources (local/external)
- Agro-pastoral competition
- Land grabs (external contest over agricultural land)
- Patterned inequality in land distribution
- Land reform
- Agricultural production as funding source for conflict

Unidirectional: impacts of conflict on agriculture
Agricultural land use
- Abandonment (due to degradation, unexploded ordinance, etc.)
- Expansion (in proximity to or far from conflict)
- Recultivation of previously abandoned areas
- Modification of existing agricultural systems (e.g. collapse of irrigated agriculture systems)

Agricultural population
- Death of agricultural workforce (combatants and civilians)
- Displacement of agricultural workforce (combatants and civilians)
- Acceleration of rural–urban migration

Agricultural livelihoods and incomes
- Destruction of traditional livelihoods
- Disruption of coping mechanisms
- Decline in cash crop production
- Decline in agricultural investment
- Displacement/destruction of agro-industry

Agricultural trading patterns
- Displacement of existing trade routes
- Disrupted access to export markets (due to transportation disruption or sanctions)

The 'land grab' phenomenon is one particular type of contest over agricultural land, which gained attention from the mid-2000s and has been subject to extensive academic and public scrutiny in part for its role in enhancing competition and sometimes triggering conflict between local populations, government figures and national or international investors. Large-scale commercial investments in agricultural production, typically undertaken in areas of the global south,[15] may lead to dispossession and direct displacement of rural households, and serve to increase competition for scarce resources including land and water. In the most extreme cases, including incidents reported in Ethiopia and Mali, the result has been violence and armed protest (Cotula, 2013).

Patterns of land distribution in agricultural contexts may also underpin conflict, according to Midlarsky's assessment of access to land, land distribution patterns and the incidence of conflict: quantitative analysis of data on land distribution from Latin America in the 1980s demonstrated statistical correlation with the incidence of political violence, leading to the conclusion that patterned inequality in land distribution is a predictor (but not strict determinant) of conflict in agrarian societies (Midlarsky, 1988). Midlarsky argues that changes in relative land distribution – average size of landholdings among the poor in contrast to the rich – are the result of long-standing processes, and possibly linked to related phenomena such as income inequality (Muller et al., 1989).

The financial returns to agricultural commodity production and trade may provide an attractive incentive to be gained or controlled through violent conflict, as well as a source of funds to finance conflict (Cramer, 2006). For example, Chouvy (Chapter 5, this volume) highlights the role of production of opium poppy, coca and cannabis in funding and sustaining conflict, which in turn provides a fertile context for expanded production. Jaafar and Woertz (2016) estimate the considerable rents obtained by ISIS from agricultural production in Syria.

Scholarship exploring the contribution of climate change and extreme weather events to conflict may also have implications for agriculture (see Smidt and Theisen, Chapter 3, this volume, for a fuller discussion of climate/weather/conflict interactions). Per Koren and Bagozzi (2016a), '[A] major mechanism by which climate change increases the likelihood of conflict is through its effects on food supplies', via agricultural production (p. 1000). As for extreme weather, such shocks have been linked to conflict through a loss of agricultural employment opportunities, reducing the economic prospects of individuals who may find fighting to be an attractive alternative (Brinkman and Hendrix, 2011). Growing-season drought, particularly among agriculturally dependent groups and politically excluded groups in least developed countries, may contribute to sustaining conflict and thereby play part of a cyclical interaction between environmental shock and violent conflict (von Uexkull et al., 2016). However, others cite contradictory evidence that good weather conditions foster greater agricultural productivity, which offers both incentives – crops, livestock and productive land – and opportunities for conflict (Brinkman and Hendrix, 2011, citing Burke et al., 2009, Buhaug, 2010, Hendrix and Salehyan, 2010 and Ciccone, forthcoming; Harris et al., 2013). Adano et al. (2012) report evidence of increased conflict in pastoral areas during periods of abundance, and explain this as an interaction of climate with layered social cooperation and governmental institutions. Harris et al. (2013) find that natural disasters tend to exacerbate conflict, but have in selected cases reduced the drivers of conflict in favour of peace-building and conflict resolution.

Somalia appears to be a case where extreme weather events function as a driver of conflict. Somalia is a fragile country dominated by agriculture, with more than 60% of the population classified as pastoralists or semi-pastoralists who derive their incomes from animal husbandry (FAO, 2016b). One study has concluded that an increase in the length and intensity of drought increases the likelihood of conflict: drought prompts herders to sell livestock, reducing prices and herder incomes and therefore the opportunity cost of engaging in violent conflict (Maystadt and Ecker, 2014).

Separately, studies have sought to understand the role of food insecurity as a driver of conflict globally (Pinstrup-Andersen and Shimokawa, 2008) and within specific regions (Maystadt et al., 2014). The relationship between access to food (derived from agricultural products) and conflict has been recorded and theorized for some time, but quantitative research investigating the relationship between food prices and political stability is a recent area of academic study. Evidence suggests that rising food and food import prices may cause food riots and social (especially urban) unrest (Bellemare, 2015; Weinberg and Bakker, 2015), particularly in democratic regimes (Hendrix and Haggard, 2015). Similarly, the FAO (2016b) has reported that food insecurity can trigger conflict, through channels including sharp increases in food prices, reductions in real incomes due to cuts in food and fuel subsidies, dispossession of assets and reduced access to food, but is never the single factor driving a conflict. For example, McMichael (2010) roots violence associated with the 2007–2008 world food crisis to the structure of the current food regime. Bush and Martiniello (2016) link food riots not only to the price or accessibility of staple foods, but to the more complex political economy of food systems.

Impact of conflict on agriculture

According to the FAO (2016b, p. 2), 'Most conflicts mainly affect rural areas and their populations. This is particularly true for civil conflicts, which have tripled in recent years, and which today are the most common form of armed conflict, and increasingly prolonged'. A geographical analysis of armed conflict events over recent decades indicates that agricultural land systems are particularly strongly affected by armed conflict (Baumann and Kuemmerle, 2016).[16]

The effect of conflict on agriculture includes impacts on the wider economy, the physical environment and resources, agricultural populations and composite outcomes including food security: 'Conflict often affects the ability to produce, trade and access food' (Brinkman and Hendrix, 2011, p. 12). These impacts may be multiple, compounding, simultaneous, and complex (FAO et al., 2017).

Conflict typically damages economies in the form of lower economic production and reduced employment (including within the agricultural sector), eroded public finances and investment (including for agricultural purposes), trade disruption (including reduced import capacity and disruption to export channels) and price inflation (FAO *et al.*, 2017).

Direct impacts of conflict on agriculture include damage to agricultural infrastructure, disruption of food (crop and animal) production systems and loss of assets (whether plundered or destroyed). Crop production may be reduced directly, or through damage to supporting infrastructure such as irrigation systems (e.g. Jaafar *et al.*, 2015). Abandonment of agricultural lands may reduce crop production in areas of particular proximity to conflict (Jaafar, Chapter 6, this volume; Eklund and Lange, Chapter 9, this volume). Livestock production systems typically experience negative effects on animal health, access to livestock products (milk, meat, blood) and animal ownership (FAO, 2016b); grazing patterns of pastoralists may be affected, making systems less self-sufficient; and in some cases livestock may be a direct target of warfare (FAO *et al.*, 2017). The effects of conflict may be observed along the entire food-value chain (input supply, production, harvesting, processing, transportation, financing and investment, and marketing as well as import/export channels). While such effects tend to be negative, conflict may also create opportunities for agricultural trade or new market structures (FAO *et al.*, 2017; Chouvy, Chapter 5, this volume). Protracted crises in particular may increase competition for remaining resources among different livelihood groups, even those not previously in conflict: 'As the economy contracts (and freedom of movement may also contract during a conflict), livelihoods have come under increasing pressure' (FAO, 2010, p. 19). Household coping strategies in response to conflict – including diversification of land holdings and crop cultivation, shifting to short-term yields, multi-year grain storage, and sale of productive assets like cattle and land (FAO *et al.*, 2017) – may reshape agricultural systems.

Conflict may trigger population movement and migration, thereby affecting employment and livelihood patterns in the agricultural sector and others. According to the FAO (2010, p. 18), 'People living in protracted crises are often forced to make radical adjustments to their livelihoods, including relocation from rural areas to the relative safety of population centres. This can disrupt traditional livelihoods and coping mechanisms, either temporarily or permanently, but can also present new livelihood opportunities if properly supported.' In the short- to medium-term, farmers and agricultural producers may shift their areas of production (e.g. see Eklund and Lange, Chapter 9, this volume), alter trade routes or abandon certain types of production in favour of new crops or animals; the term 'maladaptation' may describe cases where these changes cause overexploitation of resources or environmental damage. Over the longer term, protracted crises commonly accelerate the process of rural–urban migration. Additionally, Cramer (2006, p. 217) describes this displacement as 'the separation of people from their means of production', with IDPs and refugees increasingly forced to rely on wage labour markets (agricultural and non-agricultural) to earn a living.

The cumulative effect of disruption to agricultural production systems and the loss of assets and incomes is frequently food insecurity and malnutrition (acute and chronic) (FAO, 2016b), with mixed evidence as to whether these effects are restricted to the conflict period or persist post-conflict (Gates *et al.*, 2012). Recent incidence of the most extreme form of food insecurity, famine, has been reported in countries or regions currently in (northern Nigeria, south Sudan, Yemen) or emerging from (Somalia) conflict (Sengupta, 2017; FAO *et al.*, 2017).[17] As of 2016, approximately 60% of all undernourished people and 75% of stunted children under the age of 5 years were living in countries affected by conflict, violence and fragility (FAO *et al.*, 2017).

Bidirectional relationship of conflict and agriculture

The geographic relationship between agriculture and conflict is non-random. Since 'most civil conflicts are rural wars, fought primarily in rural areas by predominantly peasant armies' (Kalyvas, 2004, p. 161), conflict should be more likely to affect rural areas, which are often largely agricultural (Koren and Bagozzi, 2016a). Moreover, by equating food availability with supply and food access with demand, Koren and Bagozzi (2016a) argue that widespread limitations to either may affect the location of armed conflict in developing countries: in a context of otherwise low food access, agricultural production may attract conflicting parties – and thus conflict – towards agricultural and rural areas, in an effort to secure access to food resources for their forces or to seize those resources as a prize of war.

As noted above, conflict may erupt in agricultural communities as a result of competition for increasingly scarce resources in the context of protracted conflict (FAO, 2016b).[18] In fact, there may be a cyclical relationship between environmental shocks and conflict, specifically among groups dependent on agricultural production for their income and livelihoods. von Uexkull et al. (2016, p. 12394) explain the cyclical relationship as follows: 'Sustained fighting and insecurity deter investments, trigger capital flight, undermine public goods delivery, and have negative health implications, all of which may decrease the local population's ability to cope with increased environmental hardship and increase their incentives to sustain ongoing resistance'.

The FAO (2016b) summarizes the relationship between agriculture and conflict as such: '[T]he causal relationship between natural resources, weak governance, conflict and poor development remains unclear, and points to the need for a deeper analysis of the particular elements and circumstances in which they are relevant' (FAO, 2016b, p. 19).

The current state of knowledge around conflict reveals that there is no single-cause theory of conflict and that it may be futile to search for such simplistic explanations in the context of the 21st century. In line with the evidence reviewed above, we accept that conflicts may have multi-layered causes, including geo-political and economic circumstances, agrarian transformation and immediate shocks such as climate or natural disaster, and that these forces combine in different ways in different circumstances. This layering and complexity explain why we do not find linear, unidirectional relationships to explain conflict (Figs 1.1–1.3). Moreover, the complicated and bidirectional relationships between agriculture and conflict – which have not yet been fully explored – indicate the merits of efforts to expand our understanding and to consider a range of theoretical frameworks so as to gain a more complete understanding of agrarian conflict or conflict in agrarian contexts. To that end, we now consider the agrarian question as one framework that can potentially help us to understand the relationship between agriculture and conflict in the 21st century.

The Agrarian Question and Conflict

The agrarian question is a prominent theoretical framework used to connect agricultural transition to social change, and which accounts for resource competition, modernization of agricultural labour and migration, and household and social accumulation. Cramer and Richards (2011) have called for a renewed focus on agrarian dynamics in the study of conflict, and we explore whether the agrarian question can be applied to help us understand the causal relationship between agriculture and conflict. We believe that the agrarian question may be relevant to consider given the evidence of overlap of conflict with highly agricultural contexts, as demonstrated in the opening section of this chapter. The following section defines and briefly discusses terms central to the topic, presents an abbreviated history of the agrarian question and its most prevalent contemporary interpretations, and illustrates its application to the relationship between agrarian transition and contemporary conflict through selected cases.

Agrarian and agrarian transition

Agrarian activity is that which is 'concerned with cultivation (any kind of care for and production of crops or domestic animals), and typically requires access to land, as well as labour, capital and technology' (Cramer and Richards, 2011, p. 280).

The agrarian transition refers to the process by which capitalist commodification enters into the traditional (subsistence) model of agricultural production and labour. This penetration has the effect of undermining, supplanting or co-opting peasant modes of production, and provokes 'the rise of an attendant class structure, the development of new production techniques, and the emergence of urban-industrial societies' (Kelly, 2011, p. 482).

The agrarian transition disrupts individuals with regard to both economic relationships and physical location. The emergence of agrarian capitalism transforms petty commodity producing peasants into wage labourers, through complex forces of dispossession (Akram-Lodhi and Kay, 2010b). Wage labourers whose ties to the land are disrupted may migrate to other areas. This migration (rural–urban or rural–rural), may contribute to urbanization of populations and act as a driver of further economic, social, cultural and political change in rural areas (Kelly, 2011).

The classical agrarian question

The agrarian question was first posed in the 18th century, in the context of early industrialization, by Marxist political theorists: Marx laid the foundations of the agrarian question, with different dimensions expanded upon by Engels (political), Kautsky and Lenin (social), and Preobrazhensky (economic) (Bernstein, 2006; Moyo et al., 2013). Kautsky defined the agrarian question as, 'whether, and how, capital is seizing hold of agriculture, revolutionizing it, making old forms of production and property untenable and creating the necessity for new ones' (Akram-Lodhi and Kay, 2010a, p. 179, citing Kautsky, 1988). The agrarian question principally addressed three 'problematics': first, production, or the structure and dynamics of the rural production process; second, accumulation, or the extent to which agriculture supplies surpluses to sustain industrialization and structural transformation; and third, rural politics, or the response of the rural population to changes in the structure of rural production and agrarian accumulation, including class dynamics and the possibility of an alliance between peasants and (urban) workers to advance socialist political goals[19] (Akram-Lodhi and Kay, 2010a; 2010b; Hendricks, 2014). For some, the classical agrarian question has become 'synonymous with analysis of class transformations in the countryside, from the point of view of the capitalization of land' (McMichael, 2013, p. 62). Though agrarian transition may lead to social revolution, according to Marxian theorists, Marx himself acknowledged that social revolution does not necessarily lead to violent conflict (Moshiri, 1991).

Contemporary agrarian questions

The agrarian question has evolved over time, taking on new meaning and emphasizing new dimensions, largely in response to the changes in contemporary agriculture provoked by capitalism (Hendricks, 2014). McMichael (2013) notes particular changes as of the mid-20th century, as the agrarian question shifted from its original, socialist problematic addressing the political role of the European peasantry in revolution, to an imperial problematic concerned with the demobilization of post-colonial peasantries and the defence of landholding elites against reform movements, and attributes the continued evolution of the agrarian question to ongoing structural changes in global agriculture. The emphasis of the agrarian question has also shifted over time, as can be seen by Brenner's late 20th-century reassertion of 'the centrality of class structure, class relationships, and class struggle as the central dynamic variables in understanding processes of development and change, including the resolution of the agrarian question through a capitalist agrarian transition' (Akram-Lodhi and Kay, 2010a, p. 198). New considerations have been introduced as well; for example, gender and ecology entered the debate amidst the national liberation movements of the 20th century ('The Agrarian Question: Past, Present and Future' 2012). The topic of population migration was largely ignored in discussions of the agrarian question even into the late 20th century (Kearney, 2002).

This evolution of the agrarian question has generated multiple variants, each rooted in different circumstances and historical context, and the result has been complication and confusion within the academic debate. This complexity makes it difficult to suggest a definitive typology; however, it is possible to identify contemporary approaches through which the agrarian question has been considered (Hendricks, 2014). Akram-Lodhi and Kay (2010b) identify seven contemporary variants of the agrarian question (AQ), each adopting the balance of forces between capital and labour as the critical analytical variable; the following section draws primarily from Akram-Lodhi and Kay (2010b) unless otherwise indicated.

The *class forces variant (AQ1)*, principally derived from the work of Byres (1991a; 1991b; 2003; 2006), emphasizes the role of class forces (including peasant class differentiation) in shaping transformations in agrarian production systems towards rural capitalism. Class forces influence the process of agrarian transition such that it may follow multiple paths: '[A]grarian transition will be highly contingent, subject to substantive diversity, and is embedded within historical trajectories of variation that reflect processes of differential and uneven incorporation into the capitalist mode of production operating both within capitalist social formations and on a global scale' (Akram-Lodhi and Kay, 2010b, p. 266).

The *path-dependent variant (AQ2)*, derived largely from the work of Warren (1980), considers the creation of labour power commodification (or wage labour) to be central to the development of agrarian capital and rural capitalism, and capitalist industrialization. Per AQ2, contest over the terms and conditions of access to wage labour are both an outcome of the establishment of rural capitalism, and a source of socio-economic and spatial differentiation. AQ2 concludes that the emergence of agrarian capital and rural capitalism is inevitable (Akram-Lodhi and Kay, 2010b).

The *decoupled variant (AQ3)* draws principally from the work of Bernstein (2004; 2006), who divides the matter into an agrarian question of capital and an agrarian question of labour. Transnational capital has become decoupled from national labour regimes, which in turn are fragmenting and increasingly unable to provide viable livelihoods. Bernstein argues the agrarian question of capital has been resolved through the globalization of capital, and the emergence of agrarian capital matters only insofar as it shapes political struggles over resources, production and accumulation (Akram-Lodhi and Kay, 2010b; Thomson, 2011). The agrarian question of labour persists, however, as labour struggles to earn a livelihood in the face of the productive forces of capital: its central concerns are the struggle over rural livelihoods, a crisis of reproduction among fragmented labour classes and a growing reserve army of labour (Akram-Lodhi and Kay, 2010b).

The *global reserve army of labour variant (AQ4)* is derived largely from the work of Araghi (2009), who views the current period of neoliberal globalization as an extension of 19th-century liberal imperialism. The 'enclosure food regime' has favoured capital over labour; established subsidized consumption in the global north; led to processes of dispossession by displacement in agriculture; and generated a reserve army of migratory labour. Under AQ4, the agrarian question centrally concerns the reproduction of agrarian labour, whose value is continually reduced under the enclosure food regime. The struggle between a globalizing capitalism and dispossessed peasants is part of the establishment of the law of value and an effort to reduce the value of labour power in order to generate greater surplus for capital (Akram-Lodhi and Kay, 2010b).

The *corporate food regime variant (AQ5)* is based principally on the work of McMichael (2006, 2013), who defines the contemporary agrarian question within and through the lens of world history but finds importance in distinct world-historical moments. AQ5 reframes the global agrarian crisis as an agrarian question of food, and its central consideration is how peasant households and rural labour can reproduce when progressively excluded from the corporate food regime. Amidst the neoliberal globalization of capital and labour, the agrarian question of food centres on the struggle over rural livelihoods (Akram-Lodhi and Kay, 2010b).

The *gendered agrarian question (AQ6)* as elaborated by O'Laughlin (2009) emphasizes the gender dynamics of production, accumulation and rural politics. Gender must be a central consideration of social change in contemporary rural settings, because gender relations both influence and are influenced by the class forces and the formal and informal social and political institutions produced by those class forces (Akram-Lodhi and Kay, 2010b).

The *ecological agrarian question (AQ7)* described by Blaikie (1985), Weis (2007), Bellamy Foster (2009) and Watts (2009) rests on the argument that there are ecological dynamics within the rural production process, agrarian accumulation and rural politics, and therefore processes of class formation. The agrarian question serves to explain social change in contemporary rural settings only if it critically assesses ecological relationships and the ways in which they affect a potential resolution of the agrarian question, and addresses contradictions of class and ecology. In sum, AQ7 suggests that the political ecology of agrarian change and the biophysical contradictions in capitalism exert a mutual influence, and the latter are integral to a proper conception of the agrarian question (Akram-Lodhi and Kay, 2010b).

Much of the scholarship on the agrarian question comes from the global north and northern dialogues, while the effects of contemporary agrarian transformation are felt in the global south. For that reason, we highlight what might be termed an *agrarian question of the south (AQ8)*. For example, Moyo *et al.* (2013) reject Bernstein's conception (AQ3) to propose their own bifurcation of the agrarian question into an agrarian question of industrialization, focused on developed or metropolitan countries, and an agrarian question of national liberation focused on developing countries. Elsewhere, Amin (2012) argues that the incorporation of the peasantry across the south into a global agri-food system must inevitably lead to marginalization, particularly as mass emigration is no longer a viable remedy. Similarly, Patnaik (2013) views today's financial and trade relationships as a contemporary form of colonialism, with exploitation of developing by developed countries leading to increasing pressures on and displacement of peasants, who will eventually turn to resistance. Amin (2012) is optimistic but non-specific in predicting that politicized social movements will arise to channel this resistance.

Alternative definitions of the agrarian question continue to be put forth, adding to the variants described above. We anticipate continued debate and further elaboration of the agrarian question. Nevertheless, and despite important distinctions in focus, all variants place a central focus on the effect of capitalism on agricultural production and the role and position of agricultural labour.

Is the agrarian question relevant in the 21st century?

Debate remains as to whether the agrarian question remains meaningful in the 21st century and, if so, in which contexts. While some academics argue for its continued relevance, others argue that the agrarian question – in one form or another – is effectively dead.

Akram-Lodhi and Kay (2010b) claim a renewed relevance of the agrarian question in the 21st century, citing evidence including:

> the remarkable stability in the absolute number of peasant farmers over the last 40 years; the continued importance of smallholder food production to rural livelihoods in much of the South; the deepening of the market imperative and the law of value across the world capitalist economy under neoliberal globalization. …; the expanded commodification of natural resources, including land, labour power and genetic resources; the strong spatial specificities to these processes, as cross-border megaregions transcend the state in driving substantial shares of global capital accumulation; and the global resurgence, in response, of peasant movements. (Akram-Lodhi and Kay, 2010b, p. 279)

Similarly, Hendricks (2014) points to the persistence of peasants to argue for the continued relevance of the agrarian question, and calls for 'a critical interrogation of their current role under conditions of advanced capitalism', both between advanced industrial countries of the north and underdeveloped countries of the south, and also between rural and urban-dwellers within countries. Amin (2012) anticipates that the resolution of the agrarian question in developing or periphery countries is outstanding and will only be achieved through a 'genocide' of the billions of peasant farmers engaged in smallholder farming in those areas. McMichael (2013) also argues for the continued relevance of the agrarian question, provided that it expand to address the impacts of transition and the barriers to developing healthy, just and sustainable agri-food systems.

These authors counter others, like Byres and Bernstein, who argue that the agrarian question is no longer relevant, in some formulation or in a specified geography such as the global north. 'For Byres the agrarian question ceases to have relevance in the North with the overwhelming dominance of industry and the urban bourgeoisie' (Hendricks, 2014, p. 285). Bernstein (2004; 2006) argues that the agrarian question of capital has been resolved globally, if not yet completely in the global south. However, Moyo *et al.* (2013) respond that precisely because industrialization in the West relied and continues to rely on capital accumulation made possible by exploitation of colonies and later the periphery, the agrarian question has not been resolved even for developed countries. Patnaik (2013) applies the agrarian question to understand financial and trade relationships between developed and developing countries, to explain worsening food insecurity and crisis among Indian farmers since the mid-1990s to today.

Practical evidence of the relevance of the agrarian question and specifically its application to understand conflict is mixed. For example, issues of agrarian change, violent conflict and crisis remain relevant – if not predominant – in the academic debate: Academic journals continue to publish dozens of articles addressing agrarian change, agriculture, violent conflict and crisis.[20] Thomson (2011) notes that, while it is not atypical for agrarian transitions to be conflictual, contemporary conflict studies have tended to exclude treatment of agrarian transition within the process of capitalist development from their analysis. 'In fact, mainstream development and conflict literature mostly ignore the concept of "capitalism" outright. In overlooking the development of capitalism, these analyses arguably have only a partial view of the dynamic processes that influence violent conflict' (Thomson, 2011, p. 329). The splintering of the classical agrarian question into multiple, contemporary approaches to the agrarian question mirrors the diversity of the theories of causes of conflict. This diversity within the agrarian question is perhaps one reason that it is not widely applied as a theoretical framework to understand conflict.

Targeting this gap, Cramer (2006) argues that the longer-term process of transition to capitalism is central to understanding the context of conflict and crisis in many contemporary conflicts. For Cramer, the transition to capitalism is inherently conflictual, but does not in all cases result in violent conflict. Understanding the agrarian transition – and therefore the agrarian question – is important for a more complete analysis of conflicts. An explanation of why conflict occurs in some cases but not others must accordingly combine analysis of the agrarian transition with other explanatory theories. However, others point precisely to the failure of the agrarian question to offer a comprehensive explanation of the incidence of conflict: writing specifically on revolutions as one form of crisis and conflict, Goldstone (2014) argues that modernization – the process by which 'people encounter free markets for goods and services, inequality rises, and traditional religious and customary patterns of authority lose their power. . . . [and] people demand new, more responsive political regimes and turn to force to create them' – is not a singular process that results in common outcomes or that operates on a singular timeline, and therefore there is no consistent relationship between modernization and the onset of revolution (Goldstone, 2014, pp. 11–12). Cramer (2006) counters that the diversity in how violence manifests does not mean that the agrarian question embedded within a transition to capitalism is therefore without explanatory power.

The preceding discussion leads us to conclude that the agrarian question does not provide a robust theoretical framework to explain contemporary violent conflict: neither the classical nor the contemporary agrarian question provides a compelling explanation of when violent conflict will erupt, nor whether it will persist in agrarian contexts. However, we may consider the agrarian question as a useful tool to illuminate the connections between agriculture and conflict, and vice versa. Rather than attempting to definitively prove that the agrarian question offers a compelling explanation of conflict, we apply the agrarian lens to explore the dimensions between crisis, conflict and agriculture.

Case Study Illustrations: Agrarian Dynamics in 21st-Century Conflicts

Selected cases are briefly discussed below to illustrate the relevance of the agrarian question to understand conflict in the contemporary world. However, other cases indicate that the agrarian question may not provide a compelling explanation of crisis or conflict.

Agrarian concerns – for example, over the exploitation and displacement of agricultural labour and competition for scarce resources including land – have served as primary motivation for several recent conflicts. For example, the inter-linked civil wars that affected Liberia and Sierra Leone from the 1990s to the early 2000s have their roots in deep-rooted agrarian tensions, reflecting institutionalized abuses tied to domestic slavery dating to those countries' settler and colonial periods, respectively. Richards (2005) cites post-war interviews with and studies of ex-combatants to conclude that the majority came from rural and agricultural areas neglected by the state and still subject to patrimonial rule and exploitation, and were motivated by agrarian concerns including ineffective land tenure and the forced division of labour on estates and plantations. Similarly, competition over land contributed to conflict in Côte d'Ivoire from 2002–2007 (first civil war) and renewed crisis following a disputed presidential election in 2010 (second civil war). The pro-governmental mobilization of rural young people to fight in that conflict reveals the importance of the agrarian roots of the conflict, including questions of access to land, the value of labour supplied by rural youth and rural–urban mobility. Notably, rising competition for land between local and settler communities transformed into a nationalist ideology held by Ivoirians against outsiders, used to mobilize collective action and violence (Chauveau and Richards, 2008); these tensions have persisted, including in areas dominated by cocoa production, and appear to underpin violent clashes reported as recently as late 2017 (Adele, 2017). The role of land is even clearer within Nepal's recent conflict: the Communist Party of Nepal-Maoist launched an insurgency in 1996 that lasted until 2006, with a primary objective to revolutionize agriculture and address issues of landlessness,[21] unfair rural labour practices and poverty through comprehensive land reform (Bray et al., 2003).

Colombia is one case in which the agrarian question is particularly relevant to understand a prolonged conflict.[22] The internal conflict in Colombia appears to have its basis at least partially in questions of unequal access to land by peasants and landowners reaching back to the 19th century, exacerbated by financial resource inflows, natural resource capture and illicit drugs (World Bank, 2008 citing Deininger et al., 2007 and World Bank, 2002). Indeed, Thomson (2011) traces the history of conflict in Colombia throughout the 20th century, paying particular attention to the development of the agricultural sector (production of crops for commercial export, introduction and expansion of coca production, and conversion of cropland to less labour-intensive livestock production) within a wider process of capitalist economic development; competition for control of land, leading to a series of land reforms; displacement of peasants and rural labourers to both urban and marginalized rural areas; and the increasingly class-based conflict between landowners, peasants and their respective armed counterparts fuelled by the illicit drug economy. A counter-reform process jointly undertaken by the Colombian government, paramilitaries and large-scale agribusinesses demonstrates how the integration of Colombia into the global capitalist system has shaped the internal conflict. Echoing the argument that the agrarian transition is inherently conflictual and may lead to violent outcomes (Cramer, 2006), Thomson (2011, p. 352) argues that, 'Contrary to the claim that "the armed conflict in Colombia constitutes the central and greatest obstacle to development," capitalist development may actually be an obstacle to peace in the country'.

An agrarian lens also helps to understand the protracted, violent conflict in the southern Philippine archipelago of Mindanao that began in the 1970s and for which a 2014 peace agreement between the government and rebels has yet to be implemented. The conflict has generally been viewed along a Muslim–Christian divide or through the lens of the 'war on terror'. This approach limits understanding of the conflict by obscuring the complexity of the situation and diverting attention from issues related to the local economy, state weakness and fragile democratic institutions. Vellema et al. (2011) argue that the conflict in Mindanao, which has been driven by highly unequal patterns of land ownership and control as well as contested institutions governing land property, and that the violent conflict is a result of an ongoing socioeconomic transformation in the Philippines, 'embedded in a new and evolving division of labour driven by global agrarian modernization affecting group structures and regulatory ties in agrarian communities' (Vellema et al., 2011, p. 300).

In terms of crisis, the agrarian question (specifically the ecological agrarian question) provides a useful explanation for the tensions and conflict that followed the 2010 floods in Southern Punjab, Pakistan. These tensions were triggered by environmental events, but rooted in a history of internal colonization, politically influenced agricultural practices, and ethno-nationalism that serves to explain the way in which protest manifested (Nobil Ahmad, Chapter 14, this volume).

Conversely, tensions in Vietnam and Cambodia linked to agrarian transformation, particularly competition over land in both rural and peri-urban areas, have not led to widespread violent conflict in the contemporary period (Gironde, Chapter 16, this volume). The shifting identity of peasants and the growing importance of concerns beyond the agrarian space may reflect the process of agrarian transition. In this case, however, the agrarian question appears relevant but is not sufficient to explain the (absence of) conflict.

In some cases of conflict within agrarian communities, the contribution of the agrarian question remains to be determined. For example, Yemen's agricultural sector is in crisis as a result of both long-term processes of deterioration and outright war (Ajl, Chapter 10, this volume), but the agrarian dimensions of the Houthi rebellion are as-yet unclear. In other cases, the agrarian question may be relevant but does not serve as the primary explanatory model or factor. Tensions among agro-pastoral communities of the Ferghana Valley of Central Asia are rooted in collective, rather than capitalist, disruption, and appear to be principally driven by coordination failures, exacerbated by the imposition of national borders and dispersed natural resource management systems (Murzakulova and Mestre, Chapter 13, this volume).

These case illustrations point to the preliminary conclusion that the agrarian question is helpful in exposing historical tensions linked to the agricultural sector and thereby in analysing some cases of conflict, but not independently sufficient to fully explain conflict in the agrarian world.

Conclusions

This chapter has demonstrated that modern, 21st-century conflicts overlap significantly with and have strongly adverse effects on global agriculture, which include remaking the agricultural sector and agricultural trading patterns.

We then highlighted key theoretical approaches to explaining conflict, and concluded that no single theory put forth to date offers a convincing, single-cause explanation for all conflicts. Next, we explored the connection of selected theories to agriculture and the agrarian world, as well as the impact of conflict on agrarian contexts. Evidence on the causal relationship between agriculture and conflict is less clear, leading us to consider the agrarian question in an effort to better understand that relationship. The agrarian question is appropriate to consider in the 21st century, particularly insofar as conflict remains prevalent and insofar as it predominantly takes place in the context of agricultural populations and lands.

The agrarian question anticipates that the spread of capitalism within agrarian societies will be in all cases conflictual, but not necessarily violent; it does not explain under what circumstances violence will result. Indeed, application of the agrarian question to the study of conflict and crisis does not *a priori* guarantee that agrarian transformation is a factor or driver of conflict or crisis. Rather, such an approach considers both the spatial (rural versus urban) distribution of conflict and crisis, as well as 'how access to and control of land and labour, as well as financial capital, is shaped by social structures and relations, including class, gender and age' and imposes pressure that can transform into conflict and/or crisis (Cramer and Richards, 2011, p. 280).

Parallel consideration of prominent theories of conflict and the agrarian question exposes diversity in each, such that the diversity within the contemporary agrarian question reflects the diversity of theories of causes of conflict. The lack of a single, coherent definition of the agrarian question may initially deter researchers from applying it; however, just as the evidence reviewed above demonstrates the futility of seeking simplistic, single-cause explanations or unidirectional relationships between cause and effect of conflict, we should not expect the agrarian question to offer a simplistic, narrow application to the intersection of conflict and agriculture or to prove satisfactory in analysing all cases of crisis or conflict. Indeed, both the quantitative and qualitative evidence presented above demonstrate that the relationship of conflict and agriculture is bidirectional and merits thoughtful consideration through a range of lenses. Nevertheless, the agrarian question does not offer an independently compelling theory of conflict. Instead, we recognize its utility in exploring the underlying context in which conflict takes place, and so suggest that it be applied in combination with theories of conflict to identify and understand causal interactions.

Preliminary evidence from selected case studies suggests that agricultural transition underpins the incidence of (some but not all) conflict; that the unresolved agricultural transition may serve as a causal factor driving (some but not all) conflict; and therefore that the agrarian question as a theoretical framework can help us to understand (some but not all) conflict today. The agrarian question is useful to illuminate – but not fully or independently explain – the relationship between agriculture and conflict.

Why should the agrarian question be considered when analysing conflict and crisis? This matters both conceptually and practically. Conceptually, a refusal to consider the agrarian question in assessing conflict may lead us to overlook important patterns, so long as there are agricultural societies in the world: if the introduction or expansion of capitalist, market-based relationships within the agricultural sector is assumed to be the desired and normal (even inevitable) path to economic development, then we should seek to understand the implications of the agrarian transition underpinning this economic model. Such an understanding should ideally point us to policies that support development while managing its social and political implications. The danger of ignoring these implications is to be unprepared for or unable to mitigate their consequences – including crisis and (potentially) violent conflict. Practically and more immediately, an understanding of the agricultural roots of conflict and crisis can and should inform efforts to resolve conflict and rebuild in agricultural settings, with concrete impacts on the lives of millions.

Acknowledgements

The authors thank Archana Prasad and Eckart Woertz for careful review of and valuable comments on a draft of this chapter.

Notes

[1] Various metrics are used to define small farms (e.g. size of land holding, extent of livestock or assets, and gross sales), reflecting that small is a relative concept in light of diversity in farming systems across regions. The most common definition of a small farm is land area of less than 1–2 ha. Small farms are defined differently from family farms, though there is considerable overlap in these categories (Lowder et al., 2016).

[2] This analysis relies on the World Bank classification of fragile and conflict-affected states for the sake of expediency. The Organisation for Economic Co-operation and Development and the Failed States Index use similar but different data and methods to identify FCS countries (Simmons, 2017). Of the 20 of the G7+ countries – a self-identified group of fragile countries – 17 are listed among the World Bank classification; exceptions are Guinea, Sao Tome and Principe, and Timor-Leste (G7+, 2017).

[3] The figures are calculated from the dataset of Koren and Bagozzi (2016b), which in turn draws on data from the PRIO-Grid dataset, the Globcover 2009 project and Nordhaus (2006). These spatial data provide information at the 0.5 by 0.5 decimal degree resolution, or a geographic squared 'cell' level of approximately 55 km by 55 km at the equator (decreases at higher latitudes); the spatial data exclude Antarctica and the Arctic. 'Conflict' and 'civil conflict' (intrastate conflict) are defined on the basis of a war with at least 25 intentional deaths from combatants; this definition excludes deaths due to collateral damage or civilian casualties.

[4] 'Cropland' includes post-flooding and irrigated cropland, rain-fed cropland and combined cropland–vegetation systems (Koren and Bagozzi, 2016a).

[5] The percentage of sub-national administrative units in which the average farm size is less than 5 ha is 100% (Haiti), 72% (Liberia), 32% (Sierra Leone), 13% (Sudan), 2% (Mali) and 0% (South Sudan) (Samberg et al., 2016 – supplementary data).

[6] For example, the FAO (2009) reports measures of inequality including the Gini coefficient of land for 52 of 184 total countries, only one of which is an FCS country.

[7] Using the 2017 classification of 35 countries, as presented above.
[8] Trade data considered across Harmonized Tariff Schedule codes (two-digit level) 01–24, 41 and 50–53.
[9] One exception is for cocoa and cocoa-based products, for which Côte d'Ivoire provides a significant share of global exports, suggesting stronger integration with global commodity markets.
[10] Elbadawi et al. (2008) point to 1993 as a pivot point in the incidence of recurrent as compared to new conflict.
[11] Alternatively, Breisinger et al. (2015) summarize literature on the root causes of conflict as including ethnic tension, religious competition, real or perceived discrimination, poor governance and state capacity, competition for land and natural resources, population pressure and rapid urbanization, and economic factors including poverty, youth unemployment and food insecurity.
[12] Writing on revolutions in the 20th century, Goldstone (1991) notes that revolutionary crisis is the result of both the explanatory factors themselves and their ability to combine in different ways.
[13] Baumann and Kuemmerle (2016) conclude that 'conflicts occur predominantly in areas most valuable for humans, possibly prompted by conflicts over access to these lands or its resources' (p. 680).
[14] Earlier, notable cases of inter-state war motivated by demographic pressures arising from scarcity of arable land include German aggression in eastern Europe and Japan's invasion of China and elsewhere in Southeast Asia (Brinkman and Hendrix, 2011).
[15] Cotula (2013) observes the systematic connection between land grabs in sub-Saharan Africa and capitalist production models in agriculture: 'The origins of today's land rush can be traced to long-standing, unequal relations that have cast Africa as a supplier of commodities to the outside world' (Cotula, 2013, p. 8).
[16] The most conflict-prone land systems are, in order: urban and densely settled systems; extensive cropland systems; medium cropland systems; grassland systems with livestock; intensive agricultural systems; bare lands; forest systems; and grassland systems without livestock (Baumann and Kuemmerle, 2016).
[17] According to de Waal (2015), 'When famine or acute hunger occurs today, it is usually the result of armed conflict'.
[18] While not focusing specifically on agriculture, Hendrix and Brinkman (2013) offer a clear explanation of the circular linkages between food insecurity and conflict.
[19] Marxian analysis defines class on the basis of relations to the mode of production, and applies it as an analytical tool. Marx considered class consciousness as a prerequisite for proletariat-led social revolution (Moshiri, 1991).
[20] A search of the Web of Science Core Collection for academic articles containing topics 'agrarian change' or 'agrarian transformation' and 'violent conflict' or 'crisis' published over the period 2000–2016 yielded 39 results. A search of the Web of Science Core Collection for academic articles containing topics 'agrarian' or 'agrarian change' or 'agrarian transformation' and 'violent conflict' or 'crisis' published over the period 2000–2016 yielded 444 results. A search of the Web of Science Core Collection for academic articles containing topics 'agrarian' or 'agrarian change' or 'agrarian transformation' or 'agriculture' and 'violent conflict' or 'crisis' published over the period 2000–2016 yielded 2542 results (searches conducted 13 January 2017).
[21] Approximately 25% of Nepali households and 17% of Nepali agricultural labourers were landless in 2001.
[22] Castro Osorio and Culma (Chapter 17, this volume) also acknowledge the unresolved agrarian problems underpinning Colombia's armed conflict.

References

Adano, W.R., Dietz, T., Witsenburg, K. and Zaal, F. (2012) Climate change, violent conflict and local institutions in Kenya's drylands. *Journal of Peace Research* 49(1), 65–80. DOI: 10.1177/0022343311427344.
Adele, A. (2017) Land clashes test Côte d'Ivoire's fragile security. *IRIN News*, 25 October. Available at: www.irinnews.org/news/2017/10/25/land-clashes-test-cote-d-ivoire-s-fragile-security (accessed 6 November 2017).
Akram-Lodhi, A., and Kay, C. (2010a) Surveying the agrarian question (Part 1): unearthing foundations, exploring diversity. *Journal of Peasant Studies* 37(1), 177–202.
Akram-Lodhi, A. and Kay, C. (2010b) Surveying the agrarian question (Part 2): current debates and beyond. *Journal of Peasant Studies*, 37(2), 255–284. DOI: 10.1080/03066151003594906.
Amin, S. (2012) Contemporary Imperialism and the Agrarian Question. *Agrarian South: Journal of Political Economy* 1(1), 11–26.

Araghi, F. (2009) The invisible hand and the visible foot: peasants, disposession and globalization. In: Akram-Lodhi, F. and Kay, C. (eds) *Peasants and Globalization: Political Economy, Rural Transformation and the Agrarian Question*. Routledge, London, pp. 111–147.

Ballentine, K. (2003) Beyond greed and grievance: reconsidering the economic dynamics of armed conflict. In: Ballentine, K. and Sherman, J. (eds) *The Political Economy of Armed Conflict: Beyond Greed and Grievance*. Lynne Rienner Publishers, Boulder, Colorado, pp. 259–283.

Baumann, M. and Kuemmerle, T. (2016) The impacts of warfare and armed conflict on land systems. *Journal of Land Use Science* 11(6), 672–688. DOI: 10.1080/1747423X.2016.1241317.

Bellamy Foster, J. (2009) *The Ecological Revolution: Making Peace with the Planet*. Monthly Review Press, New York.

Bellemare, M. (2015) Rising food prices, food price volatility, and social unrest. *American Journal of Agricultural Economics* 97(1), 1–21. DOI: 10.1093/ajae/aau038.

Bernstein, H. (2004) Changing before our very eyes: agrarian questions and politics of land in capitalism today. *Journal of Agrarian Change* 4(1–2), 190–225.

Bernstein, H. (2006) Is there an agrarian question in the 21st century? *Canadian Journal of Development Studies* 27(4), 449–460.

Blaikie, P. (1985) *The Political Economy of Soil Erosion in Developing Countries*. Longman, London.

Breisinger, C., Ecker, O. and Trinh Tan, J.F. (2015) Conflict and food insecurity: how do we break the links? In: *2014–2015 Global Food Policy Report*. International Food Policy Research Institute, Washington, DC. Available at: http://ebrary.ifpri.org/utils/getfile/collection/p15738coll2/id/129072/filename/129283.pdf (accessed 21 April 2017).

Brinkman, H.-J. and Hendrix, C. (2011) Food insecurity and violent conflict: causes, consequences, and addressing the challenges. *Occasional Paper No. 24*. World Food Programme, Rome, Italy. Available at: http://documents.wfp.org/stellent/groups/public/documents/newsroom/wfp238358.pdf?_ga=1.232466971.546340923.1487918749 (accessed 24 February 2017).

Brückner, M. and Ciccone, A. (2007) Growth, democracy, and civil war. *CEPR Discussion Paper No. 6568*. Centre for Economic Policy Research, London. Available at: www.antoniociccone.eu/?p=240 (accessed 24 February 2017).

Bray, J., Lunde, L. and Mansoob Murshed, S. (2003) Nepal: economic drivers of the Maoist insurgency. In: Ballentine, K. and Sherman, J. (eds) *The Political Economy of Armed Conflict: Beyond Greed & Grievance*. Lynne Rienner Publishers, Boulder, Colorado, pp. 107–132.

Bush, R. and Martiniello, G. (2016) Food riots and protest: agrarian modernizations and structural crisis. *World Development* 91, 193–207.

Byres, T. (1991a) Agrarian question. In: Bottomore T. (ed) *A Dictionary of Marxist Thought*, 2nd edn. Blackwell, Oxford, UK, pp. 9–11.

Byres, T. (1991b) The agrarian question and differing forms of capitalist transition: an essay with reference to Asia. In: Breman, J. and Mundle, S. (eds) *Rural Transformation in Asia*. Oxford University Press, Delhi, India, pp. 3–76.

Byres, T. (2003) Paths of capitalist agrarian transition in the past and in the contemporary world. In: Ramachandran, V.K. and Swaminathan, M. (eds) *Agrarian Studies: Essays on Agrarian Relations in Less-Developed Countries*. Zed Books, London, pp. 54–83.

Byres, T. (2006) Differentiation of the peasantry under feudalism and the transition to capitalism: In defence of Rodney Hilton. *Journal of Agrarian Change* 6(1), 17–68.

Chauveau, J.-P. and Richards, P. (2008) West African insurgencies in agrarian perspective: Côte d'Ivoire and Sierra Leone compared. *Journal of Agrarian Change* 8(4), 515–552.

Collier, P. and Hoeffler, A. (1998) On economic causes of civil war. *Oxford Economic Papers* 50, 563–573.

Collier, P. and Hoeffler, A. (2004) Greed and grievance in civil war. *Oxford Economic Papers* 56, 563–595.

Collier, P., Elliott, V.L., Hegre, H., Hoefler, A., Reynal-Querol, M. and Sambanis, N. (2003) *Breaking the Conflict Trap: Civil War and Development Policy*. World Bank and Oxford University Press, Washington, DC.

Cotula, L. (2013) *The Great African Land Grab? Agricultural Investments and the Global Food System*. Zed Books, London.

Cramer, C. (2006) *Civil War is Not a Stupid Thing: Accounting for Violence in Developing Countries*. Hurst & Company, London.

Cramer, C. and Richards, P. (2011) Violence and war in agrarian perspective. *Journal of Agrarian Change* 11(3), 277–297.

de Waal, A. (2015) Armed conflict and the challenge of hunger: is an end in sight? In: von Grebmer, K., Bernstein, J., de Waal, A., Prasai, N., Yin, S., and Yohannes, Y. (eds) *2015 Global Hunger Index: Armed*

Conflict and the Challenge of Hunger. Welthungerhilfe, International Food Policy Research Institute (IFPRI), and Concern Worldwide Bonn, Washington, DC, and Dublin. Available at: http://ebrary.ifpri.org/utils/getfile/collection/p15738coll2/id/129685/filename/129896.pdf (accessed 17 March 2017).

Elbadawi, I., Hegre, H. and Milante, G. (2008) The aftermath of civil war. *Journal of Peace Research* 45(4), 451–459.

Food and Agriculture Organization (FAO) (2009) FAO statistical yearbook 2007–2008 – Table F.5. FAO, Rome.

Food and Agriculture Organization of the United Nations (FAO) (2010) The State of Food Insecurity in the World 2010. Food and Agriculture Organization of the United Nations, Rome. Available at: http://www.fao.org/docrep/013/i1683e/i1683e03.pdf (accessed 12 January 2017).

Food and Agriculture Organization of the United Nations (FAO) (2015) The impact of disasters on agriculture and food security. FAO, Rome. Available at: www.fao.org/3/a-i5128e.pdf (accessed 1 February 2017).

Food and Agriculture Organization of the United Nations (FAO) (2016a) FAOSTAT. Available at: http://ref.data.fao.org/web/guest/statistics (accessed 29 March 2018).

Food and Agriculture Organization of the United Nations (FAO) (2016b) Peace and food security: investing in resilience to sustain rural livelihoods amid conflict. FAO, Rome. Available at: www.fao.org/3/a-i5591e.pdf (accessed 12 January 2017).

FAO, IFAD, UNICEF and WHO (2017) The state of food security and nutrition in the world 2017: building resilience for peace and food security. FAO, Rome. Available at: www.fao.org/3/a-I7695e.pdf (accessed 25 October 2017).

G7+ (2017) Who we are. G7+. Available at: www.g7plus.org/en/who-we-are (accessed 29 March 2017).

Gates, S., Hegre, H., Nygard, H.M. and Strand, H. (2012) Development consequences of armed conflict. *World Development* 40(9), 1713–1722. DOI: 10.1016/j.worlddev.2012.04.031.

Goldstone, J. (1991) Introduction. In: Goldstone, J., Gurr, T.R. and Moshiri, F. (eds) *Revolutions of the Late Twentieth Century.* Westview Press, Boulder, Colorado, pp. 1–3.

Goldstone, J.A. (2014) *Revolutions: A Very Short Introduction.* Oxford University Press, Oxford, UK.

Harris, K., Keen, D. and Mitchell, T. (2013) When disasters and conflicts collide: improving links between disaster resilience and conflict prevention. Overseas Development Institute, London, UK. Available at: https://www.odi.org/sites/odi.org.uk/files/odi-assets/publications-opinion-files/8228.pdf (accessed 30 October 2017).

Hegre, H. and Sambanis, N. (2006) Sensitivity analysis of empirical results on civil war onset. *Journal of Conflict Resolution* 50(4), 508–535. DOI: 10.1177/0022002706289303.

Hegre, H., Ellingsen, T., Gates, S. and Gleditsch, N.P. (2001) Toward a democratic civil peace? Democracy, political change, and civil war, 1816–1992. *American Political Science Review* 95(1), 33–48.

Hendricks, F. (2014) Class and nation in the agrarian questions of the south: notes in response to Moyo, Jha and Yeros. *Agrarian South: Journal of Political Economy* 3(2), 275–293.

Hendrix, C. and Brinkman, H.-J. (2013) Food insecurity and conflict dynamics: causal linkages and complex feedbacks. *Stability: International Journal of Security and Development* 2(2), 26. DOI: http://dx.doi.org/10.5334/sta.bm.

Hendrix, C. and Haggard, S. (2015) Global food prices, regime type, and urban unrest in the developing world. *Journal of Peace Research* 52(2), 143–157.

Herrero, M., Thornton, P., Power, B., Bogard, J., Remans, R. *et al.* (2017) Farming and the geography of nutrient production for human use: a transdisciplinary analysis. *Lancet Planet Health* 1, e33–e42.

Internal Displacement Monitoring Centre (IDMC) (2016) *Global Report on Internal Displacement 2016.* Norwegian Refugee Council. Available at: www.internal-displacement.org/globalreport2016 (accessed 1 February 2017).

International Labour Organization (ILO) (2014) *Global Employment Trends 2014: Risk of a Jobless Recovery?* International Labour Office: Geneva. Available at: www.ilo.org/wcmsp5/groups/public/---dgreports/---dcomm/---publ/documents/publication/wcms_233953.pdf (accessed 5 January 2017).

Jaafar, H.H. and Woertz, E. (2016) Agriculture as a funding source of ISIS: a GIS and remote sensing analysis. *Food Policy* 64, 14–25.

Jaafar, H.H., Zurayk, R., King, C., Ahmad, F. and Al-Outa, R. (2015) Impact of the Syrian conflict on irrigated agriculture in the Orontes Basin. *International Journal of Water Resources Development* 31(3), 436–449. DOI: 10.1080/07900627.2015.1023892.

Kalyvas, S. (2004) The urban bias in research on civil wars. *Security Studies* 13(3), 160–190. DOI: 10.1080/09636410490914022.

Kearney, M. (2002) Transnational migration from Oaxaca, the agrarian question and the politics of indigenous people. *Österreichische Zeitschrift für Geschichtswissenschaften* 13(4), 7–21.

Kelly, P. (2011) Migration, agrarian transition, and rural change in Southeast Asia. *Critical Asian Studies* 43(4), 479–506. DOI: 10.1080/14672715.2011.623516.

Koren, O. and Bagozzi, B.E. (2016a) From global to local, food insecurity is associated with contemporary armed conflicts. *Food Security* 8, 999–1010.

Koren, O. and Bagozzi, B.E. (2016b) Replication data for: from global to local, food insecurity is associated with contemporary armed conflicts. *Harvard Dataverse*, 1. DOI: 10.7910/DVN/5OGHBE. Available at: https://dataverse.harvard.edu/dataset.xhtml?persistentId=doi:10.7910/DVN/5OGHBE (accessed 4 January 2017).

Lowder, S., Skoet, J. and Raney, T. (2016) The number, size, and distribution of farms, smallholder farms, and family farms worldwide. *World Development* 87, 16–29. DOI: http://dx.doi.org/10.1016/j.worlddev.2015.10.041.

Martin, P., Mayer, T. and Thoenig, M. (2008) Civil wars and international trade. *Journal of the European Economic Association* 6(2–3), 541–550.

Maxwell, J. and Reuveny, R. (2000) Resource scarcity and conflict in developing countries. *Journal of Peace Research* 37(3), 301–322.

Maystadt, J.F. and Ecker, O. (2014) Extreme weather and civil war: does drought fuel conflict in Somalia through livestock price shocks? *American Journal of Agricultural Economics* 96(4), 1157–1182.

Maystadt, J.F., Trinh Tan, J.F. and Breisinger, C. (2014) Does food security matter for transition in Arab countries? *Food Policy* 46, 106–115.

McMichael, P. (2006) Reframing development: global peasant movements and the new agrarian question. *Canadian Journal of Development Studies* 27(4), 471–483.

McMichael, P. (2010) The world food crisis in historical perspective. In: Madoff, F. and Tokar, B. (eds) *Agriculture and Food in Crisis: Conflict, Resistance, and Renewal*. Monthly Review Press, New York.

McMichael, P. (2013) *Food Regimes and Agrarian Questions*. Fernwood Publishing, Halifax, Canada.

Midlarsky, M. (1988) Rulers and the ruled: patterned inequality and the onset of mass political violence. *American Political Science Review* 82(2), 491–509.

Moshiri, F. (1991) Revolutionary conflict theory in an evolutionary perspective. In: Goldstone, J., Gurr, T.R., and Moshiri, F. (eds) *Revolutions of the Late Twentieth Century*. Westview Press, Boulder, Colorado, pp. 4–36.

Moyo, S., Jha, P. and Yeros, P. (2013) The classical agrarian question: myth, reality and relevance today. *Agrarian South: Journal of Political Economy* 2(1), 93–119.

Muller, E., Seligson, M., Fu, H. and Midlarsky, M. (1989) Land inequality and political violence. *American Political Science Review* 83(2), 577–596.

O'Laughlin, B. (2009) Gender justice, land and the agrarian question in southern Africa. In: Akram-Lodhi, K. and Kay, C. (eds) *Peasants and Globalization: Political Economy, Rural Transformation and the Agrarian Question*. Routledge, London, pp. 190–213.

Patnaik, U. (2013) Some aspects of the contemporary agrarian question. *Agrarian South: Journal of Political Economy* 1(3), 233–254.

Pinstrup-Andersen, P. and Shimokawa, S. (2008) Do poverty and poor health and nutrition increase the risk of armed conflict onset? *Food Policy* 33, 513–520.

Richards, P. (2005) To fight or to farm? Agrarian dimensions of the Mano River conflicts (Liberia and Sierra Leone). *African Affairs* 104(417), 571–590.

Ross, M. (2004) What do we know about natural resources and civil war? *Journal of Peace Research* 41(3), 337–356.

Samberg, L., Gerber, J., Ramankutty, N., Herrero, M. and West, P. (2016) Subnational distribution of average farm size and smallholder contributions to global food production. *Environmental Research Letters* 11(124010). DOI: 10.1088/1748-9326/11/12/124040. Available at: http://iopscience.iop.org/article/10.1088/1748-9326/11/12/124010/pdf and http://iopscience.iop.org/1748-9326/11/12/124010/media/erl124010_suppdata.pdf.

Sengupta, S. (2017) Why 20 million people are on the brink of famine in a 'world of plenty'. *The New York Times*, 22 February. Available at https://www.nytimes.com/2017/02/22/world/africa/why-20-million-people-are-on-brink-of-famine-in-a-world-of-plenty.html?_r=0 (accessed 23 February 2017).

Simmons, E. (2017) *Recurring Storms: Food Insecurity, Political Instability, and Conflict*. Center for Strategic and International Studies, Washington, DC.

The agrarian question: past, present and future (2012) Editorial. *Agrarian South: Journal of Political Economy* 1(1), 1–10. DOI: 10.1177/227797601200100101.

Thomson, F. (2011) The agrarian question and violence in Colombia: conflict and development. *Journal of Agrarian Change* 11(3), 321–356. DOI: 10.1111/j.1471-0366.2011.00314.

United Nations (2017) UN comtrade database. Available at: https://comtrade.un.org (accessed 9 February 2017).

United Nations Development Programme (2015) Income Gini coefficient. Available at: http://hdr.undp.org/en/content/income-gini-coefficient (accessed 2 September 2017).

Vellema, S., Borras, S. and Lara, F. (2011) The agrarian roots of contemporary violent conflict in Mindanao, southern Philippines. *Journal of Agrarian Change* 11(3), 298–320. DOI: 10.1111/j.1471-0366.2011.00311.

von Uexkull, N., Croicu, M., Fjelde, H. and Buhaug, H. (2016) Civil conflict sensitivity to growing-season drought. *Proceedings of the National Academy of Sciences of the USA* 113(4), 12391–12396.

Warren, B. (1980) *Imperialism: Pioneer of Capitalism*. Verso, London.

Watts, M.J. (2009) The southern question: agrarian questions of capital and labour. In: Akram-Lodhi, K. and Kay, C. (eds) *Peasants and Globalization: Political Economy, Rural Transformation and the Agrarian Question*. Routledge, London, pp. 262–287.

Weinberg, J. and Bakker, R. (2015) Let them eat cake: food prices, domestic policy and social unrest. *Conflict Management and Peace Science* 32(3), 309–326.

Weis, T. (2007) *The Global Food Economy: The Battle for the Future of Farming*. Zed Press, London.

World Bank (2008) World development report 2008: agriculture for development. World Bank Group, Washington, DC. Available at: http://siteresources.worldbank.org/INTWDRS/Resources/477365-1327599046334/8394679-1327606607122/WDR_00_book.pdf (accessed 12 January).

World Bank (2011) World development report 2011: conflict, security, and development. World Bank Group, Washington, DC. Available at: http://siteresources.worldbank.org/INTWDRS/Resources/WDR2011_Full_Text.pdf (accessed 12 January 2017).

World Bank (2016a) Fragile and conflict affected situations. World Bank Group, Washington, DC. Available at: http://data.worldbank.org/region/fragile-and-conflict-affected-situations (accessed 15 September 2016).

World Bank (2016b) World databank – world development indicators. Available at: http://databank.worldbank.org/data/reports.aspx?source=world-development-indicators (accessed 15 September 2016).

Appendix. Selected statistics on agriculture in fragile and conflict-affected states.

Country	Agricultural land (% of land area)[a]	Number of farms <2 ha[b]	Prevalence of farms <2 ha (% of total holdings)[b]	Rural population (% of total population)[a]	Rural poverty headcount ratio (% of rural population)[a]	Agricultural value added (% of GDP)[a]	Employment in agriculture (% of total employment)[a]	Unemployment (% of total labour force)[c]	Gini coefficient – income[d]
World	37.7	475,000,000	84	46.1	N/A	3.9	19.8	5.9	N/A
Fragile/conflict-affected states	33.4	N/A	N/A	59.2	N/A	20.6	N/A	7.9	N/A
Afghanistan	58.1	N/A	N/A	73.3	38.3	21.7	N/A	9.1	27.8
Burundi	79.2	N/A	N/A	87.9	68.8	43.0	92.2	6.9	33.36
Central African Republic	8.2	N/A	N/A	60.0	69.4	42.4	N/A	7.4	56.24
Chad	39.7	N/A	N/A	77.5	52.5	52.4	83.0	7.0	43.32
Comoros	71.5	N/A	N/A	71.7	48.7	33.6	N/A	6.5	55.93
Democratic Republic Congo	11.6	4,351,000	97	57.5	64.9	20.6	N/A	8.0	42.10
Côte d'Ivoire	64.8	629,366	56	45.8	56.8	20.2	N/A	4.0	43.18
Djibouti	73.4	1,135	100	22.7	N/A	3.9	N/A	N/A	44.13
Eritrea	75.2	N/A	N/A	79.0	N/A	14.5	N/A	7.2	N/A
The Gambia	59.8	N/A	N/A	40.4	73.9	21.1	31.5	7.0	47.33
Guinea-Bissau	58.0	73,929	88	50.7	75.6	43.7	N/A	6.9	50.66
Haiti	66.8	N/A	N/A	41.4	74.9	N/A	50.5	6.8	60.79
Iraq	21.3	N/A	N/A	30.5	30.6	N/A	23.4	16.4	30.9
Kiribati	42.0	N/A	N/A	55.7	N/A	23.0	22.1	N/A	37.61
Kosovo	52.4	N/A	N/A	N/A	31.5	13.5	4.6	N/A	26.71
Lebanon	64.3	169,028	87	12.2	N/A	4.8	N/A	6.4	N/A
Liberia	28.0	N/A	N/A	50.3	67.7	N/A	46.5	3.8	36.48
Libya	8.7	42,867	24	21.4	N/A	1.9	N/A	19.2	N/A
Madagascar	71.2	N/A	N/A	64.9	81.5	25.6	75.3	3.6	42.65
Mali	33.8	1,135	100	60.1	50.6	41.0	66.0	8.1	33.04
Marshall Islands	63.9	N/A	N/A	27.3	N/A	14.7	11.0	N/A	N/A
Federated States Micronesia	31.4	N/A	N/A	77.6	N/A	28.2	N/A	N/A	42.46
Myanmar	19.4	1,972,070	57	65.9	N/A	26.7	62.7	3.3	N/A

Papua New Guinea	2.6	N/A	N/A	87.0	41.6	35.8	72.3	2.5	43.88
Sierra Leone	54.7	N/A	N/A	60.1	66.1	61.3	68.5	3.3	33.99
Solomon Islands	3.9	N/A	N/A	77.7	N/A	34.7	N/A	3.9	46.10
Somalia	70.3	N/A	N/A	60.4	N/A	65.4	N/A	6.9	N/A
South Sudan	N/A	N/A	N/A	81.2	55.4	N/A	N/A	N/A	46.34
Sudan	28.7	N/A	N/A	66.2	57.6	39.3	44.6	14.8	35.39
Syrian Arab Republic	75.8	N/A	N/A	42.3	36.9	17.9	13.2	10.8	35.8
Togo	70.2	247,033	29	60.0	68.7	40.7	54.1	6.9	46.02
Tuvalu	60.0	N/A	N/A	40.3	27.5	21.7	N/A	N/A	41.1
West Bank and Gaza	49.5	N/A	N/A	24.7	19.4	4.5	10.5	26.2	34.46
Yemen	44.6	989,785	84	65.4	40.1	9.5	24.7	17.4	37.7
Zimbabwe	41.8	N/A	N/A	67.6	84.3	13.4	65.8	5.4	43.15

[a]Source: World Bank, 2016b; figures reported are latest available per country or aggregate.
[b]Source: Lowder et al., 2016, Web appendix tables; figures are taken from 1990 and 2000 census rounds.
[c]Modelled ILO estimate. Source: World Bank, 2016b; figures reported are latest available per country or aggregate.
[d]Source: World Bank, 2016b; figures reported are latest available per country or aggregate. Data for Afghanistan, Iraq, Syrian Arab Republic and Yemen are taken from UNDP, 2015, citing World Development Indicators 2013.

2 Geopolitics, Food and Agriculture

Eckart Woertz*

CIDOB (Barcelona Centre for International Affairs) Barcelona, Spain, and Kuwait Chair, Sciences Po, Paris, France

Introduction

In academic debates the subject of agriculture and conflict is mostly associated with domestic development issues such as the disenfranchisement and dislocation of traditional farming populations during a process of 'primitive accumulation' (Marx) or class-based conflicts surrounding the 'agrarian question', once commodification of the agricultural sector and of the larger economy have been achieved (see Bahn and Zurayk, Chapter 1, this volume). Yet food supplies and agriculture also have geopolitical importance. They have occupied planners and strategists, who sought to exploit them to further sovereign interests. In doing so they interacted with the domestic development dimension of agricultural conflict on various levels.

Food supplies are a necessary precondition of urbanization and the rise of the city as the locus of state power. Initially such supplies were local, but the first examples of long-distance grain trade occurred during the Roman Empire. However, by the time of Napoleon's Continental System such international grain trade still constituted a niche supply; using it as a tool of coercion never came to Napoleon's mind during his blockade of Great Britain.

The massive rise of international grain trade during the first food regime changed this strategic landscape from the 1870s onwards. Steam power reduced transportation costs and grain supplies from settler states fed the growing industrial workforce of the UK and continental Europe. Achieving and denying access to long-distance food supplies became a decisive factor during World Wars I and II, engraining strategic self-sufficiency consideration in post-war reconstruction efforts. The second food regime of the post-war decades saw the rise of subsidized intensive 'grain–livestock complexes' in developed countries (Weis, 2007). They produced structural surpluses that their governments sought to dump in developing countries via export promotion in order to stabilize prices. Export support such as the 'Food for Peace' programme shaped agriculture and development options in the developing world, and the US exploited such programmes as a foreign policy tool, especially in the 1970s when global food trade was highly politicized.

During the 1980s and 1990s commercial priorities prevailed, but the global food crisis of 2008 put agriculture and international food supplies again in the geopolitical spotlight. Food-deficient countries in Asia and the Middle East announced large-scale agro-investments in often food-insecure countries, propelling Africa to the centre of attention as the continent with

---

* Email: ewoertz@cidob.org

the largest available land bank and substantial unused water resources, but also with the highest population growth worldwide. Reliable external food supplies, support of domestic agricultural production and the sustainability of water consumption practices will continue to be of strategic importance and a vital state interest in the 21st century.

Urbanization and Food Supplies

Food supplies are a necessary precondition of urbanization. Until the advent of railroads their transportation beyond waterways was difficult. First city-states in Mesopotamia and Egypt emerged in river valleys downstream of grain-producing areas that allowed for convenient transportation. Much of the time of their kings and their literate administrative class was spent on ensuring agricultural surpluses on favourable terms; priests legitimized it and warriors enforced it (Lees, 2015). The grain production relied on slavery and other forms of coerced labour, establishing a close link between agricultural production and conflict in these early city-states. The surplus appropriation had to be in the form of grains. No other food had the proprieties that made grains so accessible for taxmen. They have a high energy density and can be easily stored, divided, measured and transported. They also have to be harvested during a particular season and cannot be hidden underground to be dug up at a later point in time, as some roots and tubers. Hence these states were grain states. There have not been cassava, lentil or chickpea states in history (Scott, 2017, pp. 128–137). However, water management was less centralized than Wittfogel's simplistic notion of 'hydraulic civilizations' suggests and was also dependent on decentralized agency of the cultivators (Hassan, 1997).

The food supplies were local and relied on hinterlands in relative vicinity. Bulk commodities like timber and grain could be provided efficiently only over a certain distance as the German geographer Heinrich von Thünen described in his Thünen circles (1826) (Altvater and Mahnkopf, 1996, p. 221). Production of perishable goods like dairy products and fruit and vegetables would typically need to be located still closer to cities. Livestock could come from further away, but beyond this outer Thünen circle there was an economic 'wilderness' that could not be harnessed for agricultural supplies to cities. Modern transportation and cooling chains have changed this equation. Even in earlier times there were differentiations depending on availability of water transport, variations of topography, soil fertility and changing demand prices, but overall geography constituted a powerful barrier to long-distance transportation of foodstuffs.

The Roman Empire was the first to rely on long-distance trade of grains to supply its armies and some of its cities, namely Rome, which was uniquely large by the standards of its time. It grew from a population of 300,000 in 150 BCE to 700.000–1.2 million in the 2nd century CE. Grain supplies for such a large city were important for political stability ('bread and circuses') and required a hinterland of unprecedented size that not only encompassed the south of Italy and Sicily, but also the Cyrenaica in today's Libya and Egypt. Given the needs of large standing armies, there was a seasonality of warfare. Fresh food supplies were unavailable in the winter and the storms in the Mediterranean precluded shipping from November to April. Inland warfare was severely restricted when waterways for grain transportation were unavailable (Erdkamp, 1998, p. 19f and pp. 152–154).

Apart from Rome and its armies, the need for self-sufficiency prevailed. The long-distance grain was a tax that was raised in kind; it was not freely traded on markets. Access to it depended on social and political rights and allocation mechanisms, namely for citizens and slaves. Only workmen and foreigners procured grain via markets (Erdkamp, 2005).

The long-distance grain trade was only possible because of Rome's unifying role (currency, measures, rules) and its role as guarantor of security that lowered transaction costs (Erdkamp, 2005, p. 207). With the Barbarian invasions that started in the 4th century this framework came apart, together with the cities that were the demand base for the grain trade. Rome was sacked in 410 CE. By 700 CE the once proud city only had 50,000 inhabitants, declining further to 35,000 at the beginning of the 11th century (Lees, 2015, p. 30f).

The relative importance of the Roman long-distance grain trade during its heyday remained unsurpassed until the 17th/18th centuries (Erdkamp, 2005, p. 203). Cities had witnessed renewed growth since 1500, but even in the Mediterranean with its long coastlines and well-connected waterways, long-distance grain in the 16th century only represented about 8% of consumption (Braudel, 2000, Vol. 2, p. 62).

Long-distance grain trade continued to be a niche market as late as the early 19th century. Napoleon attributed no strategic importance to it during the Continental System (1806–1814) when he sought to inflict an economic blockade on Britain. There was a small 5% share of grain imports in British consumption, half of which came from the Baltics via Danzig. Yet Napoleon only targeted British manufactured exports, never its food imports! His food policies were governed by France's domestic concerns about supply security and the 'policy of plenty'. Commercial consideration usually did not enter this equation. He declared, 'The corn question is for sovereigns the most important and the most delicate of all. The first duty of the prince in this question is to hold to the people, without listening to the sophisms of the landowners' (Heckscher, 1922, p. 341).

In contrast, Napoleon followed a typical mercantilist policy when it came to manufactured goods. Here, improving the trade balance and maximizing profits were the orders of the day (Milward and Saul, 1979). Only when abundant domestic grain supplies were assured after a good harvest as in 1809–1810, grains were included in this mercantilist logic. In these years Napoleon allowed corn exports from France, Flanders and Holland to the UK to improve prices for domestic farmers and benefit from export duties and an improved trade balance.

Beside the limited role of long-distance grain trade and the non-commercial 'policy of plenty' disposition, Napoleon never contemplated a grain boycott because he thought that the UK would have been able to source alternative supplies from Canada and its supplies from Ireland were off limits to him due to British naval supremacy. He also doubted that he could sufficiently control Prussia and Poland to prevent grain exports from his realm, which his mercantilist logic deemed 'more natural' than imports on which his blockade focused (Heckscher, 1922).

However, relative British wheat prices rose as a result of trade disruptions of the Continental System and preceding trade wars between France and the UK (O'Rourke, 2006). Although the grain trade was never directly targeted and only a niche supply there was reason for concern, as evidenced in British counter measures (Olson, 1963). The subsequently repealed Brown Bread Act of 1800 prohibited the baking of bread with unmixed fine bolted wheat flour, the use of grains for the production of spirits and starch was curtailed, and awareness campaigns exhorted the public to consume less bread (Heckscher, 1922, p. 339). The impact lasted beyond the Continental System: The UK protected its domestic agricultural producers with the Corn Law of 1815 that would only be repealed in 1846.

The First Food Regime and the Rise of Long-Distance Trade

If the UK remained largely self-sufficient in grains in the early 19th century, this changed after the abolition of the Corn Laws in 1846. They had provided protection to domestic producers and were favoured by landlords, but were opposed by the nascent bourgeoisie and industrialists who were interested in cheap food for the growing urban working class. Long-distance grain trade grew steadily. Transportation costs fell with the advent of steam ships and railroads that opened hitherto inaccessible interiors to the grain trade. By 1870 the UK imported about half of its cereal supplies. Cheap grain supplies from settler states such as Canada, the US and Australia, but also Russia and India fed the workforce of industrialization. They were a cornerstone of the first food regime that lasted from the 1870s to the 1930s (Friedmann and McMichael, 1989; McMichael, 2009). Large grain-trading houses such as Cargill, Dreyfus and Bunge emerged that would continue to play a major role in global food trade in the 20th and 21st centuries (Morgan, 1979; Murphy et al., 2012).

The growth in foreign grain supplies was accompanied by growing domestic production. Both accommodated steep population growth during Britain's 'second agricultural revolution' that occurred between 1830 and 1880 (Foster, 1999, pp. 373–375). The UK's population roughly doubled from 11 to 21 million people between 1800 and 1850, and then almost doubled again to 37 million by the end of the century. If the first agricultural revolution had improved agricultural productivity since the 17th century via improved techniques of crop rotation, drainage, manuring and livestock management, the 19th century saw the increased importance of fertilizers and an advanced understanding of soil chemistry as pioneered by the German chemist Justus von Liebig (1803–1873). Until the 16th/17th century urbanization was still limited by bioregional constraints. It had to rely on an immediate hinterland and its ecological capacities. The necessary nutrients for food production were of organic origin. As cities separated the production and consumption of agricultural products spatially, nutrition cycles as they occur in rural communities were interrupted. Only some of the waste products of consumption were given back to the land, via recycling of urban night soils and other measures (Harvey, 1996, p. 410f).

Once cities grew in size, this 'metabolic rift' (Marx) and the accompanying soil exhaustion became a major concern of advanced capitalist societies in the 19th century (Foster, 1999). Much of Marx' theorizing in this matter focused on a possible recycling of nutrition flows from the city back to the land, but instead it was the increased application of fertilizers that was used to bridge the nutritional gap that the metabolic rift had created. Demand was so high that European farmers raided Napoleonic battlefields to grind bones and use them as fertilizer. Bone imports of Britain increased more than 17-fold between 1823 and 1837, and in 1835 first imports of Peruvian guano arrived in Liverpool, a trade that the UK monopolized for strategic reasons. The US in turn sought to secure its share of the global guano business that encompassed places like Chile, Bolivia, Peru, West Africa and Pacific islands such as Nauru. The bird faeces that had accumulated over millennia could be mined and provided a precious source of the macronutrients nitrogen (N), phosphorus (P) and potassium (K). The Guano Islands Act of 1856 allowed US citizens to take possession of unclaimed islands with guano deposits and authorized the US president to use military force to protect such interests. In the 1860s, when Peruvian guano was largely exhausted, inorganic nitrates from Chile became a major source of fertilizers in the UK and elsewhere.

Overseas possession were 'spatial fixes' (Harvey, 1996) to address ecological constraints of accumulation and urbanization. The advent of the global grain trade provided a larger hinterland for the growing cities of the centre, while fertilizer imports from overseas improved the productivity of domestic agriculture. The commodification of agricultural relations in colonies undermined traditional livelihoods and made vast parts of the global south vulnerable to the famines of the 'Late Victorian Holocausts' of the late 19th century (Davis, 2001). Capitalism's 'robbing of the soil' (Marx) went beyond agriculture and included the mining of fertilizer inputs that were crucial in addressing the ecological challenges it had created. A fertilizer treadmill of steadily increasing inputs was needed to address the structural unsustainability. The first food regime thus left the UK and parts of western Europe dependent on imports of grains and fertilizers. This strategic vulnerability would manifest itself in the two World Wars of the 20th century.

World Wars I and II

During World War I Entente forces implemented a naval blockade against the German Reich, Austria and the Ottoman Empire with disastrous effects (Vincent, 1985; Offer, 1989). The blockade lasted from August 1914 and was only lifted in July 1919 after the Treaty of Versailles. By 1915 Germany had to introduce rationing and by 1916 staple foods such as grains, potatoes, meat and dairy products had become so scarce that people had to subsist on subsidiary products of limited nutritional value, such as *Kriegsbrot*, which used potato flour, turnips and even sawdust (National Archives, 2017). Occupied territories felt the impact first: food exactions from Austrian-occupied Serbia caused famine-related deaths of 365,000 people in 1916. At the end of the war in October 1918, calorie provisions in Germany had declined by half and supplies of proteins by 80%, with the urban population and women particularly affected in comparison to rural self-suppliers and the military. Malnutrition made people susceptible to diseases such as scurvy, tuberculosis, influenza and dysentery. Almost half a million people succumbed to famine and related diseases in Germany alone (Howard, 1993).

In the political turmoil between the armistice in November 1918 and the Treaty of Versailles, German command structures partially collapsed. More than 10,000 Soldiers' Councils filled the vacuum. During the following withdrawal they played a crucial role in alleviating famine in Germany by repatriating food from stores of the German army in occupied territory in France, Belgium, Poland and Ukraine. The continued blockade was used by the Entente as a stick in negotiations ahead of the Treaty of Versailles. Relief aid of the Entente was seen as a tool to push back Bolshevist groups and was denied to regions that were under their influence. There was a degree of collusion with the remnants of the old regime and the German High Command in that matter. With an internal blockade the latter sought to punish dissident regions that refused to dissolve the Workers' and Soldiers' Councils in early 1919 by withholding food supplies (Howard, 1993). Nationalist circles in Germany blamed the 'hunger blockade' for domestic unrest and military defeat. Although it was impending military collapse on the Western Front that sealed this fate, the questionable interpretation would inform the conspiratorial 'stab in the back legend' and German hunger warfare in the east in the following World War II.

The Ottoman Empire used to be a net grain exporter before World War I and regained this position in the 1930s, but during the war the Entente naval blockade caused massive famine in Greater Syria (Schatkowski-Schilcher, 1989). Over half a million people succumbed to starvation. Mount Lebanon was particularly affected, as it did not have cereal cultivation. Mulberry trees and silk production for export dominated its agriculture (Schatkowski-Schilcher, 1992). Further south the Ottoman Empire used grain deliveries to assure the loyalty of tribes in Jordan. The British on the other hand lifted the blockade of the Hejaz to allow grain exports when the Sherif of Mecca, Hussein, agreed to launch an Arab revolt against the Ottomans (Kostiner, 1993).

The importance of food and its strategic utilization grew in magnitude in World War II. The US was the only major war party that did not fight on empty stomachs. Famines in the UK were only averted by deliveries from the US. Sixty per cent of Japanese casualties were not caused by military action, but by starvation when the US Navy was able to cut off maritime supply lines in the Pacific (Collingham, 2011). The US also targeted Japanese fertilizer supplies. Japan's main source of fertilizer was French North Africa. The Allied invasion of Morocco and Allied control of sea-lanes cut off this supply. Japan had to substitute it with imports from the Pacific island of Nauru. Once the US found out, it bombarded the island, compromising Japan's domestic food production (Farago, 1954, p. 25f).

The most pernicious application of the food weapon occurred in the Soviet Union. Nazi Germany planned to feed its army off the land and subjugate people by eradicating their history, culture and means of survival. To this end the so-called *Hungerplan* aimed to cut off food supplies to cities, which had grown steeply as a result of Stalin's industrialization drive, and starve 30 million people to death (Tooze, 2006). Implementation remained limited due to the changing war fortunes, but where it occurred, its consequences were grave. During the siege of Leningrad over a million people perished.

Driven by the hunger blockade experience of World War I, the German leadership was determined to prevent food shortages at home. To this end occupied territories were looted and extorted with gusto. Food shortages only occurred again *after* the war in the harsh winter of 1946/47, when this option was not available any more (Aly, 2007). France and Denmark were particularly important for German civilian supplies. Butter, pork and beef from Denmark constituted up to one-fifth of domestic German consumption (Beevor, 2012, pp. 425, 432). This was particularly important as food supplies were rationed and focused on less cost-intensive grains (Aly, 2007, p. 170).

The Ukraine as the breadbasket of the Soviet Union played a central role in this strategic planning. Trade between Nazi Germany and the Soviet Union had shrunk to a trickle since 1932, but with the Hitler–Stalin Pact in 1939 deliveries of grains and raw materials soared and provided critical assurances ahead of the invasion of Poland. Hitler argued that Germany would need Ukraine 'so that no one is able to starve us again as they did in the last war' (Frankopan, 2015, p. 364). During his later attack on the Soviet Union he insisted on making a southern turn towards the grain fields and coalmines of Ukraine, overriding his General Staff, who had argued in favour of a direct and decisive thrust towards Moscow. However due to scorched earth tactics of the retreating Soviets, war damage, depopulation, ruthless occupation policies and partisan activity, agricultural production collapsed and Ukrainian supplies never became as important as the Germans had hoped for (Beevor, 2012, p. 418). Still, widespread looting, bartering of low-quality products against agricultural wares and their purchase with artificially overvalued German currency by German soldiers also contributed to civilian supplies at home (Aly, 2007, p. 114f; Beevor, 2012, p. 208).

The Politicization of the Food Trade and the Crisis of the Second Food Regime in the 1970s

After World War II the second food regime emerged. Grain production moved back to the centre of industrialized countries, where large grain–livestock complexes catered to an increased meatfication of diets. The growing consumption of packaged foods required new input factors, such as soybean oil. Synthetic fibres and corn syrup reduced the importance of tropical commodity imports, which had been central to the first food regime, namely cotton and sugar (Weis, 2007; McMichael, 2009).

Farm subsidies, the growing application of mineral fertilizers and mechanization of agriculture led to productivity gains and structural surpluses that weighed on prices, first in the US and, after reconstruction, also in western Europe. One way to cope with this dilemma was export promotion. Programmes such as the US Public Law 480 of 1954 encouraged subsidized food aid deliveries to the developing world that were aggressively pushed by agricultural attachés in the embassies of northern grain exporters. The meatification of diets in the developed world was accompanied by a wheatification of diets in developing countries that crowded out traditional staple foods, such as cassava, rice, corn and beans (Morgan, 1979; McMichael, 2009).

The commercial interests behind the export promotion were complemented by a developmental agenda. In his State of the Union address in 1941 President Roosevelt had characterized the 'freedom from want' as one of four essential freedoms. With the establishment of the Food and Agriculture Organization of the United Nations (FAO) after the war, food security moved to the top of the emerging international development agenda (Shaw, 2007). The Kennedy administration added a development component to the P.L. 480 programme and renamed it 'Food for Peace'. Recipient countries in the developing world welcomed the subsidized grain inflows as a cheap input to feed the growing work force of their drive towards import substituting industrialization (ISI), which was the predominant development paradigm at that time.

The growing grain trade to developing countries at subsidized costs had a geopolitical aspect as well. It created dependencies and vulnerabilities. It could be used as a carrot, but also as a stick. Close US allies were the main recipients. Nixon's 'food for war' programme allowed south Vietnam to use proceeds from P.L. 480 sales to purchase arms. It was convenient for the Nixon administration. Because of domestic farming interests, food aid was much easier to push through Congress than direct military aid (Morgan, 1979, p. 258; Wallerstein, 1980).

Adversarial behaviour in contrast was castigated with withdrawal of food aid. This happened to the Goulart and Allende governments in Brazil and Chile, to Cuba and to North Korea because of their leftist leanings. In the cases of Pakistan and India and Malaysia and Indonesia it was done to sanction inter-state belligerence. In Ceylon and Honduras food aid was withdrawn in reaction to attempts to nationalize US property (Nelson, 1968; Wallensteen, 1976, p. 292; Morgan, 1979, p. 259f).

Egypt was a prominent example of these geopolitical food aid machinations. It was the largest per capita recipient of US food aid in the world between 1958 and 1965. The US hoped to lure Nasser out of the orbit of the Soviet Union after Iraq had left the Baghdad Pact in the wake of General Qasim's coup d'état in 1958. By 1964 Egyptian wheat imports accounted for 18.6% of total imports, up from virtually nothing in 1955 (Woertz, 2013, p. 112). Almost all of these imports (91%) were P.L. 480 deliveries (Burns, 1985, pp. 119, 150). Yet when the Egyptian president showed no inclination to mellow his revolutionary rhetoric and actions, President Johnson decided to put Nasser on a 'short leash' in 1965. P.L. 480 deliveries were suspended and completely discontinued in 1966 (Burns, 1985, p. 159f). They would only resume in 1974, conditioned on Egyptian peace negotiations with Israel. By 1978 Egypt again received the lion's share of American P.L. 480 deliveries, with 30% of the total (Wallerstein, 1980, p. 46; Burns, 1985, p. 193).

By that time the global food system was in profound crisis. In the early 1970s the US, Canada and Australia sought to reduce production to stabilize prices on higher levels, right before the Soviet Union appeared as a large grain buyer on international markets in 1972. To offer a similar meatfication of diets to its citizens as in the capitalist world the Soviets had embarked on an ambitious livestock programme at home, but the required feedstock overwhelmed domestic production capacities. The so-called 'great grain robbery' caused rising prices in the US and dismayed local consumers and livestock producers. Other contributing factors to the complex crisis were strong El Niño weather conditions, high food demand at the end of an economic boom period and higher oil prices that increased input costs for agriculture.

The rising prices ushered in the world food crisis of 1972–1975. US farmers now found ample takers for commercial exports and lost interest in the export promotion of P.L. 480. Reduced food aid hurt the grain-importing developing countries. Only the few oil exporters among them were able to match rising food import costs with growing revenue streams. To alleviate the plight the US tried to stimulate agricultural production in developing countries. To this end it sought cooperation with oil exporters by offering US leadership in agricultural production growth against oil price moderation on the part of Organization of the Petroleum Exporting Countries (OPEC). This cooperation would later lead to the establishment of the International Fund for Agricultural Development (IFAD) in 1977 that was jointly funded by Organisation for Economic Co-operation and Development (OECD) and OPEC countries to spur agricultural projects in developing countries (Woertz, 2013).

The politicization of food trade reached new heights in the 1970s. This time the arsenal of the food weapon did encompass food embargoes, beside the deliberate withdrawal of food aid. The US contemplated unilateral food embargoes in 1973, in retaliation to the Arab oil boycott and against Iran in 1979 in the wake of the Iranian hostage crisis. Such moves spurred self-sufficiency programmes for cereal production in the Middle East and a failed attempt to develop Sudan as an Arab breadbasket. Other embargoes were motivated by domestic politics. To protect American livestock producers against high feedstock prices and reduce food price inflation at home, soybean exports were embargoed in 1973 and grain sales to the Soviet Union and Poland were temporarily suspended in 1975 (Ruttan, 1996; Woertz, 2013, chapter 4). The US grain embargo against the Soviet Union in 1980 that followed the Soviet invasion of Afghanistan marked a turning point. It failed. Competing grain exporters in Argentina, Europe and Australia happily picked up the slack and delivered to the Soviet Union, which only suffered minor consequences and did not modify its foreign policy. Meanwhile, American farmers were up in arms over lost export revenues (Paarlberg, 1980).

After this sobering experience, food trade in the US was depoliticized. It has prioritized commercial considerations ever since. Overproduction, high interest rates and a strong dollar led to the farm crisis of the 1980s. The US was now anxious to secure export markets and repair its damaged reputation as a reliable supplier of food. The 'Reagan Doctrine on agricultural trade' disavowed food embargoes and called for agricultural trade liberalization (Woertz, 2013, p. 131). The importance of national agencies in international food aid has declined since then. It has been increasingly administered by multilateral institutions such as the World Food Programme (WFP) (Paarlberg, 2010).

The multilateral UN embargo against Iraq that lasted from Iraq's temporary occupation of Kuwait in 1990 to the US invasion of 2003 was an exception to the general trend of depoliticization. Imports of food and medicine were formally exempted, but Iraq's ability to purchase them was severely hampered by the inability to export oil, financial sanctions and freezing of its foreign assets. In contrast to earlier embargo episodes the Iraq embargo was multilateral. Iraq was unable to source alternative trading partners. The Oil for Food programme brought a modicum of relief after 1996 as it allowed for monitored oil exports for the purchase of food and medicine, but the effect of the embargo on food security was still tremendous. Estimates of excess deaths as a result of sanctions hover around 500,000, with some reaching over 800,000. Most of them were children who suffered in particular from malnutrition and water-borne diseases (Dyson, 2006; Gordon, 2010; Marr, 2012).

The Global Food Crisis of 2008 and the Land Grab Debate: 'Africa Rising'?

In 2007/08 global food prices rose steeply and some agricultural exporter nations such as Russia, Vietnam, India and Argentina announced export restrictions out of concern for their own food security. Trust in global food markets and supply chains waned. The crisis had a geopolitical component. Food-importing nations in Asia and the Middle East that are often water scarce developed an interest in farmland investments abroad with the goal of gaining privileged bilateral access to food production. Much of this interest focused on countries in Africa with insecure land rights and problematic governance (Deininger et al., 2011). Together with a few countries in Latin America a majority of the global unused land bank is located on the continent. Sub-Saharan Africa also has substantial blue water reserves that could be harnessed, if the necessary investments in irrigation infrastructure were undertaken. It mostly has an economic water shortage as opposed to the physical water shortage that is prevalent in parts of the Middle East and North Africa (MENA) countries, Central Asia and South Asia.

Agricultural productivity gains mainly drove production growth in the post-war decades, but have declined since the 1990s. This was one contributing factor to the global food crisis of 2008 (Piesse and Thirtle, 2009; OECD-FAO, 2010). Against this backdrop Africa has been identified as a 'last frontier' of agriculture, as the Green Revolution has largely bypassed the continent. Agricultural investments in Africa are regarded as crucial to foster continuous global production growth. Although still a food net importer, it might one day be able to cater to its own growing population and satisfy changing diets of new middle classes in emerging markets alike, so the argument goes (Thurow, 2010).

China and the oil-rich Gulf countries have played a prominent role in the land investment drive. Like for India, grain self-sufficiency is a strategic concern for China (Weis, 2007). Overt reliance on imports is not an option for the country because of its population size. However, in recent years it has modified this stance by allowing a modest import component for grains for human consumption. Agricultural outsourcing on a larger scale has occurred in the case of animal feedstock such as soybeans and non-food agricultural products such as rubber and timber (Brautigam, 2009; 2015; Gironde et al., 2016).

China became a food net importer in 2004 and imports play a role in its supply mix. Yet it also has seen strong agricultural production growth at home since the reform era began in 1978 (Huang and Rozelle, 1996; Ravallion, 2009; Ash, 2010). Like the UK and the US in the 19th century China's interest has encompassed agricultural input factors such as fertilizers as the aborted take-over bid of Chinese state-owned Sinochem for Canadian Potash Corp. showed in 2010 (Massot, 2011). As a result of water scarcity, soil erosion and limited availability of arable land, domestic production growth will be more difficult to achieve in the future.

The Gulf countries on the other hand are located in the MENA region, which is the world's largest grain net importer, although its population size is much smaller than the one of Asia. Population growth in the region will only level out in the second half of the 21st century, while agricultural production will stagnate at best due to limited resources of water and land. Saudi Arabia, for example, had to phase out its large programme of subsidized wheat production between 2008 and 2016 as it caused a depletion of fossil water aquifers (Woertz, 2013).

The announcements of agricultural investments have been huge, but not many of them have been implemented, either by the Chinese, or by the Gulf countries. Lack of infrastructure, commercial viability issues, political uncertainties and corruption were major impediments to implementation. Thus, in contrast to some media images, most global food imports continue to come from developed agro-producers in the Americas, Australia, Europe and Asia, rather than from land investments in food-insecure developing countries (Oya, 2013; Scoones et al., 2013; Woertz, 2013; Brautigam, 2015, Woertz and Keulertz, 2015).

It remains to be seen to what extent commercial agriculture might proliferate in Africa in the future, how it will interact with local development ambitions and which business models it will pursue, ranging from large-scale plantations with on-farm processing to contract farming and outgrower schemes, to medium-scale commercial farms (Hall *et al.*, 2017).

When investments have occurred, their focus has not been primary production in sub-Saharan Africa, but rather further up the value chain in food processing and distribution, mostly in more developed agro markets. Saudi government owned company SALIC teamed up with international grain trader Bunge to acquire a majority stake in the privatized Canadian wheat board, and Asian food traders such as Chinese Noble or Singapore-based Olam and Wilmar have developed into major global players (Keulertz and Woertz, 2015). As such, agro-food entities from the Middle East and Asia have tried to muscle their way into increasingly corporatized global food value chains that are characteristic of what some have called a third food regime that has emerged since the 1970s (Burch and Lawrence, 2009).

The investment drive has not gone unchallenged in producer countries such as Brazil, Thailand or Australia that have implemented or discussed ceilings on foreign land ownership in an effort to keep a strategic industry national. Russia has sought to leverage its regained position as a grain net exporter by establishing a national grain-trading house. In contrast to the bottom-up resistance by farmers and holders of customary land rights to agro-investments in developing countries, such measures were instigated by government bureaucracies and parts of the domestic business class.

Conclusion

Ensuring food supplies to armies and the civilian population is an essential aspect of statecraft, often with a narrow-minded national focus. Sometimes uneven development and warfare have prevented famine at home, by causing it elsewhere. With the partial exception of the city of Rome long-distance trade of grains remained a niche supply until the 19th century. But it acquired growing importance in European supply mixes during the first food regime that emerged in the 1870s. The 19th century also saw increased application of fertilizers and improved agricultural productivity in the UK and other countries of western Europe. During the two World Wars the dependency on imported grains and fertilizers was a strategic vulnerability. It caused considerable strain and at times famine in Europe, the Middle East and Japan. The second food regime after World War II saw the advent of structural export surpluses from the north, export promotion and strategic use of food aid. More recently, the 2008 global food crisis prompted food importers in the Middle East and Asia to undertake investments in the value chains of an increasingly corporatized and globalized third food regime. Contrary to widespread media reports, so far investments in primary production in developing countries have been rare in comparison.

The geopolitical importance of food trade will likely increase in the 21st century. Growing and increasingly urban populations that often consume more varied diets will require food to be imported, often across political boundaries and delivered through international trading networks. At the same time further production growth will be challenging to achieve. Traditional exporter nations face slowing productivity improvements, soil erosion and climate change weigh on agriculture and success in hitherto underappreciated regions such as Africa is uncertain. Conflict can also take its toll on agricultural production capacities.

Paradoxically, the growing role of global food trade has gone hand in hand with efforts to spur domestic agricultural production, for example, in China and India. In contrast to the World Bank's prescription of agricultural trade liberalization and trade-based approaches to food security, food self-sufficiency concerns are widespread, especially in the developing world (Clapp, 2017). A degree of food self-sufficiency can make sense for countries with volatile export earnings, underused agricultural potential or high exposure to import dependence with simultaneously high levels of food insecurity. The same applies to countries with large populations or at risk of war and other geopolitical supply disruptions. Although food and medicine are formally excluded from international sanction regimes, financial sanctions and curtailed export earnings (e.g. oil) can affect the ability to pay for food imports, as experiences in Iraq from 1990 to 2003 and more recently in Syria and Iran have shown.

Yet often self-sufficiency aspirations will remain unfulfilled, given considerable scarcity of water and arable land in many countries that harbour them. Many of them also still show considerable population growth, although most countries worldwide, with the exception of sub-Saharan Africa, are now amidst a demographic transition and have significantly reduced birth rates. The monetary value of global net food trade has roughly tripled since 2000. Only 42 countries globally are food net exporters in calorie terms, with the US, Canada, Brazil, Argentina, Australia, Russia, Ukraine, Kazakhstan, New Zealand, Thailand, Indonesia, France and parts of eastern Europe providing the most significant quantities (*Knoema World Data Atlas*, 2017). They derive financial and strategic benefits from this position, while the vast majority of countries are food net importers in calorie terms and need to manage this dependency carefully.

Acknowledgements

A Marie Curie grant of the European Commission (RUDEFOPOS-IRAQ) supported research on this chapter.

References

Altvater, E. and Mahnkopf, B. (1996) *Grenzen der Globalisierung. Ökonomie, Ökologie und Politik in der Weltgesellschaft*. Westfälisches Dampfboot, Munich, Germany.
Aly, G. (2007) *Hitler's Beneficiaries. Plunder, Racial War, and the Nazi Welfare State*. Metropolitan Books, New York.
Ash, R. (2010) The Chinese economy after 30 years of reform: perspectives from the agricultural sector. *Copenhagen Journal of Asian Studies* 28(1), 36–62.
Beevor, A. (2012) *The Second World War*. Weidenfeld and Nicolson, London.
Braudel, F. (2000) *The Mediterranean and the Mediterranean World in the Age of Philip II*. Translation of the second revised edition, 1966 ed. 3 vols. The Folio Society, London.
Brautigam, D. (2009) *The Dragon's Gift : The Real Story of China in Africa*. Oxford University Press, Oxford, UK.
Brautigam, D. (2015) *Will Africa Feed China?* Oxford University Press, Oxford, UK.
Burch, D. and Lawrence, G. (2009) Towards a third food regime: behind the transformation. *Agriculture and Human Values* 26(4), 267–279. DOI: 10.1007/s10460-009-9219-4.
Burns, W.J. (1985) *Economic Aid and American Policy Toward Egypt, 1955–1981*. State University of New York Press, Albany, New York.
Clapp, J. (2017) Food self-sufficiency: making sense of it, and when it makes sense. *Food Policy* 66, 88–96. DOI: http://dx.doi.org/10.1016/j.foodpol.2016.12.001.
Collingham, E.M. (2011) *The Taste of War: World War Two and the Battle for Food*. Allen Lane.
Davis, M. (2001) *Late Victorian Holocausts: El Niño Famines and the Making of the Third World*. Verso, London.
Deininger, K., Byerlee, D., Lindsay, J., Norton, A., Selod, H. and Stickler, M. (2011) *Rising Global Interest in Farmland. Can It Yield Sustainable and Equitable Benefits? In Agriculture and Rural Development*. World Bank, Washington DC.
Dyson, T. (2006) On the death toll in Iraq since 1990. Crisis States Occasional Papers, No. 1. London School of Economics, London.
Erdkamp, P. (1998) *Hunger and the Sword: Warfare and Food Supply in Roman Republican Wars (264–30 B.C.)*. Dutch monographs on ancient history and archaeology v. 20. Gieben, Amsterdam.
Erdkamp, P. (2005) *The Grain Market in the Roman Empire: A Social, Political and Economic Study*. Cambridge University Press, Cambridge, UK.
Farago, L. (1954) *War of Wits: The Anatomy of Espionage and Intelligence*. Funk & Wagnalls, New York.
Foster, J.B. (1999) Marx's theory of metabolic rift: classical foundations for environmental sociology. *American Journal of Sociology* 105(2), 366–405. DOI: 10.1086/210315.
Frankopan, P. (2015) *The Silk Roads. A New History of the World*. Bloomsbury, London.

Friedmann, H. and McMichael, P. (1989) Agriculture and the state system. the rise and decline of national agricultures, 1870 to the present. *Sociologia Ruralis* XXIX(2), 93–117.

Gironde, C., Golay, C. and Messerli, P. (eds) (2016) *Large-Scale Land Aquisitions. Focus on South-East Asia*. International Development Policy Series. Brill/Nijhoff, Leiden, Germany.

Gordon, J. (2010) *Invisible War: The United States and the Iraq Sanctions*. Harvard University Press, Cambridge, Massachusetts.

Hall, R., Scoones, I. and Tsikata, D. (2017) Plantations, outgrowers and commercial farming in Africa: agricultural commercialisation and implications for agrarian change. *Journal of Peasant Studies* 44(3), 515–537. DOI: 10.1080/03066150.2016.1263187.

Harvey, D. (1996) *Justice, Nature, and the Geography of Difference*. Blackwell Publishers, Cambridge, Massachusetts.

Hassan, F. (1997) The dynamics of a riverine civilization. A geoarchaeological perspective on the Nile Valley, Egypt. World *Archaelogy* 29(1), 51–74.

Heckscher, E.F. (1922) *The Continental System: An Economic Interpretation*. Clarendon Press, Oxford, UK.

Howard, N.P. (1993) The social and political consequences of the allied food blockade of Germany, 1918–19. *German History* 11(2).

Huang, J. and Rozelle, S. (1996) Technological change: rediscovery of the engine of productivity growth in China's rural economy. *Journal of Development Economics* 49(2), 337–369.

Keulertz, M. and Woertz, E. (2015) States as actors in international agro-investments. *International Development Policy* 6(1).

Knoema World Data Atlas (2017) Role of trade – role in food consumption of: food net-trade (based on FAO Food Security Data, June 2012) 2017. Available at: https://ban.knoema.org/atlas/topics/Food-Security/Role-of-Trade/Role-in-food-consumption-of-food-net-trade?type=maps (accessed 12 May 2017).

Kostiner, J. (1993) *The Making of Saudi Arabia, 1916–1936: From Chieftancy to Monarchical State*. Studies in Middle Eastern History. Oxford University Press, New York.

Lees, A. (2015) *The City. A World History*. Oxford University Press, Oxford, UK.

Marr, P. (2012) *The Modern History of Iraq*, 3rd edn. Westview Press, Boulder, Colorado.

Massot, P. (2011) Chinese state investments in Canada: lessons from the potash saga. Canada-Asia Agenda no. 16.

McMichael, P. (2009) A food regime genealogy. *Journal of Peasant Studies* 36(1), 139–169. DOI: 10.1080/03066150902820354.

Milward, A. and Saul, S.B. (1979) *The Economic Development of Continental Europe 1780–1870*, 2nd edn. Routledge, Abingdon, UK; New York.

Morgan, D. (1979) *Merchants of Grain*. Viking Press, New York.

Murphy, S., Burch, D. and Clapp, J. (2012) Cereal secrets: the world's largest grain traders and global agriculture. Oxford, Oxfam Research Reports.

National Archives (2017) Spotlights on history: the blockade of Germany 2017. Available at: www.nationalarchives.gov.uk/pathways/firstworldwar/spotlights/blockade.htm (accessed 17 April 2017).

Nelson, J.M. (1968) *Aid, Influence, and Foreign Policy*. Government in the Modern World Series. Macmillan, New York.

O'Rourke, K.H. (2006) The worldwide economic impact of the French Revolutionary and Napoleonic Wars, 1793–1815. *Journal of Global History* 1(1), 123–149.

Organisation for Economic Co-operation and Development (OECD)-Food and Agricultural Organization of the United Nations (FAO) (2010) *Agricultural Outlook 2010–2019*. OECD, Paris.

Offer, A. (1989) *The First World War, an Agrarian Interpretation*. Clarendon Press, Oxford University Press, Oxford, UK.

Olson, M. (1963)*The Economics of the Wartime Shortage: A History of British Food Supplies in the Napoleonic War and in World Wars I and II*. Duke University Press, Durham, North Carolina.

Oya, C. (2013) Methodological reflections on 'land grab' databases and the 'land grab' literature 'rush'. *Journal of Peasant Studies* 40(3), 503–520. DOI: 10.1080/03066150.2013.799465.

Paarlberg, R. (2010) *Food Politics: What Everyone Needs to Know*. Oxford University Press, Oxford, UK.

Paarlberg, R.L. (1980) Lessons of the grain embargo. *Foreign Affairs* 59(1), 144–162.

Piesse, J. and Thirtle, C. (2009) Three bubbles and a panic: an explanatory review of recent food commodity price events. *Food Policy* 34(2), 119–129.

Ravallion, M. (2009) Are there lessons for Africa from China's success against poverty? *World Development* 37(2), 303–313.

Ruttan, V.W. (1996) *United States Development Assistance Policy: The Domestic Politics of Foreign Economic Aid*. The Johns Hopkins studies in development. Johns Hopkins University Press, Baltimore, Maryland.

Schatkowski-Schilcher, L. (1989) Die Weizenwirtschaft des Nahen Ostens in der Zwischenkriegszeit: Der Einfluß der Ökonomie auf die Politik am Beispiel Syriens. In: Schatkowski-Schilcher, L. and Scharf, C. (eds) *Der Nahe Osten in der Zwischenkriegszeit, 1919–1939. Die Interdependenz von Politik, Wirtschaft und Ideologie*. Franz Steiner, Stuttgart, 241–259.

Schatkowski-Schilcher, L. (1992) The famine of 1915–1918 in Greater Syria. In: John Spagnolo (ed.) *Problems of the Modern Middle East in Historical Perspective: Essays in Honor of Albert Hourani*. Ithaca Press (for Garnet Publishing Ltd.), Reading, UK, pp. 229–258.

Scoones, I., Hall, R., Borras, S.M., White, B. and Wolford, W. (2013) The politics of evidence: methodologies for understanding the global land rush. *Journal of Peasant Studies* 40(3), 469–483. DOI: 10.1080/03066150.2013.801341.

Scott, J.C. (2017) *Against The Grain: A Deep History of the Earliest States*. Yale University Press, New Haven, Connecticut.

Shaw, D.J. (2007) *World Food Security: A History Since 1945*. Palgrave Macmillan, New York.

Thurow, R. (2010) The fertile continent: Africa, agriculture's final frontier. *Foreign Affairs* 89(6), 102–111.

Tooze, J.A. (2006) *The Wages of Destruction: The Making and Breaking of the Nazi Economy*. Allen Lane, London.

Vincent, C.P. (1985) *The Politics of Hunger: The Allied Blockade of Germany, 1915–1919*. Ohio University Press, Athens, Ohio.

Wallensteen, P. (1976) Scarce goods as political weapons: the case of food. *Journal of Peace Research* 13(4), 277–298.

Wallerstein, M.B. (1980) *Food for War – Food for Peace: United States Food Aid in a Global Context*. MIT Press, Cambridge, Massachusetts.

Weis, T. (2007) *The Global Food Economy: The Battle for the Future of Farming*. Zed Books, London.

Woertz, E. (2013) *Oil for Food. The Global Food Crisis and the Middle East*. Oxford University Press, Oxford.

Woertz, E. and Keulertz, M. (2015) Food trade relations of the Middle East and North Africa with tropical countries. *Food Security* 7(6), 1101–1111. DOI: 10.1007/s12571-015-0502-5.

3 Climate Change and Conflict: Agriculture, Migration and Institutions

Martin Smidt[1,]* and Ole Magnus Theisen[2†]
[1]*Department of Sociology and Political Science, Norwegian University of Science and Technology (NTNU), Trondheim, Norway;* [2]*Department of Sociology and Political Science, NTNU, Trondheim, Norway and Peace Research Institute, Oslo, Norway*

Introduction

Climate change and its social consequences are increasingly 'hot' topics, which is evident in the quantitative research on the subject matter. No less than four journal special issues (*Geopolitics* 2014, *Journal of Peace Research* 2012, *Political Geography* 2007 and 2014) have been published on the climate–conflict connection in social science journals in the past 10 years. Security in a broader sense even got its own chapter in the Intergovernmental Panel on Climate Change's (IPCC) AR5 (IPCC, 2014).

Food production, migration and institutions are often pointed to as central factors in understanding how climate change might relate to conflict. While the role of food production is increasingly analysed, we believe in line with Buhaug (2015), Bernauer *et al.* (2012) and Linke *et al.* (2015), that the most pressing knowledge gap concerns the role of institutions and migration. This chapter, therefore, does not aim for a full literature review of the burgeoning quantitative climate anomalies–conflict literature (see Theisen *et al.*, 2017), nor a much-needed review of the case study literature on the subject matter. Rather, we focus on the role of institutions and migration as potential mechanisms between weather shocks and conflict and potential data sources for investigating these links.

In the following, we briefly discuss direct social effects of climate change before we give an overview of the quantitative literature on the subject matter pointing to certain limitations in how conflict is currently analysed. Thereafter, we start with a general discussion of suggested linkages between climate change and conflict, before we turn to a more focused discussion on the role of migration and institutions as intermediate factors within the climate–conflict link. We end with a cautionary note on the ethical aspects of very detailed data on mobility and recommendations to better integrate insights from studies using qualitative methods.

Likely Consequences of Climate Change for Society

The IPCC (2013) projects significant rises in ocean temperatures, air temperature and the mean sea level, the latter being amplified by larger and more common storm surges (IPCC, 2013). In latitudes far from the equator, winters will be shorter and milder, facilitating the arrival of invasive species and extending the growing

---

* Email: martin.smidt@ntnu.no
† Equal authorship implied. Theisen's work was supported by the European Research Council grant 648291.

season (IPCC, 2013, p. 1140). Areas that cycle between wet and dry seasons will see more extreme seasonal contrasts (IPCC, 2013, p. 1032), with more floods and droughts. Dry regions will likely experience more short-term droughts (IPCC, 2013, 2014, p. 247f). Flood protection has deteriorated in many parts of the world, mainly due to rapid population growth in exposed areas. Combined with a projected increase in flood hazards, this will likely cause economic harm, and adaptation measures will often be insufficient to compensate for the effects of climate change (IPCC, 2014, p. 232).

While our understanding of the physical effects of climate change has improved considerably (notwithstanding substantial uncertainties), the subsequent social impact remains much more uncertain. We can predict with some certainty how ecosystems and the availability of arable land will be affected (Hegre et al., 2016) as well as some effects on food production. The combined effects of changes in temperatures and precipitation patterns will severely affect access to freshwater in certain regions. Seven per cent of the global population will see their water supplies reduced by at least one-fifth for each centigrade of warming. This will be especially noticeable in dry subtropical regions (IPCC, 2014, p. 232) including parts of the Sahel and East Africa, regions where rain-fed agriculture and pastoralism dominate. Warming has already reduced the Sahel's growing period and subsequently its agricultural potential (IPCC, 2007, p. 9). Equatorial regions in general are at risk of decreased agricultural production with temperature increases of 1–2°C (IPCC, 2007, p. 11). Heavily populated areas, such as South Asia, could see a potential drop in production as high as 30%. Simultaneously, total population and even more the urban population is expected to rise, making many areas highly vulnerable to food shortages (IPCC, 2007, p. 13). Such reductions can be dramatic, especially for households that rely heavily on subsistence farming. These effects will subsequently also influence local wealth, as earnings from agricultural production will drop. Where cash crops are produced, local industries may lose their supply of raw materials. Areas that rely on tourism where vulnerable species are central may lose their main source of income. A reduction in wealth is likely to have wider ramifications for the community as essential services often rely on taxes collected locally.

The reduced access to water will also directly hit industry and services that require water to operate (IPCC, 2014, p. 232).

The potentially negative effects of environmental change on migration are covered extensively in the case literature (Gemenne, 2011). Only a handful of cross-national studies of internal or international migration exist, arguably due to the lack of data. Reuveny and Moore (2009) find that immigration to OECD countries is correlated with land pressure and food scarcity in the country of origin. Neumayer (2005) finds no connection between drops in food production or natural disasters in sub-Saharan Africa (SSA) and the number of asylum seekers arriving in Europe. The journey requires planning, thus sudden and unforeseen events like natural disasters are unlikely to lead to increased migration. While Naudé (2009) finds natural disasters to drive migration from SSA countries, his 2010 study (Naudé, 2010) fails to. Using urbanization rates as a proxy for internal migration, Barrios et al. (2006) find a correlation between drought and urbanization in developing countries.

Linking Climate Change, Climate Variability and Conflict

Is the weather a useful proxy?

While the current climate–conflict literature's use of variables capturing weather has been criticized for being 'about weather, not climate' (Gleditsch, 2012, p. 7, see also Selby, 2014), such criticism risks throwing out the baby with the bathwater (this critique would apply to the migration studies above as well). Busby et al. (2012, p. 10f) argue that if future temperatures (and precipitation) remain normally distributed, increasing the average will also drastically increase the chance of current extreme levels (see Fig. 3.1 for an illustration from the Fourth Assessment Report of the IPCC). A counterargument is that the change may be sufficiently gradual to enable ecosystems and societies to adapt. However, societies today capable of coping with today's extreme events will not necessarily manage this as extremes by today's standard increase in frequency and the rarer events become stronger.

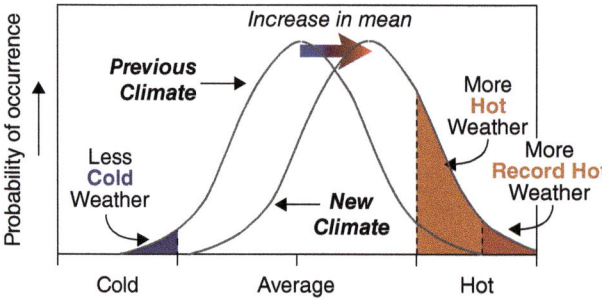

Fig. 3.1. Extreme weather events. Schematic showing the effect on extreme temperatures when the mean temperature increases, for a normal temperature distribution. (From: Solomon et al., 2007, p. 53).

Conceptualizing conflict

In this section, we focus on quantitative studies of the climate–conflict nexus that study internal armed conflict, that is, violent contests between two organized actors fighting over some incompatibility. This field is dominated by a small number of conflict datasets, most prominently the Uppsala Conflict Data Program/Peace Research Institute Oslo (UCDP/PRIO) Armed Conflict Dataset (ACD) covering civil conflict (Gleditsch et al., 2002; Melander et al., 2016), and the UCDP Non-State Conflict Dataset (NSCD) (Sundberg et al., 2012) covering violence between groups without direct state involvement. Both datasets comprise country–year aggregates with 25 battle-related deaths as the inclusion threshold. Neither is well suited to examine hypotheses where cause and effect are geographically and/or temporally close. They have therefore been integrated into the UCDP Georeferenced Event Dataset (GED) (Sundberg and Melander, 2013), which records all fatal incidents and geocodes them to enable disaggregated analyses. Another geocoded dataset is the Armed Conflict Location and Event Dataset (ACLED) (Raleigh et al., 2010), which covers similar forms of violence to the GED, but also non-fatal events (Raleigh et al., p. 656). These two event datasets have the advantage of giving more exact timing and location of violence, making them better suited for analysing the effects of shocks like floods or droughts.

Most datasets gather information primarily from media sources, potentially causing bias both in terms of areas covered and when attempting to include minor events. Areas that rely on self-help both in terms of the provision of food and security are pointed to as particularly relevant in the weather–conflict literature (Suliman, 1999). These areas also have poorer media coverage, which may lead to conflicts being under- or unreported. This can be accentuated when using data on single fights (events) with no lower casualty threshold, as events with fewer casualties are less likely to be reported. Datasets using a minimum number of deaths as thresholds for inclusion are therefore less likely to suffer from a reporting bias, as larger events and conflicts that see repeated clashes are more likely to reach the headlines irrespective of where they occur. A related potential problem with events data is that they become less likely to be mentioned in media sources the further back in time you go. Eck (2012) points out that event datasets suffer from issues like events being reported multiple times and inaccurate geocoding. As a result, the accuracy gained by using disaggregated data may be reduced by lower data quality. The ideal would, of course, be to have data on violence collected on the ground, but such studies are extremely rare (though see Adano et al., 2012).

Moreover, few researchers theorize about 'events' as such (Eck, 2012, p. 126). Rather, theories usually concern the outbreak, duration, intensity, spread, etc., of conflicts. Nevertheless, many studies on weather and violence rely on event data for their tests.[1] Getting beyond this approach where every news report weighs the same and instead applying theoretically derived operationalizations of aspects of conflicts is critical for research on weather and conflict to move forward, but also in order to get reliable measures of the multiple inter-related phenomena that make up a conflict.

Quantitative research and its gaps

The literature on climate and conflict has seen a number of reviews over the past few years

(Bernauer et al., 2012; Gleditsch, 2012; Scheffran et al., 2012; Hsiang et al., 2013; IPCC, 2014; Gemenne et al., 2014; Salehyan, 2014; Selby, 2014; Buhaug, 2015; Theisen et al., 2017). Most of these conclude, like Bernauer et al. (2012), that no general 'systematic, causal relationship' has been found. Buhaug (2015, p. 270) is even more sceptical, citing 'a disturbing disconnect between underlying theoretical arguments' and the testing of these. He particularly criticizes the lack of properly specified mechanisms and, as a consequence, similar data being used to evaluate quite different mechanisms. Scheffran et al. (2012, p. 9) are more confident, claiming that 'a significant part of the current literature supports the argument that climate change has an influence on violent conflict in at least some regions of the world'. Despite this, they conclude that the literature suffers from a number of weaknesses that cast doubt on the validity of the research, especially its ability to differentiate between climate change and other trends over time. The most enthusiasm for current research can be found in a metastudy by Hsiang et al. (2013). This met with sharp criticism from a large number of scientists (Buhaug et al., 2014).

While we agree with the majority conclusion that there so far is no clear-cut climate–conflict effect, there are signs of convergence on certain sub-topics. More often than earlier, recent studies use disaggregated data, both for conflict variables and environmental factors, arguably leading to more plausible tests, but with a caveat on data quality. Analyses of more narrow phenomena or limited to certain contexts fashioned to more plausibly catch the effects put forward by theory may sacrifice some generalizability, but this loss is outweighed by improved interpretability.

Theoretical mechanisms[2]

The simplest explanation for why resource scarcity may spur violence is derived from evolutionary theory where violence arises from a scramble over scarce resources (Gat, 2009, p. 577) causing selection of genes that predispose for violence. Similarly, hot temperatures have been linked to individual aggression, including violent crime and riots (Anderson and Delisi, 2011). A strand within ecological anthropology holds that war is an adaptation to scarcity as recurring warfare levels out resource distribution and reduces population pressure (Vayda, 1976). Generally worsening environmental conditions have led people in traditional societies to blame and attack outgroups such as witches (during the Little Ice Age in early modern Europe) or Jews (the Plague of the 14th century) (Behringer, 1999).

While most disaster–conflict studies assume Hobbesian responses to disasters, these are contrasted by findings of behaviour in the immediate disaster aftermath, where antisocial and opportunistic behaviour was reduced (Fritz, 1996), and altruistic behaviour increased, creating 'communities of sufferers' (Drury et al., 2009, p. 502), although only temporarily.

Direct applications of biological and psychological models to explain collective behaviour remain common in popular predictions on the effects of climate change (Schwartz and Randall, 2003). Their applicability to collective violence, which requires a minimum of organization and planning, in a world where states (mostly) provide a minimum of monopoly of force is uncertain at best, and has met criticism (see Raleigh et al., 2014). Any serious study aiming at understanding how resource scarcity may spur collective violence must account for its organization and the role of the state. Tilly (1978) contends that conditions favouring the organization of discontent are necessary for collective violence to occur. The option to free ride negates the individual's motivation to rebel, and for insurgencies to succeed providing selective benefits for soldiers is essential (Gates, 2002). Individual-level arguments are, therefore, arguably more relevant for violent contests between small groups or bands of fighters where individual and collective incentives are closer, rather than for rebel movements and civil wars.

Two other individual-level mechanisms – the opportunity cost (OC) and relative deprivation (RD) argument – are common in the literature and both often include a group aspect. OC holds that anything causing falling incomes also causes a relative drop in recruitment costs to rebel armies or other criminal enterprises, increasing the appeal of rebel groups' selective benefits (Grossman, 1991). This criminal cum rebel assumption has met criticism: insurgent movements do not mirror criminal organizations as norms and ideas – factors that inhibit the growth of criminal groups – are necessary for rebel groups to expand (Sanín, 2004).

RD postulates that instability is more likely when expectations are not met. However, *interpersonal* (vertical) economic inequalities are not correlated with civil violence (Fearon and Laitin, 2003; Collier and Hoeffler, 2004), but RD arguments emphasizing inter-group inequalities have found systematic links to civil and less intense forms of conflict (Østby, 2008; Fjelde and Østby, 2014). Relatedly, if distinct groups are deprived of political influence at the national level, this can give rise to resentment along ethnic lines facilitating collective violence (Buhaug et al., 2008). Consequently, it has been argued, for resource scarcity to result in civil violence, some form of alliance or shared interest between the deprived and elites is required (Goldstone, 1991), or at least the combination of high ethnic cleavages and exclusive political institutions (Kahl, 2006). While Theisen et al. (2011/12) and von Uexkull et al. (2016) find that political exclusion for groups in combination with drought do not increase the risk of civil conflict onset, the latter finds politically excluded groups hit by agricultural shocks are more likely to be involved in ongoing civil conflicts.

In general, the arguments emphasizing non-individual factors hold that contexts already at risk of conflict are the most likely candidates to experience scarcity-induced violence (see Buhaug et al., 2010 for a fuller argument). Broadly speaking, four contextual factors are underscored in the literature:

1. Low level of economic development/lack of economic diversification outside agriculture (Baechler, 1999; Homer-Dixon, 1999; Kahl, 2006), making large sections of the population directly dependent on renewable resources.
2. Weak states, which have: (i) less efficient counter-insurgency capabilities; (ii) weak infrastructure networks; and (iii) proportionately less income from mobile capital, making extortion more feasible.
3. Disrupted, unstable (Homer-Dixon, 1999) or exclusionist government institutions (Kahl, 2006).
4. Strong ethnic cleavages (Homer-Dixon, 1999; Kahl, 2006).

Several specific mechanisms pertaining to how resource scarcities may cause collective violence have been put forward. One argument holds that when the quality and/or quantity of a resource falls at the same time as populations grow, elites are likely to adjust institutions in order to restrict resource access to gain from its increased value. Deprived groups are led into further marginalization that foments grievances increasing the risk of conflict (Homer-Dixon, 1999; Kahl, 2006). Kahl (2006) suggests an alternative mechanism where increasing pressure on renewable resources hampers economic growth, in turn hurting regime legitimacy. The regime may respond by targeting oppositional groups and promising regime supporters a piece of the resource pie.

In terms of resource shocks, some studies that find negative rainfall growth to increase the risk of civil conflict, both as an instrument for economic growth (e.g. Miguel et al., 2004) and directly (Hendrix and Glaser, 2007), indicate that lower economic growth reduces a state's ability to redistribute resources as well as maintain law and order. A weakness in the argument is that it is unclear why the rebels' fighting capacity is left relatively unaffected while state capacity is almost instantly reduced. Far from all studies investigating the topic find a link, however.[3]

The role of the state in natural disasters has also been argued to increase the risk of collective violence (e.g. Nardulli et al., 2015). Olson and Gawronski (2010) discuss factors that affect the likelihood that grievances arise and the extent to which they will be directed towards the authorities. These include the extent and efficiency of mobilization, whether relief is distributed fairly, the consistency and reliability of information provided and whether proper preparations were in place prior to disaster.

Grievances rising from inappropriate state response require a nominal expectation of state responsibilities, an assumption that may not hold. For instance, de Waal (2005, p. 204) argues that during the mid-1980s famine-stricken farmers in Darfur were surprised to see the state provide food relief. Assumptions on grievances over state neglect require the expectation of a nominally working modern state.

Having underscored the importance of state function and reaction to environmental change, we now address what we see as perhaps the two most important gaps in the literature, namely the role of institutions as a crucial mediator and migration as both an adaptive option and potential catalyst of conflict.

The Way Forward

Institutions

Much statistical research on resource scarcity and violence suffers from an overreliance on either individual-level and/or state-centric explanations. Although undoubtedly representing central mechanisms, they fail to give a satisfactory explanation of a meso-level phenomenon: why in the face of resource stress do some *groups* take up arms to fight the state or another group? As Fig. 3.2 illustrates, for a conflict to spill over into violence many conditions at several steps in the causal chain have to be present, leaving the large majority of disputes over scarce resources non-violent either because proper institutions are in place or because those involved simply chose another strategy (stand-off, avoidance, cooperation, out-migration, etc.). However, much of the quantitative literature on weather and violence tends to downplay the role of institutions, often leaving the reader with a quite deterministic impression of the mechanisms at hand (Buhaug, 2015, p. 3). Only a handful of large-n studies have theorized elaborately about what kind of institutions make organized violence stemming from resource scarcity more or less likely (see Gizelis and Wooden 2010; Theisen *et al*., 2011/12; Linke *et al*., 2015; von Uexkull *et al*., 2016). All cross-national studies on the subject matter focus on national-level formal institutions, even though Ostrom (1990) in her seminal contribution has demonstrated the effectiveness and durability of local informal institutions. Tellingly, her work is rarely built upon in the quantitative climate–conflict literature.

We believe that the role and type of institutions will differ between the type of society and conflict at hand, and gaining a fuller understanding of this is critical. The relevant form of institutions in a civil war setting – for instance, the ethnic-exclusionist regime in Syria – is likely to be quite different from violent cattle-theft episodes in rural Tanzania. Moreover, we believe that the current quantitative literature on peace and conflict in general, but also the climate–conflict literature, has underplayed the crucial role of institutions at the sub-national level. Violence between groups where the state is not an active party is a form of violence frequently pointed to as particularly likely to be driven by scarce renewable resources (Suliman, 1999). A long tradition of research has held that people in the peripheries of weak states, where near-subsistence livelihoods dominate and the state is less able to keep the peace and provide its (nominal) citizens with food, are more likely to fight when they are starving (Vayda, 1976; Tornay, 1979). Recently several studies have investigated this link using statistical material most often testing an unmediated link between drought and violent events. However, the relationship is not as straightforward as could be expected (Theisen *et al*., 2017). Due to the theoretical relevance of such conflicts, we focus on the role of institutions in these kinds of conflicts below.

Institutions can be defined as 'stable, valued, recurring patterns of behavior' (Huntington, 1965, p. 394). A central distinction can be drawn between formal (state-made) and informal (made by non-state actors) institutions. Contrary to the belief that the absence of formal institutions creates acute scarcity and overuse (Hardin, 1968), self-governing common property institutions such as informal rules over common grazing grounds and locally developed irrigation channels can be quite prevalent and enduring despite no other coordinating force than the users themselves (see, in particular, Ostrom, 1990, p. 58ff). Case studies from SSA indicate that if a resource dispute arises (box II in Fig. 3.2), locals prefer to turn first to friends, neighbours and relatives, before resorting to traditional authorities like village elders or a chief (Turner *et al*., 2012, p. 749f). Formal institutions are at this stage often shunned. They are seen as less in touch with the local context, thus making inflexible judgements; being more costly and corrupt; and creating long-standing grievances between families.[4] Corroborating this finding, Linke *et al*. (2015) conclude from a household survey of three counties in Kenya that informal institutions reduce support of violence, whereas formal institutions have no effect.

Customary or informal institutions have their clear limitations, however, in particular when it comes to containing violence. In the western Sahel where members of different communities quite often live mixed in the same settlements, Turner *et al*. (2012) point out that, in contrast to other disagreements, those that turn violent are often taken directly to the public

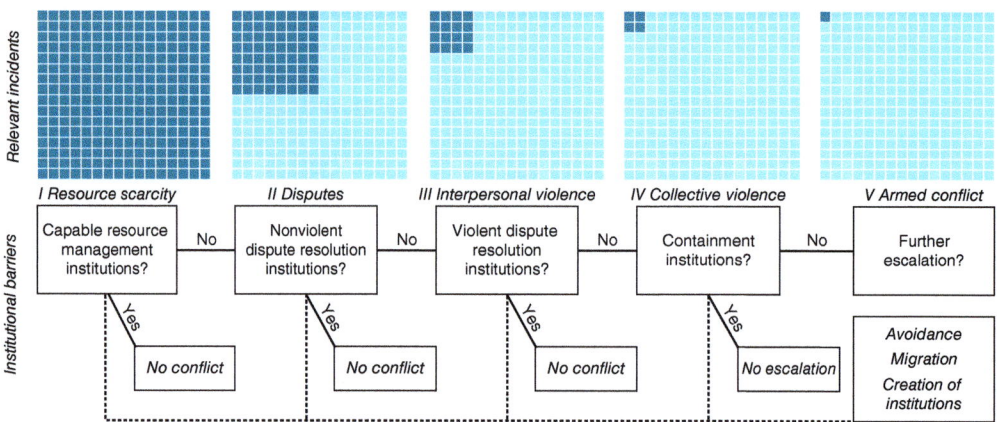

Fig. 3.2. The role of institutions regulating resource conflicts. The quadratic boxes at the top show the number of cases with a problem (dark) for simplicity assuming 25% of cases escalate for each step vs 75% are resolved (light) for each of the steps (I–V) in the causal chain at the second line.

authorities (but involving chiefs, box III in Fig. 3.2). This is quite similar to what villagers in five districts in Northern Tanzania told us during fieldwork conducted in October 2016 – if someone is hurt or killed the police are brought in. In areas where the state is less efficient, resolution of violent conflicts is frequently left to local traditional conflict resolution mechanisms, where there is no neutral arbiter to mediate or enforce peace should things get out of hand (box IV in Fig. 3.2). Suliman (1999) argues that resource-regulating and conflict-resolving institutions in Darfur and the Borana-dominated areas of Ethiopia have broken down due to increasing resource pressure caused by population growth and agricultural expansion at the expense of pastoralism. While most of the time being able to resolve conflicts relatively peacefully, these institutions rely on the very real threat of deadly violence in itself in order to be credible (Fearon and Laitin, 1996; Eaton, 2008). In areas where feuding has remained an accepted way of resolving disputes between communities, large-scale inter-ethnic violence is more likely to occur (Witsenburg and Zaal, 2012) as the state is unable to 'contain fear' in an effective and unbiased manner (Lake and Rotchild, 1996). Despite the lack of focus on local institutions in the climate–conflict literature, innovative use of existing data sources in SSA have revealed interesting patterns between local informal institutions and armed conflict and formal and/or national-level institutions. Using data on an ethnic group's degree of pre-colonial hierarchization, Wig (2016) finds that the more hierarchically organized they are, the less likely ethnic groups deprived of political power in the modern state are to experience civil conflict. Wig and Kromrey (forthcoming) find communal violence to be more likely in areas where the same hierarchization was low.[5] This relates to the argument that more centralized institutions in the past have left modern states stronger as they have more to build on. The argument is reinforced by studies using Afrobarometer surveys as a data source on local institutions. These show that trust in traditional institutions translates into trust in modern institutions (Logan, 2009). This relation arguably also goes the other way, as informal institutions are more fragile if not recognized by the state (Ostrom, 1990).

Communities living together over time often establish institutions for dispute settlement and conflict resolution (Hagberg, 1998). When a new group enters an area, such inter-ethnic institutions are not present and have to be built. Although this can be successful (Bogale and Korf, 2007; Adano et al., 2012), it is far from certain. Newcomers might challenge local notions of property rights (Feyissa, 2011) or might have less secure claims to land (Turner et al., 2012, p. 202). In a comparative study of 11 cases of inter-group conflict in arid or semi-arid areas of SSA, Seter et al. (2017) find that in-migration of new ethnic groups without previous relations can create problems over renewable resources.

It is not increased resource scarcity per se that causes conflict, but little previous experience with resource sharing arrangements and conflict management. Contrasting the peaceful, long-standing co-existence between pastoralist Fulbe and Mossi farmers in the Central Plateau of Burkina Faso with the recent in-migration of Fulbe into northern Côte d'Ivoire causing intense violence, Breusers et al. (1998, p. 375) suggest that '...the extent to which the ethnic groups involved have had a "common" history is probably of the utmost importance' for how conflicts are resolved.

Migration

In the previous section, we suggested that migration is an important relief valve when resources become scarce and disputes arise, and as an example of a problem institutions can mitigate. Figure 3.3 illustrates how institutions are just one of several factors influencing the probability that in-migration will lead to conflict. Reuveny (2007) suggests several channels. The first is a Malthusian mechanism where the host location is already experiencing a resource shortage, or migrants are perceived to cause one. Second, autochthones can feel threatened by arriving migrants, but as argued above, this is less likely where the relevant groups have a history of interaction (formal or informal institutions). Third, autochthonous groups may feel their position is threatened if the migrant group is large enough relative to the autochthonous group to alter the demographic balance, and consequently the local power relationship (as seen in Gambella in Ethiopia; Feyissa, 2011). Finally, if there are latent tensions between groups then the influx of migrants could reinforce these. Migration itself does not necessarily increase chances of conflict, but it can do so under certain conditions. A location experiencing shortages is more vulnerable regardless of migrant characteristics, but differences in ethnic compositions are expected to amplify or create problems regardless. Problems can be mitigated by different institutions, such as those listed in the previous section.

A few recent studies have investigated the migration–conflict link. One study of Indian states found that state-to-state migration of males increases the risk of riots (Bhavnani and Lacina, 2015), while another study found that the effect was only present when migrants were relatively deprived compared with the urban populace (Østby, 2016). Bhavnani and Lacina (2015) do not find that the effect is stronger for states with high unemployment. In a related study, Ghimire et al. (2015) find that the higher the number of people displaced by flooding, the higher the risk of civil conflict incidence but not onset. Again, institutions are found to be the more important factor, measured here in terms of local political alignment with national governments. Lastly, a study of urban growth by Buhaug and Urdal (2013) finds no correlation between urban growth and unrest, but points out a problem with their population data in that they do not differentiate between in-migration and natural growth and in-migration.

Overall, the migration/displacement link is theoretically sound and has support from the three large-n studies that have investigated the link, although migration increases the probability of violence only under certain conditions in one study. One clear priority for future research is to further investigate links between migration and conflict (see below). An obstacle to such efforts is the lack of accurate and comprehensive data on migration. While conflict datasets often

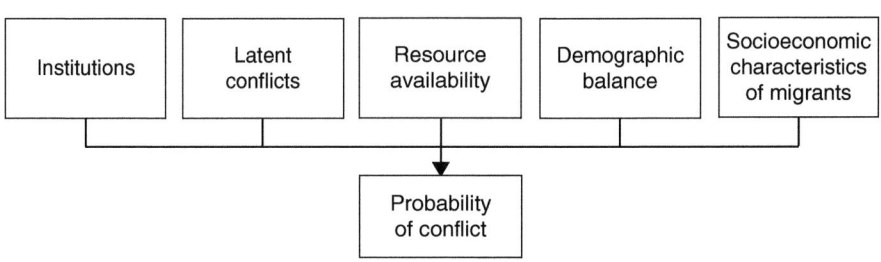

Fig. 3.3. Conditions for migrants increasing the probability of conflict.

can pinpoint conflict events down to the village–day level, data on migration are far from as extensive or detailed, and the many different forms of mobility itself (e.g. permanent or cyclical, pre-planned or distress, etc.) makes it challenging to create cross-national datasets.

Population datasets that can be used to infer in-migration come mainly from censuses and related estimations. Censuses may be quite precise for the year they are conducted, but interpolated data will be useless for studying effects on a smaller timescale and their geographical resolution may also be poor. Datasets based on satellite imagery are a recent improvement, with more frequent and precisely positioned data points. Unfortunately, these estimates suffer from imprecisions that make them unsuitable for testing many mechanisms, but their global coverage going back several decades allows for broad analyses. Both types of data are unable to distinguish internal population growth from migration, a problem that can be solved through surveys; for example, where respondents are asked about the duration of residence and whether their origin is urban or rural (Østby, 2016). Estimating in-migration from such data gives a relatively valid measure compared to other population data, but is limited by the number of locations surveyed.

Another promising data source is data from mobile phone service providers. Even in less developed countries, mobile phone ownership is increasingly common. Mobile phone data have already been applied to conflict studies by Berger *et al.* (2014, unpublished results), who find that call traffic volumes on individual cell towers coincides with violent events in Côte d'Ivoire. Bharti *et al.* (2015) use mobile phone data in tandem with night light data from satellite imaging to track population movement around the 2010 election crisis in Côte d'Ivoire, and Lu *et al.* (2016a) use mobile phone data from Bangladesh to track mobility patterns relating to a cyclone. The latter study demonstrates high resolution tracking of individuals as well as showing how such data can be aggregated to show larger trends. Combining these mobile phone usage datasets with weather and conflict datasets could boost our ability to test migration–conflict mechanisms.

Datasets can be constructed to fit most time scales useful for correlating climatic anomalies and social turmoil. Tracking individuals in a flood-stricken area in the immediate aftermath will tell us whether the flood leads to only temporary relocation or to more long-term migration. Tracking the flow from an area during a drought will tell us whether it causes migration, and when and where in the course of the drought people decide to move.

The problem with these data, as Bharti *et al.* (2015) point out, is that we are limited to data the service providers are willing, able, or allowed to release. For example, this means long-term studies back in time are impossible, as the data are no longer stored. Lastly, deciphering the data requires more time and technical skill than traditional data, making them a more demanding source (Lu *et al.*, 2016b).

There are ethical issues regarding the use of detailed mobility data on individuals, or even aggregated data on movement. As an example, is it morally defensible to study conflict (or riots) using mobile phone data? Countries subject to civil conflict and communal violence are often ruled by regimes of questionable moral virtue, and these may be inspired to use the methods for their own ends. Data that allow insight into every social aspect of conflict and migration also make it harder to publish results without compromising the safety of the subjects.

Conclusion

This chapter has focused on two central gaps in understanding possible connections between climate change and armed conflict. There is a clear need for improving the reliability and validity of conflict data. Furthermore, institutions as an intermediate variable must be taken seriously and some data sources for large-n investigations into this already exist. Likewise, using mobile phone data to track mobility down to the individual is a promising way of understanding potential links between climate shocks, mobility and conflict, noting the potential ethical issues described above.

Case studies have pointed out that there are important contextual differences in how and why resource scarcity translates into conflict (see, for example, Adano *et al.*, 2012). Future work should therefore try to rely more on in-depth case studies from fields such as anthropology and geography. While we concur with

Raleigh *et al.* (2014, p. 77) that in understanding why certain situations turn out violent or not 'it is probably more critical to understand "the nature of the state" than the "state of nature"', we emphasize that in contexts of weak states local and/or informal institutions are highly important. Despite their importance, however, institutions are no panacea. As Turner *et al.* (2012) state, mutual trust is important in making institutions work. For this, the interplay with the state is central: if it enforces its monopoly of violence in a relatively unbiased manner, it should reduce the security dilemma that tends to plague societies where groups have to rely on self-defence mechanisms.

Climate change will undoubtedly affect human societies and, in particular, it will have predominantly negative effects for the agrarian world. Although many studies find a correlation between weather anomalies and conflict in the recent past, we should be wary of alarmist predictions for the future. An important qualifier is that we need to understand the extent to which findings using current anomalies in weather are transferable to a future with a changing climate. Relatedly, we need to better understand whether the social impacts of shocks of the same magnitude are comparable between different contexts today. Similarly, we need to explore whether changes in *climate* have comparable social impacts under different social circumstances. What is uncontested in the literature, however, is the very negative joint impact of climate anomalies/change and armed conflict on vulnerability (see, for example, Busby *et al.*, 2014). In consequence, there is likely a mutually reinforcing relationship between vulnerability to climatic stress and vulnerability to conflict. However, vulnerability in the agrarian world, as elsewhere, is a dynamic process involving adaptation or the lack thereof. This is a blind spot for much research on conflicts, and should be a future priority (Javeline, 2014).

Notes

[1] This is not to point fingers – indeed that would have implied pointing fingers at one of the authors. Rather, it is to point out a weakness of the literature as a whole.
[2] This section builds on Theisen (2012). We only consider water scarcity in how it affects the outbreak of conflict.
[3] See Theisen *et al.* (2017) for a review.
[4] Corroborated by interviews in Northern Tanzania conducted in October 2016.
[5] Kromrey and Wig also use data from an expert survey conducted by one of the authors (data not released), which provide information on today's customary institutions across ethnic groups in Africa.

References

Adano, W.R., Dietz, T., Witsenburg, K. and Zaal, F. (2012) Climate change, violent conflict and local institutions in Kenya's drylands. *Journal of Peace Research* 49(1), 65–80.
Anderson, C.A. and Delisi M. (2011) Implications of global climate change for violence developed and developing countries. In: Forgas, J.P., Kruglanski, A.W. and Williams, K.D. (eds) *The Psychology of Social Conflict and Aggression*. Psychology Press, New York, pp. 249–265.
Baechler, G. (1999) *Violence through Environmental Discrimination*. Kluwer Academic, Dordrecht, the Netherlands.
Barrios, S., Bertinelli, L. and Strobl, E. (2006) Climatic change and rural-urban migration. *Journal of Urban Economics* 60(3), 357–371.
Behringer, W. (1999) Climatic change and witch-hunting. *Climatic Change* 43(1), 335–351.
Bernauer, T., Böhmelt, T. and Koubi, V. (2012) Environmental changes and violent conflict. *Environmental Research Letters* 7(1), 1–8.
Bharti, N., Lu, X., Bengtsson, L., Wetter, E. and Tatem, A.J. (2015) Remotely measuring populations during a crisis by overlaying two data sources. *International Health* 7(2), 90–98.
Bhavnani, R.R. and Lacina, B. (2015) The effects of weather-induced migration on sons of the soil riots in India. *World Politics* 67(4), 760–794.

Bogale, A. and Korf, B. (2007) To share or not to share? *Journal of Development Studies* 43(4), 743–765.
Breusers, M., Nederlof, S. and Rheenen, T.V. (1998) Conflict or symbiosis? *Journal of Modern African Studies* 36(3), 357–380.
Buhaug, H. (2015) Climate–conflict research. *WIREs Climate Change* 6(3), 269–275.
Buhaug, H., Cederman, L.E. and Rød, J.K. (2008) Disaggregating ethno-nationalist civil wars. *International Organization* 62(3), 531–551.
Buhaug, H., Gleditsch, N.P. and Theisen, O.M. (2010) Implications of climate change for armed conflict. In: Norton, A., Cameron, E. and Mearns R. (eds) *Social Dimensions of Climate Change*. World Bank, Washington, DC, pp. 75–102.
Buhaug, H. and Urdal, H. (2013) An urbanization bomb? *Global Environmental Change* 23(1), 1–10.
Buhaug, H., Nordkvelle, J., Bernauer, T., Böhmelt, T., Brzoska, M. and Busby, J.W. et al. (2014) One effect to rule them all? *Climatic Change* 127(3), 391–397.
Busby, J.W., Gulledge, J., Smith, T.G. and White, K. (2012) Of climate change and crystal balls. *Air and Space Power Journal Africa and Francophonie* 12(3), 4–44.
Busby, J.W., Smith, T.G. and Krishnan, N. (2014) Climate security vulnerability in Africa mapping 3.0. *Political Geography* 43, 51–67.
Collier, P. and Hoeffler, A. (2004) Greed and grievance in civil wars. *Oxford Economic Papers* 56(4), 663–695.
Drury, J., Cocking, C. and Reicher, S. (2009) Everyone for themselves? *British Journal of Social Psychology* 48(3), 487–506.
Eaton, D. (2008) The business of peace. *African Affairs* 107(427), 89–110.
Eck, K. (2012) In data we trust? *Cooperation and Conflict* 47(1), 124–141.
Fearon, J.D. and Laitin, D.D. (1996) Explaining interethnic cooperation. *American Political Science Review* 90(4), 715–735.
Fearon, J.D. and Laitin, D.D. (2003) Ethnicity, insurgency, and civil war. *American Political Science Review* 97(1), 75–90.
Feyissa, D. (2011) *Playing Different Games*. Berghan Books, Oxford, UK.
Fjelde, H. and Østby, G. (2014) Socioeconomic inequality and communal conflict. *International Interactions* 40(5), 737–762.
Fritz, C.E. (1996) Disasters and mental health. University of Delaware Disaster Research Center. Available at: http://dspace.udel.edu:8080/dspace/handle/19716/1325 (accessed 28 October 2010).
Gat, A. (2009) So why do people fight? *European Journal of International Relations* 15(4), 571–599.
Gates, S. (2002) Recruitment and allegiance. *Journal of Conflict Resolution* 46(1), 111–130.
Gemenne, F. (2011) Why the numbers don't add up. *Global Environmental Change* 21, S41–S49.
Gemenne, F., Barnett, J., Adger, W.N. and Dabelko, J.D. (2014) Climate and security. *Climatic Change* 123(1), 1–9
Ghimire, R., Ferreira, S. and Dorfman, J.H. (2015) Flood-induced displacement and civil conflict. *World Development* 66, 614–628.
Gizelis, I. and Wooden, A.E. (2010) Water resources, institutions, and intrastate conflict. *Political Geography* 29(8), 444–453.
Gleditsch, N.P. (2012) Wither the weather? *Journal of Peace Research* 49(1), 3–9.
Gleditsch, N.P., Wallensteen, P., Eriksson, M., Sollenberg, M. and Strand, H. (2002) Armed conflict 1946–2001. *Journal of Peace Research* 39(5), 615–637.
Goldstone, J. (1991) *Revolution and Rebellion in the Early Modern World*. University of California Press, Los Angeles, California.
Grossman, H.I. (1991) A general equilibrium model of insurrections. *American Economic Review* 81(4), 912–921.
Hagberg, S. (1998) Between peace and justice. PhD thesis, University of Uppsala, Uppsala.
Hardin, G. (1968) The tragedy of the commons. *Science* 162, 1243–1248.
Hegre, H., Buhaug, H., Calvin, K.V., Nordkvelle, J., Waldhoff, S.T. and Gilmore, E. (2016) Forecasting civil conflict along the shared socioeconomic pathways. *Environmental Research Letters* 11(5), DOI: 10.1088/1748-9326/11/5/054002.
Hendrix, C.S. and Glaser, S.M. (2007) Trends and triggers. *Political Geography* 26(6), 695–715.
Homer-Dixon, T.F. (1999) *Environment, Scarcity, and Violence*. Princeton University Press, Princeton, New Jersey.
Hsiang, S.M., Burke, M. and Miguel, E. (2013) Quantifying the influence of climate on human conflict. *Science* 341(6151), 1235367. DOI: 10.1126/science.1235367.
Huntington, S.P. (1965) Political development and political decay. *World Development* 17(3), 386–430.

IPCC (2007) *Climate Change 2007 – The Physical Science Basis Climate Change* 2007. Contribution of Working Group I to the Fourth Assessment Report of the Intergovernmental Panel on Climate Change. Cambridge University Press, Cambridge, UK.

IPCC (2013) *Climate Change 2013 – The Physical Science Basis*. Contribution of Working Group I to the Fifth Assessment Report of the Intergovernmental Panel on Climate Change. Cambridge University Press, Cambridge, UK.

IPCC (2014) *Climate Change 2014: Impacts, Adaptation, and Vulnerability*. Contribution of Working Group II to the Fifth Assessment Report of the Intergovernmental Panel on Climate Change. Cambridge University Press, Cambridge, UK.

Javeline, D. (2014) The most important topic political scientists are not studying. *Perspectives on Politics* 12(2), 420–434.

Kahl, C. (2006) *States, Scarcity, and Civil Strife in the Developing World*. Princeton University Press, Princeton, New Jersey.

Lake, D.A. and Rotchild, D. (1996) Containing fear. *International Security* 21(2), 41–75.

Linke, A.M., O'Loughlin, J., McCabe, T.J., Tir, J. and Witmer, F.D.W. (2015) Rainfall variability and violence in rural Kenya. *Global Environmental Change* 34, 35–47.

Logan, C. (2009) Selected chiefs, elected councillors and hybrid democrats. *Journal of Modern African Studies* 47(1), 101–128.

Lu, X., Wrathall, D.J., Sundsøy, P.R., Nadiruzzaman, M., Wetter, E. *et al.* (2016a) Unveiling hidden migration and mobility patterns in climate stressed regions. *Global Environmental Change* 38, 1–7.

Lu, X., Wrathall, D.J., Sundsøy, P.R., Nadiruzzaman, M., Wetter, E. *et al.* (2016b) Detecting climate adaptation with mobile network data in Bangladesh. *Climatic Change* 138(3–4), 505–519.

Melander, E., Pettersson, T. and Themnér, L. (2016) Organized violence, 1989–2015. *Journal of Peace Research* 53(5), 727–742.

Miguel, E., Satyanath, S. and Sergenti, E. (2004) Economic shocks and civil conflict. *Journal of Political Economy* 112(4), 725–753.

Nardulli, P.F., Peyton, B. and Bajjalieh, J. (2015) Climate change and civil unrest. *Journal of Conflict Resolution* 59(2), 310–335.

Naudé, W. (2009) Natural disasters and international migration from sub-Saharan Africa. *Migration Letters* 6(2), 167–175.

Naudé, W. (2010) The determinants of migration from Sub-Saharan African countries. *Journal of African Economies* 19(3), 330–356.

Neumayer, E. (2005) Bogus refugees? *International Studies Quarterly* 49(3), 389–409.

Olson, R.S. and Gawronski, V.T. (2010) From disaster event to political crisis. *International Studies Perspectives* 11(3), 205–221.

Østby, G. (2008) Polarization, horizontal inequalities and violent civil conflict. *Journal of Peace Research* 45(2), 143–162.

Østby, G. (2016) Rural–urban migration, inequality and urban social disorder. *Conflict Management and Peace Science* 33(5), 491–515.

Ostrom, E. (1990) *Governing the Commons*. Cambridge University Press, New York.

Raleigh, C., Linke, A., Hegre, H. and Karlsen, J. (2010) Introducing ACLED. *Journal of Peace Research* 47(5), 651–660.

Raleigh, C., Linke, A. and O'Loughlin, J. (2014) Extreme temperatures and violence. *Nature Climate Change* 4(2), 76–77.

Reuveny, R. (2007) Climate change-induced migration and violent conflict. *Political Geography* 26(6), 656–673.

Reuveny, R. and Moore, W.H. (2009) Does environmental degradation influence migration? *Social Science Quarterly* 90(3), 461–479.

Salehyan, I. (2014) Climate change and conflict. *Political Geography* 43, 1–5.

Sanín, F.G. (2004) Criminal rebels? *Politics and Society* 32(2), 257–285.

Scheffran, J., Brzoska, M., Kominek, J., Link, M.P. and Schilling, J. (2012) Climate change and violent conflict. *Science* 336(6083), 869–871.

Schwartz, P. and Randall, D. (2003) An abrupt climate change scenario and its implications for United States national security. *Global Business Network Report*. Available at: www.dtic.mil/get-tr-doc/pdf?AD=ADA469325 (accessed 1 May 2017).

Selby, J. (2014) Positivist climate conflict research. *Geopolitics* 19(4), 829–856.

Seter, H., Theisen, O.M. and Schilling, J. (2017) All about water and land? Resource-related conflicts in East and West Africa revisited. *GeoJournal* 83(1), 169–187. DOI: 10.1007/s10708-016-9762-7.

Solomon, S., Qin, D., Manning, M., Alley, R.B., Berntsen, T. *et al*. (2007) Technical Summary. In: Climate Change 2007: The Physical Science Basis. Contribution of Working Group I to *The Fourth Assessment Report of the Intergovernmental Panel on Climate Change* (Solomon, S., Qin, D., Manning, M., Chen, Z., Marquis, M. *et al*. (eds)). Cambridge University Press, Cambridge, UK.

Suliman, M. (1999) Conflict resolution among the Borana and the Fur. In: Suliman, M. (ed.) *Ecology, Politics and Violent Conflict*. Zed Books, London, pp. 286–289.

Sundberg, R. and Melander, E. (2013) Introducing the UCDP georeferenced event dataset. *Journal of Peace Research* 50(4), 523–532.

Sundberg, R., Eck, K. and Kreutz, J. (2012) Introducing the UCDP non-state conflict dataset. *Journal of Peace Research* 49(2), 351–362.

Theisen, O.M. (2012) Renewable resource scarcity, natural disasters, and the possibility of collective violence. PhD thesis, Norwegian University of Science and Technology, Trondheim.

Theisen, O.M., Holtermann, H. and Buhaug, H. (2011/12) Climate wars? *International Security* 36(3), 79–106.

Theisen, O.M, Gleditsch, N.P. and Buhaug, H. (2017) Climate change and armed conflict? In: Stohl, M., Lichbach, M.I. and Grabosky, P. (eds) *States and Peoples in Conflict*. Routledge, New York, pp. 113–129.

Tilly, C. (1978) *From Mobilization to Revolution*. Random House, New York.

Tornay, S. (1979) Armed conflict in the Lower Omo Valley, 1970–1976. In: Fukui, K. and Turton, D. (eds) *Warfare among East African herders*. National Museum of Ethnology, Osaka.

Turner, M., Ayantunde, A.A., Patterson, K.P. and Patterson III, E.D. (2012) Conflict management, decentralization and agropastoralism in dryland West Africa. *World Development* 40(4), 745–757.

von Uexkull, N., Croicu, M., Fjelde, H. and Buhaug, H. (2016) Civil conflict sensitivity to growing-season drought. *Proceedings of the National Academy of Sciences of the USA* 113(44), 12391–12396.

Vayda, A. (1976) *War in Ecological Perspective*. Plenum, New York.

de Waal, A. (2005) *Famine that Kills*, 2nd edn. Oxford University Press, Oxford, UK.

Wig, T. (2016) Peace from the past. *Journal of Peace Research* 53(4), 509–524.

Wig, T. and Kromrey, D. (2018) Which groups fight? *Journal of Peace Research* advance online publication. DOI: 10.1177/0022343317740416.

Witsenburg, K. and Zaal, F.M. (2012) *Spaces of Insecurity*. African Studies Collection University of Leiden, Leiden.

4 Water, Agriculture and Conflict: Global, National and Local Analysis of Conflict in MENA, sub-Saharan Africa and the United States

Martin Keulertz*
American University of Beirut, Beirut, Lebanon

Introduction

The agrarian question is deeply shaped by conflict. The struggle of the global peasantry has been a prime concern in capitalist societies for most of the 19th and 20th century. The livelihood systems of peasants have always been negatively affected by their state of development and the enclosure of global capitalism of the agricultural sector. Conflict over water is a topic of growing concern in international and national politics adding a new dimension to the struggle of peasants. Due to population growth in various parts of the world, increased living standards and urbanization, the Food and Agriculture Organization of the United Nations (FAO) projects food production to increase by 70% to meet demands of a global society reaching 9.1 billion human beings by 2050 (FAO, 2009). Annual cereal production will need to rise to about 3 billion tonnes from 2.1 billion in 2009, and annual meat production will need to rise by over 200 million tonnes to reach 470 million tonnes (FAO, 2009). These demands place a tremendous burden on the world's farmers, who are seen as the primary stewards of water resources (Allan *et al.*, 2015). Farmers will have to use available resources more efficiently. At the same time, water is no longer abundant but increasingly a bottleneck for economic development. Competition over water resources could increase as a result of economic pressures on farmers.

While conflict over water is not a new topic, this chapter provides a different perspective on the scales of conflict. It sets out an analysis on three levels: (i) global; (ii) national/bilateral; and (iii) sub-national/local. The chapter argues that different levels have different facets of water conflict. While the global level is subject to very few tensions over virtual water, nation states have used water as a tool in bilateral foreign policy to acquire access to strategic natural resources, yet no all-out water war has occurred to date. However, the most acute arena for conflict over water resources is the sub-national and local level, where water is managed by farmers. The world's water question presents an immense future challenge for agrarians due to increasingly limited water resource availability. Agrarians in dry areas will be under particular pressure. The chapter will demonstrate why water resources cannot be made more productive by 'efficiency gains' but that evapotranspiration must be reduced in order to save water. In other words, farming will have to be reduced or even abandoned in several world regions, which currently

---
* Email: mk219@aub.edu.lb

produce for local and export markets. Without a drastic shift in food and water systems, several world regions will not only risk running out of water but also experience more conflict over water resources. This emerging hydro-agrarian question will significantly impact the global south with profound political, social and economic ramifications in the developing world. Water and the agrarian question is an emerging topic, which will ultimately decide upon the environmental and social future of the planet. In order to show why water is such a crucial issue in the agrarian question, water resources management needs to be illustrated first.

The Water Challenge

By 2030, the United Nations (UN) have predicted that half of the global population will be subject to water scarcity (Riegels, 2016), which will particularly affect agriculture as it both withdraws and consumes the lion's share of global fresh water resources. Fresh water is also termed 'blue water', which is either surface or groundwater (Falkenmark and Rockström, 2006). In terms of withdrawals per sector, agriculture uses 70% of blue water, industry 20% and domestic users 10%. However, industry and domestic users return much of their water into the system through, for example, water recycling. When return flows and water recycling are taken into account, agriculture consumes up to 93% of blue water resources, industry around 4% and domestic users 3% (Hoogeveen et al., 2015). Blue water in agriculture is predominantly used for irrigation, especially to produce animal feed, and to a lesser extent for watering livestock (Allan, 2011). Conflict over water resources more often concerns blue water than other types of water, as farmers compete over allocation and access to it – in particular in water-scarce dryland areas (Huppert and Wolff, 2002; Warner and Zeitoun, 2006).

Conflict over water resources is not a new concern. Bellicose quotes about water resources have been attributed to thinkers and politicians for more than a century. While Mark Twain reportedly put it that, 'Whiskey is for drinking, water is for fighting over', Egypt's former president Anwar el-Sadat suggested that water 'is the only matter' for which his country could enter a war (Kameri-Mbote, 2011; USBR, 2017). His fellow countryman and later UN General Secretary Boutros Boutros-Ghali went further by predicting, 'The next war in our region will be over the waters of the Nile, not politics' (Ricks, 2011). These gloomy predictions were quickly taken up by academics such as Peter Gleick, who predicted future 'water wars' in world regions such as the Middle East (Gleick, 1993). However, the level of analysis remained on the national level. Allan refuted the claims about all-out water wars by pointing at the nexus between water, food and trade, or what he called 'virtual water' (Allan, 2011). Through his research, Allan showed that most blue and green water is used in food production, which in turn is subject to global agricultural trade. Trade of agricultural commodities allows even water-scarce countries to enjoy 'a version of food security' because they import 'virtual water' through agricultural commodities (Allan, 2003; Aldaya et al., 2009). He concluded that water wars are unlikely as long as food is shipped around the world through agricultural trade. Allan's most important contribution to water and conflict has been the importance of trade to alleviate local water shortages. In addition, he pointed at the crucial nexus of water and food to highlight the overarching role of agriculture in water use.

Concerns of conflict over blue water have long been associated with the basin level to highlight the inter-linkages and power asymmetries of riparian countries, who seek to use water for agricultural, industrial and domestic purposes (Waterbury, 2002; Tropp and Jagerskög, 2006; Thuo, 2013). Hydropolitics emerged as an increasingly popular academic sub-discipline of international relations from these basin-wide approaches to analyse how countries (with particular reference to the developing world) use water in political decision-making. For this purpose, Warner and Zeitoun developed the 'hydro-hegemony' framework to provide an analysis of the tactics of countries and their control over water resources (Warner and Zeitoun, 2006). This helped to bring decision-making in river basins to the attention of global policy makers. River basin institutions such as the Nile Basin Initiative were established by the World Bank and other Western donors with several million US dollars in order to mediate political tensions over water resources in riparian contexts.

To date, no conflict over water has emerged in even fully allocated basins. Contrary to the fears of some analysts, cooperation over water has in fact increased between countries, hence on the national level (Zeitoun and Mirumachi, 2010). As a result, this chapter argues that water is by nature a local resource potentially causing conflict on a much lower level than the global level. The real arena of conflict over water resources is the sub-national or local level in drylands, where many small farmers struggle with decreasing water availability. This is especially evident in Asia and Africa. For example, in 2016, police in drought-stricken Bundelkhand in India reported several very violent clashes over water between farmers and government security officials (Pacific Institute, 2017). In Darfur, 70 casualties were reported in early 2017 when herders and farmers clashed over water resources (Pacific Institute, 2017).

It is often suggested that agriculture must become more water-efficient in areas facing water scarcity. Farmers are advised to introduce water-saving irrigation methods to increase crop yields by up to 100% thanks to high-tech irrigation methods and careful irrigation scheduling (AFED, 2008). The development industry has taken up this issue by investing millions into schemes that promote smallholder high-tech irrigation (drip, localized sprinkler, sub-surface drip) technology. The introduction of drip irrigation is associated with development, modernity and progress, and its use is widely perceived as helping solve problems of water scarcity (Venot et al., 2014). However, research in sub-Saharan Africa has shown smallholders find it difficult to adapt to this new technology due to poor water conditions on the ground and high maintenance costs (Venot et al., 2014). More importantly, research in California has shown drip irrigation methods do not actually 'save' water but lead to more evapotranspiration and ultimately greater water extraction from the watershed. This is because high-tech irrigation is designed to improve the water service to the plant, and to maximize the quantity of water consumed by the crop, yet to minimize return flows to aquifers and drains. Per hectare, consumption increases in parallel with yield, thus any additional 'savings' are usually converted to extra irrigated area and expanded agricultural production (Perry, 2018).

The only solution to actually manage water more efficiently in water-scarce areas is to introduce strict allocation rules either set by the local community or the state (Perry, 2017). Water allocation is a deeply political subject as it leads to local issues over water rights and distribution of agricultural water to farmers. Since an increasing number of areas in the world are subject to water scarcity, distribution issues over scarce resources will inevitably become a very significant topic. However, issues over water allocation and water availability already exist, yet conflict occurs on different levels and scales. Recent conflict over water is therefore an acute topic in how conflict has been being playing out around the world (Table 4.1).

What is Conflict?

Conflict is a term often used without a clear definition. In order to provide a framework of the different forms of conflict in relation to water and agriculture, conflict is defined here to set out the analysis for the global, national and local levels. For good reasons, conflict is not universally defined. It involves a political, economic and

Table 4.1. Three levels of water conflict.

Level	Scale	Contested issues (illustrative)
3	Global	No known global conflicts over water resources despite previous agricultural trade wars such as between the US and the Soviet Union in 1980 and the embargo on Iraq. However, (virtual) water not the reason/tool for/of conflict
2	Bilateral	Political conflict in Nile basin between Egypt and Ethiopia; Israeli use of military against Syria in 1964 to control the headwaters of the Jordan
1	Sub-national/local	California farmer protests over water legislation drawn up by the US federal government; in 2007, conflicts in Burkina Faso, Ghana and Côte d'Ivoire led to severe clashes over water between farmers and herders

social dimension that cultural groups around the world perceive differently. According to Nicholson, conflict is 'an activity that takes place between conscious, but not necessarily rational, beings' (Nicholson, 1992, p. 11). Conflict exists when 'two people wish to carry out acts which are mutually inconsistent'. A conflict is resolved when some mutually compatible set of actions is worked out (Nicholson, 1992, p. 11). The definition can be extended to more parties and to several levels (global, national, sub-national, local). Conflict is also often a process that has developed between conflicting parties over time from past interactions. However, the key issue to understand about conflict is its subjectivity.

If conflict escalates into further violent action, it is war. In international law, war is defined as 'the use of violence and force between two or more states to resolve a matter of dispute' (Bledsoe and Boczek, 1987, p. 12). The definition is similar if, instead of states, ethnic groups are involved in using violence and force to resolve a matter of dispute. All three definitions are highly relevant for the analysis of water, agriculture and conflict. There are three levels of conflict: global, bilateral and local.

Level 3: water conflict on the global level

Although there is no global watershed, there is still the concept of globally available water resources. These are not a real resource but the water that flows around the globe is 'virtual water'. Virtual water is the water embedded in food. For example, it requires 1300 m^3 to produce a tonne of wheat and approximately 16,000 m^3 are embedded in a tonne of beef (Allan, 2011). Virtual water is silently traded through global agricultural commodities. The virtual water 'trade' has significantly increased in the past three decades. For example, Asia has almost tripled its virtual water 'imports' in the period from 1986 to 2007 with 97 km^3 and 261 km^3, respectively (Dalin *et al.*, 2012). Virtual water serves as a silent mechanism to balance out water deficits. Countries faced by population growth and limited water and/or land resources for food production can rely on the global food market to balance out their water deficits. This is especially true for the grain and staple food market. Staples such as wheat, soy, sugar cane, maize and other grains are primarily produced in the global food bowls in North and South America, eastern Europe and Australia. Although only 12% of globally produced cereals are traded, they are key for keeping the world at peace. Without trading these staples, meat and dairy production would be next to impossible in water and land-scarce regions (Allan, 2013).

Global staple food production and thus virtual water flows are currently controlled by the so-called 'corporate food regime'. In 2003, 73% of the global grain trade was facilitated by five companies – all headquartered in North America and Europe. Archers Daniels Midlands, Bunge, Cargill, Louis Dreyfus and Glencore (the ABCD+G) facilitate trade by providing storage and trading facilitates (Sojamo *et al.*, 2012). The current corporate food regime is the most recent of several distinct food regimes, as identified and elaborated by Harriet Friedmann and Philip McMichael. Food regime theory illustrates the tectonic shifts in global food politics. There have been three food regimes in place since the 19th century. The first, British-led, colonial food regime lasted from the 1860s until the 1930s. It largely imported luxury food such as cocoa, sugar, coffee and tea from British colonies. It disintegrated with the economic crisis in the 1930s (Bernstein, 2003). After World War II, a second, American-led, highly political food regime emerged. It used surplus food production in the United States as a geopolitical tool through, for example, Public Law 480 – a foreign policy tool to deliver food aid in exchange for political control of recipient countries (Woertz, 2013). With the global economic crisis in the 1970s, the second food regime saw its final breath, fundamentally impacted by the tectonic shifts of the 1970s including the collapse of the international monetary order, and a sudden shortage of food on the world market causing sky-high food prices (Bernstein, 2003). This opportunity enticed small farmers in the US to increase borrowing, which paved the way for the US farm crisis in the 1980s. The US peasantry was further diminished, yet thanks to larger farm size units and economies of scale, production levels were further expanded (Bernstein, 2003). Increasing farm sizes and high levels of production led to another phenomenon in global agriculture: the emergence of a corporate food regime.

The corporate food regime has been in place since the 1980s. The Western traders together with food processors and retail giants have since prospered and provided the world with an absence of violent conflict over food. They contributed to a context of little or no conflict over silently traded virtual water. To date, there have been no cases of 'mutual inconsistent acts' or global conflict due to contest over water. This can be explained because the corporate food regime is active in the water-abundant regions in North and South America, which have readily available water and land resources for surplus agricultural production (Allan, 2013). However, in recent years Asian governments have stepped up to challenge Western hegemony over agricultural trade and thus their virtual water hegemony. Especially China and Singapore have invested several billion US dollars to establish commodity traders of their own (Keulertz and Woertz, 2015). China's food arm of its Sovereign Wealth Fund (SWF), COFCO, has purchased the agricultural trading company Noble in 2014. The SWF of Singapore has bought the majority shares of another agricultural trader, Olam, also in 2014 (Keulertz and Woertz, 2015). This can be interpreted as an act to counter Western hegemony in the corporate food regime by investing in strategic crop trading houses. However, it takes place on a very subtle level that has not led to any inconsistent acts thus far.

Level 2: national/bilateral water conflict

The second level on the international scale is the inter-state, political level. Countries that share transboundary waters such as rivers or lakes have had several occasions when water became a source of political tensions, yet no prolonged all-out conflict/war has thus far occurred. For example, the Nile basin has seen several political threats, mainly from Egypt to take action against the southern riparian Ethiopia. Ethiopia has for decades demanded to obtain an increase of its share of the Nile River. The Nile Water Agreement of 1959 between Egypt and Sudan allowed the entire average flow of the Nile River to be shared by Egypt and Sudan with 55.5 billion cubic metres (bcm) (Egypt) and 18.5 bcm (Sudan) (Waterbury, 2002). All further upstream riparians did not get a share to use Nile water for any economic purpose (Keulertz, 2013). In recent years, upstream countries such as Uganda and in particular Ethiopia have called upon Egypt for a revised agreement to allow other riparians to use the Nile for economic growth. Egypt has for a long time resisted these claims by sending out bellicose messages to Ethiopia. Egyptian presidents Sadat and later Morsi have repeatedly stressed their willingness to take military action against Ethiopia (Keulertz, 2013). However, despite these announcements and further Ethiopian use of the Blue Nile for hydropower purposes, no military action has ever been taken. The possibility of a water war cannot be fully ruled out, but the risk is at best low due to Egypt's decreasing economic health and a lack of support for Egypt from the US to wage a water war in the Nile basin.

However, water conflict has also been used as a (foreign) policy tool. After the establishment of Israel in 1948, it quickly became clear that Israel required access to water in order to thrive economically. As a result, US President Eisenhower appointed Eric Johnston as a special ambassador in 1953 to develop the so-called 'Johnston Plan' on how to share the water resources of the Jordan basin between Israel, Jordan and Syria. The initial plan gave preference to in-basin use of the Jordan basin's waters (excluding the Litani River in Lebanon). Johnston proposed the following annual quotas: 774 million m³ for Jordan, 394 million m³ for Israel and 45 million m³ for Syria. However, the plan was quickly rejected by all riparians, each demanding different quotas and strategies. Israel demanded the inclusion of the Litani River within the water plan and to use the Sea of Galilee as the main storage facility. Moreover, Israel demanded a doubling of its water allocation to 810 million m³. Jordan and Syria countered with a proposal rejecting storage in the Sea of Galilee and demanding the inclusion of Lebanon as a riparian state while excluding the Litani from the plan. The League of Arab States proposed very different quota allocations (Israel 200 million m³, Jordan 861 million m³, Syria 132 million m³ and Lebanon 35 million m³ per year). Negotiations were settled with an agreement based on the source of water from rivers of *wadis* (valleys). In the end, the agreement borne out of the Johnston Plan was to allocate Jordan 720 million m³, Israel 400 million m³,

Syria 132 million m³ and Lebanon 35 million m³ (Gat, 2003; Sosland, 2007).

Despite the Jordan basin water plan being concluded by all riparian parties including the League of Arab States in 1955, water became a subject of political conflict in the next decade. After the Suez Canal crisis, the Arabs' stance towards Israel hardened, arguing that further Israeli economic growth would be a threat to the Arab world (Shlaim, 2001). Meanwhile, Israel and Jordan completed large-scale water infrastructure projects by building the National Water Carrier (Israel) and the East Ghor (Jordan) water canals. In 1964, the Arab states decided to reduce the flow of water by 11% into the National Water Carrier, prompting Israel to deploy air strikes against Syria and operating deep into foreign territory (Shlaim, 2001). The water conflict was also seen as a contributing factor to the 1967 Arab–Israeli war, yet it was by no means the only factor. However, the clashes over water in the Middle East were a first major occurrence of conflict over water between states.

The bilateral level of international politics has therefore not been immune to conflict over water. However, despite the incidents in the Nile and the Jordan basins, conflict has not erupted into outright war. In both the Nile and the Jordan basins, conflict was about water as a strategic resource. Conflict was not overtly over agricultural water resources. It would therefore be more apt to speak of water as a tool in bilateral foreign policy. The level that has seen most conflicts over water for agricultural purposes is the sub-national/local level.

Level 1: the sub-national/local level

A large body of literature in anthropology, development and political science has been devoted to water and conflict in pastoralist areas, where water is seasonal and often scarce (Scoones and Graham, 1994; Omosa, 2005; Le Meur, 2006; White et al., 2012). Pastoralism is the branch of agriculture concerned with the raising of livestock, in particular animal husbandry such as of camels, goats, cattle, yaks, llamas and sheep. The majority of pastoralists globally live in dryland areas and hold informal land and water rights. Culture and tradition has shaped their livelihoods for centuries. However, when water resources are scarce during times of drought, pastoralist farmers frequently clash over access to water for their livestock. The drylands of sub-Saharan Africa are a notorious hotspot for pastoralist clashes over water. Newspapers in Ethiopia, Sudan, Kenya and Tanzania frequently report of herder conflicts in the rural areas of their countries (Mankoye, 2014; Langat, 2015). Pastoralists often carry weapons to protect their herds and obtain access to natural resources. Clashes are not only frequent but often underreported. At best, clashes over water make it into national newspapers or development news websites. However, pastoralists are the prime concern for conflict over water resources needed for agricultural production. This conflict could potentially have geopolitical impacts as rural dwellers could be forced to cities and eventually migrate to other regions including Europe, North America, Asia and Oceania. Moreover, it could lead to less political stability, which could ultimately lead to so-called failed states (Stewart, 2007; Selby and Hoffmann, 2012).

A study on the Volta basin in West Africa shows that two of the key risks for conflicts over water and land resources are political strategies conceived since the immediate post-colonial era to integrate pastoralists into the economies of Burkina Faso, Ghana and Côte d'Ivoire, which have been unsuccessful. Farmers and pastoralists have been clashing over water resources since the 1960s with increasing intensification (Cabot, 2017). The most violent clashes so far reportedly occurred in the Volta basin in 2007 when 2000 farmers had to escape from conflict with pastoralists (Anonymous, 2007). Inadequate land and water rights together with discriminatory policies against pastoralists have become especially evident during times of drought. Due to the ongoing effects of climate change, pastoralist–farmer conflicts have become more frequent and are likely to intensify if no policies are designed to support agrarians in the Volta basin (Cabot, 2017). International organizations name education and technological innovations as key solutions for farmer–pastoralist conflicts (Anonymous, 2007). However, such proposals fail to acknowledge the deeper problems, which are owed to the ongoing struggle over land and water rights of small farmers and pastoralists in sub-Saharan Africa.

However, conflict over water is not a topic that is restricted to countries in sub-Saharan Africa with low economic development. In recent years, the water crisis has also reached one of the richest countries in the world: the United States of America. One of the hydrological hotspots is California, which has seen a breathtaking historical trajectory from a near desert state to one of America's key food bowls in the past 80 years. Triggered by Roosevelt's New Deal in the 1930s, Californian decision-makers built large-scale water infrastructure to divert water from the wet northern part to the dry southern part, and from the Colorado River to southern California. The engineering miracle allowed California to become one of the most prolific agricultural producers in the world. However, a 15-year drought, climate change, environmental protection policy and overuse of natural resources have caused California's water resources to quickly deplete in recent decades. Groundwater supplies have been accessed at an unprecedented rate by farmers and urban consumers in response to the drought, causing rapid depletion and pollution. Despite water-use emergency reductions issued by the state governor, which forced urban residents to decrease their water use by 25% in 2015, the agricultural side has remained largely untouched by measures intended to restrict water use (Keulertz et al., 2016).

As 80% of California's water is used by agriculture, farmers in California's Central Valley experienced the drought first-hand by decreasing natural rainfall. Most of the water available to farmers in the Central Valley is diverted from the Sacramento-San Joaquin River Delta. The State Water Project and Central Valley Project built since the 1920s altered the natural flows of the delta with negative impacts on the ecosystem and in particular fish stocks. As a response to ecosystem risks, Californian decision-makers under the leadership of Governor Brown introduced the Bay Delta Conservation Plan in 2009, which has its legal roots in the Endangered Species Act of 1973 conceived by the Nixon Administration (Keulertz et al., 2016). The intent of the state and federal government was to allow ecosystem rehabilitation through a 50-year habitat conservation plan. Less water is therefore allowed to be pumped from the delta to the Central Valley, where the majority of farmers cultivate food crops. As of 2015, California had experienced a 14-year drought that intensified competition for water among farmers, ecosystem uses and cities. Farmers exposed to water cuts have raised vocal opposition to the state and federal governments. Farming lobby groups such as Families Protecting the Valley see the roots of the water crisis by overly regulated 'big government' in Washington, DC (Keulertz et al., 2016). Unsurprisingly, farmers in drought-affected areas in California enthusiastically supported Donald Trump in the 2016 presidential elections, as he promised to deliver more water to farmers once elected to the White House (Anonymous, 2016).[1]

Conflict in California has played out through a democratic process. Family farmers who grow food on small farming plots in the western part of the United States have protested through visible signposts along inter-state highways in California to put pressure on the federal and state governments. Moreover, they have taken the water crisis to the ballot box to support a populist candidate, who promised a water solution based on anti-scientific facts. Conflict between farmers and government is according to Nicholson 'an activity that takes place between conscious, but not necessarily rational, beings' (Nicholson, 1992, p. 11). It has not led to violence so far but severe disagreement. For now, the 'magic pill' has been a populist president, who may or may not be able to solve the water crisis. If President Trump proves unsuccessful in dealing with climate change and overuse of water, the democratic 'magic pill' may not provide the remedy needed by California's farmers.

The Agrarian Question and Conflict

This brings the chapter to the overarching question of the importance of addressing water conflict in the agrarian question in the 21st century. Conflict over agriculture is a longstanding topic in the academic literature. The recurring agrarian question has shaped European, Asian and Latin American politics for most of the 20th century. With its archaic feudal, merchant capitalist system, the agricultural sector has like no other exploited billions of people to date. Classical Marxists noted the key contradiction of class conflict in agriculture as it was unclear where peasants stood. Were the rural *lumpenproletariat* a strong ally in the struggle against capitalism or would they resist socialism due to their specific cultural ties to feudal landlords? For Marx, peasant farmers were under no compulsion to realize the average rate of profit on capital. Thus, grain prices would be low where peasant proprietorship predominated (*Capital*, Vol. III: 805–806). The agrarian question in the 19th century contributed to the rise of communism in eastern Europe, Asia and Latin America in the 20th century. Land was nationalized and feudal landowners were dispossessed of their assets. Conflict in the agrarian context meant the class struggle of the peasantry in the so-called 'West' (comprising North America and Europe) and the rest of the world in which the state played a key role. While the communist project nationalized agriculture, the Western model introduced subsidies and technologies to leverage peasants out of poverty by supporting their livelihoods. The capitalist Western world reshaped its agricultural policies before and in the aftermath of World War II. While Roosevelt in the US introduced the Farm Bill in 1933 providing payments to farmers to subsidize their production, Europe followed suit in 1962 introducing a sophisticated set of subsidies to European farmers through the Common Agricultural Policy. Subsidies in Europe and America were not only designed to support the livelihoods of farmers but also promoted overproduction to dominate global agricultural trade (Keulertz and Woertz, 2015).

Peasants in the West were thus fostered by financial incentives and technological innovation such as seeds and fertilizers. In the US, their increasing production capacities were used for strategic foreign policy, using food as a tool to dump into developing world markets and increase dependency of the global south on the Western world (Bernstein, 2003). In the Soviet Union and the wider communist world, self-sufficiency drove the political agenda. However, thanks to the different speed of agricultural development, US farmers increasingly supplied grain to the Soviet Union, making food a Western catalyst of its hegemony in the days of the Cold War (Paarlberg, 1982).

The agrarian question of the 21st century is defined by the contradictions of capitalism. While the food regime has kept the global level largely food-secure, it has mainly tapped into areas of the world where water is readily available and not scarce. However, other world areas with low water availability are not penetrated by those multinationals. These are especially located in the global south: in the Middle East, sub-Saharan Africa and the drylands of Asia. What has been left untouched by the corporate food regime will be the key arena for the agrarian question faced by water scarcity. Climate change with its effects on floods and droughts will only aggravate the agrarian question. This brings the chapter to the hydro-agrarian question.

Water will expand the agrarian question to another domain. While Roseberry (1993) saw the agrarian question primarily through an economic lens defined by national politics, McMichael asserted the role of global structures such as globalization as a key point of concern (McMichael, 1997). However, water scarcity and uncertainty over access to water due to the tragedy of the commons and exacerbated by climate change add another challenging layer to the agrarian question. Water is a localized and highly variable resource, which is beyond national and international power structures to govern. While land tenure and rights are socially and politically constructed and thus manageable by local and national governments, water availability will be subject to nature and therefore forces beyond the control of policy makers. In sum, then, the hydro-agrarian question means that class struggles in the global south may be aggravated by the availability of water resources to small farmers. The agrarian class is also growing fast, with Africa outstripping all other continents in terms of population growth by 2100: the African continent will host 4.1 billion people (from 1.2 billion today) of which approximately 30–40% are going to live in rural areas (UNDESA, 2015). Acute crises triggered by strong population growth, natural disasters such as droughts and floods, and an overuse of water resources could occur frequently, leading to real concern over conflict that may no longer remain merely local but could have geopolitical impacts due to the ongoing trends of out-migration in rural societies in the global south (Nyantakyi-Frimpong and Bezner Kerr, 2017). If agrarians lack access to water resources, there may be an acute risk for out-migration to urban areas, which would pose significant challenges to overall governance of countries including providing more employment in cities to absorb the newly arriving rural population.

Water scarcity and climate change mean that agrarians will be further marginalized to allow countries in the global south to produce enough food for more than three times the current population, such as will be necessary in sub-Saharan Africa. The recent surge for land in Africa has signalled which direction the continent will take in the coming decades. Triggered by investors from urban areas and the periphery of the corporate food regime, the enclosure of African land was greeted by farmer protests and violent resistance (Greco, 2013). The hydro-agrarian question of the 21st century will therefore only intensify existing class struggles with very significant impacts on the stability of countries in the global south. At the same time, the primary mechanism for this class struggle will be population growth and rural out-migration to cities and potentially foreign countries.

Conclusions

This chapter has covered the nexus of water, agriculture and conflict. It has analysed the importance of understanding water resources and argued that there is no silver bullet other than strict water allocation policies to share available resources. Three levels for analysis have been identified: (i) the global level; (ii) the bilateral/national level; and (iii) the sub-national/local level. The global level has not seen major conflicts over water resources so far as the silent 'virtual water trade' has acted as a stabilizing factor. However, the global power asymmetries of the corporate food regime act as the more compelling reason why there has been no conflict: Prime land with water is in the hands of the corporate food regime, which exerts a major power force in the global economy. The bilateral/national level has seen some conflict in the past such as in the Nile basin or in the Jordan basin. However, water was used more as a foreign policy tool in political power games. Agriculture and farmers never played an important role in either of these two analysed basins.

The chapter has shown that currently the real level of concern is the sub-national/local level, where conflict over water and food resources is a subject of increasing importance. Especially pastoralist farmers are subject to decreasing water resources and the enclosure of land. Moreover, conflict over water resources is deeply connected to inadequate water rights and climate change, which serve as aggravating factors that lead to an emerging hydro-agrarian question in the global south. However, not only farmers in the global south are affected by the nexus of water, agriculture and conflict. Farmers in the most powerful nation in the world, the United States, are also experiencing low-level (albeit non-violent) conflict over water resources. The current national leadership has promised easy solutions based on unscientific facts; hence water, agriculture and conflict could soon hit the United States, too.

The hydro-agrarian question of land tied to abstract water resources for the production of food offers yet another contradiction in world capitalism. Unlike other contradictions, the hydro-agrarian question is perhaps one of the most neglected topics in global politics. Understanding the challenges ahead is therefore mandatory to avoid conflict and potentially further migration and radicalization in the global south. The hydro-agrarian question needs to be incorporated into analyses on the future of the agrarians, which is likely to become an acute issue of global significance in future decades.

Note

[1] As Trump said during a 27 May 2016 meeting with farmers in Fresno, 'We're going to solve your water problem – you have a water problem that is so insane, it is so ridiculous, where they're taking the water and shoving it out to sea. And I just met with a lot of the farmers who are great people, and they're saying, we don't even understand it... They have farms up here, and they don't get water. I said, "Oh, that's too bad. Is it a drought?" "No, we have plenty of water... We shove it out to sea..." The environmentalists don't know why. They're trying to protect a certain kind of three-inch fish... My environmental standard is very simple... I want clean air and clean water' (Hitzik, 2016, p. 1).

References

Aldaya, M.M., Allan, J.A. and Hoekstra, A. (2009) Strategic importance of green water in international crop trade. *Ecological Economics* 69, 887–894.
Allan, J.A. (2003) *The Middle East Water Question*. I.B. Tauris, London.
Allan, J.A. (2013) The food-water value chain. In: Antonelli, M. and Greco, F. (eds) *L'acqua che mangiamo: Cos'è l'acqua virtuale e come la consumiamo*. Edizione Ambiente, Milan, pp. 48–64.
Allan, T. (2011) *Virtual Water: Tackling the Threat to Our Planet's Most Precious Resource*. I.B. Tauris, London.
Allan, T., Keulertz, M. and Woertz, E. (2015) The water–food–energy nexus: an introduction to nexus concepts and some conceptual and operational problems. *International Journal of Water Resources Development* 31(3), 301–311.
Anonymous (2007) Innovation and education needed to head off water war. *Irinnews*. Available at: www.irinnews.org/report/74308/burkina-faso-innovation-and-education-needed-head-water-war (accessed 22 February 2017).
Anonymous (2016) California presidential election results. *Politico*. Available at: www.politico.com/2016-election/results/map/president/california (accessed 22 February 2017).
Arab Forum for Environment and Development (AFED) (2008) *Water Efficiency in Agriculture*. Amman: AFED.
Bernstein, H. (2003) Farewell to the peasantry. *Transformation* 52, 1–19.
Bledsoe, R.L. and Boczek, B.A.S.E.-C. dictionaries in political science (1987) *The International Law Dictionary*. ABC-Clio, Santa Barbara, California.
Cabot, C. (2017) Case study: farmer–herder conflicts in Burkina Faso, Côte d'Ivoire and Ghana. *Climate Change, Security Risks and Conflict Reduction in Africa: A Case Study of Farmer-Herder Conflicts over Natural Resources in Côte d'Ivoire, Ghana and Burkina Faso 1960–2000*. Springer, Berlin, Heidelberg, Germany, pp. 113–155.
Dalin, C., Konar, M., Hanasaki, N., Rinaldo, A. and Rodriguez-Iturbe, I. (2012) Evolution of the global virtual water trade network. *Proceedings of the National Academy of Sciences of the USA* 109(16), 5989–5994.
Falkenmark, M. and Rockström, J. (2006) The new blue and green water paradigm: breaking new ground for water resources planning and management. *Journal of Water Resources Planning and Management* May/June, 129–132.
Food and Agricultural Organization of the United Nations (FAO) (2009) How to feed the world in 2050. Insights from an expert meeting at FAO, 2050(1), 1–35. Available at: www.fao.org/wsfs/forum2050/wsfs-forum/en (accessed 22 February 2017).
Gat, M. (2003) *Britain and the Conflict in the Middle East, 1964–1967: The Coming of the Six-Day War*. Praeger, Westport, Connecticut.
Gleick, P. (1993) Water and conflict: fresh water resources and international security. *International Security* 18(1), 79–112.
Greco, E. (2013) Struggles and resistance against land dispossession in Africa: an overview. In: Allan J.A., Keulertz, M., Sojamo, S. and Warner, J. (eds) *Handbook of Land and Water Grabs: Foreign Direct Investment and Food and Water Security*. Routledge, Abingdon, pp. 456–465.
Hitzik, M. (2016) California's drought: how Trump's blistering caricatured a genuine crisis. *LA Times*. Available at: www.latimes.com/business/hiltzik/la-fi-hiltzik-trump-westlands-20160606-snap-story.html (accessed 22 February 2017).
Hoogeveen, J., Faurès, J.M., Peiser, L., Burke, J. and van de Giesen, N. (2015) GlobWat – a global water balance model to assess water use in irrigated agriculture. *Hydrology and Earth System Sciences* 19, 3829–3844.
Huppert, W. and Wolff, B. (2002) 'Principal Agent Problems' in irrigation – inviting rent-seeking and corruption. *Zeitschrift für Bewässerungswirtschaft* 37(2), 179–198.
Kameri-Mbote, P. (2011) *Water, Conflict and Cooperation*. Wilson Center, Washington, DC.
Keulertz, M. (2013) *Drivers and Impacts of Farmland Investment in Sudan: Water and the Range of Choice in Jordan and Qatar*. King's College London, London.
Keulertz, M. and Woertz, E. (2015) State actors in international agro-investments: the role of China, Russia and Gulf. *Development Policy* 6(1), 30–52.
Keulertz, M., Sowers, J., Woertz, E. and R. Mohtar (2016) *The Water-Energy-Food Nexus in Arid Regions*, Available at: http://oxfordhandbooks.com/view/10.1093/oxfordhb/9780199335084.001.0001/oxfordhb-9780199335084-e-28 (accessed 22 February 2017).
Langat, W. (2015) As water falls short, conflict between herders and farmers sharpens. *Reuters*, 23 November 2015. Available at: www.reuters.com/article/kenya-climatechange-conflict/as-water-falls-short-conflict-between-herders-and-farmers-sharpens-idUSL8N13D4G420151123 (accessed 22 February 2017).

Le Meur, P.-Y. (2006) Conflict over access to land and water resources within sub-Saharan dry lands. *Underlying Factors, Conflict Dynamics and Settlement Processes. Roma: FAO/Lead report* (September). Available at: www.fao.org/fileadmin/templates/lead/pdf/tanzania/conflict.pdf (accessed 22 February 2017).

Mankoye, K. (2014) Herders fight farmers Tanzania water. *Al Jazeera*. Available at: www.aljazeera.com/indepth/features/2014/01/herders-fight-farmers-over-tanzania-water-201411610756420189.html (accessed 22 February 2017).

McMichael, P. (1997) Rethinking globalization: the agrarian question revisited. *Review of International Political Economy*, 4(4), 630–662.

Nicholson, M. (1992) *Rationality and the Analysis of International Conflict SE – Cambridge Studies in International Relations 19*. Cambridge University Press, Cambridge, UK.

Nyantakyi-Frimpong, H. and Bezner Kerr, R. (2017) Land grabbing, social differentiation, intensified migration and food security in northern Ghana. *Journal of Peasant Studies* 44(2), 421–444.

Omosa, E. (2005) *The Impact of Water Conflicts on Pastoral Livelihoods: The Case of Wajir District in Kenya*. IISD, Winnipeg.

Paarlberg, R. (1982) Food as an instrument of foreign policy. *Proceedings of the Academy of Political Science* 34(3), 12–33.

Pacific Institute (2017) Water conflict chronology. *World Water*. Pacific Institute, Oakland, California.

Perry, C. (2018) The water crisis: which solutions will work? In: Allan, T., Bromwich, B., Keulertz, M. and Colman, A.J. (eds) *Handbook of Water, Food and Society*. Oxford University Press, New York, pp. 30–48.

Ricks, T. (2011) The future of water wars. *Foreign Policy* 756–759. Available at: http://foreignpolicy.com/2011/05/05/the-future-of-water-wars (accessed 22 February 2017).

Riegels, N. (2016) Options for decoupling economic growth from water use and water pollution: A report of the Water Working Group. UNEP, Nairobi.

Roseberry, W. (1993) Beyond the agrarian question in Latin America. In: Cooper, F., Isaacman, A.F., Mallon, F.E., Roseberry, W. and Stern S.J. (eds) *Confronting Historical Paradigms. Peasants, Labor and the Capitalist World System in Africa and Latin America*. University of Wisconsin Press, Madison, pp. 318–70.

Scoones, I. and Graham, O. (1994) New directions for pastoral development in Africa. *Development in Practice* 4(3), 188–198.

Selby, J. and Hoffmann, C. (2012) Water scarcity, conflict, and migration: a comparative analysis and reappraisal. *Environment and Planning C: Government and Policy* 30(6), 997–1014.

Shlaim, A. (2001) *The Iron Wall: Israel and the Arab World*. W.W. Norton & Company, New York.

Sojamo, S., Keulertz, M., Warner, J. and Allan, J.A. (2012) Virtual water hegemony: the role of agribusiness in global water governance. *Water International* 37(2), 169–182.

Sosland, J.K. (2007) *Cooperating Rivals: The Riparian Politics of the Jordan River Basin*. SUNY Series in Global Politics. State University of New York Press, Albany, New York.

Stewart, P. (2007) 'Failed' states and global security: empirical questions and policy dilemmas. *International Studies Review* 9(4), 644–662.

Thuo, S. (2013) *The Nile in Black*. GWP, Entebbe.

Tropp, H. and Jagerskög, A. (2006) *Water Scarcity Challenges in the Middle East and North Africa (MENA)*. UNDP, New York.

United Nations Department of Economic and Social Affairs (UNDESA) (2015) *World Population Prospects*. UNDESA, New York.

United States Bureau of Reclamation (USBR) (2017) *Reclamation and Arizona*. Available at: https://www.usbr.gov/lc/phoenix/AZ100/1940/index.html (accessed 22 February 2017).

Venot, J.-P., Zwarteveen, M., Kuper, M., Boesveld, H., Bossenbroek, L., Van Der Kooij, S., Wanvoeke, J., Benouniche, M., Errahj, M., De Fraiture, C. and Verma, S. (2014) Beyond the promises of technology: a review of the discourses and actors who make drip irrigation. *Irrigation and Drainage* 63(2), 186–194.

Warner, J. and Zeitoun, M. (2006) Hydro-hegemony – a framework for analysis of trans-boundary water conflicts. *Water Policy* 8, 435–460.

Waterbury, J. (2002) *The Nile Basin: National Determinations of Collective Actions*. Yale University Press, New Haven, Connecticut.

White, B., Borras, S.M., Hall, R., Scoones, I. and Wolford, W. (2012) The new enclosures: critical perspectives on corporate land deals. *Journal of Peasant Studies* 39(3–4), 619–647.

Woertz, E. (2013) *Oil for Food: the Global Food Crisis and the Middle East*. Oxford University Press, Oxford, UK.

Zeitoun, M. and Mirumachi, N. (2010) Transboundary water interaction: reconsidering conflict and cooperation. In: Wegerich, K. and Warner, J. (eds) *The Politics of Water: A Survey*. Routledge, London, pp. 12–28.

5 Illegal Drug Plant Cultivation and Armed Conflicts: Case Studies from Asia and Northern Africa

Pierre-Arnaud Chouvy*
Centre national de la recherche scientifique (CNRS), Paris, France

In Asia and other continents, the internal peace of a number of countries has been affected, sometimes even conditioned, by the existence of illegal agricultural production and the ensuing illegal trade (Chouvy and Laniel, 2007). However, through loss of politico-territorial control, the armed conflicts that have afflicted certain states have made possible and even encouraged the development of such agricultural production and trafficking. Indeed, significant systemic effects have long existed between guerrilla and civil war economies on the one hand, and the economies resulting from illegal activities on the other hand. As demonstrated by Alfred McCoy as early as 1972 in his seminal book *The Politics of Heroin* (McCoy, 1972; re-edited in 1991 and in 2003), war economies and drug economies have a long common history, in Asia and elsewhere. Asia will be the main focus of this chapter since it holds a special importance in the history of illegal drugs, notably because the opium trade has long been part of various armed conflicts and geopolitical rivalries (from the Sino-British opium wars to the more recent Afghan and Burmese wars); and because, as a consequence of the economic, political and historical importance of the opium trade in the region, this is where 'the symbiotic relationship between trafficker and politician that has become the dominant feature of the contemporary drug trade has its roots' (Meyer and Parssinen, 1998, p. 12), and this is also where the global prohibition of certain drugs originated.

In Afghanistan and in Burma the illegal opium economy has been partly responsible for financing the ongoing war efforts of some of the opposing factions, since the Soviet–Afghan War (1979–1989) in Afghanistan and since the independence (1948) of Burma (where the world's longest civil war is still going). But if opium has clearly been one of the sinews of war for some of the Afghan and Burmese guerrillas and factions, it often subsequently has become one of the stakes of war. Understandably, the strong synergies existing between civil war economies and illegal drug economies, especially agriculture-based drug economies, have therefore weighed upon the two countries' potential for political and economic development. As well as allowing and even encouraging prolongation of conflict and making any resolution of crises all the more difficult, the conflict/drug 'synergy' has also laid the foundations for the criminalization of the peace economies in both countries, so potentially compromising their internal peace and security.

Through these connections with the war economy, the opium economy has certainly had a destabilizing effect in the recent histories of Afghanistan and Burma. But it is important to stress that, while the opium economy has surely

* Email: pachouvy@geopium.org

helped perpetuate the Afghan and Burmese conflicts, it did not cause them. Also, the ongoing politico-territorial and economic crises that exist in both countries do not result from their illegal drug economies – at least, not directly. Nor did the opium economy simply bankroll some of the parties at war to a greater or lesser extent. In fact, it enabled many Afghan and Burmese farmers to survive as best as they could during long periods of economic depression. As Jonathan Goodhand writes, 'opium is simultaneously a conflict good, an illicit commodity and a means of survival' (Goodhand, 2005, p. 211).

Synergies between war economies and drug economies are nothing new, although the oldest documented case of an Asian civil war to have been financed, at least to some extent, by opium proceeds is China in the late 1910s. Yuan Shikai, the second president (1912–1916) of the Republic of China, had perpetuated the opium suppression campaign started before him but, after his death in 1916, and the subsequent breakdown of the central government, former warlords-turned-military governors split into countless factions, and warlordism emerged again. During this period, 'narcotics provided a means to finance the expensive arms and ammunition required to survive as a warlord' and 'opium revenue became a major financial resource for warlords, mainly through "fines" on cultivation, trafficking, selling, and smoking' (Meyer and Parssinen, 1998, p. 143; Zhou, 1999, p. 40).

In the late 1920s, 'the escalating cost of warfare forced even the most reluctant and high-minded politicians to turn to the opium business for revenues' and even Chiang Kai-shek, despite his hostility to morphine and heroin, was forced to 'acknowledge opium's significance' in order to consolidate his power in the country (Meyer and Parssinen, 1998, pp. 158, 154). Later, in the mid-1930s, the Nationalists, confronted with increasing international and national pressure, launched a nationwide antidrug campaign that eventually failed. However, as historian Zhou Yongming stresses, such a move was strategically motivated as it was in part designed to 'consolidate the power of the central government nationwide by cutting off the revenue sources of regional powers' (Zhou, 1999, p. 78). Then, as historians Kathryn Meyer and Terry Parssinen explain:

It was in this hothouse, created by China's disintegration and the League [of Nations]'s successes, that gangsters and politicians molded the modern international narcotics trafficking industry. The symbiotic relationship between trafficker and politician that has become the dominant feature of the contemporary drug trade has its roots in Asia in the early twentieth century. The men in the shadows succeeded because they structured their careers with webs of smoke at the point where profits and power converge.

(Meyer and Parssinen, 1998, p. 12)

However, symbiosis between drug traffickers, politicians and other power holders, and synergies between war economies and drug economies only fully developed during the Cold War.

The Cold War and the Rise of Opium Production in Asia

The Cold War played a direct and prominent role in the production and trafficking of illegal drugs. Indeed, the financing of many anti-communist covert operations, such as those led by the Central Intelligence Agency (CIA), resorted to the drug economy that existed in various proxy states where drug trafficking was often condoned and even encouraged. Specific historical cases illustrate how the anti-communist agenda of the CIA played a decisive role in spurring the global illegal drug trade by condoning it and even facilitating it. These include the French Connection and the role of the Corsican mafia against communists in France and in Southeast Asia (Laos and Vietnam), the propping up of the defeated Kuomintang (KMT) in northern Burma, the Islamic mujahideen resistance in Afghanistan, and, on another continent, the Contras in Nicaragua, as extensively and very convincingly documented and demonstrated by Alfred McCoy (McCoy, 2003; Chouvy, 2007).

The United States, as the leader of the global struggle against communism, extensively used its special services and intelligence agencies to conduct covert operations worldwide. In the global struggle to contain communism, local aid was needed and widely found in local criminal organizations. It is in Southeast Asia, in Southwest Asia and in Latin America that the CIA most significantly influenced the illegal drug trade. Its anti-communist covert operations benefited from the participation of some drug-related combat units who, to finance their own struggle, were directly involved in illegal drug production and trafficking. Considering the involvement of different groups in the drug trade (e.g. the Hmong in Laos, the Kuomintang (KMT) in Burma and the mujahideen in Afghanistan), their CIA backing implied that the agency condoned the use of drug proceeds and the increase of opiate production in Asia. However, no evidence has surfaced to suggest that the CIA condoned or facilitated the exportation of heroin to the US or Europe, as clearly happened with the Nicaraguan Contras (McCoy, 2003; Chouvy, 2007).

In October 1949, the communists defeated the KMT in China, and in the years that followed they cracked down on what was then the world's largest opium production network. Opium production then shifted to the mountainous and frontier areas of Burma, Laos and Thailand, where KMT remnants had fled and became deeply involved in drug trafficking. Beginning in 1951, the CIA supported the KMT in Burma in an unsuccessful effort to assist it in regaining a foothold in China's Yunnan province. Arms, ammunition and supplies were flown into Burma from Thailand by the CIA's Civil Air Transport (CAT), later renamed Air America and, still later, Sea Supply Corporation, created to mask the shipments. The Burmese Army eventually drove KMT remnants from Burma in 1961, but they later resettled in Laos and northern Thailand and continued to run most of the opium trade (McCoy, 2003; Chouvy, 2007).

Following the 1954 French defeat in Indochina, the US gradually took over the intelligence and military fight against communism in both Laos and Vietnam. It also took over the drug trafficking business developed by the French by buying the opium produced by the Hmong and Yao hill tribes to enlist them in counterinsurgency operations against the Viet Minh. To meet the costs of this war, the French secret intelligence service, then called the SDECE (*Service de documentation extérieure et de contre-espionnage*), allied itself with the Corsican syndicates trafficking opium from Indochina to Marseille to take over the opium trade that the colonial government had outlawed in 1946. The CIA ran its secret army in Laos, composed largely of Hmong tribesmen led by General Vang Pao. Air America would fly arms to the Hmong and fly back their opium to the CIA base at Long Tieng, where Vang Pao had set up a large heroin laboratory. Some of the heroin was then flown to south Vietnam, where part of it was sold to US troops. After the Americans pulled out of Vietnam in 1975, Laos became the world's third largest opium producer and retained this rank until the mid-2000s (McCoy, 2003; Chouvy, 2007).

However, Vietnam was not the only battleground of Cold War drug operations. The CIA launched a new major covert operation in Southwest Asia in the early 1980s to support Afghanistan's mujahideen guerrillas in their fight against Soviet occupation. US President Ronald Reagan was determined to counter what he viewed as Soviet hegemony and expansionism, a goal shared by his CIA director, William Casey. To support the mujahideen with arms and funds, the CIA resorted to one of Pakistan's intelligence services, the Inter-Services Intelligence (ISI), which chose which Afghan leaders to back and used trucks from Pakistan's military National Logistics Cell (NLC) to carry arms from Karachi to the Afghan border. However, the ISI not only chose Gulbuddin Hekmatyar, an important Afghan opium trafficker, as its main beneficiary, but also allowed NLC trucks to return from the border loaded with opium and heroin. After the Soviet withdrawal from Afghanistan in 1989, US aid to the mujahideen stopped, and the internecine conflict that ensued in the country favoured an increase in opium production in order to maintain rival warlords and armies. Afghanistan eventually became the world's biggest opium-producing country and has remained such since (McCoy, 2003; Chouvy, 2009).

As Jill Jonnes puts it:

> In the years before World War II, American international narcotics policy had been extremely straightforward. The United States was righteously against anything that promoted or sustained the nonmedical use of addicting drugs. But the Cold War created not only new national security policies, but a new shadow world that accepted a far more ambivalent attitude toward drugs and drug trafficking.
>
> (Jonnes, 1996, pp. 164–165)

Illegal drug production and trafficking increased during the Cold War. During this period, the US government was less interested in waging the 'war on drugs' begun in 1971 by Richard Nixon than in using drug traffickers to support its wars and proxies abroad. Indeed, had the CIA cracked down on drug trafficking during the Cold War, it would have forgone valuable intelligence sources, political influence and much-needed funding for its covert, and sometimes illegal, operations. Ironically, there is no evidence that the Soviet Union or its secret intelligence agency, the KGB, resorted to drug sales to fund activities during the Cold War. Therefore, after the modern international narcotics trafficking industry emerged in pre-World War II China, and after communism enabled the People's Republic of China to suppress local opium production, trafficking and consumption, the fight against communism implied by the Cold War justified that opium production and trafficking be resorted to in order to finance covert operations and secret wars.

In the third edition of *The Politics of Heroin*, Alfred McCoy writes:

> Rhetoric about the drug evil and the moral imperative of its extirpation has been matched by a paradoxical willingness to subordinate or even sacrifice the cause for more questionable goals. The same governments that seem to rail most sternly against drugs, such as Nationalist China in the 1930s and the United States since the 1940s, have frequently formed covert alliances with drug traffickers.
> (McCoy, 2004, p. 459)

In his effort to reveal the extent of the 'CIA complicity in the global drug trade' McCoy then explains that 'nowhere is this contradiction between social idealism and political realism more evident than in the clash between prohibition and protection during the cold war' (McCoy, 2004, p. 459). However, the end of the Cold War was not to reduce illegal opium production in Asia, as the end of foreign subsidies to warring Afghan factions largely spurred opium poppy cultivation in Afghanistan. During most of the 20th century, wars and conflicts fostered illegal opium production and made peace-building more difficult as war economies and drug economies fed each other in a vicious circle.

War and Illegal Agricultural Drug Production

Illegal agricultural drug production does not only occur in Asia: vast expanses of illegally cultivated coca bushes and cannabis plants, but also of opium poppies, exist in the Americas (including large-scale outdoor cannabis cultivation in the US and Canada, and opium poppy cultivation in Mexico and Colombia) and in Africa (widespread, though never properly estimated, cannabis cultivation). Of course, illegal cash crops are usually, though not always, more profitable than local food crops or even other possible, legal, cash crops, and it is tempting to explain that people resort to them simply due to economic considerations. But illegal cash crops proliferate above all in contexts of armed conflict (Afghanistan, Burma, Colombia) or open or rampant social, economic or political crisis (economic and political crises in sub-Saharan Africa, Morocco and Peru; social and political tensions in Bolivia) that compromise the politico-territorial controls necessary to impose the rule of law in a given country (Chouvy and Laniel, 2007). Illegal cash crops are not only and perhaps not mainly the result of economic problems, but instead thrive in political contexts marked by the use and consequences of force and violence, and by complex and often transnational power struggles.

This is precisely how opium production was first commercially developed in Asia: in the context of colonialism and of early globalization. War played a very early role in the spread of opium production and consumption in Asia as two 'opium wars' (1839–1842 and 1856–1860) were waged in the 19th century by the British in order to impose their opium trade upon Imperial China. Later, opium economies largely contributed to making wars more viable and even profitable and, in turn, wars and the political and territorial disruptions they caused made illegal opium production easier. In Asia, opium production clearly thrived in the two countries that underwent the continent's longest wars: Afghanistan and Burma (Chouvy, 2009).

Therefore, it seems that war best explains the success of Afghanistan's and Burma's illegal drug economies. In fact, to simply pretend that illegal drug crops prosper on the ruins of underdevelopment proves much harder. If low levels of economic development and (even more so) a strong incidence of poverty can logically and legitimately, yet not legally, urge certain populations, mainly rural and often peripheral and marginal, to resort to an agricultural illegal drug economy (as indeed do many Afghan and Burmese farmers), nevertheless the drugs–underdevelopment correlation does not seem to possess the validity that we very often wish to lend it. It is indeed easy to observe that economic underdevelopment and poverty are not burdens endemic to areas of illegal agricultural drug production and that they cannot explain its emergence and perpetuation in a systematic way: in Asia as well as in Latin America (mostly Colombia and Mexico) illegal opium production is much more concentrated and localized than poverty. Concurrently, illegal agricultural production is far from being restricted to the developing world or to countries at war, since Canada and the US are among the world's very first illegal producers of marijuana (both indoor and outdoor cultivation). The market value of US-produced marijuana reportedly exceeded US$35 billion in 2005, a time when it was still completely illegal, more than the country's most profitable staple crop, maize (US$23 billion) (Gettman, 2006). Now that cannabis cultivation has become legal in a few select US states, the data would need to be reevaluated.

The overlap that can be observed between areas of underdevelopment and areas of agricultural production of illegal drugs does not explain the resort to the latter in a satisfactory way. Indeed, many more countries would resort to the production of illegal drugs, especially in Asia where history and ecology would make Central Asian, Chinese and Indian production all the more easy, if economic underdevelopment were the main cause of illegal opium production (Chouvy, 2009).

In the same way, it does not seem that the other variable frequently advanced to explain the appeal of the opium economy – that of an economic strategy designed by dominated and marginalized ethnic minorities – has any more validity than that of poverty. While in Asia opium production is almost exclusively undertaken by tribal people, opium farmers differ greatly in Burma, in Laos and in Thailand, where they all belong to ethnic minorities, and in Afghanistan, where it is the largely dominant (politically, culturally and demographically) Pashtun people, one the world's largest tribal groups, who resort to opium poppy cultivation (yet, other ethnic/tribal groups also produce opium). In Pakistan's Khyber Pakhtunkhwa province, previously (1955–2010) known as the North West Frontier Province (NWFP), where opium production had almost been suppressed until it reappeared in the early 2000s, poppies are not cultivated by the country's dominant group, but Pakistan's poppy farmers are, like in Afghanistan, Pashtun people, and are not really dominated as much as Southeast Asia's tribal group may be (although they are far from being dominant the way the ethnic Thais or Burmans are) (Chouvy, 2009).

Therefore, economics and ethnicity may be pushed aside as unique causal factors leading to illegal agricultural production. But even ecological or more classic geographical factors fail to explain why opium production is resorted to by some specific groups or peoples in some countries and why it is not more widespread. Of course, in Southeast Asia, opium production takes place in very different ecological milieus and geographical environments than in Afghanistan (Chouvy, 2011). The heavily rainfed highlands of Southeast Asia where opium is produced lie in the far peripheries and borderlands of Burma, Laos and Thailand, while the main opium-producing areas of Afghanistan are largely, but not exclusively, located in much drier lowlands, irrigated or not. Ecological and geographical constraints are definitely worse in Southeast Asia's hills and mountains, where most people rely almost exclusively on rainfed agriculture, suffer from a lack of access to regional markets and do not have many other cash crop opportunities than the one offered to them by opium (Chouvy, 2009).

But what changed the 'hill tribe economy from subsistence agriculture to cash-crop opium farming' (McCoy, 1991, p. 119) in Southeast Asia in the 1940s is similar to what spurred large-scale commercial opium production in Afghanistan in the 1980s and 1990s. War, whether through strategic use of opium and opium producers (the Hmong in Laos and the mujahideen in Afghanistan), or through physical destruction (orchards, irrigation channels, landmines in Afghanistan), or both (added physical destruction and economic disruption caused by decades of war in Burma and Afghanistan), has turned opium production into a funding (or enriching) resource for military commanders and warlords confronted with finance shortages, and into a coping mechanism for farmers confronted with a new war-driven market (based on the strategic and therefore economic value of opium) and with war-induced physical and economic disruptions.

Therefore, there clearly exists a strong correlation between war economies and drug economies, most notably in Afghanistan and in Burma. Although opium production predated the Afghan and Burmese conflicts, the wars and internecine conflicts that plagued both countries clearly spurred illegal opium production. In return, opium production helped perpetuate the Afghan and Burmese conflicts by making them economically viable. However, as Gaston Bouthoul warned – long before the World Bank economist Paul Collier argued that 'greed considerably outperforms grievances' in triggering and perpetuating civil wars – 'one should not confuse the economic aspect of conflicts with their necessity or their economic fatality' (Collier and Hoeffler, 2001; Bouthoul, 1991, p. 226). Indeed, the Afghan and Burmese conflicts obviously did not start because of opium production, for their causes were much more complex and deep rooted. The key causes of large-scale illegal opium production in Afghanistan and in Burma lie in the pre-existence of opium production to war in both countries, in the transnationalization (mainly by the Soviet Union, the CIA, China, the Communist Party of Burma and the KMT) of their conflicts, and finally in the necessity of both countries' belligerents to find alternative financial resources after foreign subsidies and support were cut off.

War alone, however, cannot satisfactorily explain the emergence or the development of illegal opium production. While the cost of war may explain such recourse in Burma and in Afghanistan, where many warring factions clearly resorted to the opium economy to fund their operations during decades, the case of illegal cannabis production urges some caution. Cannabis cultivation and hashish production have developed considerably in Morocco during the last decades of the twentieth century until, interestingly, the 2003 crop equalled that of opium poppy in Afghanistan in 2004: 134,000 ha of cannabis were reportedly cultivated in 2003 in Morocco (UNODC, 2003), while opium poppies covered 131,000 ha in Afghanistan a year later. The comparison is all the more striking since both countries hold almost the same area of arable land (Afghanistan holds 12% of arable land (7.8 million ha) and Morocco, whose territory is smaller, holds 19% of arable land (8.5 million ha)) (Chouvy, 2009).

Yet, hashish production in Morocco differs greatly from opium production in Afghanistan or in Burma, or even from coca production in Colombia, as no armed conflict challenges the writ of the Cherifian kingdom over its territory (Chouvy and Laniel, 2007). Although cannabis cultivation is illegal in Morocco, a complex set of colonial, political and economic factors has resulted in an entrenched tolerance of hashish production in the northern region of the country, the Rif Mountains. The Rif is one of the poorest regions in Morocco and its tribal Berber population has long resisted foreign and even Arabic rule, eventually obtaining a *de facto* tolerance of cannabis cultivation by the Moroccan state that obviously saw the large hashish economy as an alternative to regular economic development (Chouvy, 2005). Both ecologically and economically, cannabis cultivation and its rapid increase in the Rif Mountains during the past decades are understandable. The Rif is densely populated and is one of the most unsuitable regions of Morocco for intensive agricultural production: a rugged relief of steep slopes and poor soils, combined with heavy but irregular rainfall compounded by a lack of irrigation infrastructure, make most crops other than cannabis not worth the labour invested.

War clearly cannot explain Morocco's large-scale illegal agricultural drug production but other features previously identified in Afghanistan and in Burma are also present in Morocco: poverty, geographical and ecological constraints, and problematic inter-ethnic relations make a complex set that is highly favourable to the production of an illegal cash crop. But although Moroccan cannabis cultivation has not developed in an armed conflict context, it is nevertheless, at least to some extent, the consequence of tense and violent relations and even of full-blown wars (the Rif War, 1921–1926) between the Cherifian state and the Riffian Berbers. What the Moroccan example shows is that a crop whose production benefited both economically and strategically the French and the Spanish Moroccan Protectorates (cannabis cultivation was only really prohibited in 1954 in the French Protectorate and in 1956 in the Spanish Protectorate, at independence) became entrenched in one of the poorest and most restive areas of the country (Chouvy, 2005).

Long-term consequences of war and low-intensity conflicts have made the Rif perhaps the largest region of hashish production in the world. Without an ongoing war in the Rif that could explain its current existence and importance, illegal cannabis cultivation appears to be tacitly tolerated by the state, whatever the reason: either because the state benefits economically and strategically from it, or because it does not have the means to control its own territory and to impose its writ over it. Most likely, the reality lies somewhere in between: since independence the Moroccan state has not had the political and economic means to prohibit and/or eradicate cannabis cultivation in the Rif, nor has it had the means to promote economic alternatives to a fast-growing and profitable hashish industry. Corruption has of course played a large role in the development of cannabis cultivation, and many drug-related scandals have rocked the state's administration in the past years. In the end, what explains the huge extent of illegal cannabis cultivation in a country at peace such as Morocco is the failure of the state to control its territory, that is, by economic development, political integration and law enforcement. Most likely, the Moroccan state has lacked the authority, legitimacy and capacity to impose the rule of law on its entire territory, to formulate adequate strategies and to carry out reforms (Chouvy, 2009).

In Afghanistan and in Burma, war has played a fundamental role in the development of illegal opium production and it is highly probable that their respective productions would not have reached such levels if they had not been at war for so long. In fact, although underfunded wars clearly favour the resort to informal and illegal economies (both by civilians and the military), it is not war per se that makes large-scale illegal agricultural production possible, but the lack of state territorial control that war implies. While opium production preceded war in Afghanistan and in Burma and developed as it financially helped to perpetuate war, it does seem, according to the cases of Afghanistan, Burma and Morocco, that illegal agricultural production has a tendency to outlive war and to complicate transitions from war economies to peace economies. Worse, it is likely that forced suppression of illegal agricultural productions without adequate compensation and alternatives may well threaten old status quos, as in Morocco, or compromise peace-building and state-building, as in Afghanistan and in Burma.

Drugs and War: the 'War on Drugs'

Throughout modern history, illegal agricultural drug production has been spurred on or even imposed by wars, among other things because it has helped make wars viable or even profitable for some belligerents, and sustainable for some civilians. Yet, wars have not only been waged by way of and for drug proceeds: they have also been waged against drugs. In fact, US President Richard Nixon (1969–1974) launched a 'war on drugs' in 1971 that, after it successfully attacked Turkish opium production, 'defined the character of subsequent drug wars by applying the full coercive resources of the United States government to eradicate narcotics production at its source' (McCoy, 2004, p. 47). However, during four decades the extremely expensive US-led war on drugs denounced at length by McCoy (2003, 2004) and many others not only produced many 'unintended consequences' (Tullis, 1996) and failed to achieve both US and UN objectives of global interdiction and suppression of *certain* drugs, it also proved counterproductive. As McCoy stresses, 'after 30 years of failed eradication, there is ample evidence to indicate that the illicit drug market is a complex global system, both sensitive and resilient, that quickly transforms suppression into stimulus' (McCoy, 2004, p. 96). Reduction and even suppression of drug supplies in producer countries has been the guiding line as well as the ultimate goal of the global prohibition regime and of the war on drugs (Bewley-Taylor, 2001). However, almost 40 years of war on drugs have in fact accompanied, if not encouraged, expansion not only of illegal opium poppy cultivation (in Asia as well as Latin America) but also of coca (in South America) and cannabis cultivation (worldwide).

In fact, as many observers have noted, 'the Drug war has achieved a self-perpetuating life of its own' for 'rather than reassess the failure of US prohibition policies, [US] federal officials blame smaller countries with meagre resources for the problems in their inner cities and suburbs' (Blumenson and Nilsen, 1998, p. 38; Davenport-Hines, 2001, p. 348). This is why drug war politics have been described as 'politics of denial'

(Bertram et al., 1996). Despite its gigantic yet unmeasured global cost (around US$50 billion spent annually by the US alone in the 2000s), the war on drugs has not only failed to reduce surface areas dedicated to illegal drug crops and quantities produced, it has also expanded and dispersed these illegal crops worldwide, while doing much to contribute to the militarization of many countries and areas of production (Chouvy and Laniel, 2007).

The steady increase in global opium production observed since the early 1970s occurred in spite of the many efforts deployed by the international community to suppress or reduce illegal opium poppy cultivation worldwide. Illegal opium production has increased despite countless forced eradication campaigns and in spite of many crop substitution and alternative development programmes. It can even be argued, as I have done elsewhere (Chouvy, 2009), that the increase of illegal opium production is due, at least to some extent, to the counterproductivity of forced eradication campaigns. Of course, the reasons for such a global failure are many and complex, rooted in the long history and politics of Asia and of the poppy.

First and foremost, opium production has clearly benefited from the turmoil of Asian history and geopolitics. The 19th-century opium wars, the 20th-century Cold War and its many local conflicts waged by proxy in Burma, Laos and Afghanistan, and even the 21st-century War on Terrorism in Afghanistan and Pakistan, have all spurred the continent's illegal opium production. Illegal drug economies and war economies share a long common history and have affected many territories in Asia and elsewhere.

Yet, illegal opium production has not benefited only from synergies between war economies and drug economies. It has also thrived on economic underdevelopment and poverty, whether war-induced or not: it is now widely acknowledged that the vast majority of Asian opium farmers grow poppies in order to cope with poverty and, above all, food insecurity (Chouvy, 2009). But despite the fact that the vast majority of Asian opium farmers are among the poorest of the poor, too many observers and policy makers still doubt that they engage in illegal opium production out of need – and not out of greed. In 2007, even the United Nations Office on Drugs and Crime bluntly argued that Afghan opium production was not linked to poverty – 'much to the contrary'. In fact, as shown by history and geography, illegal opium production never thrives better than when war and poverty overlap, as is the case in both Afghanistan and Burma. Part of the problem, in both Afghanistan and Burma, is that illegal opium production is likely to outlive their respective and successive wars. Obviously, peace-building is a difficult task and peace is hard to obtain and sustain. But war often transformed political and economic realities and dynamics to such an extent that time is needed for war-torn countries to achieve transition from war economies to peace economies. To bring an end to illegal opium production has proven as, if not more, difficult as ending wars – and maybe poverty – in the countries where poppies are illegally grown. In predominantly rural countries such as Afghanistan and Burma, whose conflicts have lasted for decades and have stalled economic growth and development, it seems that the suppression of illegal opium production can only follow – and proceed from – the establishment of peace and the initial reconstruction of the state and of the economy.

References

Bertram, E., Blachman, M., Sharpe, K. and Andreas, P. (1996) *Drug War Politics. The Price of Denial*. University of California Press, Berkeley, California.

Bewley-Taylor, D.R. (2001) *The United States and International Drug Control, 1909–1997*. Continuum, London.

Blumenson, E. and Nilsen, E. (1998) Policing for profit: the drug war's hidden economic agenda. *University of Chicago Law Review* 65, 35–114.

Bouthoul, G. (1991) (1951) *Traité de polémologie. Sociologie des guerres*. Payot, Paris.

Chouvy, P.-A. (2005) Morocco said to produce nearly half of the world's hashish supply. *Jane's Intelligence Review* 17(11), 32–35.

Chouvy, P.-A. (2007) Drug trafficking, In: Tucker, S.C. (ed.) *The Encyclopedia of the Cold War: A Political, Social, and Military History*. ABC-CLIO, Santa Barbara, California, pp. 339–441.

Chouvy, P.-A. (2009) *Opium. Uncovering the Politics of the Poppy*. I.B. Tauris, London.

Chouvy, P.-A. (2011) Finding an alternative to illicit opium production in Afghanistan, and elsewhere. *International Journal of Environmental Studies* 68(3), 373–379.

Chouvy, P.-A. and Laniel, L. (2007) Agricultural drug economies: cause or alternative to intra-state conflicts? *Crime, Law and Social Change* 48(3–5), 133–150.

Collier, P. and Hoeffler, A. (2001) *Greed and Grievance in Civil War*. Economics of Civil War, Crime, and Violence Research Project, Policy Research on the Causes and Consequences of Conflict in Developing Countries. World Bank, Washington, DC.

Davenport-Hines R. (2001) *The Pursuit of Oblivion. A Social History of Drugs*. Phoenix Press, London.

Gettman, J. (2006) Marijuana Production in the United States (2006). *Bulletin of Cannabis Reform* 2, 1–28.

Goodhand, J. (2005) Frontiers and wars: the opium economy in Afghanistan. *Journal of Agrarian Change* 5(2), 191–216.

Jonnes, J. (1996) *Hep-Cats, Narcs, and Pipe Dreams. A History of America's Romance with Illegal Drugs*. Johns Hopkins University Press, Baltimore, Maryland.

McCoy, A.W. (1972) *The Politics of Heroin in Southeast Asia*. Harper & Row, New York.

McCoy, A.W. (1991) *The Politics of Heroin. CIA Complicity in the Global Drug Trade*. Lawrence Hill Books, New York.

McCoy, A.W. (2003) *The Politics of Heroin. CIA Complicity in the Global Drug Trade (Afghanistan, Southeast Asia, Central America, Colombia)*. Lawrence Hill Books, New York.

McCoy, A.W. (2004) The stimulus of prohibition. A critical history of the global narcotics trade. In: Steinberg, M.K., Hobbs, J.J. and Mathewson, K. (eds) *Dangerous Harvest. Drug Plants and the Transformation of Indigenous Landscapes*. Oxford University Press, New York, pp. 24–111.

Meyer, K. and Parssinen, T. (1998) *Webs of Smoke. Smugglers, Warlords, Spies, and the History of the International Drug Trade*. Rowman & Littlefield, Lanham, Maryland.

Tullis, L. (1996) *Unintended Consequences. Illegal Drugs and Drug Policies in Nine Countries*. Lynne Rienner Publishers, Boulder, Colorado.

United Nations Office on Drugs and Crime (UNODC) (2003) Maroc. Enquête sur le cannabis 2003. United Nations, Vienna.

Zhou Yongming (1999) *Anti-Drug Crusades in Twentieth-Century China. Nationalism, History, and State Building*. Rowman & Littlefield, Lanham, Maryland.

6 Remote Sensing and GIS-based Technologies for Assessing the Impact of Conflict on Agricultural Production

Hadi Jaafar*
Department of Agriculture, American University of Beirut, Beirut, Lebanon

Introduction

Agriculture is one of the most affected sectors in conflicts for several reasons. Conflicts destabilize countries and communities by disrupting supply chains and availability of inputs required for agricultural production (Witmer and O'Loughlin, 2009). Quantifying the changes in agricultural production in relation to conflicts is an important measure towards food security monitoring schemes necessary in food-insecure regions of the world. Very often, conflicts create zones that cannot be accessed even by humanitarian agencies or personnel, making it difficult and hazardous to conduct ground-truth assessment of the effect of the conflict on agriculture and rural livelihoods (De Soysa et al., 1999).

Geographic information systems (GIS) are computer information systems that can collect, store, analyse and relate all types of information within a geographical and a database concept (Zeiler, 1999) (Longley, 2005). They present a helpful modelling tool for assessing impact of conflicts from space. Within the context of food insecurity threats amongst regions of conflict, remote sensing techniques can also provide a prompt and relatively accurate assessment of agricultural activities and productivity in those regions (Brink and Eva, 2009). Remote sensing is a process by which information about an object is collected using sensors that are not in direct contact with that object. An example would be a satellite orbiting the earth and recording electromagnetic radiation that is reflected from the earth's surface (Jensen, 2009). Remote sensing techniques work in synergy with GIS and can be used as change detection tools that enable the analyser to compare satellite data from different times to assess agricultural damage from disasters like conflicts, droughts, floods or changes in human activity due to other reasons (Campbell and Wynne, 2011). Remote sensing and GIS-based technologies can help to test and benchmark assumptions about changes in agricultural production in conflict settings. Separately, this information can in some instances be used to inform humanitarian or food security interventions. GIS becomes a must to geographically reference agricultural census data (Pierce and Clay, 2007). Agricultural census data are not usually geographically referenced. By combining satellite data and global positioning systems (GPS) ground observations and census data, scientists and practitioners can document and quantify such changes. GIS and remote sensing can be used to relate remotely sensed indices to census yield datasets and provide useful relationships

---

* Email: hj01@aub.edu.lb

from which future yield data could be derived. Effects of conflicts on agricultural livelihoods could be analysed using what is called change detection methods. Considerable efforts have been spent in developing change detection methods using remotely sensed data; a thorough review of these methods can be found in Singh (1989).

These methods have been recently used in remote sensing of satellite imagery used for civil and agricultural applications. Their uses have been increasing due to the abundance of remotely sensed products and the recent advances in software availability and reasonable ease of use for such analysis. There are many examples in the scientific literature on the use of satellite imagery in studying the effects of war on changing landscape and food security. Such examples include imagery-based crisis identification that have been used by various agencies like the United Nations (UN) and other non-governmental agencies (Marx and Loboda, 2013), mainly for assessment of violation of human rights. Other examples include the case of the Bosnia massacre (Witmer, 2008), land use changes in the Caucasus (Baumann et al., 2015) and in Iraqi Kurdistan (Eklund et al., 2016), the effect of the conflict in Sudan on eco-scarcity (Brown, 2010) and, more recently, the Syrian conflict (Jaafar et al., 2015) and the Syrian refugee crisis (Jaafar et al., 2016). Other research applications include conflict-caused rural abandonment of agricultural lands in the 2-year war within Bosnia (Witmer, 2008) and in Darfur (Prins, 2008). Another active area of research focuses on determining crop yields using remotely sensed indices that measure vegetation health in general (Jaafar and Ahmad, 2015a; Lobell et al., 2015; Sun et al., 2017) and within irrigated ecosystems (Jaafar and Ahmad, 2015b), and how these indices could be used to estimate biomass and consequently agricultural productivity. Such indices include the Normalized Difference Vegetation Index (NDVI) and the Enhanced Vegetation Index (EVI) (Huete, 1988). A recent application of the EVI was in Jaafar and Woertz (2016), where the authors used GIS and a remotely sensed estimate of EVI in territories controlled by the terrorist group Islamic State in Iraq and Syria (ISIS) to determine the income generated by the group from agriculture.

The objective of this chapter is to present some methodological approaches that utilize GIS and vegetation-related remotely sensed indices as a helpful tool in detecting the effects of conflicts on rural livelihoods and agricultural production in affected regions. Methods to disentangle drought/wet year effects from conflict effects will be discussed. Moreover, approaches to derive crop yields from remote sensing data will be described and then illustrated using a recent case study from Syria.

Materials and Methods

The major input required for remotely detecting agricultural production is the remotely sensed image of the site of interest. Imagery could be aerial (i.e. from drones or aeroplanes) or satellite-derived. Whereas aerial photography could acquire much higher image resolutions and could be customized to fit an area of interest, satellite imagery capture larger areas of land and most imagery suitable for agricultural studies are freely available. However, the recent developments of unmanned air vehicles (UAV) and their affordable price facilitated the acquisition of aerial photography (Bergo et al., 2010).

There are several satellite imagery products with various spatial and temporal resolutions. Spatial resolution of satellite imagery determines the ability of the imagery to resolve details of objects. Temporal resolution is the time it takes a satellite to return to the same point on earth. Some imagery is freely available (NASA products (Heinsch et al., 2003), Copernicus (Gascon et al., 2014)) and others are not (private companies' higher resolution satellites like Worldview (Jawak and Luis, 2013) and Quick Bird (Yang et al., 2005)). Today there are many existing satellites that orbit the earth and capture images of the earth's surface in various temporal frequencies. High temporal frequency (i.e. daily) observations come at the disadvantage of low resolution (1 km for instance). Resolution defines the size of the smallest feature that can be possibly detected, usually in terms of the pixel dimension (the smallest unit area of an image that could be discernable). Until recently, the highest spatial resolution of freely available imagery was that of NASA's Landsat mission

(15 m and 30 m, respectively) (Roy *et al.*, 2014). The launch of the Copernicus mission brought it down to 10 m with the Sentinel-2 satellite (Drusch *et al.*, 2012). The enhanced resolution is useful to separate agricultural fields from natural vegetation and to detect changes in small-scale farming practices.

Data requirements – satellite imagery

There is an abundance of satellite imagery that is freely available to the public. Table 6.1 shows a list of freely and commercially available imagery and the characteristics of the image from which vegetation indices and production-related parameters could be derived and/or estimated. This imagery in addition to any other imagery that has red and infrared bands could be used to derive the major vegetation indices. Bands are ranges of the electromagnetic spectrum of light that are registered on remote sensing instrument sensors.

Spatial resolution of satellite imagery is very important in classification and feature identification. For example, at 30 m pixel resolution, geographic features such as shorelines, rivers and mountains can be easily separated, whereas the 0.3 m resolution imagery is enough to identify small vehicles and irrigation canals bigger than 60 cm in width. High-resolution imagery that is commercially available could be of high value in detecting small fields and farming activities, as well as in establishing base maps that can be used to derive the high-level land use/cover maps. Destruction of large open-channel irrigation systems (transmission canals) can be only detected using high-resolution imagery (0.5 m or better). Burnt areas can be also delineated using either changes in vegetation indices or high-resolution imagery.

To illustrate how satellite imagery can be first visually interpreted to detect changes in agricultural production, we use the freely available Google Earth historical imagery platform. Figure 6.1 shows Google Earth historical imagery for the Syrian/Turkish border in Al-Hassake governorate in Syria (August 2005 vs August 2015). Over 10 years, agriculture in the Turkish region has flourished, while the agricultural fields in the conflict-affected Al-Hassake region are much less productive. A similar situation is noticed in other Syrian governorates. The Ghab Valley in Idlib governorate, for example, shows a drastic decrease in vegetation between 2010 (pre-conflict) and 2015. Visual image interpretation remains a robust tool in image interpretation and can easily locate areas of interest to be studied more deeply (Jensen, 2009; Lillesand *et al.*, 2014). Despite the obviousness of the decrease in agricultural activity (that could be directly or indirectly attributed to the conflict) as in Figs 6.1 and 6.2, it remains crucial to quantify these changes in order to come up with an estimate of the decrease in production within the affected lands.

Vegetation detection

Detecting vegetation using satellite imagery is based on the specific responses of healthy leaves to incident light. Vegetation indices are based on calculating the difference in the surface reflectance of light wavelengths (spectrum) that are received by vegetation. In simple terms, sunlight reflected from the terrain is separated into several wavelengths, or spectral bands. Detectors on the satellite sensors sense the reflected energy and convert it into an electronic signal that is recorded and transmitted as image data. The information of the reflected radiation is registered in numbers that are later translated into an image that needs to be corrected for interferences of atmospheric particles (dust, water vapour, and other objects or material). Reflectance is a measurement of the percentage of incoming or incident energy that a surface reflects. It is a fixed characteristic of an object or its state. Reflectance is a key for differentiating objects or states of objects. Vegetation as any other matter interacts with the electromagnetic energy of light, and satellites usually record the reflectance of the surface onto their sensors.

Many indices are available to detect vegetation vigour. As plants grow, canopy and biomass increase. Green leaves absorb more red and blue light than any other light bands, and they reflect near-infrared (NIR) radiation (Fig. 6.1); near-infrared is a portion of the electromagnetic spectrum that is not visible to humans. Some vegetation indices are based on the difference between NIR and red reflectance. 'Greenness' is

Table 6.1. Satellite imagery sensors that are mostly used in agriculture along with their temporal and spatial resolutions and dates of availability.

Satellite sensor	Ownership	Launch date	Swath width (km)	Resolution (m), multispectral	Resolution (m), visible	Spectral bands	Revisit time
Worldview-4	Digital Globe	Nov-16	13.1	1.24	0.31	Pan, R, G, B, NIR	1 day
Worldview-3	Digital Globe	Aug-14	13.1	1.24	0.31	Coastal-B, B, G, Y, R, R-edge, NIR1, NIR2	1 day
Worldview-2	Digital Globe	Oct-09	16.4	2.4	0.46	Coastal-B, B, G, Y, R, R-edge, NIR1, NIR2	1.1 days
Geoeye	Digital Globe	Sep-08	15.2	1.84	0.46	Pan, B, G, R, NIR	3 days
Worldview-1	Digital Globe	Sep-07	17.6	3	0.5	Pan, R, G, B, NIR	1.5
Pleaides-1A	AirBus Defense and Space	Dec-11	20	2	0.5	Pan, B, G, R, NIR	Daily
Pleaides-1B	AirBus Defense and Space	Dec-12	20	2	0.5	Pan, B, G, R, NIR	Daily
KOMPSAT-3A	Korean Aerospace Research Institute	Mar-15	12	MS: 2.2 m, IR: 5.5 m	0.55	Pan, B, G, R, NIR, MWIR	3 days
QuickBird	Digital Globe	Oct-01	16.8–18	2.62	0.65	Pan, B, G, R, NIR	1–3.5 days
Gaofen-2	Beijing Space Eye Innovation Technology	Aug-14	45	3.2	0.8	Pan, B, G, R, NIR	60 days
TripleSat	Twenty First Century Aerospace Technology	Jul-15	23.4	1	0.8	Pan, B, G, R, NIR	Daily
IKONOS	Digital Globe	Sep-99	11.3	3.2	0.82	Pan, B, G, R, NIR	3 days
SkySat	SkyBox Imaging	Nov-13	2	2	0.9	Pan, B, G, R, NIR	Multiple daily visit
SkySat2	SkyBox Imaging	Jul-14	2	2	0.9	Pan, B, G, R, NIR	Multiple daily visit
SPOT-6	AirBus Defense and Space	Sep-12	60	6	1.5	Pan, B, G, R, NIR	Daily
SPOT-7	AirBus Defense and Space	Jun-14	60	6	1.5	Pan, B, G, R, NIR	Daily
FORMOSAT	AirBus Defense and Space	May-01	24	8	2	Pan, B, G, R, NIR	Daily
ALOS	JAXA	Jan-06		10	2.5	Pan, B, G, R, NIR, DEM	46 days
SPOT-5	AirBus Defense and Space	May-02	60	SWI: 20 m, MS: 10 m	5	Pan, B, G, R, NIR, SWIR	2–3 days
RapidEye	BlackBridge	Aug-08	77	5	NA	B, G, R, R-Edge, NIR	5.5 days

Continued

Table 6.1. Continued.

Satellite sensor	Ownership	Launch date	Swath width (km)	Resolution (m), multispectral	Resolution (m), visible	Spectral bands	Revisit time
Sentinel-2A	ESA and Airbus Defense and Space	Jun-15	290	10	NA	Aerosol, B, G, R, R-Edge1&2&3, NIR1&2, water vapor, cirrus, SWIR1&2	10 days
Sentinel-2B	ESA and Airbus Defense and Space	Jul-16	290	10	NA	Aerosol, B, G, R, R-Edge1&2&3, NIR1&2, water vapor, cirrus, SWIR1&2	10 days
LANDSAT 7	NASA	Apr-99	185	30	15	Pan, B, G, R, NIR, SWIR1&2, TIR	16 days
LANDSAT 8	NASA	Feb-13	185	30	15	Pan, coastal-B, B, G, R, NIR1, SWIR1&2, cirrus, TIR1&2	16 days
ASTER	NASA	Dec-99	60	VNIR: 15, SWIR: 30, TIR: 90	NA	VNIR, SWIR, TIR	16 days
MODIS_terra	NASA	Dec-99	2330	250, 500, 1000	NA	36 NIR bands	1–2 days
MODIS_aqua	NASA	May-02	2330	250, 500, 1000	NA	36 NIR bands	1–2 days
EO-Hyperion	NASA	2000–2017	37	30	NA	Hyperspectral	On demand

Data have been extracted from the website of the corresponding company or agency.

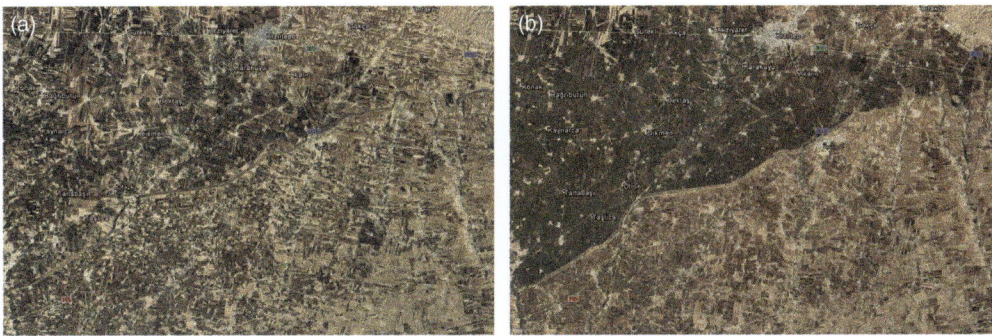

Fig. 6.1. Google Earth historical imagery comparison between the Syrian (below) and Turkish (above) border in (a) 2005 (pre-conflict) and (b) 2015 (conflict). (Map data: Google, Landsat, Copernicus.)

associated with plant photosynthetic activity, biomass and fractional cover of vegetation.

The red to NIR simple ratio (SR) index is the first true vegetation index. It takes advantage of the inverse relationship between chlorophyll absorption of red radiant energy and increased reflectance of NIR energy for healthy plant canopies (Birth and McVey, 1968). The ratio has

Fig. 6.2. Google Earth historical imagery comparison between the northern Ghab Valley in the Syrian governorate of Idlib (2010: pre-conflict, and 2015: conflict). (Map data: Google, Landsat, Copernicus.)

been found to be sensitive to biomass and changes in leaf area (Huete et al., 2002).

Another index that is widely used is the NDVI (Rouse, 1974), which was eventually used to estimate plant yields by quantifying the photosynthetic plant capacity. NDVI is calculated by using the red and NIR bands of spectral satellite imagery. The importance of the use of NDVI in vegetation analysis lies in the fact that it can be used to detect the different growth stages of crops. This is crucial for aiding in crop classification studies and vegetation mapping. For example, a wheat crop could be distinguished from spring and summer crops by simply knowing the growing season of wheat and summer crops in a region. As a winter crop, wheat turns yellow in late spring and hence the NDVI value for a wheat field would be very low in May, whereas other crops are in their vegetative stage in spring and it will have a high NDVI.

Figure 6.3 shows a hyperspectral plot of four 30-m pixels image taken by EO-1 Hyperion NASA's satellite (decommissioned as of 22 February 2017) on 1 July 2003. Note the low reflectance of healthy vegetation in the red spectrum (670 nm) and the high reflectance in the NIR (700–900 nm and 1000 nm). Water can be easily distinguished by its high reflectance in the blue band (450 nm and high absorbance in the remaining region of the electromagnetic spectrum).

Steps for identification of changes in agricultural production

The procedure to quantify changes in agricultural production using remote sensing and GIS can be conceptualized in the following steps:

Step 1. Identify the location of interest and acquire the image for the location of interest by defining the geographic coordinates of the rectangle corners for the study area.

Step 2. Subset satellite images to regions of interest.

Step 3. Use geo-spatial analysis software to display the image. To be of greatest value, the original remotely sensed data must usually be calibrated in two distinct ways:

(i) It should be calibrated in space and in reflectance so that remotely sensed data obtained on different dates and satellite systems can be compared with one another.

(ii) The remotely sensed data must usually be calibrated (compared) with what is on the ground in terms of biophysical (e.g. leaf-area-index, biomass) or cultural characteristics (e.g. land use/cover, population density) (Jensen, 2009).

Step 4. Extract the bands of interest, for example NIR, red, blue and shortwave infrared (SWIR). Thermal bands could be also of interest in some applications (indirectly differentiating rainfed from irrigated lands and also detecting water stress and estimate consumptive use of vegetation; see, for example, Anderson et al., 1997; AghaKouchak et al., 2015).

Step 5. Convert the digital number of the band into radiance and top of atmosphere (TOA) reflectance. This step is necessary to make the image time independent (i.e. remove the effect of the day of the year, the sun azimuth and the band range from the calculated reflectance).

Step 6. Perform atmospheric correction on the image to convert the TOA reflectance into surface reflectance. This is necessary to remove

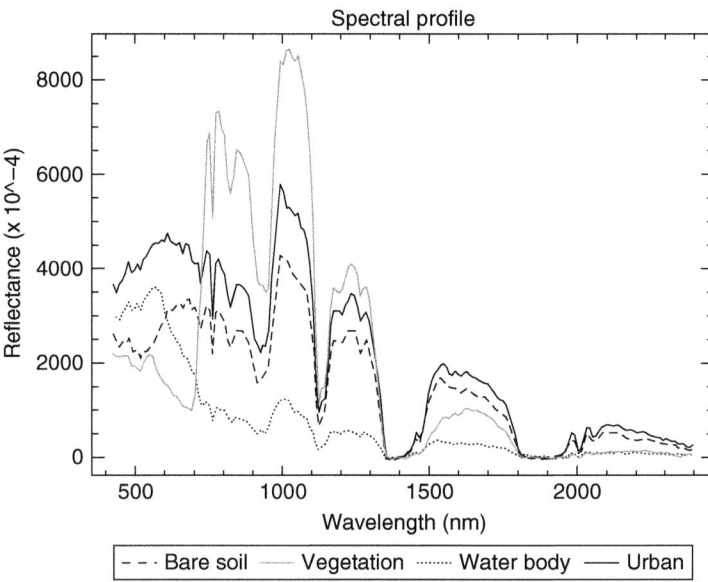

Fig. 6.3. Spectral plot of a vegetated pixel, a bare soil pixel, an urban pixel and a water body pixel, Beqaa, Lebanon, 1 July 2003. (Hyperspectral imagery EO-1 Hyperion satellite.)

the effects of aerosols and water vapour from the reflectance.

Step 7. Calculate the target vegetation index, for example the NDVI or the EVI that corrects for some soil reflectance interference and some aerosols using the blue band.

Step 8. Delineate the agricultural fields, or use an area of interest for the fields or a high-resolution image delineation of agricultural fields from a pre-conflict image. A land use/land cover map should be used to identify agricultural areas.

Step 9. Where possible, identify the crops being planted. Normally this could be attained via existing high-level land use/land cover maps, or through the use of image classification techniques. Hyperspectral imagery might be available for the site of interest.

Step 10. Perform spatial analysis on the data overlaid by the area of interest. Zonal statistics or of the vegetation index on each of the delineated fields or the target areas can be performed. The zonal statistics calculate the sum, mean, standard deviation, minimum and maximum of the VI for each feature in the shapefile of interest or for the whole file. This could be done in GIS software such as QGIS (open-source), IDRISSI, Global Mapper, ArcMap or remote sensing software (ENVI, ERDAS).

Step 11. Repeat this process for each image available (depending on the satellite revisit date for the study area).

Step 12. Determine a temporal average of the vegetation index or the productivity for the fields in the study area.

Step 13. Perform raster operations[1] in a GIS software by subtracting the value of the parameter for the time of interest from the temporal average to identify deviations from the mean. This step could also make use of the less subjective standardization of the difference by creating Z-scores for EVI. This could be done by subtracting the mean of the value and dividing by the temporal standard deviation on a pixel-by-pixel basis. The standardization offers a more rigorous way to identify areas of highest variation, and also could be used to compare EVI to NDVI. This approach has been used in deriving drought indices (Vicente-Serrano *et al.*, 2010; Mu *et al.*, 2013), EVI (Jaafar and Ahmad, 2015a) and agricultural productivity (Jaafar and Ahmad, 2015b).

An initial time investment is needed to set up a running model for the above steps, but once the model is up and running an almost real-time monitoring system can be easily implemented, with a lag time of 1–2 weeks following imagery acquisition depending on the platform. Given the recent advances in computational technology and computer storage and processing abilities, the disadvantages of the data burden would be overruled by the nearly impossible traditional analysis of agricultural production in conflict-affected zones.

In the case where no census data are available to relate remotely sensed estimates to anomalies of agricultural production, the latter can be inferred from a remotely sensed analysis. Some knowledge of planted crops in the area of interest is necessary to give those anomalies a meaningful value. If this also is lacking, commercial higher resolution imagery can be tasked and crop identification (machine-based or using visual interpretation) could be resorted to. This requires an additional investment due to the sometimes expensive cost of such imagery.

Deriving biomass estimates

There are multiple ways to estimate agricultural biomass from remote sensing and GIS data. This sub-section presents three of the most common estimation methods. Less interested or non-technical readers may skip this section.

First, the plant biomass can be estimated from what is referred to as net primary productivity (NPP), a measure of the net amount of carbon uptake by vegetation within an area of land (Monteith, 1972), commonly constituting what is referred to as the biomass. Once the daily biomass production is calculated, it is summed over the crop growing season. The fresh yield can be estimated by multiplying the crop harvest index by the dry biomass and dividing the result by the dry matter fraction.

Another option is to use the readily available primary productivity datasets that are generated by MODIS (Heinsch et al., 2003; NASA, 2015), such as gross primary productivity (GPP), NPP and net photosynthesis (P_{snet}). These products have been found effective in deriving remotely sensed evapotranspiration estimates in many countries like the US (Lobell et al., 2002) and recently Syria (Jaafar and Ahmad, 2015a; 2015b). However, they might be challenged on field-scale level calculations because of their coarser resolution.

Once the NPP is calculated, the economic yield per hectare (EY) can then be estimated using Eqn 6.1 (Lobell et al., 2002; Jaafar and Ahmad, 2015b), among others:

$$EY_i = (NPP_i \times HI_i \times f_{AG})/(DF_i \times C) \quad \text{(Eqn 6.1)}$$

Where EY_i is the metric tonnes of economic yield per hectare of crop i; NPP_i is the net primary production of crop i; HI_i is the harvest index of crop i; f_{AG} is the fraction of production allocated above-ground, directly related to the root:shoot ratio; DF_i is the dry proportion of the economic yield i; and C is the carbon content (0.45 gC (g dry matter)).

The calculated yield per field per crop should be multiplied by the cropped area per crop for the region of interest for that year. A total production figure (in tonnes) will then be derived using the remotely sensed estimates of production. Ideally this should be done for a sufficiently long period of record (10–15 years) immediately preceding the beginning of the conflict.

In conflict situations, it is sometimes impossible to distinguish between different field crops due to the unavailability of data in that area and also the lack of hyperspectral imagery for the locations and times of interest. Because crops look similar on medium- to low-resolution vegetation imagery, it is sometimes possible to adopt a total production approach rather than a crop yield approach (Jaafar and Ahmad, 2015a). In other words, instead of estimating the yield per crop, a total production figure for all crops (e.g. summer crops) could be lumped together and used as an indication of the percentage reduction in agricultural production.

In some cases, agricultural spring GPP would give lower root mean square error (i.e. better prediction) than the NPP or EVI index (Jaafar et al., 2015; Jaafar and Ahmad, 2015a; 2015b). GPP datasets that are obtained by the Terra MODIS and Aqua MODIS instruments can be used to derive the GPP estimates (gC m^{-2}) for the croplands (NASA, 2015). NASA publishes these datasets from 2000 up to the present. The GPP algorithm assumes a linear relationship between the amount of absorbed photosynthetically active radiation (FPAR) and biomass production. Like NPP, GPP has been used to estimate crop yields by converting GPP values to biomass estimates and then using an averaged crop harvest index to derive the yield portion in the biomass in Jaafar and Ahmad (2015a). GPP was also used in the case study presented below to estimate winter crop production for the crop lands that are normally planted with wheat and barley. MODIS GPP is valid for providing production estimates at the regional scale due to the trade-off between spatial and temporal resolution. Production could be related to cumulative GPP over the main growing season of winter crop growth in the northern hemisphere (February, March, April and sometimes May) where GPP is summed over these months. The process would be repeated for all years and plotted against the winter crop production. Care should be taken to determine the planted areas. Crop land areas could be either derived from existing spatial coverage (GIS shapefiles or polygon features from crop databases) or from freely available land cover products (e.g. NASA's MCD12 land cover product; NASA, 2013). The land cover product provides a 500-m resolution grid of land cover mapping as developed by the University of Maryland Department of Geography (UMD) classification (Friedl et al., 2010). The product describes land cover properties derived from observations spanning a year's input of Terra and Aqua MODIS data. The released versions provide a temporal coverage on land cover mapping for the period of record (2000–present). The MCD12 land cover product incorporates five different land cover classification systems.

Similar to the GPP product, MCD12 does not differentiate among crop types. This drawback could be overcome if it agrees well with the totalized government-reported planted areas and also if the derived regression relationships between the spring GPP and the winter crop production is significant. When there is a major change in croplands due to conflict or drought, which could also have happened during the conflict years, the GPP for those lands would be lower. When imagery analysis from cross-referenced datasets such as Landsat and Sentinel-2 shows that new areas are cultivated and others abandoned, all of the areas should be included in the spatial GIS analysis. In such cases, global 30-m Landsat-derived land cover product (Chen et al., 2015) could be used to extract the new cultivated lands. The dataset comprises ten types of land cover, including forests, artificial surfaces and wetlands for the years 2000 and 2010, extracted from Landsat and Chinese HJ-1 satellite images derived. The layer should be spatially subset to include only cultivated lands as per the classification scheme used within the dataset for the study area of interest.

Jaafar and Woertz (2016) developed a new spatio-temporal wheat detection index (WDI) that identifies planted areas in the winter and spring seasons. The index was based on temporally normalizing the 30-m Landsat-derived EVI. This was done by differencing raster EVI values from northern Syria for early April (peak of green cover of wheat) and late June (at peak yellowness/harvest) and dividing by the sum (Eqn 6.2):

$$WDI = \left(EVI_{April} - EVI_{June}\right)/\left(EVI_{April} + EVI_{June}\right)$$

(Eqn 6.2)

WDI identifies a threshold that can detect winter crops by noting the difference in EVI due to wheat and barley phenology (maximum greenness in early spring and yellowness following harvest). The areas with that threshold would then be classified as the planted areas, and the spatial mean of spring GPP over the extracted coverage clipped over governorate boundaries would be determined.

Controlling for rainfall

In Syria and Iraq, and other countries with similar rainfall seasonality, rainfall affects winter crop yields, because higher yields are correlated with higher rainfall (up to a limiting point, where excess water can cause yield drops especially in waterlogged lands) (Doorenbos and Kassam, 1979). A detailed discussion on how to control for rainfall can be found in Jaafar and Woertz (2016). Rainfall data may be obtained from several sources (including ground stations, government reports, climatic reports, among others) and also, more recently, remotely sensed observations (Hou et al., 2014).

Although both drought and conflict can result in lower agricultural production, it remains desirable for some analysts to disentangle the effects of the two on agricultural production. In some situations conflict might lead to a shift in agricultural activities and cropping patterns, and sometimes it can lead to a sustained agricultural production. It is useful to first study the relation (if any) between crop yields and rainfall in an area in non-conflict years. A no-conflict counterfactual scenario can then be constructed that incorporates the rainfall–yield relation to which the conflict situation can be fairly compared.

Estimating agricultural production

The results of the biomass estimates should be plotted against local production figures, usually obtained from concerned agencies (agriculture ministries, census agencies, government portals, etc.). Governments usually provide agricultural production figures at the district level or governorate level. The cropped area per district should be determined and, where possible, checked against the areas delineated using satellite imagery for every year of record. The yield or production data reported by the government should be highly correlated with the remotely sensed estimate of the production figures.

Generalized linear regression models should be used and tested for significance (F-statistic) and goodness of fit. Akaike information criterion (AICc) is normally calculated and then used to determine the best model (Sakamoto et al., 1986).

In case the calculated correlation of the government-reported production estimate and the remotely sensed biomass estimate is not statistically significant, a year-by-year analysis should be conducted to remove outliers from the relation; this could be done by a user-defined criterion using any statistical software program. Care should be taken to exclude certain anomaly years. An example would be a situation in which a wheat crop looks healthy on a satellite image but then prior to harvest it was devastated by a flood or a rust disease, as happened in Syria in 2010 (Solh, 2010). An appropriate relationship (usually linear or quadratic in nature) should be then determined that relates the remotely sensed biomass estimate to the government yield production figure. The relationship would be used to bias-correct the remotely sensed estimate against the government production estimates. Finally, the relationship should be used to determine yields when no government data are available.

Case Study: Syria Wheat and Barley Production in 2016

To illustrate how remote sensing data can be used to derive crop production in conflict zones with little ground reference information, we run the analysis described above over the country and governorates of Syria to determine the wheat and barley production in 2016. We do this using regression analysis of EVI-estimated production and GPP-estimated production and compare the results. The reported governorate winter crop production is calculated as the weighted mean for each zone, based on its respective proportion of the total area cropped. The calculated values are then regressed against the mean cumulative spring GPP as derived from zonal statistical analysis within a GIS coverage. For each year within this period, croplands were extracted from this dataset and merged to generate the file used in this analysis.

Syria has been under the effect of devastating conflict since March 2011 (Gleick, 2014). Following the beginning of the conflict in 2011, reported wheat and barley harvests were less than reported in 2013 (possibly due to difficulties in collecting field data by government officials). As in prior conflict years, agricultural production in Syria was reported to be severely impacted by the conflict in 2016, with the country reaching its lowest production of wheat in 27 years (El Dahan and Olga, 2017).

Both estimates (EVI and GPP) show a low production in Syria (less than 3 million metric tonnes of wheat and barley combined), and the EVI estimate is only 2.2 million metric tonnes of wheat and barley (Table 6.2). Note that pre-conflict deviation of both estimates from the reported government data on cereal production is within 15% (+/−), except for the year 2010 when production of cereals was reduced by yellow rust (a cereal disease) (Fig. 6.4). In that year, productivity in the northeastern part of Syria dropped by an average of 40% (Solh, 2010).[2]

The low cumulative spring EVI and spring GPP for 2016 indicate a severe drop in the wheat and barley production in Syria. This is evident in Fig. 6.5a and b, which show the spring GPP and EVI anomalies of 2016 as compared to pre-conflict average GPP and EVI. Green areas show higher than average production while red areas show lower than average production. All Syrian governorates show severe drops in productivity, except the Kurdish-controlled northern region of Al-Hassake (which witnessed decreased, albeit not severe, average reductions), in contrast to the highly productive Turkish territory just north

Table 6.2. EVI and GPP predicted wheat and barley production as compared to government-reported production.

Syria	Wheat and barley production (1000 metric tonnes)			difference from reported (%)	
	Reported	GPP-derived	EVI-derived	GPP-derived	EVI-derived
2000[a]	3317	3243	3141	−2.3	−5.3
2001	6700	6206	6185	−7.4	−7.7
2002	5695	5617	5044	−1.4	−11.4
2003	5992	5390	6389	−10.0	6.6
2004	5065	5447	5451	7.5	7.6
2005	5065	4550	4756	−10.2	−6.1
2006	6134	5174	5147	−15.6	−16.1
2007	4826	4074	4839	−15.6	0.3
2008[a]	2400	2756	2497	14.8	4.0
2009	4547	5198	4837	14.3	6.4
2010[b]	3763	5285	5002	40.5	32.9
2011[c]	4525	5089	4742	12.5	4.8
2012	4337	4387	4329	1.2	−0.2
2013	4093	6171	5744	50.8	40.3
2014[a]	2624	2999	1820	14.3	−30.6
2015	–	6031	4582		
2016	–	2819	2227		

[a] Drought years; [b] wheat crop hit by diseases; [c] beginning of conflict (March 2011).

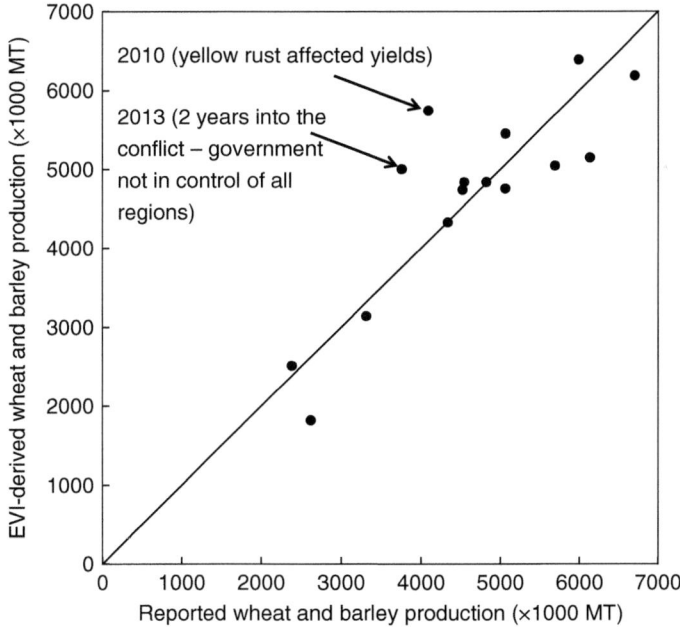

Fig. 6.4. Estimated versus reported wheat and barley production in Syria for the period (2000–2014).

of Al-Hassake. Al-Hassake itself, where most of the wheat in Syria is produced, shows recovery in the northeastern 'Jazira' region (due to the relative stability of these Kurdish-controlled regions). To the west and to the south, the decline in production is evident. Southern Al-Hassake was under ISIS control in 2014 and 2015. Aleppo shows a similarly severe decline in winter crops,

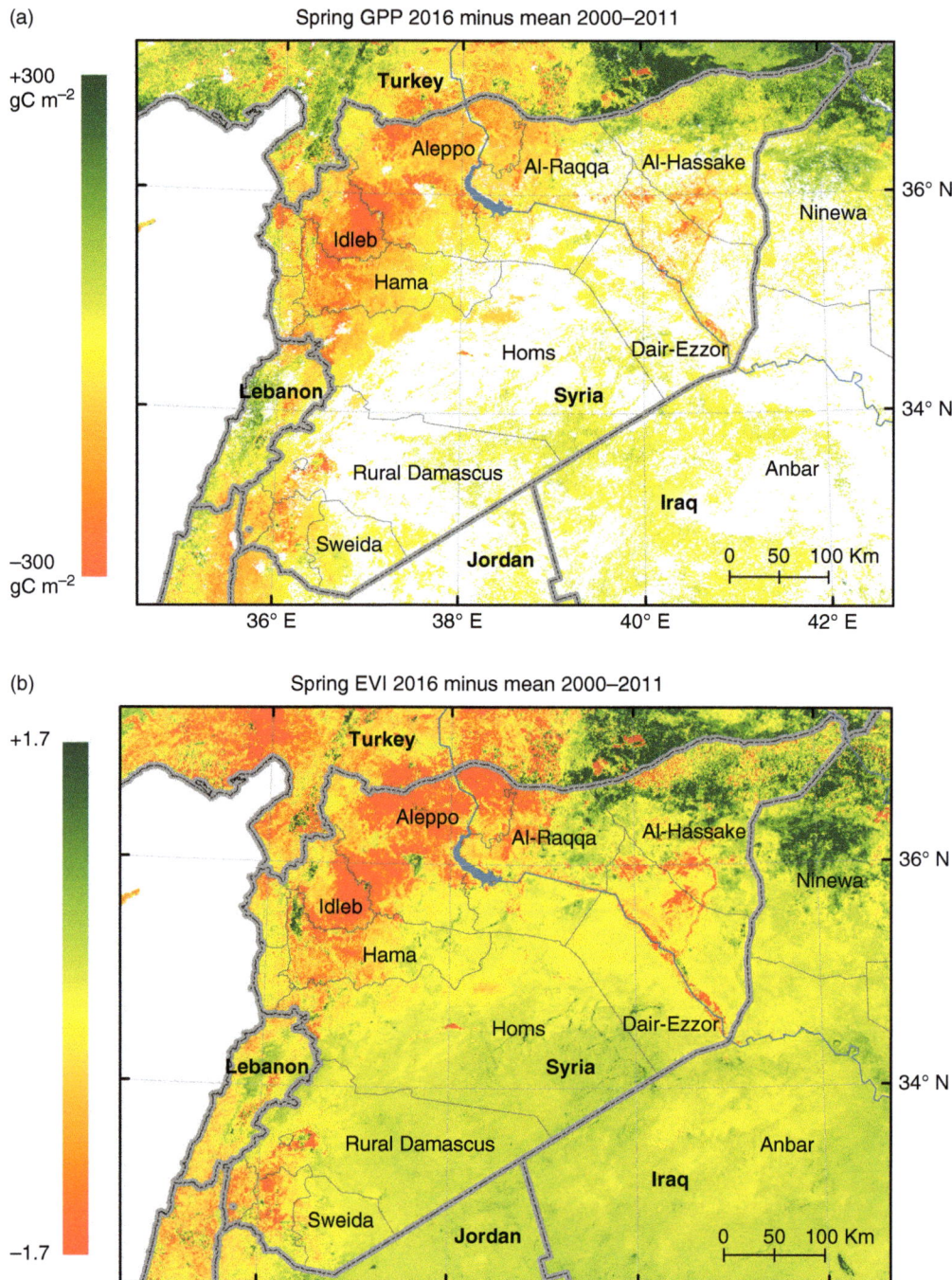

Fig. 6.5. (a) 2016 anomaly of cumulative spring gross primary productivity (GPP) of Syria, an indicator of agricultural production; and (b) 2016 cumulative spring EVI anomaly z-scores.

as do the bordering Turkish areas, possibly due to the abandonment of agricultural lands due to the increase in the Turkish military operations.

In contrast to Al-Hassake, other parts of Syria do not show any recovery in production, with vegetation indices well below pre-conflict levels. It is also evident that drought had hit the region in 2000, 2008 and 2013, when average rainfall in Syria was 284, 330 and 227 mm, respectively, which is less than 60% of optimal wheat water requirements in that region (Ministry of Agriculture and Agrarian Reform in Syria, 2014). Due to the presence of supplemental irrigation in some governorates, rainfall cannot be used to estimate cereal production except in zones where no irrigation is practised.

Figure 6.6 shows the 1:1 fit of the remotely sensed estimates and the reported estimates of wheat and barley production by Syrian governorate for the years 2000–2014. The reported estimates and remotely sensed data in Syria align, demonstrating that the latter is an excellent indication of agricultural production and can be used in absence of other information about crop yields and agricultural production. The robustness of the estimate at the finer resolution (i.e. district-level vs country-level estimates) is a promising finding and is an incentive to always collect data at the smallest scale possible.

The results are comparable to those reported by Lobell et al. (2002), specifically the positive correlation of remotely sensed estimates of productivity and census data at the county level within the US.

Conclusions

In this chapter we have described remote sensing and GIS methodologies related to detecting agricultural productivity and yields from agricultural lands. We have highlighted methods that aid in determining crop yields from vegetation variables as measured from satellite systems. Methods to control for rainfall (separate the rainfall effect on yield from the conflict effect) were also discussed. Finally, we also used available land cover products and ecosystem production variables to derive estimates of agricultural production in the conflict-affected country of Syria for 2016. Remotely sensed vegetation indices prove to be of high value in agricultural applications, primarily in detecting changes in agrarian practices and consequently agricultural production. These indices could be used for deriving primary production using biophysical equations as well as empirical models, in standard as well as in conflict-affected contexts.

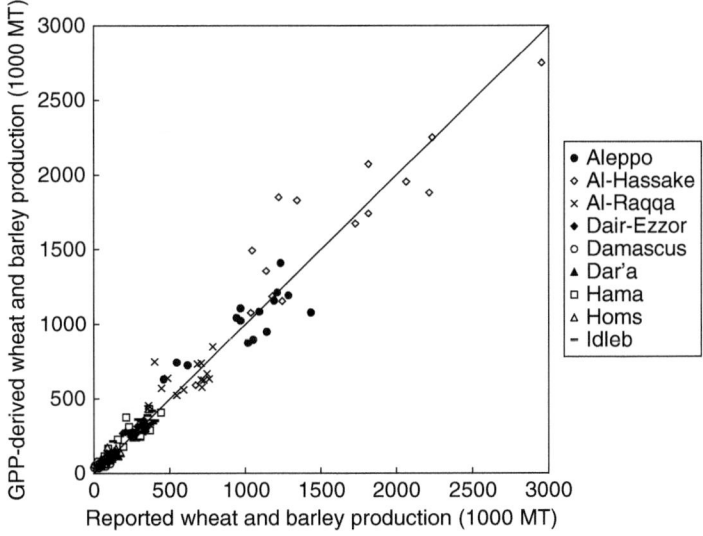

Fig. 6.6. Remotely sensed GPP estimates versus reported wheat and barley production estimates in Syrian governorates for the period 2000–2014; the northern governorate of Al-Hassake is the breadbasket of Syria, consisting of more than one-third of total Syrian production.

Conflicts disrupt human activities and can significantly impact agriculture. The abundance of satellite imagery in space and in time, as well as the recent advancements in modelling ecosystem variables using remote sensing techniques, has led to an increased use of these variables in monitoring effects of conflict on agriculture. Agricultural production estimates would be valuable in helping assess the socio-economic impact of conflict on rural livelihoods. It could be also helpful in analysing the agricultural resources available to support the campaigns of various parties to the conflict.

Notes

[1] Raster data is a type of spatial data storage in which values are stored in grids/cells.

[2] The detailed methodology for quantifying the effect of the conflict on agriculture in the Orontes basin in Syria can be found in Jaafar et al. (2015). Other studies extend the analysis to the governorates of the Euphrates River basin (Jaafar and Ahmad, 2015a) as well as the governorates that partly lie within the Yarmouk basin (Jaafar and Ahmad, 2015b). Jaafar and Woertz (2016) extended the methodology to areas in Iraq and Syria that were under the control of ISIS, and showed how remotely sensed estimates of production in those areas could be used to estimate the amount of funding ISIS could generate from agriculture. The authors also provided a remotely sensed analysis of the Euphrates Lake levels in the context of the hydro-hegemony practised by ISIS.

References

AghaKouchak, A., Farahmand, A., Melton, F., Teixeira, J., Anderson, M., Wardlow, B. D. and Hain, C. (2015) Remote sensing of drought: progress, challenges and opportunities. *Reviews of Geophysics* 53, 452–480.

Anderson, M., Norman, J., Diak, G., Kustas, W. and Mecikalski, J. (1997) A two-source time-integrated model for estimating surface fluxes using thermal infrared remote sensing. *Remote Sensing of Environment* 60, 195–216.

Baumann, M., Radeloff, V.C., Avedian, V. and Kuemmerle, T. (2015) Land-use change in the Caucasus during and after the Nagorno-Karabakh conflict. *Regional Environmental Change* 15, 1703–1716.

Bergo, M., Calleri, A. and Vescovi, F. (2010) Precision agriculture deploying Unmanned Aerial Vehicle (UAV). *Mondo Macchina* 19, 44–50.

Birth, G.S. and McVey, G.R. (1968) Measuring the color of growing turf with a reflectance spectrophotometer. *Agronomy Journal* 60, 640–643.

Brink, A.B. and Eva, H.D. (2009) Monitoring 25 years of land cover change dynamics in Africa: a sample based remote sensing approach. *Applied Geography* 29, 501–512.

Brown, I.A. (2010) Assessing eco-scarcity as a cause of the outbreak of conflict in Darfur: a remote sensing approach. *International Journal of Remote Sensing* 31, 2513–2520.

Campbell, J.B. and Wynne, R.H. (2011) *Introduction to Remote Sensing*. Guilford Press, New York.

Chen, J., Chen, J., Liao, A., Cao, X., Chen, L., Chen, X., He, C., Han, G., Peng, S. and Lu, M. (2015) Global land cover mapping at 30m resolution: a POK-based operational approach. *ISPRS Journal of Photogrammetry and Remote Sensing* 103, 7–27.

De Soysa, I., Gleditsch, N.P., Gibson, M. and Sollenberg, M. (1999) *To Cultivate Peace: Agriculture in a World of Conflict*. International Peace Research Institute, Oslo.

Doorenbos, J. and Kassam, A. (1979) Yield response to water. *Irrigation and Drainage Paper* 33, 257.

Drusch, M., Del Bello, U., Carlier, S., Colin, O., Fernandez, V., Gascon, F., Hoersch, B., Isola, C., Laberinti, P. and Martimort, P. (2012) Sentinel-2: ESA's optical high-resolution mission for GMES operational services. *Remote Sensing of Environment* 120, 25–36.

Eklund, L., Persson, A. and Pilesjö, P. (2016) Cropland changes in times of conflict, reconstruction, and economic development in Iraqi Kurdistan. *Ambio* 45, 78–88.

El Dahan, M.P. and Olga, P. (2017) Syria's million-ton Russian wheat deal in jeopardy. *The Daily Star – Lebanon* 26 January 2017.

Friedl, M.A., Sulla-Menashe, D., Tan, B., Schneider, A., Ramankutty, N., Sibley, A. and Huang, X. (2010) MODIS Collection 5 global land cover: algorithm refinements and characterization of new datasets. *Remote Sensing of Environment* 114, 168–182.

Gascon, F., Cadau, E., Colin, O., Hoersch, B., Isola, C., Fernández, B.L. and Martimort, P. (2014) Copernicus sentinel-2 mission: products, algorithms and Cal/Val. SPIE Optical Engineering+ Applications, 2014. International Society for Optics and Photonics, 92181E-92181E-9.

Gleick, P.H. (2014) Water, drought, climate change, and conflict in Syria. *Weather, Climate, and Society* 6, 331–340.

Heinsch, F.A., Reeves, M., Votava, P., Kang, S., Milesi, C., Zhao, M., Glassy, J., Jolly, W.M., Loehman, R. and Bowker, C.F. (2003) GPP and NPP (MOD17A2/A3) products NASA MODIS land algorithm. *MOD17 User's Guide*, 1–57.

Hou, A.Y., Kakar, R.K., Neeck, S., Azarbarzin, A.A., Kummerow, C.D., Kojima, M., Oki, R., Nakamura, K. and Iguchi, T. (2014) The global precipitation measurement mission. *Bulletin of the American Meteorological Society* 95, 701–722.

Huete, A.R. (1988) A soil-adjusted vegetation index (SAVI). *Remote Sensing of Environment* 25(3), 295–309.

Huete, A., Didan, K., Miura, T., Rodriguez, E.P., Gao, X. and Ferreira, L.G. (2002) Overview of the radiometric and biophysical performance of the MODIS vegetation indices. *Remote Sensing of Environment* 83, 195–213.

Jaafar, H. and Ahmad, F.A. (2015a) Crop yield prediction from remotely sensed vegetation indices and primary productivity in arid and semi-arid lands. *International Journal of Remote Sensing* 36, 4570–4589.

Jaafar, H. and Ahmad, F.A. (2015b) Relationships between primary production and crop yields in semi-arid and arid irrigated agro-ecosystems. *International Archives of the Photogrammetry, Remote Sensing and Spatial Information Sciences – ISPRS Archives* 27–30.

Jaafar, H. and Woertz, E. (2016) Agriculture as a funding source of ISIS: a GIS and remote sensing analysis. *Food Policy* 64, 14–25.

Jaafar, H., Zurayk, R., King, C., Ahmad, F. and Al-Outa, R. (2015) Impact of the Syrian conflict on irrigated agriculture in the Orontes basin. *International Journal of Water Resources Development* 31, 436–449.

Jaafar, H., King-Okumu, C., Haj-Hassan, M., Abdallah, C., El-Korek, N. and Ahmad, F. (2016) Water resources within the Upper Orontes and Litani Basins: a balance, demand and supply analysis amid the Syrian refugees crisis. IIED Working Paper. IIED, London.

Jawak, S.D. and Luis, A.J. (2013) Improved land cover mapping using high resolution multiangle 8-band WorldView-2 satellite remote sensing data. *Journal of Applied Remote Sensing* 7, 073573–073573.

Jensen, J.R. (2009) *Remote Sensing of the Environment: An Earth Resource Perspective*, 2nd edn. Pearson Education, Harlow, UK.

Lillesand, T., Kiefer, R.W. and Chipman, J. (2014) *Remote Sensing and Image Interpretation*. John Wiley and Sons, Chichester, UK.

Lobell, D.B., Hicke, J.A., Asner, G.P., Field, C.B., Tucker, C.J. and Los, S.O. (2002) Satellite estimates of productivity and light use efficiency in United States agriculture, 1982–98. *Global Change Biology* 8, 722–735.

Lobell, D.B., Thau, D., Seifert, C., Engle, E. and Little, B. (2015) A scalable satellite-based crop yield mapper. *Remote Sensing of Environment* 164, 324–333.

Longley, P. (2005) *Geographic Information Systems and Science*. John Wiley and Sons, Chichester, UK.

Marx, A.J. and Loboda, T.V. (2013) Landsat-based early warning system to detect the destruction of villages in Darfur, Sudan. *Remote Sensing of Environment* 136, 126–134.

Ministry of Agriculture and Agrarian Reform in Syria (2014) Statistical datasets 2013 (in Arabic). Damascus. Available at: www.moaar.gov.sy/site_ar/agristat/2013/stat2013.rar (accessed 15 December 2015).

Monteith, J. (1972) Solar radiation and productivity in tropical ecosystems. *Journal of Applied Ecology* 9, 747–766.

Mu, Q., Zhao, M., Kimball, J.S., McDowell, N.G. and Running, S.W. (2013) A remotely sensed global terrestrial drought severity index. *Bulletin of the American Meteorological Society* 94, 83–98.

NASA (2013) MCD12Q1, 051. NASA EOSDIS Land Processes DAAC, USGS Earth Resources Observation and Science (EROS) Center. NASA, USGS, Sioux Falls, South Dakota. Available at: https://lpdaac.usgs.gov (accessed 5 January 2015).

NASA (2015) MOD17A2H, MODIS/TERRA Gross Primary Productivity 8-Day L4 Global 500 m SIN Grid V006. NASA EOSDIS Land Processes DAAC, USGS Earth Resources Observation and Science (EROS) Center. NASA, USGS, Sioux Falls, South Dakota. Available at: https://lpdaac.usgs.gov (accessed 5 January 2015).

Pierce, F.J. and Clay, D. (2007) *GIS Applications in Agriculture*. CRC Press, Boca Raton, Florida.

Prins, E. (2008) Use of low cost Landsat ETM+ to spot burnt villages in Darfur, Sudan. *International Journal of Remote Sensing* 29, 1207–1214.

Rouse, J.R. (1974) Monitoring the vernal advancement and retrogradation (green wave effect) of natural vegetation. NASA/GSFC Type III Final Report, Greenbelt, Maryland, 371pp.

Roy, D.P., Wulder, M., Loveland, T., Woodcock, C., Allen, R., Anderson, M., Helder, D., Irons, J., Johnson, D. and Kennedy, R. (2014) Landsat-8: Science and product vision for terrestrial global change research. *Remote Sensing of Environment* 145, 154–172.

Sakamoto, Y., Ishiguro, M. and Kitagawa, G. (1986) Akaike information criterion statistics. D. Reidel, Dordrecht, The Netherlands.

Singh, A. (1989) Review Article: Digital change detection techniques using remotely-sensed data. *International Journal of Remote Sensing* 10, 989–1003.

Solh, M. (2010) Tackling the drought in Syria. *Nature Middle East*. DOI: 10.1038/nmiddleeast.2010.206.

Sun, L., Gao, F., Anderson, M.C., Kustas, W.P., Alsina, M.M., Sanchez, L., SAMS, B., McKee, L., Dulaney, W. and White, W.A. (2017) Daily mapping of 30 m LAI and NDVI for grape yield prediction in California vineyards. *Remote Sensing* 9, 317.

Vicente-Serrano, S.M., Beguería, S. and López-Moreno, J.I. (2010) A multiscalar drought index sensitive to global warming: the standardized precipitation evapotranspiration index. *Journal of Climate* 23, 1696–1718.

Witmer, F.D.W. (2008) Detecting war-induced abandoned agricultural land in northeast Bosnia using multispectral, multitemporal Landsat TM imagery. *International Journal of Remote Sensing* 29, 3805–3831.

Witmer, F.D. and O'Loughlin, J. (2009) Satellite data methods and application in the evaluation of war outcomes: abandoned agricultural land in Bosnia-Herzegovina after the 1992–1995 conflict. *Annals of the Association of American Geographers* 99, 1033–1044.

Yang, Z.-Y., Kang, J.-Y. and Zhao, G.-D. (2005) Renewing urban large-scale topographic maps by using Quick Bird panchromatic remote sensing images [J]. *Engineering of Surveying and Mapping* 2, 009.

Zeiler, M. (1999) *Modeling our World: the ESRI Guide to Geodatabase Design*. ESRI, Inc., Redlands, California.

Part 2

Case Studies on Agriculture, Crisis and Conflict

7 The 'Arab Spring' in North Africa: Egypt and Tunisia

Ray Bush*
School of Politics and International Studies (POLIS), University of Leeds, Leeds, UK

Introduction

It is now more than 30 years since the editor of essays focusing on the Middle East and North Africa's (MENA) agrarian question concluded:

> Unless the Middle Eastern and North African countries can adopt 'equitable growth' strategies, the situation of those at the bottom of the social scale, especially in the rural areas is likely to remain rather grim.
> (Richards, 1986, p. 17).

This was poignant indeed as the 2010–2011 uprisings in Tunisia and Egypt demonstrated. Yet, apart from a couple of exceptions, there remains very little published in relation to rural conflict and inequality, social differentiation and household social reproduction in North Africa and the Middle East, including Egypt and Tunisia (Ayeb, 2013; 2017; Gana, 2017). Most of the debate since 2010 has focused on urban conflict, campaigns against authoritarianism in Egypt and the limited democratic opening in Tunisia (Korany and El-Mahdi, 2012; Achcar, 2013; Alexander and Bassiouny, 2014; Beinin, 2016; Marfleet, 2016).

This chapter breaks the 'symptomatic silence' regarding rural political economy and farmer resistance to dominant patterns of agricultural modernization. The chapter does not move solely in a chronological manner. It explores instead a range of conflicts in Egypt and Tunisia's countryside in the context of broader government-declared agricultural strategy. The final section investigates examples of rural conflict including prior to the uprisings. The chapter argues that the silences of investigation, and assessment of rural conflict and the drivers for social change and transformation, mask the dynamics of rurality that have repeatedly featured in Egypt and Tunisia's historical formation.

The primary driver for rural social conflict in our case studies, and as a more common feature of rural resistance worldwide, is the desire for smallholder producers to secure sustained access to sufficient calories to avoid hunger. Access is maintained by ensuring income to enable crop production, for self-provisioning and market sales and non-financial exchanges. This is dependent upon access to land of sufficient size and quality and/or wage employment. Wages from work in the village or from labour migrancy confirm what is commonly known, although mostly ignored by policy makers: smallholder farmers have multiple occupational roles and pluriactivity provides flexibility and security especially when pressures from capitalist crisis, poor prices for outputs yet mounting input costs, prevail (Van Der Ploeg, 2009). Pluriactivity enables farmers to have greater control over life chances and finances. Farmers may also have a

---
* Email: r.c.bush@leeds.ac.uk

range of ownership and access arrangements to land. Farmers may be landowners, tenants and sharecroppers, and this variation in access offers a dimension of support and safety netting in times of economic and ecological crisis.

Unequal farmer access to land in Egypt and Tunisia has driven and sustained high levels of rural poverty. Struggles over resources more generally, like farming implements and inputs, have shaped the ways in which communities and households are able to socially reproduce themselves. Our investigation goes to the heart of what have been called agrarian questions (Akram-Lodhi et al., 2009; Bernstein, 2010; Van Der Ploeg, 2013). These relate to the impact capitalist modernity has on farmers (see Bahn and Zurayk, Chapter 1, this volume). We explore this not by reifying theoreticist debates about what constitutes a peasantry and whether they continue to exist or not: they do (FAO, 2014). Instead, we look at the concrete examples of recent conflict in rural Egypt and Tunisia and situate them in the broader crisis of capitalist accumulation in each country. In doing this we explore how the agricultural policies of the governments and the international financial institutions (IFIs), like the World Bank, exacerbate rural struggles. In doing so we ask why policy makers repeat past policy failures. Rural conflict and protest is the outcome of unresolved capitalist contradictions that reproduce displacement and rural dispossession. The uncertainty and poverty this produces challenge a rural moral economy of governability and further conflict is generated. We will see evidence of rural struggles and their contribution to uprisings in Egypt and Tunisia but we are ultimately confronted by the failures of agrarian struggle to be transformed into urban supported radical transformation. Farmers resist the consequences of the internationalization of capital at the local level (Bush and Martiniello, 2017). However, scaling this up to deliverable radical policy has been sabotaged by urban elite capture as the counter-revolution was secured. Policy makers excuse urban bias by promoting a trade-based 'food security'. They refuse to recognize the specificity and dynamics of the agrarian structures that are being undermined and reconstituted. Family farmers demand access to the key resource of land, and conflict is the outcome of a policy environment that cannot recognize the (contradictory) relationship between capital and nature (Moore, 2015).

Agriculture and the Ancien Regimes

The post-World War II era has been shaped by US production and subsidies creating a 'postwar food order, food aid and dumping'. That 'cumulatively lay foundations for the transfer of the world's peasant populations to camps of surplus labour in urban locations' (Araghi, 2010, p. 112; see also Friedmann and McMichael, 1989). The second food regime from the 1940s to 1973 was characterized by the Cold War and the use of subsidized US PL480 grain as a political weapon. Cheap wheat imports in the global south undermined local agriculture and labour surplus economies. Food dependence upon the US, and other grain exporters in the global north, intensified with grain price inflation and the onset of the third food regime in 1973. The end of the Cold War and trade liberalization effectively marked the close of the ideology of developmentalism. In the food sector neoliberal reform after 1973 was advanced by the increased power of agribusiness. Economic liberalization, structural adjustment lending and accelerated depeasantization and rural displacement drove increased austerity in the south (McMichael, 2013).

Although there are many important local specificities that distinguish contrasts in the experiences of Egypt and Tunisia – geography, class structure, role of the military and geostrategic interests – there are also important similarities. Small farmer agriculture is central to both political economies and continues to contribute significantly to production, employment and income. Its role is explored in each country, in turn.

Egypt

The Nile Valley and Delta comprises about 3 million ha of irrigated land with probably as many as 4 million farmers. Agriculture is very productive on just under 3.7% of the land area in 2014. Many farmers harvest two or three crops a year. Major crops include wheat and rice, beans and fruit and vegetables and cotton. Yet self-sufficiency reaches only 50% of consumption levels and small farmers remain poor. Land access is very unequal with half of the cultivated area held by 90% of farmers with an average size of less than 5 feddans; almost 70% of farms access less than 2 feddans (1 feddan = 0.42 ha or 1 acre). Yet 3% of farmers control 33.5% of the agricultural land averaging more than 10 feddans (Ayeb, 2010, 2012). Moreover, since economic reform began in rural Egypt in 1987, preceding the formal adjustment programme with the IFIs in 1991, and especially following Law 96 of 1992 (fully effective in 1997), which impacted deleteriously on the country's more than 1 million tenants, rural social differentiation and conflict have intensified.

Egypt's political economy has been dominated and shaped by its dependence upon imported wheat. President Nasser was astute enough to reduce the impact of the US-dominated international food regime in the 1960s by playing Washington and Moscow off against each other. This ensured adequate flows of imported wheat but later Egyptian presidents have been plagued by the price uncertainty of the global food system. Egypt produced 9 million tonnes of wheat in 2016/17 and imported 12 million tonnes (FAO, 2017).

The price of imported wheat, and the spikes in prices in 2008; food price inflation of 20% (closer to 25% in 2017); deaths in bread queues; and industrial strikes culminated in the January 2011 uprisings. In 2010 wheat prices rose by 32% and rice by 42%. The uprisings were caused by persistent attempts by the government of Egypt (GoE) to implement neoliberal market reform. That accelerated conflict and failed to shift the economy away from dependence upon rents that accrued from oil and gas sales, Suez Canal charges and labour migrant remittances, the latter falling precipitously after imperialist interventions in Iraq and Libya. The IFIs continued to applaud the economic strategy of erstwhile President Hosni Mubarak, denying that their policies had a role in his downfall. It was a 'strategy' founded upon direct foreign investment that deepened the extractive economy characterized by asset stripping by crony capitalists linked to the ruling National Democratic Party, and by poverty and inequality, especially in rural Egypt although becoming more entrenched and common in the country's towns and cities (Shenker, 2016).

Agriculture contributed about 14% of Egypt's value-added gross domestic product (GDP) in 2011 but this was down to 11% in 2015 (World Bank, 2017). Perhaps as many as 30% of the country's 95 million people remain engaged in or are linked to agriculture. Since 2000, 17% of Egypt's import bill pays for food (it is just 3% for India and 4% in Brazil) (Al-Riffai, 2015, p. 37). Food insecurity and wheat dependency remain persistent features of post-uprising Egypt. One reason for this is the recurrent and systemic character of rural conflict that stems from, among other things, the failure of the country's policy makers to think creatively and analytically about what constitutes food security. The dominant view regarding food security, among the Egyptian authorities and IFIs, is one that is trade-based: if a country produces insufficient food locally then food security is maintained by macro-economic policy to generate income to purchase food from international markets. Egypt's problem, shared by Tunisia, is that even when the national economy was growing, as it did in the 10 years prior to 2011 at about 3% per capita, poverty and inequality worsened and so too did food insecurity for the poor. This was the direct result of neoliberal modernization that promoted exports funded by a regime of borrowing, debt and farmer poverty. After the appointment of liberalizer Ahmed Nazif as prime minister in 2004, the contradictions of Egypt's strategy for capital accumulation became more pronounced. Unemployment worsened, more than half the population were deemed poor and worker and civil unrest intensified (Abdelrahman, 2015; Beinin, 2016). Egypt's agricultural strategy was to try and boost the export of high value crops and do so with incentives to large-scale producers. This received the blessing of the United States Agency for International Development (USAID), which had effectively run the Ministry of Agriculture and Land Reclamation in the 1990s. USAID encouraged the replacement of the historical role of the state in providing many agricultural inputs at controlled prices, including credit. Yet the private sector failed to boost agricultural production as the planners promised (Faris and Khan, 1993; Bush 1999; 2016; GoE, 2009).

A nuanced view of food security requires more than income generated nationally to enable the import of food to meet local needs. It also encompasses the needs of households to access revenue with which to purchase food and to grasp the ways in which food producers (who are the first to suffer food insecurity) understand

and also transform their well-being. It raises the importance of a strategy for food sovereignty.

In Egypt the agricultural investment scheme at Toshka, which built the Mubarak Pumping Station and irrigation canal linking the Nile with part of the south Western Desert, failed to attract sustained investment and migrant farmers from what the GoE described as an overpopulated Delta. It failed to generate significant export agriculture and it had a poor environmental impact. The scheme was projected to create 2.8 million new jobs and attract 16 million migrants into new towns on up to 1.5 million acres of newly reclaimed land. One estimate of the financial cost since the 1990s is that it has sucked in US$90 billion (Schilling, 2013).

Tunisia

Tunisia has also long pursued the trade-based view of food security. Yet it is not that the Egyptian or Tunisian peasants could not feed the local population, it is that the agricultural strategies pursued by both authoritarian regimes prevented them from so doing. Tunisia's trade-based view of food security, especially after the early 1970s, centred on capital intensive agriculture, which neglected investment, consultation or support for small family farmers. It promoted an environmentally deleterious agricultural strategy that has created immense regional variations and uneven capitalist development.

Small farmers dominate Tunisian agriculture and, like Egypt, there are huge inequalities in the size of land holdings. Fifty-four per cent of farmers have less than 5 ha covering about 11% of the total area. Just 3% of farmers, however, have holdings of more than 50 ha, which represents 34% of Tunisia's farmland (Jouili, 2009, p. 5).

In 2010 agriculture accounted for nearly 8% of GDP, contributing nearly one-tenth of total exports, and 20% of employment. Perhaps as much as 34% of the population is rural and 50% of these rely directly or indirectly on the agricultural sector.

Although the World Bank has noted that agriculture is 'a very important sector for growth and poverty reduction' (2014, p. 260), two processes have undermined the efficacy of small-scale family farming in Tunisia. The first was the structural adjustment program (SAP) introduced in 1986. The state implemented a policy of applying a 'true price' for agriculture subordinated to the market. Subsidies to small farmers for inputs ended, increasing vulnerability to market pressure and excluding small farmer access to credit, land and other hitherto state-provided services (Jouili, 2009, p. 2). Small land holdings fragmented. Those of less than 5 ha increased in number from 133,000 in 1961 to 281,000 in 2004. At the same time a restructuring of land occurred as larger landowners purchased land from increasingly indebted smallholders.

The second process undermining small-scale family farming was competition between the richer Sahel region in the north and the rest of the country, especially over access to land and water. A government of Tunisia (GoT) strategy accelerated a pattern of combined and uneven capitalist development as state land was converted into privatized holdings bought by coastal-based entrepreneurs (Ayeb and Bush, 2014). These investors increased the export of olive oil and vegetables from coastal areas but the production of the surpluses took place away from the shoreline and more developed and urbanized coastal enclaves. Agricultural and mining production and extraction took place in the south and centre of the country (Gana, 2012; 2013). The strategy of promoting water-intensive agriculture excluded and dispossessed local small-scale farmers. After years of collectivized agriculture (1964–1970), Tunisia's Fifth Five Year Plan 1977–1981 promoted a neoliberal agenda. The plan enabled state- and privately funded expensive and environmentally damaging irrigation schemes. In the area of Sidi Bouzid, for example, where Mohemed Bouazizi committed suicide in December 2010, whose family along with many others were coping with land dispossessions, debt and de-agrarianization, irrigated farmland increased from 2000 ha in 1958 to more than 47,000 ha in 2011 (Ayeb, 2012; Fautras, 2015).

Sidi Bouzid lies in a semi-arid zone. Farmers mostly herd sheep and camels, and grow almonds, olives and cereals. Yet while the area is highly productive, the people who produce the wealth – farmers and, in the case of Gafsa, governorate miners – remain poor: the wealth has been appropriated by Tunis and coastal-based elites. With the usual caveats about the accuracy of poverty levels Sidi Bouzid is the fourth poorest governorate in Tunisia. In 2011 as many as 42% eked an existence on less than US$2 per day (Touhami, 2012; Szakal, 2016b). Many farmers lost land as crops failed and input costs escalated. Dispossession followed indebtedness.

Interior poverty rates in 2010 were estimated to be up to 32% and unemployment rates 22%. These compared with coastal poverty and unemployment rates of 9% and 11%, respectively (Mestiri, 2016, n.p.). The uprising in 2010 began in the underdeveloped regions where poverty and food insecurity were most intense. The World Bank expressed concern that poverty-generated conflict and progress regarding the Millennium Development Goals had lagged in rural and interior regions (World Bank, 2012, pp. 6–7).

Plus ça Change

The optimism from uprisings in Egypt and Tunisia raised the hope that small farmers could voice their concerns and help facilitate alternatives to their historical marginalization and persistent rural conflict. Disappointment, however, about the lack of positive change for the rural poor in both countries has been extensive.

Egypt

The inclusion of farmer representation in the drafting of Egypt's constitution in 2014, and even the insertion of a clause affirming the importance of 'food sovereignty', did not effect change. The policy of continued farmer marginalization or more appropriately their abjection has continued (Bush, 2012; El Nour, 2015).

President Abdel Fattah al-Sisi has persisted with an agricultural strategy that rhetorically declares concern with wheat dependency, and the need for greater self-sufficiency, with open trade and land reclamation and very little mention of small farmers. Erstwhile Agriculture Minister Essam Fayed noted in January 2017 that the Ministry planned to achieve food security of strategic crops like wheat, maize and rice based on modern technology. He indicated the need for high productivity seeds in the context of limited water, improved marketing and a strategy to increase the value of agricultural products and reduce loss of crops, which he estimated to be 20% of production.

Following a cabinet reshuffle in January 2017 the new Agriculture Minister Abdel Moneim al Banna continued to emphasize the need to reduce the food gap of strategic crops in order to reduce import dependence (Mitwally, 2017). He reiterated concern about import dependency and also included the need to develop contracts with farmers to harness greater efficiency in farm cycles, reducing the deleterious impact of middlemen, merchants and traders. He committed to the farmers' card system of support for 7 million farmers including the need to reduce the amount of poor quality and overpriced fertilizer (Hassan, 2017). He also promised the introduction of farmer health insurance and reiterated the need for water rationalization (Al-Noubi, 2017).

Ministry statements of intent, perhaps to allay continued accusations of corruption and government ineptitude, sit alongside the continued GoE preoccupation with pharaonic-style, project-led growth. In December 2015 President al-Sisi announced a plan to reclaim 1.5 million feddans of desert land in the New Valley Governorate near Farafra. He had scaled down an earlier announcement to reclaim 4 million feddans for agricultural use. The idea that boosting agricultural production requires more land is a recurrent feature of Egyptian politics. There seems to have been little learnt from historical failures in meeting targets and aims to reclaim land and resettle farmers – as Nasser had tried in Beheira in the 1960s and Mubarak in the late 1990s attempting to reclaim 500,000 feddan in Toshka. This latter is a project President el-Sisi is now reviving. The 1.5 million feddan scheme, as it has become known, met all the necessary environmental impact assessments, according to the president, but it is likely to suck up 25% of Egypt's available groundwater – 2 billion cubic m^3. The cost to purchase and prepare the land was to be organized by the Egyptian Rural Development Company, which was given LE8 billion capital (approximately US$440 million or £345 million) (Mada Masr, 2015; Mukhtar, 2016). Yet there seems to have been very little detailed discussion of how land will be distributed and to whom. What will the balance be between small farmers and large companies and how will the early farming activities of new young farmers be financed? There were mixed government signals on this. In February 2016 the head of the Egyptian Rural Development Company said it was negotiating with the Central Bank to provide funding for small-scale farmers and young people, but the expected interest rate was likely to be an unrealistic 5% repayable over 8 years (Ramadan, 2016). This latter vexed the farmers' syndicate as they protested the increased cost of credit that would follow from plans to convert the Agricultural Credit and Development Bank into an investment bank (Dardeer, 2017).

The persistence with large-scale scheme agriculture followed President el-Sisi's 'Egypt the Future' donor conference in March 2015. The conference was the opportunity to show the Western world just how important the president was to maintaining political order in Egypt and more broadly in the struggle against Islamic fundamentalism. The 'rebranding' of Egypt (Shenker, 2015) provided the moment for the world's 'greatest' to congratulate the president on his 'leadership' (Tony Blair) and the 'journey to economic growth…' (Christine Lagarde). In winning praise and donor support for his project-driven development, including a new administrative capital city, el-Sisi noted ominously, 'we don't approve of violation of Human Rights but this is an exceptional state in Egypt's history' – and his foreign guests agreed. To persuade Western dissenters, the Egyptian state embarked on an unprecedented strategy of paying two US public relations companies US$1.8 million a year to highlight Cairo's role in managing regional risk (Rohan, 2017). Far from focusing on the needs of Egypt's farmers, and recognizing and then understanding why rural conflict persists, el-Sisi has focused on building the security state (Abdelrahman, 2015).

Tunisia

After interminable months of constitutional wrangling and elections in Tunisia the Ministry of Agriculture announced in February 2017 a new Five Year Plan for agriculture. There were to be seven areas of focus yet there seemed to be very little endeavour to explore how the consequences of historical underdevelopment and marginalization of poor farmers and regions would be redressed. While there is recognition of the need to promote crops important to small and family farmers, and strengthening agriculture's role in rural development, there is very little analysis of why Tunisia's small farmers have become so poor and why and how rural poverty continues to be reproduced. There is too little effort made to correct and reverse the dominance of larger estate agriculture, which, as we have noted, sucked in available land and water resources to the detriment of small farmer agriculture. The new initiative read like a shopping list with little debate as to how it will be delivered and how it will be linked to the performance of Tunisia's broader macro-economic performance and reform. The declared strategy is to develop natural resources; address land problems (sic) including fragmentation; promote production systems and their sustainability; and disseminate knowledge and the rules for good governance in agriculture and fisheries (Kapitalis, 2017).

The intention of the Five Year Plan is to deliver an annual increase in agricultural growth of 3.3%. Minister of Agriculture Samir Taieb announced there would be an investment boost of 55% from 5.8 billion dinars (US$2.3 billion or £1.8 billion) in the period 2011–2015 to 9 billion in 2016–2020. The investment would be equally distributed between the public and the private sectors (Kapitalis, 2017). He said little, however, regarding how that investment would be delivered and how if left to the market, a more balanced agricultural system could emerge.

This strategy is likely to receive at least a warm endorsement from the World Bank, which has recently criticized earlier Tunisian agricultural policy (World Bank, 2012; 2014). For a generation the World Bank advanced and defended the interests of dictator Ben Ali and the land-owning *colons* from Tunis and the north. Recognizing that the World Bank among other IFIs had 'heralded Tunisia as a role model for other developing countries' it recently noted, however, that there were 'serious flaws' in Tunisia's economic strategy. These included weak job creation and regional inequalities in an economy 'frozen in low-value added activities' (World Bank, 2014, p. 16).

The World Bank now criticizes Tunisia's historical agricultural policy and its trade-based view of food security. The IFI advocates a strategy to 'unleash' agricultural growth in Tunisia's interior regions. It argues that state intervention has 'repressed the agricultural sector' by distorting production away from Mediterranean products in which Tunisia has a natural comparative advantage, like olive oil and tomatoes, towards continental products like cereals, beef and milk in which Tunisia is not very competitive but which are key to food security (World Bank, 2014, p. 260).

Tunisia's agricultural competitiveness lies in its comparative advantage in crops with 'greater labor intensity and a disadvantage with high land intensity'. This reflects an understanding that Tunisia has an abundance of cheap labour but a scarcity of water and arable land. The policy implication that follows from this is that Tunisia should focus more on producing olives, tomatoes, and vegetables and out-of-season crops for export. Soft wheat production may be profitable in the north and northwest, when there is good rainfall, but there should be less wheat, barley, beef and milk production because it is inefficient and subsidized. Tunisia should instead improve its export of arboriculture to the European Union. The country remains the world's second largest producer of olive oil, but since 2000 production stagnated despite an increase in global demand.

The World Bank's analysis raises concerns that are at the heart of small farmer anxiety: what might future cropping patterns look like and how can uneven spatial development be redressed? Unfortunately the policy suggestions made by the IFIs fall into time-weary tropes of neoliberal ideology. They also absolve the IFIs of creating the type of Tunisian state that has overseen economic decay and environmental crisis. Structural adjustment in Tunisia promoted privatization of state farms, reductions in farm subsidies that impacted mostly the poor as prices were also liberalized and opportunity for accessing farm credit diminished, at least for smallholders. Economic reform also encouraged transformation of land rights, accelerated social differentiation and increased access to water, land and credit for larger landholders, that led to the stripping of Tunisia's topsoil.

The World Bank's recent analysis and recommendations echo IFI commentary in 2006 when it called for a greater role for the private sector, reduced state intervention and 're-engineering' of institutions especially in cereal production and land titling (World Bank, 2006). In value terms, agricultural production increased by 67% from 1990–2010. Seventeen per cent of this growth by value was in beef and milk production, areas the World Bank says are uncompetitive but have received strong state support. Subsidies, price support and trade protection have resulted in an inefficiently inflated sector that has benefited from transfer of resources paid for by taxpayers, according to the World Bank (2014, p. 265). For the World Bank the strategy is simple:

> To unleash the potential of agriculture, the state needs to play a different role in agricultural markets. The state should allow markets to freely establish prices and should refrain from direct intervention in the market, focussing instead on providing a regulatory framework and public goods to support the development of the sector.
> (World Bank, 2014, p. 273)

The World Bank's suggestions for Tunisia's agricultural development make little mention of small farmers or the rural conflict that helped topple dictatorship. The IFIs' focus on macro-economic policy and a positive role for the market, which they see not as a market where economic power is exercised by powerful landed and coastal elites accelerating rural dispossession, displacement and social differentiation, but instead as a market that has been hindered by too much state intervention, which has driven inefficiency. Successful agriculture is dependent upon an end to price and input support, an end to state marketing, and a boost to infrastructure with research, extension, irrigation and land registry. A simplification of access to land with land registration will, moreover, 'facilitate large investments in agriculture' (World Bank, 2014, p. 274). Despite rhetoric to reduce regional inequality, the World Bank says little about how small farmers can be empowered to deliver the IFI-favoured policy or reduce the tensions and conflict in Tunisia's periphery. The policy emphasis remains focused on larger landowners whose historical strategy of mining Tunisia's environment and exploiting small farmer livelihoods has created the farming crisis.

Farmers and Resistance

Egypt

Independent trade unions in MENA have been invested with the importance of throwing off the yoke of state-run federations that have been used to co-opt workers and farmers and suppress workplace democracy. Independent trade unions have been more successful in Tunisia than Egypt. Egypt's General Federation of Trade Unions (FGSTE), founded in 1957, was used by successive political regimes to limit farmer representation and suppress demands for reducing input costs and raising farm gate prices. The FGSTE was challenged by the spread and activism of independent trade unions (Beinin, 2016) and optimism that accompanied the Declaration on the Freedom of Association in March 2011. After the uprising, independent and locally convened farmer associations were represented by the Egyptian Federation of Independent Trade Unions. However, the difficulty remained of transforming protest and representation from Egypt's countryside into robust and effective political representation. If the January 2011 uprising had been successful, precisely because of its spontaneity (following decades of crisis and mobilizations), its demise, first with conservative Islamist President Morsi and then with the shameless return to dictatorship with Abdel Fattah al-Sisi, was easy because of the political failings to organize, structure and institutionalize protest and protestor rights.

Egyptian small farmers mobilized around grievances protesting especially around land access, input prices and merchant power, but they did not have an effective mechanism to deliver their protest by converting it into policy change. Accurate accounting of farmer protests is difficult because of the often remote spatial and geographical context in which rural resistance to dispossession, struggles over access to land and the consequences of market liberalization take place. Nevertheless, there are indications that protest and unrest have for many years been a persistent feature of Egypt's rural life. The human rights advocacy group Sons of the Soil estimated that between 2009 and 2010, the year preceding the 25 January uprising, there were at least 180 sit-ins, 132 demonstrations and 6 strikes in rural Egypt. And the protests they reported met with violent responses by the police and security forces. In 2010 alone, these responses caused a recorded 1700 arrests, 297 fatalities and 1451 injuries (Bush, 2011, p. 400). Similar numbers of incidents are reported by the Cairo-based Land Centre for Human Rights (LCHR). In 2014, another Cairo-based human rights organization noted 64 protests in the agricultural sector, from a recorded 2274 worker protests (El-Mahrousa, 2014). The recorded number of rural disputes is less than those for industrial worker conflicts but we might expect that. Rural resistance is often informal and unrecognized by national media, and disputes around irrigation and land boundaries are often proxies for small farmer resistance to being marginalized by the state, investors and larger landowners. Rural conflict also takes the form of opposition to price hikes, often by local middlemen and traders who are frequently accused of selling out-of-date and overpriced seeds and fertilizers (author interviews with farmers in Dakahlia, May 2013; LCHR, 2014).

Rural conflict in Egypt has been especially evident as private landowners, or the state, try and dispossess tenants from their land. This has occurred far more often after the full implementation in 1997 of Law 96 of 1992, which impacted at least 1 million tenant farmers and their families. That legislation fully liberalized land markets and sales, purchases and rentals. In some cases rents rose by 400% and many tenants were dispossessed (Bush, 2002b; Saad, 2002; Ayeb, 2012, p. 80). The legislation gave land owners the confidence to try and quash the impact of Nasser's 1960s legislation, which had given many farmers with tenancies rights to land in perpetuity. There are many cases where landowners dispossessed tenants illegally. Nasser's land reform gave tenants rights to farm while the land became state land. Yet the Nawar family in Sarando in Egypt's Delta near Damanhour, for example, has persistently waged a violent struggle to dispossess farmers claiming back land that had been farmed by tenants for more than 40 years (Human Rights Watch, 2005; Williams, 2005).

In what has become another celebrated case (widely reported since 2015), a local businessman, landowner and erstwhile National Democratic Party figure, Farid Al-Masry, chased farmers from land that they had been allocated as veterans from the Yemen war in 1962 and farmed peacefully from 1967–1996. Conflict since 1997 has been caused by Al-Masry, who defied court orders and violently intimidated villagers. He claimed, on a disproven technicality, that he should never have lost the land after Nasser's 1961 land reform that limited individual holdings to less than 100 feddan. It is perhaps a measure of just how overdetermined politically land access issues have become that legal battles have been intense and the violence against farmers and their families so systematic and systemic for more than 20 years. In 2015 more than 20 farmers were imprisoned, later released, after the Agricultural Reform Agency in Mansura refused to return land to villagers even though the farmers had won a court order in their favour. For a while after the January 2011 uprising the farmers retook their land in the confidence that the political turmoil would settle in their favour. It did not. Farmers have faced repeated harassment, crops damaged by thugs and police intimidation and equipment stolen, reportedly stored on the grounds of the businessman's ceramics factory. The Spanish company Roca now has a 50% stake in local Gravena ceramics and the farmers assert that their land was targeted by Al-Masry for a factory extension – even though it is illegal to build on farmland. As one of the oldest farmers noted, 'These are the worst days. You let the dogs eat the peasants. It's important to tell people that the country is rotten from top to bottom.' Referring to President Sisi, he added, 'If he doesn't come out and make sure people are treated well, the entire country will go up in flames' (Esterman, 2015: n.p.).

Probably the biggest source of conflict in Egypt's countryside relates to attempts by historic landowners, often linked to the state and state institutions, to displace and dispossess farmers and tenants from their land. The LCHR annual report in 2016 noted how state agencies and media reporting, or the lack of the latter, favoured big landowners under a pretext of law and neutrality attacking smallholder mobilizations as the advance of factional interests. The problem for small farmers is that their attempts to advocate defence of their land rights since 1997 have been confronted by powerful property claimants. As the LCHR has detailed there is a recurrent trend in the way that small farmer interests are challenged in conflicts with the state. There are frequent allegations of theft against small farmers; lawsuits accusing them of stealing the land on which they have farmed, often for a generation. Small farmers are forced to challenge in law decisions that have been fabricated by state representatives and new property claimants. There are many cases where peasants and their families are harassed and imprisoned: victims of torture at the hands of the local forces of law and disorder – the police (LCHR, 2002; Human Rights Watch, 2017).

Many of the complaints and conflict for Egyptian small farmers relate to having to deal with dramatically raised land rents and prices for farming inputs, seed and fertilizer (LCHR, 2014; Egypt Report, 2017). Law 96 of 1992 created openings for historic owners to try and make unlawful land claims and this was abetted by the state promoting market liberalization. The chaos for small farmers that neoliberal reform engendered was intensified by confusions that were generated after the January 2011 uprising as landholders strived to claim land and irrigation rights when policing was in chaos. Confusion spread to interministerial competition and failings for ministries with *locus standi* to limit lawlessness. The Ministry of Agriculture's oversight of rural Egypt was challenged and its erstwhile minister, Salah Helal, became ensnared in a corruption scandal. He was sacked in September 2015 for allegedly accepting bribes in return for legalizing the sale of 2500 acres of land to the businessman Ayman Gamil (El-Fekki, 2015a; 2015b). The Ministry of Agriculture was also implicated in the acrimonious declaration by Hisham Geneina, Head of Accountability State Authority, Central Auditing agency, that corruption cost the Egyptian state LE600 billion (US$76 billion) in 2015 (Al-Monitor, 2016). He later corrected this time frame to the 4 years 2011–2015, but that early imprecision and lack of clarity in defining corruption cost him his job. It was not lost on commentators that Geneina had been a supporter of erstwhile President Morsi, and that the implication for agriculture of the corruption to which he referred included the loss of agricultural land and malpractice across a range of ministries throughout the country (Al-Monitor, 2016; Middle East Eye, 2016).

Tunisia

Farmer claims, squatting and occupation of state land were strong elements in the ousting of President Ben Ali. Landless and near landless workers occupied as many as 100 farm units covering 10,000 ha in 2011 (Gana, 2017, p. 268). Probably the most celebrated case in Tunisia of farmer resistance to what can be seen as 'privatization' of state occurred in the Jemna Oasis. This is a particularly important case because it lies in an area of southern Tunisia impacted by decades of neoliberal agricultural policy and where peasants have clawed land and land rights away from the state and its policy of privatization. The farmers of Jemna have, for instance, formed the Association for the Protection of Jemna Oasis. This promotes a more socially sensitive management of agriculture and reinvestment of revenues for community development (Ayeb, 2016; Hamouchene, 2017). Famous for quality date production the Oasis was colonized by the French occupation in 1912 and after independence in 1956 by the state. Instead of returning land to previously dispossessed families the Tunisian state nationalized it for mechanized agriculture. The farmers in the Oasis, however, want the state to return to them the land that was their grandparents'. The state company that managed Jemna Oasis was bankrupt in 2002 and two 'entrepreneurs' close to erstwhile dictator Ben Ali took it over and made high profits from increasing surplus from local farmers.

The ouster of Ben Ali led to a committee of farmers reclaiming the land. The people of the Jemna Oasis 'demonstrated that the Tunisian revolution is not uniquely about narrowly defined political rights but rather a broader agenda entailing economic sovereignty (including over land), dignity and justice' (Hamouchene, 2017, n.p.).

Despite the farmer initiative, or probably because of it, the government told the local farmer association in 2015 to reorganize themselves as a 'company for agricultural promotion and development' where the state could hold a majority share (Szakal, 2016a). But farmers contest and oppose the reintroduction of neoliberal agriculture, dominance of agribusiness and strategy underpinned by agricultural economics that only prioritized cost–benefit analysis and market liberalization. Farmers also contest prioritization of export-led agriculture that requires high water input and accelerates land degradation.

The association members recalled how, before the uprising, state-owned land had been leased to private investors and the returns to capital remained with them and the benefits of local labour did not profit either the Oasis or the local town. In contrast, since the uprising, the association has facilitated investment of agricultural gains in the town, in health provision, paying workers and maintaining the ecological balance of the Oasis (Guerfali, 2015). Production doubled from 2011–2014 and 300 workers were employed compared with only 20 in 2011.

The success of independent farmer associations is uneven and, like the case of Egypt, dependent upon a unified platform or union that can represent the views of small farmers in the institutions of government. The informal and sporadic, spontaneous and organized small farmer resistances are effective and important in advancing the interests of the rural poor. However, rural protesting can become strengthened if it receives unconditional support from democratic and representative organizations that can petition government.

Small farmer dissatisfaction with the official state union, the Tunisian Union of Agriculture and Fisheries (UTAP), led in December 2014 to the creation of the independent Union of Tunisian Farmers (SYNAGRI). This union, much like its counterparts in Egypt, has failed to broaden its social base of support and political backing in Tunis. It is not exactly a union for small farmers as it also includes investors and larger landholders but it has at different times tried to raise small farmer interests (Gana, 2017).

Conclusions

This chapter has documented some recent forms of small farmer protest in Egypt and Tunisia. It has done so by situating protest in the context of the 2010 and 2011 uprisings and the enormous hope and optimism that accompanied the toppling of the two dictatorships. The uprisings in Tunisia were relatively more successful because in Egypt the military re-asserted itself having never lost its hegemony. Yet both countries share the historical amnesia as to why an important dimension of the uprisings emerged, namely small farmer and landless desperation with poverty and abjection. Both countries persist with food security strategies that externalize their economies and fail to understand the dimensions and dynamics of small farmer agriculture. Smallholder and landless resistance will continue and is unlikely to be structured and organized by trade unions. And this is the tension that can only be resolved in the practice of local political struggles. Rural producers are at their strongest when they (informally) defend their autonomy and livelihoods, squatting, blockading, seizing land and other resources stolen from them, and yet without a stronger politically organized push, perhaps with independent trade unions, there is little opportunity for rural struggles to be embraced by urban workers and pressure groups necessary for radical transformation.

Acknowledgements

Thanks to Yosra el Gendi and Giulio Iocco for research assistance. Many thanks too for comments from Habib Ayeb and the insight (and patience) of the editors. Errors are mine alone.

References

Abdelrahman, M. (2015) *Egypt's Long Revolution*. Routledge, London.
Achcar, G. (2013) *The People Want: A Radical Exploration of the Arab Uprising*. Saqi, London.
Akram-Lodhi, H., Kay, C. and Borras, S.J. (2009) The political economy of land and the agrarian question in an era of neoliberal globalization. In: Akram-Lodhi, H. and Kay, C. (eds) *Peasants and Globalization: Political Economy, Rural Transformation and the Agrarian Question*. Routledge, London, pp. 214–238.
Al-Noubi, E. (2017) The New Minister of Agriculture in his first statements: Egypt shall not progress agriculturally without scientific research. Available at: www.youm7.com/story/2017/2/14/ال-زراعة-وزير هـ-الـ جديد ى-الـ فـ أول-هـ حالـ صريـ د فوق-لـ ن-مصرـة تـ ياتـ الزراع/3102418 (accessed 20 February 2017).
Al-Monitor (2016) How calling out corruption cost this top Egyptian official his job. Available at: www.al-monitor.com/pulse/iw/originals/2016/04/egypt-corruption-report-auditing-authority-dismissed-geneina.amp.html (accessed 9 April 2017).
Al-Riffai, P. (2015) How to feed Egypt: enhancing food availability and nutrition for a bulging population. *The Cairo Review of Global Affairs* 18, 36–43. Available at: https://cdn.thecairoreview.com/wp-content/uploads/2015/09/CR18-Al-Riffai.pdf (accessed 19 April 2017).
Alexander, A. and Bassiouny, M. (2014) *Bread, Freedom, Social Justice: Workers and The Egyptian Revolution*. Zed Books, London.
Araghi, F. (2010) The invisible hand and the visible foot. In: Akram-Lodhi, H. and Kay, C. (eds) *Peasants and Globalization: Political Economy, Rural Transformation and the Agrarian Question*. Routledge, London, pp. 111–147.
Ayeb, H. (2010) *La Crise de la société rurale en Egypte: la fin du fellah?* Karthala, Paris.
Ayeb, H. (2012) The marginalisation of the small peasantry: Egypt and Tunisia. In: Bush, R. and Ayeb, H. (eds) *Marginality and Exclusion in Egypt*. Zed Books, London, pp. 72–96.
Ayeb, H. (2013) Le rural dans la révolution en Tunisie. Blog, 28 September. Available at: https://habibayeb.wordpress.com/2013/09/28/le-rural-dans-la-revolution-en-tunisie-les-voix-inaudibles (accessed 20 April 2017).
Ayeb, H. (2016) Jemna, ou la résistance d'une communauté dépossédée de ses terres agricoles. Blog, 3 October. Available at: https://habibayeb.wordpress.com/2016/10/03/jemna-ou-la-resistance-dune-communaute-depossedee-de-ses-terres-agricoles (accessed 25 March 2017).
Ayeb, H. (2017) Food issues and revolution: the process of dispossession, class solidarity, and popular uprising: the case of Sidi Bouzid in Tunisia. *The Food Question in the Middle East: Cairo Papers in Social Science* 34(4), 86.
Ayeb, H. and Bush, R. (2014) Small farmer uprisings and rural neglect in Egypt and Tunisia. *Middle East Report* 272, 2–10.
Beinin, J. (2016) *Workers and Thieves*. Stanford Briefs, Stanford, California.
Bernstein, H. (2010) *Class Dynamics of Agrarian Change*. Fernwood Publishing, Halifax; Winnipeg.
Bush, R. (1999) *Economic Crisis and Politics of Reform in Egypt*. Westview Press, Boulder, Colorado.

Bush, R. (ed.) (2002a) *Counter Revolution in Egypt's Countryside: Land and Farmers in the Era of Economic Reform*. Zed Books, London.

Bush, R. (2002b) Land reform and counter-revolution. In: Bush, R. (ed.) *Counter Revolution in Egypt's Countryside: Land and Farmers in the Era of Economic Reform*. Zed Books, London, pp. 3–31.

Bush, R. (2010) Food riots: poverty, power, protest. *Journal of Agrarian Change* 10(1), 119–129. Available at: http://onlinelibrary.wiley.com/doi/10.1111/j.1471-0366.2009.00253.x/pdf.

Bush, R. (2011) Coalitions for dispossession and networks of resistance? Land, politics and agrarian reform in Egypt. *British Journal of Middle Eastern Studies* 38(3), 391–405.

Bush, R. (2012) Marginality or abjection? The political economy of poverty production in Egypt. In: Bush, R. and Habib, A. (eds) Marginality and Exclusion in Egypt. Zed Books, London, pp. 3–13.

Bush, R. (2014) Food security and food sovereignty in Egypt. In: Babar, Z. and Mirgani, S. (eds) *Food Security in the Middle East*. Hurst, London, pp. 89–114

Bush, R. (2016) Uprisings without agrarian questions. In Kadri, A. (ed.) *Development Challenges and Solutions After the Arab Spring*. Palgrave Macmillan, Basingstoke, pp. 153–172.

Bush, R. and Ayeb, H. (eds) (2012) *Marginality and Exclusion in Egypt*. Zed Books, London.

Bush, R. and Martiniello, G. (2017) Food riots and protest: agrarian modernizations and structural crises. *World Development* 91, 193–207.

Dardeer, A.E. (2017) The Egyptian Countryside Company: 7200 application forms were distributed to small peasants and youth. 19 February. Available at: www.almasdar.com/63481 (accessed 20 April 2017).

Egypt Report (2017) In protest over the increase of the prices of fertiliser, peasants demand the dismissal of the minister of agriculture and threaten to escalate. 31 January. Available at: www.egyrep.com/ الـ فلاح-الأ سمدة-أ سعار-رفـ ع-ع لـى-اد تجاجا (accessed 4 April 2017).

El-Fekki, A. (2015a) Agriculture corruption case does not involve other ministers: Cabinet. *Daily News* 10 September. Available at: www.dailynewsegypt.com/2015/09/10/agriculture-corruption-case-does-not-involve-other-ministers-cabinet (accessed 7 March 2018).

El-Fekki, A. (2015b) Former corrupt agriculture minister faces first trial on 12 December. *Daily News*, 4 November.

El-Mahrousa (2014) Annual report on the labor movement in Egypt. Available at: http://elmahrousacenter.org/english/wp-content/uploads/2015/01/The-Annual-Report-Brief-on-Labor-Movement-in-Egypt1.pdf (accessed 29 July 2017).

El Nour, S. (2015) Small farmers and the revolution in Egypt: the forgotten actors. *Contemporary Arab Affairs* 8(2), 198–211.

Esterman, I. (2015) A fight over land rights shaped by wars, an uprising and power politics. Madamasr.com, 10 June. Available at: www.madamasr.com/en/2015/06/10/feature/politics/a-fight-over-land-rights-shaped-by-wars-an-uprising-and-power-politics (accessed 7 March 2018).

Faris, M.M. and Khan, M.H. (eds) (1993) *Sustainable Agriculture in Egypt*. Lynne Rienner, Boulder, Colorado.

Fautras, M. (2015) Land injustices, contestations and community protest in the rural areas of Sidi Bouzid. *Spatial Justice*, N. 7 (January 2015). Available at: www.jssj.org/article/injustices-foncieres-contestations-et-mobilisations-collectives-dans-les-espaces-ruraux-de-sidi-bouzid-tunisie-aux-racines-de-la-revolution (accessed 20 April 2017).

Food and Agriculture Organization of the United Nations (FAO) (2014) International year of family farming. Available at: www.fao.org/family-farming-2014/en (accessed 26 April 2017).

Food and Agriculture Organization of the United Nations (FAO) (2017) Global information and early warning system, country briefs, Egypt. 19 May. Available at: www.fao.org/giews/countrybrief/country.jsp?code=EGY (accessed 7 June 2017).

Friedmann, H. and McMichael, P. (1989) Agriculture and the state system: the rise and decline of national agriculture from 1870 to the present. *Sociologia Ruralis* 14, 93–118.

Gana, A. (2012) The rural and agricultural roots of the Tunisian revolution: when food security matters. *International Journal of Sociology and Agriculture and Food* 19(2), 201–213.

Gana, A. (2013) Aux origines rurales et agricoles de la Révolution tunisienne. *Maghreb-Mashreq* 1(215), 57–80.

Gana, A. (2017) Rural and farmers' protest movements in Tunisia and Egypt in the era of Arab revolts. In: Corrado, A., de Castro, C. and Perrotta D. (eds) *Migration and Agriculture: Mobility and Change in the Mediterranean Area*. Routledge, London, pp. 261–276.

Government of Egypt (GoE) (2009) *Agricultural Sustainable Development Strategy 2030*. Ministry of Agriculture and Land Reclamation, Cairo.

Guerfali, R. (2015) Entrepreneuriat social: la réussite remarquable de Jemna malgré les carences de la loi! Available at: http://nawaat.org/portail/2015/07/11/entrepreneuriat-social-la-reussite-remarquable-de-jemna-malgre-les-carences-de-la-loi (accessed 10 April 2017).

Hamouchene, H. (2017) Jemna in Tunisia: an inspiring land struggle in North Africa. *Open Democracy*, 13 April. Available at: https://www.opendemocracy.net/arab-awakening/hamza-hamouchene/jemna-in-tunisia-inspiring-land-struggle-in-north-africa (accessed 18 April 2017).

Hassan, K. (2017) Minister of Agriculture: I Presented to the prime minister the plan to secure food to egyptians and the mechanisms of delivering wheat. 18 February. Available at: http://gate.ahram.org.eg/News/1393034.aspx (accessed 20 February 2017).

Human Rights Watch (2005) *Egypt: Attacks by Security Forces in Sarando*. HRW, New York.

Human Rights Watch (2017) *Egypt. Events of 2016*. Available at: https://www.hrw.org/world-report/2017/country-chapters/egypt (accessed 7 June 2017).

Jouili, M. (2009) Tunisian agriculture: are small farms doomed to disappear? 111 EAAE-IAAE Seminar 'Small farms: decline or persistence', University of Kent, Canterbury, 26–27 June. Available at: http://ageconsearch.tind.io/record/52816/files/051.pdf (accessed 20 April 2017).

Kapitalis (2017) Plan de développement de l'agriculture et de la pêche 2016–2020. 9 February. Available at: http://kapitalis.com/tunisie/2017/02/09/plan-de-developpement-de-lagriculture-et-de-la-peche-2016-2020 (accessed 5 April 2017).

Korany, B. and El-Mahdi, R. (eds) (2012) *Arab Spring in Egypt. Revolution and Beyond*. The American University in Cairo Press, Cairo.

Land Centre for Human Rights, Cairo (LCHR) (2002) Farmer struggles against Law 96 of 1992. In Bush, R. (ed.) *Counter Revolution in Egypt's Countryside*. Zed Books, London, pp. 126–138.

Land Centre for Human Rights, Cairo (LCHR) (2014) Conditions of the farmers. Available at: http://www.lchr-eg.org/land-and-farmer/212-عام-ال-2014.فلاح ين-أو ضاع-ف ى - سري عقـة راءة.html (accessed 12 March 2017).

Mada Masr (2015) Sisi inaugurates 1st phase of 1.5 million feddan reclamation project. 31 December. Available at: www.madamasr.com/en/2015/12/31/news/u/sisi-inaugurates-1st-phase-of-1-5-million-feddan-reclamation-project (accessed 29 April 2017).

Marfleet, P. (2016) *Egypt. Contested Revolution*. Pluto Press, London.

McMichael, P. (2013) *Food Regimes and Agrarian Questions*. Fernwood Publishing, Halifax; Winnipeg.

Mestiri, M. (2016) Disparités régionales, etat des lieux d'une discrimination. Available at: http://nawaat.org/portail/2016/02/09/disparites-regionales-etat-des-lieux-dune-discrimination (accessed 15 April 2017).

Middle East Eye (2016) Sisi fires Egypt's top auditor known for anti-corruption drives. 29 March. Available at: www.middleeasteye.net/news/sisi-fires-egypts-key-auditor-known-his-anti-corruption-drives-321612060 (accessed 11 June 2017).

Mitwally, S. (2017) Agriculture and Irrigation (Ministers): weekly joint meetings to resolve the issues of the two ministries. 20 February. Available at: http://today.almasryalyoum.com/article2.aspx?ArticleID=535788&IssueID=4243 (accessed 25 February 2017).

Moore, J. (2015) *Capitalism in the Web of Life. Ecology and the Accumulation of Capital*. Verso, London.

Mukhtar, H. (2016) The Egyptian Countryside Company: we shall not leave the small peasants and we shall teach them how to plant the desert. Available at: www.youm7.com/story/2016/10/18/شركة-الـ ريـ ف-الـ م صرى-ن تركـلـ صغـارد- ين- المـزارع -لمهم سـ نـعـ ف يـقـو بـ زراعـتـكـ/2927331 (accessed 20 January 2017).

Ramadan, B. (2016) The head of the Egyptian Countryside Company: facilitations to youth and small peasants in the One Million Acres Project. 25 November. Available at: www.almasryalyoum.com/news/details/1047079 (accessed 20 April 2017).

Richards, A. (1986) Introduction. In: Richards, A. (ed.) *Food, States, and Peasants. Analyses of the Agrarian Question in the Middle East*. Westview, Boulder, Colorado.

Rohan, B. (2017) Mukhabarat hires Washington lobbyists to boost image. Available at: https://apnews.com/d8d55dbbcedb4e589d33555cc5fa8855/egypts-general-intelligence-registers-washington-lobbyist (accessed 7 March 2018).

Saad, R. (2002) Egyptian politics and the tenancy law. In Bush, R. (ed.) *Counter Revolution in Egypt's Countryside*. Zed Books, London, pp. 103–125.

Schilling, D. (2013) Egypt's $90 billion south valley project. *Industry Tap*, 10 January. Available at: www.industrytap.com/egypts-90-billion-south-valley-project/539 (accessed 20 April 2017).

Shenker, J. (2015) The corridors of counter revolution. Sharm el Sheikh and the international elite. Available at: www.jackshenker.net/egypt/the-corridors-of-counter-revolution-sharm-el-sheikh-and-the.html (accessed 20 April 2017).

Shenker, J. (2016) *The Egyptians: a Radical Story*. Allen Lane, London.

Szakal, V. (2016a) In Jemna, a social experiment against state policies. 27 September. Available at: http://nawaat.org/portail/2016/09/27/in-jemna-a-social-experiment-against-state-policies (accessed 7 March 2018).

Szakal, V. (2016b) Migration interne, marché de l'emploi et disparités régionales. 9 March. Available at: http://nawaat.org/portail/2016/03/09/migration-interne-marche-de-lemploi-et-disparites-regionales (accessed 7 March 2018).

Touhami, H. (2012) Seuil de pauvrete, population pauvre. Conference on 7 March 2012, la Faculté des Sciences Economiques de Tunis à l'invitation du Club des Econometres Tunisiens Economiques de Tunis. Available at: www.leaders.com.tn/uploads/FCK_files/file/SEUIL%20DE%20PAUVRETE-VDF-Leaders.pdf (accessed 19 June 2017).

Van der Ploeg, J.D. (2009) *The New Peasantries. Struggles For Autonomy and Sustainability in an Era of Empire and Globalization.* Earthscan, London.

Van der Ploeg, J.D. (2013) *Peasants and The Art of Farming. A Chayanovian Manifesto.* Fernwood Publishing, Winnipeg.

Williams, D. (2005) In Egypt's countryside, farmers' anger seen as 'silent time bomb'. *Washington Post*, 17 July.

World Bank (1986) *Poverty and Hunger: Issues and Options for Food Security in Developing Countries.* World Bank, Washington, DC.

World Bank (2006) *Tunisia: Agriculture Policy Review.* Report No. 35239-TN, World Bank, Washington, DC. Available at: http://documents.worldbank.org/curated/en/242951468114530031/pdf/352390TN.pdf (accessed 28 April 2017).

World Bank (2012) Interim strategy note for the Republic of Tunisia for the period FY13-14. Available at: http://documents.worldbank.org/curated/en/786001468173647922/Tunisia-Interim-strategy-note-for-the-period-FY13-14 (accessed 24 April 2017).

World Bank (2014) *The Unfinished Revolution. Bringing Opportunity, Good Jobs and Greater Wealth to All Tunisians.* Development Policy Review, report number 86179-TN. World Bank, Washington, DC.

World Bank (2017) Agriculture, value added (% of GDP). *DataBank: World Development Indicators*. Available at: http://databank.worldbank.org/data/reports.aspx?source=world-development-indicators (accessed 3 July 2017).

8 Degraded Capital Formation: the Achilles' Heel of Syria's Agriculture

Linda Matar*
National University of Singapore, Singapore

Introduction

During most – if not all – of its developmental trajectory, pre-conflict Syria relied heavily on its agricultural sector, characterized in the literature as being both large and productive (Metral, 1984; Hinnebusch, 1989; FAO, 2003). In 1981, in his article on Arab economies in the 1970s, Roger Owen described the agricultural sector as the Achilles' heel of most Arab countries, advising them to follow in the footsteps of Syria and Algeria and invest in the agricultural sectors that employed nearly half of their population (Owen, 1981, p. 9). Syria, of all major Arab states, has traditionally invested heavily in its agricultural sector and simultaneously prevented losses in valuable agricultural land to residential and commercial construction. For years, the agricultural sector was dependent on the government's guidance and support, a legacy inherited from the Ba'athist state-interventionist policies. The Ba'athist government (1963–1970) intensified land reform and other interventionist measures as part of its broader dirigiste strategies that aimed at installing centralized planning.[1] In the agricultural sector, it controlled the production, marketing and delivery of strategic crops that are essential for food security and export earnings. It also exercised centralized planning in cultivation, state-set procurement prices and subsidized inputs of production, all of which underpinned Syria's transition from a food-dependent to a food-sufficient economy in major food commodities (al-Hindi, 2011, p. 36). By 1995 the country had turned into a small wheat net exporter (USDA, 2017). The government also improved the system of agricultural cooperatives and guaranteed stable if not lucrative markets for peasants by not extracting a surplus from them through state marketing (Owen, 1981, p. 9; Hinnebusch, 2011, pp. 11–12).

The outcome of the agrarian reforms of the 1960s remains a matter of controversy. Hinnebusch (2009) described the land reform as one that occurred 'from above', since it was not preceded by an agrarian revolution launched by the peasants. In this regard, Perthes underestimates the economic and political results of this exercise by arguing that it was beneficial to the middle class landowners who exploited the poor peasants. Economically, land reform, according to Perthes, implicated downside effects in terms of stagnant production throughout the 1980s that was unable to meet local demand (Perthes, 1995, p. 45). Nevertheless, Hinnebusch's 1989 study demonstrated that land reform was followed by a temporary loss in production, after which the latter resumed at normal or even higher absolute

---

* Email: linda@nus.edu.sg

levels (Hinnebusch, 1989, pp. 257–259). Still, other scholars like Batatu argued that, despite the limitation of the Ba'athist land reform, the latter was anything but regressive because it enabled the vulnerable peasantry to acquire land through which they enhanced their independence, boosted their potential for initiative and improved their living conditions (Batatu, 1999). Landlords were also encouraged to invest in their reduced holdings (Hinnebusch, 2011, pp. 12–13). With regard to infrastructure, the countryside had witnessed remarkable improvements in transport, communication and electrification services, and the expansion of the educational system due to increased government investment in the agricultural sector – all of which had substantially improved rural living conditions (Perthes, 1995, p. 93).

However, it was not long until Syria, like the other developing countries, was also pushed into the neoliberal trail, gradually shifting the responsibility for resource allocation from the hands of the state to market forces. The neoliberal reforms took some time to stretch out to the agricultural sector. As Kadri states, 'where outright occupation was not the case, trade openness treaties, dislocation-laws dispossessing farmers and macro policies allocating resources away from agriculture uprooted the peasantry en masse' (Kadri, 2012, p. 4).

In the 1990s, the agricultural sector witnessed a gradual transition from a sector relying on heavy-handed state guidance to a market-dependent structure. The private sector intervened in the production and marketing of agricultural products, slowly during the 1990s and more quickly in the 2000s (al-Hindi, 2011). In the absence of a strong and mobilized peasantry that could have defended the benefits gained from the radical agrarian reforms, counter-reforms were imposed 'from above' just as the initial agrarian reforms were introduced 'from above'.

Furthermore, the Bashar al Assad government, especially in the late 2000s and, as advised by the International Financial Institutions (IFIs), deepened the austerity measures towards agriculture for the sake of ensuring fiscal consolidation and macroeconomic balancing. The intensified lifting of government support after 2005 was accompanied by severe climate conditions (a decade of warmer temperature and less rainfall combined with 4 years of extremely dry weather). During 2006–2010, nearly 60% of Syria's land suffered from severe drought that forced a million rural Syrians to move to the cities and pushed about 3 million Syrians into extreme poverty (Matar, 2016, p. 134). Rising rural poverty explains why the initial protests against the government started in the rural areas before spreading to the small towns and suburbs.

This chapter examines agriculture in pre-conflict Syria and discusses how agricultural policies enacted during the Ba'athist (1963–1970) and the pre-conflict Assad (1970–2011) governments had shaped rural development and affected small farmers' livelihoods. It focuses on the investment or capital formation question and follows its trend since the 1960s. I argue that, although Syria's agricultural policies throughout its development trajectory promoted agricultural production, starting in the 1990s the government and the private sector failed to boost the investment rate in the agricultural sector, particularly in infrastructural projects that are needed to strengthen agricultural resistance to shocks. At a time when the whole economy was experiencing a depressive cycle and investment was moving towards low productivity areas in the tertiary sector, the agricultural sector witnessed a systematic decline in the share of agricultural investment out of total investment since 1990. The lack of quality investment in infrastructural projects, especially those that are needed for the delivery of irrigation water and for modernizing the irrigation techniques, weakened Syria's ability to evade the devastating impact of the drought that haunted Syria during 2006–2010. In hindsight, the reduced share and quality of capital formation in agriculture was coupled with the intensification of austerity measures by the Bashar al Assad government, leaving the country vulnerable on the security front once a draught set in. The combined factors identified above also deepened the social rift between town and country, pre-supposing the objective conditions for the onset of the crisis. The more immediate causes of the revolt, however, remain ideological and are related to a crisis of rule or to the perception that the ruling elite can no longer govern (Kadri, 2016).

The chapter starts by giving a brief overview of the agricultural sector that sets the scene for further analysis of Syria's farming sector.

The next section summarizes the radical agricultural reforms that were undertaken during the 1960s, underpinning Syria's transformation to a self-sufficient economy in food production. The third part of the chapter covers the Hafiz and Bashar al Assad governments and their endorsement of gradual agrarian counter-reforms. The following section traces the trend in capital formation in the agricultural sector since the 1960s. Finally, the conclusion sheds light on agricultural conditions under conflict.

Setting the Scene

Syria is an agrarian economy, whose agricultural sector has accounted for 20–25% of gross domestic product (GDP) during most of Syria's development path up until 2007 (Table 8.1). Until the discovery of new crude oil reserves that massively boosted Syrian oil production between the mid-1980s and mid-1990s, the agricultural sector was the backbone of Syria's economy. Estimates show that 25.5% of Syria's land was arable in 2010 (World Bank, 2017). With Syria's positioning in the western Fertile Crescent, 70% of its cultivated area was highly dependent on rainfall and Syria's agricultural production fluctuated accordingly (al-Hindi, 2011). Syria's agriculture consumed nearly 80–90% of the country's total water withdrawals – both surface and groundwater, because of the traditional irrigation methods and the government's non-rationalizing of water use (al-Nahhas, 2011; De Châtel, 2014).[2]

Nearly 9 million people lived in rural areas before the conflict and earned their living from farming and livestock breeding (Zurayk and Gough, 2014, p. 123). In the early 1970s, the agriculture sector absorbed more than 60% of the labour force, as compared to 5% in construction and 12% in services (Lawson, 1989, p. 26). This ratio then dropped in the late 1970s to 33%, while the proportion of those employed in other sectors had risen (Lawson, 1989, p. 26). In 2005, the agricultural sector ranked second (20%) after services (27%) in terms of absorbing the labour force (al-Hindi, 2011, p. 23). In 2007, the total number of those employed in the agricultural sector amounted to 19% of total employment (World Bank, 2017). This decline was the result of agricultural counter-reforms that increased migration from the countryside to the cities, enlarging the reservoir of cheap wage labour.[3] Despite this drop in the agricultural sector workforce, agricultural production remained adequate, contributing no less than 20% of GDP (Central Bureau of Statistics, various issues). This is because of the government's deliberate attempt through its agricultural policies –

Table 8.1. GDP decomposition by economic sectors, percentage share based on constant 2000 prices, various years. (From: Central Bureau of Statistics, Syrian Statistical Abstract, various issues.)

	1990 (%)	1995 (%)	2000 (%)	2005 (%)	2006 (%)	2007 (%)	2008 (%)	2009 (%)	2010 (%)
Agriculture	25	23	25	23	24	21	18	19	16
Mining and manufacturing	26	28	30	25	24	23	23	23	24
Building and construction	3	3	3	3	4	4	4	4	4
Wholesale and retail trade	20	21	15	20	18	21	22	21	20
Transport and communication	10	11	13	11	11	11	12	12	13
Finance and insurance	3	4	4	5	5	5	5	5	5
Social and personal services	2	2	2	3	3	3	4	4	4
Government services	11	8	8	10	11	13	12	13	14

whether protectionist or liberal – to promote agricultural production, resorting to investment in land reclamation and irrigation infrastructure that increased the irrigated areas.[4] The government's long-term support to agriculture from the 1960s up until 2005 – providing loans at concessional rates, tax exemptions on production and subsidies to production inputs (fertilizers, seeds, pesticides) – was also key. Subsidies covered a large proportion of the production costs and promoted the use of better-quality seeds and chemical fertilizers, especially for wheat and cotton (al-Hindi, 2011, p. 28). Furthermore the agricultural sector had been the main employment sector for women. In 2003, the female labour force participation rate was 23% in rural areas and 15% in urban areas (Fafo, 2007, p. 12).

Agriculture had also provided raw materials for the domestic manufacturing and food industries. In the 1960s, the industries that processed agricultural inputs produced nearly 65% of total industrial output (Hinnebusch, 1989, p. 253). Syria managed to ensure food self-sufficiency – especially in food staples as well as animal products – up until 2005 and occasionally exported fruit and vegetables under favourable weather conditions (al-Hindi, 2011, p. 36; Khaddam, 2011, p. 80). It was a net exporter of wheat from 1995 until 2008 (USDA, 2017), after which it started to import wheat following production shortages as a result of the drought (IRIN, 2009) and more recently, the war damages.

By the mid-2000s, agricultural production was carried out by a large number of relatively small farm units amidst the diminishing role of state farms.[5] Production remained under the state's control, as the government planned, regulated, marketed, subsidized inputs, invested, distributed and provided the channels of credit to agriculture (FAO, 2003; 2016b, p. 10).

The following sections present a detailed review of different periods' agricultural policies, along with their implications for rural development and agrarian capital formation.

The Ba'athist Radical Policies Supporting Syria's Journey in Achieving Self-Sufficiency in Food Production

When the Ba'ath party came to power in 1963, it pursued policies of economic nationalism that promoted the national reorientation of economic resources, protected Syria's infant industry through tariff measures, and enhanced the economy's productive capacity through state-led and import-substituting investment. Radical economic reforms were undertaken to set the course for economic independence, leaving an indelible mark on Syria's development, especially in its agricultural sector. Apart from implementing extensive nationalization and taking control of the economy's productive resources, the Ba'athist government ushered in land reforms at a time when 60% of the rural population did not own any land (al-Ahsan, 1984). As a result, the government expropriated nearly one-third of the land and redistributed it to landless peasants (Hopfinger and Boeckler, 1996, p. 184). Among the major laws that it enacted and which shaped rural development were a law that prohibited the displacement of farmers from the land, a law that reduced the ceilings on private landholdings and a cooperative law (Chouman, 2005, The socialist experience in Syria, the consequences of its movement towards the market economy, and the impact of restructuring and globalization, unpublished paper). Cooperatives were also incorporated into the Peasant Union, an independent political organization formed by the Ba'ath to ensure peasant participation in corporatization and acquiescence in the agrarian plans (Hinnebusch, 1989, p. 41).

The government undertook these measures for political and economic reasons. The main objective was to uproot the remnants of the liberal model of the pre-Ba'ath ancien regime that existed in the early 1950s and to reduce the degree of political and financial dependence on the advanced capitalist world during the fragile post-independence period. From a social perspective, the aim was also to build and consolidate the regime's social base of support from its constituency of the middle class, workers and rural peasantry.

The Ba'athist government committed itself to enhancing food production, because ensuring food self-sufficiency was part and parcel of its broader strategy of ensuring national security (Zaim, 2002). The government intervened closely in the production, pricing, import and distribution activities of agricultural products. It decided on crop patterns and rotations – tailored to various regions – and on the amount of credit and quantity of fertilizers and other inputs according to a pre-determined agricultural production plan

(FAO, 2003). The Agricultural Co-operative Bank (ACB), a dispenser of farm loans, was also responsible for distributing locally produced and imported materials, especially fertilizers, either directly or indirectly (through cooperatives) to farmers. Because the government was responsible for supplying food to the citizens, it controlled the marketing of agricultural products of state farms, thereby excluding the middlemen from such activities and more importantly, from surplus extraction and appropriation (Hinnebusch, 2011, p. 11). The state-set pricing policy for agricultural products facilitated the planning of agricultural production. Both the production cost and the degree to which the government considered a crop important and aimed to encourage production determined the profit margin on crops (al-Hindi, 2011, p. 27).

The government enhanced mechanization, improved the irrigation and drainage systems, and introduced prolific strains of seeds. It pursued animal husbandry, and in some arid regions improved land irrigation and reclamation. It also set up state farms, cooperatives to sell inputs such as fertilizers to peasants and public industrial enterprises to promote agro-business industrial activities. In the Jazira region, the government relied on modern farming techniques for cultivation (Hopfinger and Boeckler, 1996).

In terms of financial support, the Ba'athist strategy provided farmers with subsidized credit through the ACB, subsidized inputs of production, extended tax exemptions on agricultural investments and agricultural warehouses, and purchased strategic products at prices that were occasionally higher than market levels (The Syria Report, 2 June 2008; al-Hindi, 2011, p. 27). The latter measure encouraged production and ensured that local farmers earned a minimum guaranteed income. Through its multiple exchange rate system, the government subsidized the cost of importing inputs that were not produced locally (al-Hindi, 2011, p. 29). Additionally, farmers, unlike other exporters, were allowed to retain 100% of their export revenues (FAO, 2003).

Against this backdrop, the above-mentioned interventionist measures that the Ba'athist government introduced and the early phase of the Hafiz al Assad rule endorsed, not only diversified but also ensured an increase in agricultural production from the mid-1970s throughout the 1980s (Hinnebusch, 1989, p. 269–270; FAO, 2003; Hinnebusch, 2011, p. 13). From a social perspective, these reforms changed the distribution of the active working population among the different social classes between 1960 and 1970. As detailed in the quantitative assessment of the class structure presented by Longuenesse, the number of small landowning peasants increased by 150%, rising from 243,460 in 1960 to 608,540 in 1970 (Longuenesse, 1979, p. 4). This was accompanied by a drop in the number of peasants working in the agricultural sector, known as the agrarian proletariat, whose share of the total active population dropped from 20.5% in 1960 to 8.9% in 1970 (Longuenesse, 1979, p. 4). To sum up this section, radical agrarian policies during the Ba'athist phase ensured 'greater equality with greater growth' (Hinnebusch, 2011, p. 14).

The Assad Government (1970–2011): from Protectionism to Neoliberalism

After Hafiz al Assad seized state control in 1970, the government introduced cautious market-driven economic reforms in the 1970s and 1980s. Hafiz was known for his pragmatic steps, such as the 'Corrective Movement' of November 1970 that aimed at gearing resources to increasing economic production through the contributions of both private and public sectors (Perthes, 1995, pp. 41–42). It was only during Hafiz's days, especially after the 1986 economic crisis, that the private sector – initially hibernating in the preceding Ba'athist state-interventionist period – was allowed to have an increasing role in resource management and allocation. In this regard, mixed-sector stock corporations were launched in the 1980s with 25% public and 75% private ownership to accelerate private sector participation (Polling, 1994). These ventures were set up in the tourism and agricultural sectors. Law No. 10 of 1986 provided the framework for the establishment of joint ventures in the agriculture sector (agro-business). Such enterprises decided on production levels without abiding by the state's central production plans, enjoyed financial privileges such as tax holidays, and imported production equipment tax-free (Polling, 1994; Hopfinger and Boeckler, 1996). However, these agro-business enterprises were of limited success. Not only did they not raise agricultural production, but they also failed to initiate innovative projects with modern techniques (Hopfinger and Boeckler, 1996).

Starting in the 1990s, the liberalization of the agricultural sector gained momentum, shifting the agricultural policies from protectionism to a more liberal orientation. The government's interventionist role changed from direct planning that set quantitative targets to indicative planning that takes into consideration signals from both state and market entities. In hindsight, one can argue that the decision to move to indicative planning – as well as other decisions such as the endorsement of the 'social market economy' paradigm in 2005[6] – were only pretexts that facilitated the government's acceleration of a gamut of neoliberal reforms. Using a bottom-up approach and guided by state-set national targets, agricultural studies were planned in governorates by the relevant agricultural authorities (al-Hindi, 2011, pp. 26–27). By 2004, the planning policy was further loosened, giving farmers the complete freedom to decide on the quantity of production of the strategic crops (lentils, chickpeas, peas, etc.) as long as their total production met the target of the category (legumes) that the central government set. Additionally, the pricing strategy also witnessed a change as the prices of major crops, such as wheat, barley, lentils, chickpeas, cotton and sugar beet, were the only ones to be administratively determined. These administered prices were not compulsory unless the farmers were to sell the strategic crops to state establishments (al-Hindi, 2011, p. 27). After 1996, the government dismissed its policy that had occasionally increased the local prices of agricultural products above international prices (al-Hindi, 2011, p. 27). Between December 2000 and December 2001, the Bashar al Assad government enacted a series of decisions, such as Decision 83 of 2000 that carried out the privatization of state farms in Syria's northeast, marking the end of the 'socialist' measures of the preceding Ba'athist phase (Ababsa, 2011).

During the Assad reigns, the private sector furthered its intervention in production, manufacturing, marketing and the processing of crops (al-Hindi, 2011). After 1990, the government ended the compulsory delivery of agricultural products to state marketing agencies, except for the strategic crops that state enterprises processed (cotton, sugar beet and tobacco). In terms of government support, it was mainly after the government endorsed the 'social market economy' paradigm as its new development strategy in 2005 that the state accelerated its lifting of supportive measures towards agriculture. It reduced credit to farmers, removed subsidies on inputs of production and decreased its investment expenditure. Initially the ACB set low interest rates on loans for agriculture, ranging between 4.5% for the public and cooperative sectors to 5.5% for the private sector. However, these rates were revised in 2007 and rose to 5–8% (al-Hindi, 2011, p. 28). The government also revised its policy for subsidies. While subsidies on fuel and on the maintenance of government irrigation projects remained, those on pesticides, fertilizers and farming machinery were phased out, forcing farmers to cultivate less or abandon their crops altogether. As a result, smallholder farming became untenable and many herders had to sell their livestock following the dwindling of vegetation in pastures and the exhaustion of feed reserves.

In their pursuit of economic liberalization, both Hafiz and Bashar put aside the interests of the farmers. The contractionary fiscal policy that springs from the neoliberal framework and its associated price-based resource allocation mechanism had negative consequences for Syria's periphery[7] because it misallocated resources away from the productive sectors (Seifan, 2012, pp. 115–116). By failing to install an interventionist investment policy that would target the productive sectors, the government through tax incentives (tax holidays) encouraged investors to stay away from long-term/low-return investments, especially in agriculture. The path of capital formation was distorted as private investors favoured short-term gains through speculative activity. The result of accelerated liberalization in Syria's agricultural sector, especially during the Bashar period, was beneficial to the new commercial bourgeois class that emerged during the al Assad rule. In the case of agriculture, the main beneficiaries of the privatization of state farms, according to Ababsa, '[were] not the traditional rural constituents of the Ba'ath party, but a re-emergent class of latifundists tied to the central state' that strengthened the existing power structures (Ababsa, 2005, p. 13). In the words of Zurayk and Gough (2014): 'the "modernisation" and liberalisation of farming and the lack of investment in family farming has led to a reduction in the resilience of the food and farming systems in Syria' (Zurayk and Gough, 2014, pp. 124–125). But what was effectively detrimental to the resilience of the farming system was the deficiency in both private and public quality investment, mainly in modernized irrigation systems and other infrastructure needed to mitigate water shortages.

Degraded Capital Formation in Syria's Agricultural Sector

Expansive state-led development that relied on substantial state-directed and import-substituting investment to enhance the economy's industrial nucleus characterized Syrian dirigisme. State-led investment was first kick-started under the United Arab Republic (UAR), after which the Ba'athist government accelerated it. Both governments gave priority to the agricultural sector. While most Middle Eastern states allocated a small proportion of their budgets (less than 10% of total expenditure) to agriculture during 1976–1981 (Owen, 1981), Syria reserved 51% and 24% of its total public investment to agriculture in its first and second Five-Year Plans of 1960–1965 and 1966–1970, respectively (Hinnebusch, 1989, p. 254). The government maintained its role in agriculture, allocating a proportion of not less than 23% of total public investment throughout the 1970s (Hinnebusch, 1989, pp. 255–256).

This policy of allocating a considerable amount of budget funds to agriculture continued during the early phase of Hafiz al Assad's rule. In the early 1970s, the government prepared an agricultural intensification plan to intensify agricultural production in order to be able to meet rising demand and reduce imports, mitigating international economic and political pressures. The actual public investment in agriculture in absolute value increased significantly throughout the 1960s and 1970s (Hinnebusch, 1989, p. 259). Apart from investing in tractors and processing factories, the Ba'athist and the Hafiz governments focused on infrastructure projects, such as hydraulic projects and land reclamation and irrigation (the Euphrates project), the setting up of electricity services in rural areas, and the import of turnkey industrial plants that were needed for import-substitution (Picard, 1988, p. 139; Hinnebusch, 1989, pp. 253–257; Hopfinger and Boeckler, 1996, p. 185). These investments were financed by the considerable inflow of foreign capital, comprising financial assistance from the former Soviet Union and funding by the Gulf States after the 1973 War (Hopfinger and Boeckler, 1996, p. 185). The expanded investment in irrigation and reclamation managed to consolidate and expand the irrigated sector and to transform some desolated areas into viable peasant communities, but failed to cut off Syria's dependency on the unpredictable rainfall (Hinnebusch, 2011, p. 12).

In hindsight, Syria witnessed a temporary decline in agricultural output during the implementation phase of the investment projects in the 1960s. Nevertheless, the sluggish outcome was reversed after the industrial and irrigation projects started operation in the early 1970s. In the words of Hinnebusch (1989, p. 256): 'agriculture seemed on the road to recovery [in the 1970s]'. The bulk of industrial expansion during the 1970s was rural-linked with agricultural industries contributing nearly 60% of the value of total industrial output (Hinnebusch, 1989, p. 253). Agricultural production then resumed its stagnation in the early 1980s due to bad weather conditions, but the sector managed to overcome this decline in the late 1980s, indicating its strength and adequate level of diversification (Hinnebusch, 1989, pp. 258–259).

The results of the intensification effort were also quite significant in terms of modern and extensive farming methods, multi-cropping, boosted harvests, a flourishing agricultural industry and the strengthening of livestock industry (Hinnebusch, 1989, p. 258; Hopfinger and Boeckler, 1996, p. 184). Against this backdrop, the rate of growth of agricultural and livestock production registered 6.3% annually during the 1980s (Hinnebusch, 1989, p. 258). Despite its occasional wastefulness and inefficiency, state planning increased capital formation, boosted agricultural production and productivity and developed the agricultural sector (Hinnebusch, 2011, p. 11).

Since the mid-1980s, the Hafiz government started to introduce tight measures on agriculture, pressured by the drying up of external funding from the Gulf States. As a result, the share of agricultural investment out of total gross fixed capital formation (GFCF) dropped from a high of 27% in 1970 to 6% in the early 1980s before recovering to 14% in the late 1980s (Hinnebusch, 1989, pp. 259–260; Central Bureau of Statistics, various issues). Although the share exceptionally increased to 23% in 1990, it did not surpass the rate witnessed in the early 1970s. Amidst market-driven economic reforms throughout the 1990s, the government deliberately cut down on its public investment in agriculture (Seifan, 2012, pp. 115–116), especially in modern irrigation projects and other infrastructural projects needed to sustain Syria's food security and sovereignty during critical times of unfavourable climate conditions. Figure 8.1 shows that agricultural investment as a percentage of total GFCF witnessed a decelerating trend in the period between 1990 and 2009. As a result the share of agricultural investment out of total GFCF dropped from 23% in 1990 to 16% in 2000, then further to 8% in 2006 (Central Bureau of Statistics, various issues). In addition, the average ratio of agricultural investment to GDP dropped from 3% in the 1990s to 2% in 2000s (Central Bureau of Statistics, various issues). Concurrently, the share of investment in manufacturing also witnessed a systematic decline, leading to a drop in manufacturing's contribution to total value-added output from 6% in 1995 to 2% in 2000 (UNIDO, 2014). This indicates that Syria's economic growth after 1990 was mainly driven by oil and short-term speculative types of investments that were conducted in the tertiary sector.

Whatever investment the Assad governments had provided to agriculture in terms of agricultural services, irrigation and livestock breeding following economic liberalization, they were not enough to enhance the resilience of the farming sector against any potential technical or environmental threat. For instance, these governments did little in terms of rehabilitating the existing government projects, modernizing the irrigation systems, increasing the use of modern irrigation techniques (drip, sprinkler, localized), and increasing the means of water management and preservation throughout the country. The areas irrigated by modern techniques amounted to only 18.5% of the total irrigated area throughout the 1990s and up until 2005 (al-Nahhas, 2011, p. 27). In comparison, this rate reached 60% in Jordan (Khaddam, 2011, p. 70).

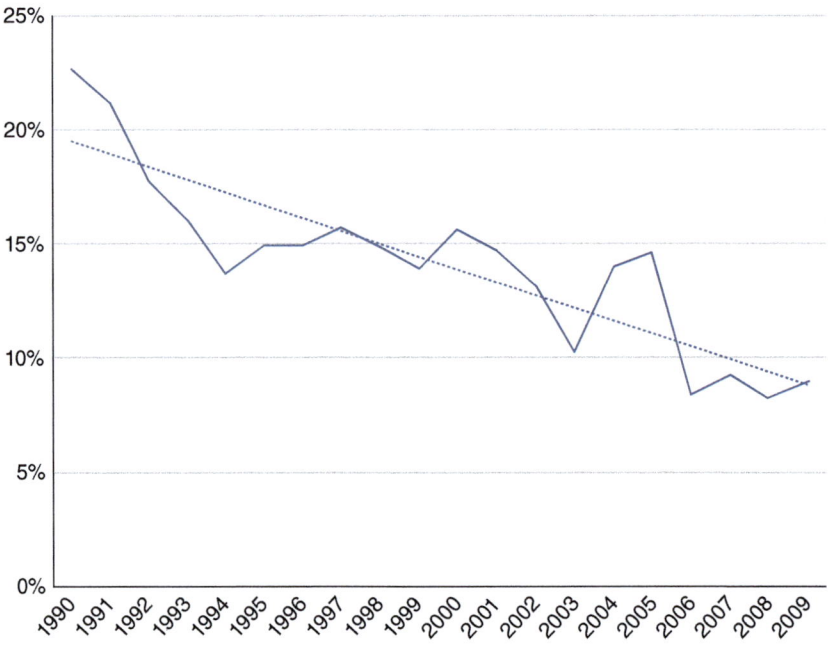

Fig. 8.1. Investment in agriculture, forestry, and fisheries as percentage of total GFCF, 1990–2009.[8] (From: Central Bureau of Statistics, *Syrian Statistical Abstract*, various issues.)

The Assad governments opted for liberalizing the agricultural sector to give more room for the private sector to intervene and promote investment in the productive sectors. However, this failed to materialize. When state-led investment tapered off following economic liberalization during the 1990s and 2000s, the private sector could not fill the gap and boost the share of agricultural investment out of GFCF, because private investors opted to invest in short-term and fast-earning ventures in the tertiary sector.

I have argued elsewhere that investment liberalization as manifested in the prominent Law No. 10 of 1991 and its later amendment Decree No. 7 of 2000 – which were major reform measures that underpinned Syria's transformation from a state-led to market-led economy – failed to enhance the economy's productive capacity (Matar, 2016). At a later stage, the Bashar government, instead of addressing the downsides of these laws (in terms of failing to prioritize the agricultural and the industrial sectors), went ahead and intensified liberalization (land ownership and the surpassing of ownership ceilings of land) through the ratification of Legislative Decree No. 8 of 2007. As a result, LD No. 8, like its predecessor Law No. 10, failed to boost investment in the productive sectors (Matar, 2016). The Syrian Investment Agency's *Annual Investment Report* published data on the *licensed* and *executed* investment projects – their total numbers and capital costs – that were approved and implemented by Law No. 10 and LD No. 8. Data show out of 176 *licensed* private sector projects in the agricultural sector, only 71 (40%) were executed during the period 1991–2007. These projects carried a total capital cost of SYP 11.5 billion (US$225 million) and created 3690 new jobs (SIA, 2007, p. 52). The majority of the 176 licensed projects (81%) were in livestock breeding and fattening, while the remaining (19%) were in agricultural production, services and irrigation (SIA, 2007, p. 52). These dismally insignificant numbers – over a period of 17 years – serve as evidence for my contention that private sector-led investment failed to boost investment in modern infrastructural projects.

In summary, the long-term depletion of investment projects in infrastructure and in modern irrigation systems needed to deliver water efficiently and to ensure the conservation of water resources coupled with the decrease in government support that was curtailed even more sharply after 2005 worsened the situation for Syria's small peasantry. When the drought hit in 2006, the farming sector was not prepared to face it, adding pressure to the hard conditions in rural areas and the rest is history. According to De Châtel, it was mainly the 'long-term mismanagement of natural resources' rather than the drought itself that 'led to the growing disenfranchisement and discontent in Syria's rural communities' (De Châtel, 2014, p. 532). However, one needs to note that this mismanagement has been part of the neoliberal policies that failed to address the fault lines in agriculture. Back in 2000, Bush warned that 'the strategy for agricultural reform pursued by the IFIs … will not lead to sustained higher rates of agricultural output and greater food security and neither will it address environmental issues like water logging, salination and soil erosion. Instead, it will lead to greater rural inequality and the possibility of social unrest, as policy makers refuse to talk with the people that are most affected by the economic reforms – the landless, near landless and tenants' (Bush, 2000, p. 235).

Concluding Remarks: Agriculture Under Conflict

There is no doubt that Syria's food security has been undermined over the past 6 years of conflict. According to the Food and Agriculture Organization of the United Nations (FAO): 'Syria's food chain is disintegrating – from production to markets – and entire livelihood systems are collapsing' (FAO, 2014, p. 1). Currently, an estimated 6.7 million people – including those internally displaced – are residing in rural areas and are relying directly or indirectly on the farming sector as a source of living (FAO, 2017, p. 1). Nevertheless, increasing production and marketing costs have dampened the net incomes of most farmers (FAO, 2017).

The agricultural sector has functioned poorly within a context of war-induced destruction, internationally imposed sanctions, and shortages of supply and pillage, making it difficult for farmers to feed a war-torn economy and maintain their own livelihoods. The country's fragmented economy along with the destruction of irrigation systems, mills and farming equipment have made matters worse, hampering agricultural production. The loss of government control over roads and territory has obstructed the transport and marketing of agricultural products from production to core markets. More importantly, economic sanctions, difficulties in transportation and the devaluation of the Syrian pound have induced shortages in primary inputs of production (fertilizers, seeds and pesticides), whose prices increased, imposing heavy losses on farmers. Of course, supply bottlenecks also raise the prices of local food production.

Despite the above, the agricultural sector contributed 26.8% and 25.4% to Syrian GDP in 2013 and 2014, respectively (SCPR, 2016, p. 24), but like the rest of economy it has recently suffered from severe losses.[9] The Syrian farming sector is fast losing its capacity to cope with the crisis, further threatening the country's food availability. As agricultural output contracts and food prices skyrocket following the lifting of government subsidies on bread, rice, sugar and water in mid-2014, Syrians can hardly afford their essential basket of commodities (IMF, 2016). According to FAO estimates, nearly 7 million people can be currently classified as food insecure (FAO, 2016a). As pointed out by Zurayk: 'The war in Syria is a calamity for the Syrian people, for the poor, for the farmers, and for every person who feels concerned by the fate of a fellow human being' (Zurayk, 2013, p. 9).

In 2016, FAO proposed a contingency plan for Syria, known as the 'Plan of Action (PoA) for Syria 2016–2017' with which FAO assists the Syrian agricultural sector to rebuild resilience, enhance its institutions and ecosystems, strengthen its capacity to adapt and recover from the conflict as well as mitigate any potential shock that can affect its food security (FAO, 2016b). FAO states that its 'resilience-based approach' combines both emergency and development efforts, and can make Syrian crop production less dependent on water by enabling farmers to adopt new techniques that can restore 'soil organic matter' (FAO, 2016b, p. 48). I argue that apart from its short-term 'emergency life-saving' interventions, some of its other proposed measures, such as promoting 'climate-smart agriculture technologies' and 'value-chain development' and 'risk-proof infrastructure and plantation' (FAO, 2016b, pp. 5 and 21) are not only overambitious but delusional, given Syria's critical security condition. With the evolving humanitarian disaster, widespread poverty, food and medicine shortages, displacement of families and homelessness, to a scale unforeseen in this century so far, the priority is on mobilizing all international forces to bring about peace and to strengthen Syria's resilience to survive the conflict. It is definitely misleading to earmark funds (regardless of whether they are internal or external) on ambitious organic and environmentally safe projects that are not even achieved in the US in times of peace.

Unfortunately, the FAO also has failed to see that the agricultural policies for Syria under conflict follow from or are subsidiary to overall macroeconomic policies. New scenarios have been at play during the war years that require attention because the way matters are handled at present may mitigate or prolong the conflict. For instance, the emergence of the war economy has enabled war-traders in the government-controlled and the 'opposition'-controlled areas to control food supplies as means of warring and simultaneously making profit. A macroeconomic strategy that curbs the growing control of warlords over food supply is necessary. Even though the war costs have depleted the civilian economy's resources, the strategy should also allocate resources to priority sectors. At any rate, the agricultural sector and its food-related security are part of national security.

However, as sound as the above strategy may seem, it cannot be made operational without an international consensus aiming at forestalling the resources that feed the war economy. At the current stage of the conflict such a consensus appears remote and farfetched. Even if we restrict the macroeconomic strategy to the areas falling under the purview of the Syrian government, the latter would not be able to carry it out, given the scarcity of resources and the crippling internationally imposed sanctions. Unless the major international players push for a political settlement that ends the war by restricting the channels that feed the war, things are set to remain the same. Judging by the growing depth of international cleavages over

Syria, especially the Sino-Russian and American rift, it is likely that more of the warlording by means of controlling food supplies may leave the Syrian population more abject than they are at present.

Acknowledgements

I am grateful to Max Ajl for providing valuable comments on this chapter; all errors, however, are my own.

Notes

[1] Land reform, which took nearly two decades to be completed, was initially introduced under the UAR (1958–1961), briefly reversed by the separatist government before it was intensively accelerated by the radical Ba'athist military officers after their coup d'état in March 1963. The project is marked as the centrepiece of the Ba'athist 'socialist' radicalization measures with land redistribution ending in 1970 while cooperatization fully completed by the end of the 1970s (Hinnebusch, 2011, p. 8).

[2] More than 80% of Syria's irrigated areas are irrigated through traditional methods. When water is transferred from the source to the fields, total water loss, either through leakage or evaporation, is estimated to range between 30 and 50% as a result of using the open government irrigation canals (al-Nahhas, 2011).

[3] Barout (2011) estimated that the total number of people who migrated from rural to urban cities between 2002 and 2008 was 600,000 people, but Shadid (2011) stated that this number reached 1.5 million by 2011.

[4] The country's irrigated area increased from 651,000 ha in 1985 to 1.35 million ha in 2010 (De Châtel, 2014, p. 529).

[5] State farms were created in the 1960s as a result of the radical 'socialist' measures that were promulgated then. In the words of Hinnebusch (1989, p. 20): 'the leading Ba'thi ideologues initially went so far as to advocate collective farms as the ultimate form of socialist ownership: they were seen as the solution to land fragmentation, as instruments of state planning and resource mobilisation in the agrarian sector'. Ababsa examines the failure of the state farms in the Euphrates Valley – in terms of corruption and decline in agricultural production – and the agrarian counter-reform that followed in that area during 2000s (Ababsa, 2005; 2011).

[6] In 2005, the Syrian government endorsed the 'social market economy' paradigm that advocates the engagement of both market forces and state intervention in the process of economic development, whereby the private sector takes control of economic activities and the state ensures that social and welfare benefits are delivered to the citizens. However, as argued by Abboud (2015), this paradigm was anything but social. It was the veneer behind which the state intensively pursued market liberalization without ensuring social safety nets for its citizens.

[7] FAO's 2003 survey reveals that the government policies (in terms of access to credit and subsidized inputs) during the 1990s and 2000s intensified rural stratification, favouring the larger and more capitalized farmers at the expense of small farmers (FAO, 2003, pp. 303–304).

[8] Investment in agriculture is carried out through agricultural cooperatives. Although agricultural cooperatives are state-led and state-financed institutions that follow the state edict in its choice of investment, these cooperatives are termed autonomous.

[9] The FAO's 2017 report estimated the total loss in terms of destroyed assets and infrastructure and depletion in production in the agricultural sector to amount to roughly US$16 billion over the past 6 years. It also estimated that the rehabilitation of the sector will cost between US$11 to 17 billion (FAO, 2017, p. 1).

References

Ababsa, M. (2005) Privatisation in Syria: State Farms and the Case of the Euphrates Project. EIU Working Paper No. 2005/02.

Ababsa, M. (2011) Agrarian counter-reform in Syria (2000–2010). In: Hinnebusch, R. *et al.* (eds) *Agriculture and Reform in Syria*. Lynne Rienner Publishers, Inc., Boulder, Colorado, pp. 83–107.

Abboud, S. (2015) Locating the 'social' in the social market economy. In: Hinnebusch, R. and Zintl, T. (eds) *Syria from Reform to Revolt: Volume 1: Political Economy and International Relations*. Syracuse University Press, Syracuse, New York, pp. 45–66.

al-Ahsan, S.A. (1984) Economic policy and class structure in Syria: 1958–1980. *International Journal of Middle East Studies* 16(3), 301–323.

al-Hindi, A. (2011) Syria's Agricultural sector: situation, role, challenges and prospects. In: Hinnebsuch, R. *et al*. (eds) *Agriculture and Reform in Syria*. Lynne Rienner Publishers, Inc., Boulder, Colorado, pp. 15–55.

al-Nahhas, A. (2011) Modern irrigation in Syria. *Damascus University Journal* 27(2), 23–42 [in Arabic].

Barout, M.J. (2011) The last decade in Syrian history [Al-'Aqd al-Akhir fi Tarikh Suriyah: Jadalliyat al-Jumud wa-al-Islah (1–4)]. Arab Centre for Research and Policy Studies, April, 2011. Arab Centre for Research and Policy Studies, Doha.

Batatu, H. (1999) *Syria's Peasantry, the Descendants of its Rural Notables, and their Politics*. Princeton University Press, Princeton, New Jersey.

Bush, R. (2000) An agricultural strategy without farmers: Egypt's countryside in the new millennium. *Review of African Political Economy* 27(84), 235–249.

Central Bureau of Statistics, *Syrian Statistical Abstract*. Various issues. Central Bureau of Statistics, Damascus.

De Châtel, F. (2014) The role of drought and climate change in the Syrian uprising: untangling the triggers of the revolution. *Middle Eastern Studies* 50(4), 521–535.

Fafo (2007) The Syrian labour market: findings from the 2003 unemployment survey. Available at: http://almashriq.hiof.no/general/300/320/327/fafo/reports/20002.pdf (accessed 15 May 2017).

Food and Agriculture Organization of the United Nations (FAO) (2003) *Syrian Agriculture at the Crossroads*. Agricultural Policy and Economic Development Series No. 8. FAO, Rome.

Food and Agriculture Organization of the United Nations (FAO) (2014) Sub-regional strategy and action plan: resilient livelihoods for agriculture and food and nutrition security in areas affected by the Syria crisis. Summary. Food and Agriculture Organization (FAO) – Regional Office for the Near East and North Africa. Available at: www.fao.org/documents/card/en/c/19477cf3-886a-4423-82b8-3ecea165ebb8 (accessed 9 June 2017).

Food and Agriculture Organization of the United Nations (FAO) (2016a) Food Production in Syria at all-time low. (15 November 2016). Available at: www.fao.org/emergencles/fao-In-action/stories/stories-detail/en/c/453428 (accessed 10 June 2017).

Food and Agriculture Organization of the United Nations (FAO) (2016b) *Plan of Action for Syria: towards Resilient Livelihoods for Sustainable Agriculture, Food Security and Nutrition, 2016–2017*. FAO, Rome.

Food and Agriculture Organization of the United Nations (FAO) (2017) *Counting the Cost: Agriculture in Syria after Six years of Crisis*. FAO, Rome.

Hinnebusch, R. (1989) *Peasant and Bureaucracy in Ba'thist Syria: The Political Economy of Rural Development*. Westview Press, Inc., Boulder, Colorado.

Hinnebusch, R. (2009) Syria under the Ba'th: the political economy of populist authoritarianism. In: Hinnebusch, R. and Schmidt, S. (eds) *The State and the Political Economy of Reform in Syria*, St Andrews Papers on Contemporary Syria. Lynne Rienner Publishers, Inc., Boulder, Colorado, pp. 5–24.

Hinnebusch, R. (2011) The Ba'th's agrarian revolution (1963–2000). In: Hinnebusch, R. *et al*. (eds) *Agriculture and Reform in Syria*. Lynne Rienner Publishers, Inc., Boulder, Colorado, pp. 3–14.

Hopfinger, H. and Boeckler, M. (1996) Step by step to an open economic system: Syria sets course for liberalisation. *British Journal of Middle Eastern Studies* 23(2), 183–202.

International Monetary Fund (IMF) (2016) Syria's conflict economy. IMF Working Paper, WP/16/123. IMF, Washington, DC.

IRIN (2009) Syria: drought blamed for food scarcity. 22 February 2009. Available at: www.irinnews.org/printreport.aspx?reportid=83069 (accessed 18 December 2012).

Kadri, A. (2012) Proletarianisation under neoliberalism in the Arab world, published in Athimar publications (18 January 2012). Available at: www.athimar.org/Article-23#sdfootnote16sym (accessed 15 May 2017).

Kadri, A. (2016) *The Unmaking of Arab Socialism*. Anthem Press, London.

Khaddam, M. (2011) Syrian agriculture between reality and potential. In: Hinnebsuch, R. *et al*. (eds) *Agriculture and Reform in Syria*. Lynne Rienner Publishers, Inc., Boulder, Colorado, pp. 57–82.

Lawson, F. (1989) History of liberalization in Syria. In: Berch, B. (ed.) *Power and Stability in the Middle East*. Zed Books, London.

Longuenesse, E. (1979) The class nature of the state in Syria: contribution to an analysis. *MERIP Reports* 77, 3–11.

Matar, L. (2016) *The Political Economy of Investment in Syria*. Palgrave Macmillan, Basingstoke.

Metral, F. (1984) State and peasants in Syria: a local view of a government irrigation project. *Peasant Studies* 11(2), 69–89.

Owen, R. (1981) The Arab economies in the 1970s. *MERIP Reports*, 100/101, 3–13.

Perthes, V. (1995) *The Political Economy of Syria Under Asad*. I.B. Tauris, London.
Picard, E. (1988) Arab military in politics: from revolutionary plot to authoritarian state. In: Dawishi, A. and Zartman, I.W. (eds) *Beyond Coercion: The Durability of the Arab State*. Croom Helm, New York, pp. 116–147.
Polling, S. (1994) Investment Law No. 10: which future for the private sector. In: Kienle, E. (ed.) *Contemporary Syria: Liberalisation Between Cold War and Cold Peace*. British Academic Press, London, pp. 14–25.
Seifan, S. (2012) Policies of income distribution and its role in the social explosion. In: *The Background of Revolution: Syrian Studies*. Arab Center for Research and Policy Studies, Doha [in Arabic].
Shadid, A. (2011) Syria's ailing economy poses a threat to Assad. *New York Times* 23 June 2011. Available at: www.nytimes.com/2011/06/24/world/middleeast/24damascus.html (accessed 25 June 2011).
Syrian Centre for Policy Research (SCPR) (2016) Confronting fragmentation: impact of the Syrian crisis report. UNDP Country Office in Syria.
Syrian Investment Agency (SIA) (2007) *The Second Annual Investment Report in Syria for the Year 2007*. Prime Ministry, Damascus.
The Syria Report (2008) Agriculture fund to streamline government's subsidies policy. The Syria Report, 2 June 2008.
United States Department of Agriculture (USDA) (2017) Production, supply and distribution online. Available at: https://apps.fas.usda.gov/psdonline/app/index.html#/app/advQuery (accessed 16 June 2017).
United Nations Industrial Development Organization (UNIDO) (2014) *Industrial Statistics Database, INDSTAT4*. UNIDO, Vienna.
World Bank (2017) *World Development Indicators (WDI)*. World Bank Group, Washington, DC.
Zaim, I. (2002) The program and strategies for economic reforms. *Economic Files*. Mazzeh: The Syrian Economic Society [in Arabic].
Zurayk, R. (2013) Civil war and the devastation of Syria's food system. *Journal of Agriculture, Food Systems, and Community Development* 3(2), 7–9.
Zurayk, R. and Gough, A. (2014) Bread and olive oil: the agrarian roots of the uprising. In: Gerges, F. (ed.) *The New Middle East Protest and Revolution in the Arab World*. Cambridge University Press, Cambridge, pp. 107–135.

9 Crisis and Agricultural Change in the Kurdistan Region of Iraq, 1980s–2010s: an Interdisciplinary Approach

Lina Eklund[1],* and Katharina Lange[2]

[1]*Lund University, Lund, Sweden; Aalborg University, Copenhagen, Denmark;*
[2]*Leibniz-Zentrum Moderner Orient, Berlin, Germany*

Introduction

The Middle East has been confronted with a number of challenges that are severely impacting local food production systems. Rapid population growth, urbanization, armed conflict and changing patterns of food consumption are weakening domestic food supply bases, making the region increasingly food insecure and reliant on food imports (Kamrava *et al.*, 2012). At the same time limitations in natural resource availability, especially water and arable land, are putting biophysical constraints on food production, and aggravating national and regional tensions. Thus, social, political and ecological factors interact in shaping complex and challenging conditions for local food production. To gain a more systematic and complete understanding of this multidimensional field, we argue, it is necessary to use integrated approaches drawing on qualitative and quantitative methods from different disciplines (Stock and Burton, 2011). However, multi-, inter- and transdisciplinary research can be challenging, and integration of different disciplines can be difficult to reach. This chapter explores possibilities and difficulties encountered when using an interdisciplinary, mixed-method approach for better understanding how crises may affect agriculture. Combining satellite image analysis with qualitative interviews, we present a case study from the Duhok province in Kurdistan Region of Iraq. By discussing our preliminary findings, we demonstrate the potential of our integrated approach while also addressing practical challenges when collaborating across disciplines.

Historical background

The Kurdistan Region of Iraq (Fig. 9.1) has experienced fundamental socio-economic, cultural, political and environmental transformations in the past decades. Once a main exporter of agricultural produce, its agrarian production – in concert with the rest of Iraq – declined significantly during the late 1970s. Nevertheless, in the 1980s, a third of all foodstuffs consumed in Iraq was still grown in the Kurdish northern provinces (Fischer-Tahir, 2003, p. 76). In the late 1980s and 1990s, agricultural production declined even further. Today, only a small proportion of produce consumed in the region is still locally grown (UNDP, 2010, p. 34; Eklund *et al.*, 2015, p. 79). The causes for this decline are widely attributed to the region's violent history.

Since Ottoman times, the town of Duhok with its hinterland had been part of Mosul province; it became a province in its own right only

* Email: lina.eklund@cme.lu.se or linae@plan.aau.dk

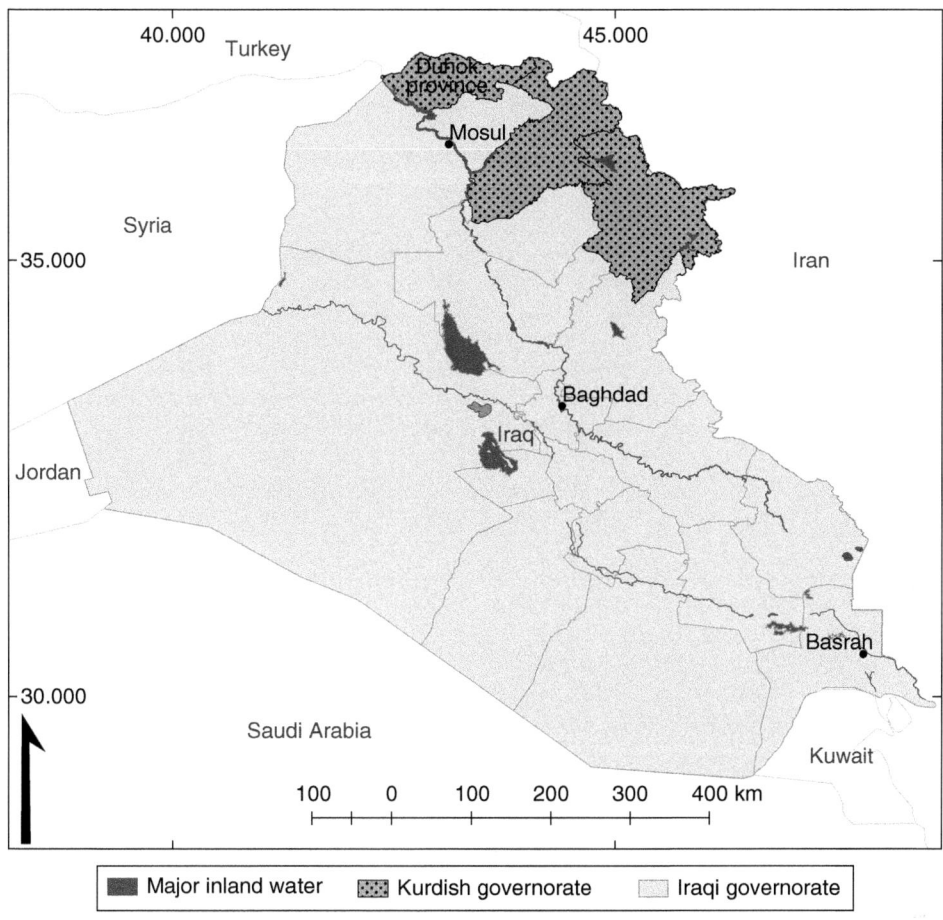

Fig. 9.1. Location of Duhok province, Iraqi Kurdistan.

in 1976. Today, it is one of the three provinces making up the Kurdistan Region of Iraq, besides Hewlêr (Erbil) and Silêmanî (Sulaymaniya). During the Kurdish uprising against Baghdad, which began in 1961 (Ibrahim, 1983, pp. 447–653; McDowall, 2004, pp. 308–313), the region was shaken by extreme military and political conflicts. Loyalties and political alliances in the course of the uprising shifted, in Duhok as in other parts of the Kurdish region, influenced by tribally structured solidarities and rivalries, political opportunities and calculations, as well as social and class-specific aspects (van Bruinessen, 1992, pp. 255–257; Wimmer, 1997; McDowall, 2004, pp. 306, 308—319). The so-called March Manifesto, negotiated between the warring parties in March 1970, raised hopes for peace as well as a relative autonomy of the Kurdish region within Iraq. The following years were characterized by relative stability but growing political tensions, and in 1974 fighting resumed. Following the Algiers Agreement between Iran and Iraq in March 1975, Iran withdrew its support for the Peshmerga (the Kurdish fighters, Peshmerga literally meaning 'those who are in front of death') under the command of Mulla Mustafa Barzani, and the Kurdish uprising was finally suppressed (Ibrahim, 1983, pp. 633–644; McDowall, 2004, pp. 323–342). Especially in rural areas, the civilian population was subjected to extreme repression (HRW, 1993; McDowall, 2004, pp, 343–357). Wide areas along Iraq's borders with Turkey and Iran were declared zones forbidden for habitation, cultivation and construction; villages were destroyed, their inhabitants forcibly resettled in so-called *Mujammaʿāt* (collective towns), or in distant parts of southern Iraq (Farouk-Sluglett and Sluglett, 1990, pp. 187–189; HRW, 1993; Fischer-Tahir, 2003, pp. 80–81). In 1988, repression culminated in the infamous genocidal *Anfal* campaign during which hundreds of thousands were killed, injured, poisoned by chemical weapons, displaced, deported and forcibly resettled in urban conglomerations (Mlodoch, 2015; Casim, 2016). Those who were able to flee sought refuge in Turkey, Iran or, eventually, Europe. The eighth and last operation of this campaign was carried out in August and September 1988 in Duhok province. In the course of just a few weeks, between 65,000 and 80,000 people from this region escaped to Turkey and Iran (Physicians for Human Rights, 1989; Hardi, 2011, pp. 84–86). Following Iraq's invasion of Kuwait, and the subsequent escalation of war between the US-led coalition and Iraq, another 1.5 million Kurds fled to the neighbouring countries in March 1991. After the unilateral establishment of a no-fly-zone by the USA, the UK and France north of the 36th parallel in April 1991, a large-scale return of refugees and reconstruction of Kurdistan began.

This massive and protracted violence had devastating psychological and physical effects (see Hardi, 2011; Mlodoch, 2015). It also led to a fundamental transformation of economic and social structures in the region of Kurdistan, an area that had hitherto been predominantly rural. Rural spaces now had been largely depopulated (Sinemillioglu, 2011, p. 1), agrarian lands were overgrown by grass and mountainous areas deforested (Dziegiel, 1981, p. 14; Mubareka and Ehrlich, 2010). The dramatic loss of human life, accompanied by the repeated destruction and devastation of villages, fields and orchards, as well as widespread experiences of flight, deportation and resettlement have all contributed to a lasting marginalization of rural life and agriculture. Besides the material dimension, these experiences have also led to massive intergenerational changes in the area of agricultural knowledge and skills, as well as symbolic valuations and lifestyle aspirations, away from rural ways of life (Lange, 2016).

After 1991, a number of non-governmental organizations (NGOs) as well as (quasi) state actors sought to rehabilitate village life (see, for instance, Sinemillioglu, 2011; Barwari, 2013), for instance by granting easily accessible loans for (re)construction of rural housing. In the years immediately after *Anfal*, agricultural activity had indeed initially resumed, not least in response to the detrimental supply situation in the region. However, the region suffered from a double embargo after 1991 (the United Nations embargoed Iraq, while the rest of Iraq boycotted the Kurdish region; trade was principally conducted through smuggling; cf. Leezenberg, 2002; 2005, p. 636; 2006, p. 162). This also impacted local agricultural infrastructure as it led to a lack of fertilizer, as well as spare parts for agricultural machinery. At the same time, a satellite study shows that the embargo led to an extensification of agricultural land use as an effort to increase production (Gibson *et al.*, 2012). In the medium term, this embargo enhanced effects of other, older political schemes that distributed revenues from oil production while effectively discouraging a sustained investment in agricultural development. The increased privatization and commercialization of agriculture, which had set in well before the 1990s; the expansion of the oil industry (first beginning in the 1930s) at the expense of the agricultural sector; and from 1996 on notably the UN *Oil-for-Food* programme, which continued and amplified earlier policies of importing foodstuffs in exchange for oil (cf. Natali, 2010; 2012; Fischer-Tahir, 2011; Woertz, 2013), all enhanced and solidified economic and demographic shifts towards a rapid urbanization (see Dziegiel, 1981; Springborg, 1986; Farouk-Sluglett and Sluglett, 1990, pp. 215–254; and Fischer-Tahir, 2003, pp. 75–79 on socio-economic transformations before the 1990s).

In 1992, the first parliament of the Kurdistan Region was elected, and administrative structures were gradually put in place, with the two dominant parties (PUK and KDP) settling on a 50:50 power-sharing agreement. However, tensions between the two parties remained, and the region was shaken by armed conflict between the militias of the two parties, as well as between PKK and KDP, between 1994 and 1998 (cf. Fischer-Tahir, 2016).

Following the overthrow of Saddam Hussein by the USA and its allies in 2003, the situation stabilized, and the general political and economic climate was coloured by increasing wealth, optimism and consumerism: compared to other parts of Iraq, security in Kurdistan was excellent, and a policy of decided economic liberalization attracted considerable foreign investment.

However, in 2014 this mood began to fade again due to a number of factors. Since 2012, the global oil price had been in a lasting decline, exacerbating already existing budgetary difficulties due to persistent political tensions between the Kurdish Regional Government (KRG) and the central government in Baghdad, which still controlled 90% of Kurdistan's budget. In this context, earlier declarations of intent by the KRG Ministry of Agriculture and Water Resources to develop the agricultural sector in order to obtain 'food sufficiency and security in the region' (Baban, 2012; Eklund *et al.*, 2017a) take on new, and increased significance.

The war against the fighters of the self-styled Islamic State, which occupied wide tracts of land immediately south of Kurdistan's boundary in summer 2014 (Eklund *et al.*, 2017b); the increased influx of refugees from Syria, from Iraq's Arab provinces, and the Sinjar Mountains' Yezidi population; and, last but not least, unresolved internal political tensions within the Kurdistan Region have all contributed to the pervasive sense of crisis that has made itself increasingly felt since 2014. The deterioration of the political-economic situation has become tangible through the protracted failure of the KRG to pay salaries to state employees, including members of the security forces (Osgood *et al.*, 2016). Growing dissatisfaction and protest against Kurdistan's political elites (whom many suspect of corruption) has been expressed through demonstrations and even riots most prominently in Silêmanî (Sulaymaniya), but also in Hewlêr (Erbil) while residents of Duhok province – though also disaffected – have largely remained more quiet (Hussein *et al.*, 2016).

The broad outlines of the political-economic development of the Kurdistan Region since the mid-20th century have been repeatedly described (Leezenberg, 2003; 2005; 2006; 2015; Stansfield, 2003; Natali, 2010). However, publications about regional and social variations and differences within these broader lines are still lacking. Qualitative research (e.g. Mlodoch, 2015; Fischer-Tahir, 2016) suggests that Kurdistan's public memory about the past decades is a highly politicized and sensitive space, not least because the precarious and tense state formation processes weigh heavily on the opportunities for open discussion of internal conflicts and contrasts. In this context, dominant historical narratives that highlight Kurdish unity and persecution by external enemies leave only little space for the articulation of internal differences that would nuance and modify the larger narrative. Moreover, Leezenberg's observation, stated more than a decade ago (2002, p. 289), is still valid: a regionally and locally differentiated discussion of the recent economic changes in the different parts of Iraq is still lacking. Furthermore, reliable data regarding Kurdistan's economic trajectory and situation are scarce (Eklund, 2015).

Nevertheless, recent findings challenge these larger narratives, suggesting that a closer look is needed to uncover and explain regional and local differences with regard to agricultural change. Thus, while satellite data show that large areas (between 64 and 83%) of cultivated land reverted to grassland after *Anfal* in the Jafati valley of Silêmanî (Sulaymaniya) province, the same kind of data indicate only marginal changes in the area under cultivation before and after *Anfal* when the regional focus shifts to the southern rim of Duhok province (Mubareka and Ehrlich, 2010). Eklund *et al.* (2015)'s satellite-based analysis of cropland changes in the Duhok province also points to local variations in the development of agriculture after *Anfal*. Another temporal reference point is 2003, when Kurdistan entered a period of intense economic growth. Here, it seems that, while overall agricultural productivity in Kurdistan continued to decline, agricultural land use in some regions and sites has increased.

Satellite data provide a useful basis to identify these variations, especially in a region such as Kurdistan where government produced data are neither always available nor necessarily reliable (Eklund, 2015). However, the causal factors behind these variations can only be analysed when the satellite-based data are combined with other methods, in particular qualitative methodology using biographical and narrative interviews. This chapter therefore proposes an interdisciplinary study (cf. Bromber and Lange, 2016; Eklund *et al.*, 2016), combining quantitative and qualitative methodologies, to scrutinize the transformations in agriculture in two sites of Duhok province in greater detail: Mangesh and Semel (Fig. 9.2).

Cropland Changes 1984–2014: a Quantitative Approach

The use of satellite data in studies of land dynamics

Satellite-based remote sensing has become a useful approach to study land use and land cover changes. It has been used to study global land cover dynamics and their effects on, for example, carbon emissions (DeFries *et al.*, 1999; Hansen *et al.*, 2000; Bartholomé and Belward, 2005; Pittman *et al.*, 2010). It can also be used to study land cover/land use dynamics at regional (Alcantara *et al.*, 2013; Estel *et al.*, 2015) or local levels (Kuemmerle *et al.*, 2009; Baumann *et al.*, 2011; 2014; Hostert *et al.*, 2011; Eklund *et al.*, 2015). Furthermore, remote sensing analyses can be very useful when studying the effects of armed or political conflicts on land systems (Ahram, 2015; Eklund *et al.*, 2015; Jaafar *et al.*, 2015; Jaafar and Ahmad, 2015; Gibson *et al.*, 2016; Jaafar and Woertz, 2016). Earth observation satellites carry instruments that record data on how the earth's surface reflects light in different wavelengths, which can help identify, for example, different vegetation types, soils and manmade structures.

Cropland changes in Duhok 1984–2014

To investigate changes to cropland extent in the Duhok province in Iraqi Kurdistan, Eklund *et al.* (2015) collected Landsat data from four different periods: A = pre-*Anfal* (1984–1987), B = post-*Anfal* (1989–1991), C = reconstruction (1998–2002) and D = economic growth (2011–2014).[1] The images were pre-processed so that all were corrected for atmospheric disturbances. Then, Normalized Difference Vegetation Index (NDVI) values, an index representing greenness or vegetation density, were calculated. For each of the four periods, two image composites were created, one that represented spring (March–June) greenness, and one that represented summer (July–September) greenness. This distinction was made based on the knowledge that the most common crop type in the Duhok province, winter crops (wheat and barley), were harvested around June every year. Thus, comparing spring greenness with summer greenness helped distinguish areas that had been harvested from areas that had maintained the vegetation (or lack thereof) over the harvest period, identifying harvested areas as winter croplands. The classification was validated using visual inspection (periods A–C) and ground truth data (period D), showing an overall accuracy of between 93 and 97%. For more detailed information on the methods used in the cropland classification, see Eklund *et al.* (2015).

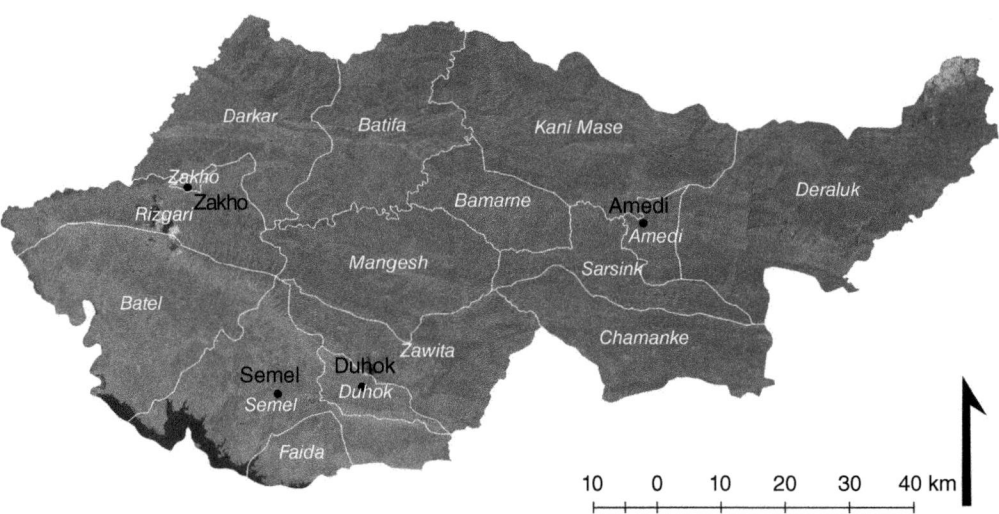

Fig. 9.2. The Duhok province, cities and subdistricts. Background image: Landsat 8 (2014, day 144).

The results of Eklund et al. (2015) show that winter cropland in the whole province had an extent of 1200 km² in the pre-*Anfal* period (A). This had decreased to about 870 km² after the *Anfal* campaign (period B), but increased again in the reconstruction period (C), to 1040 km². In the economic growth period (D), the cropland extent had increased further to 1300 km². The analysis further showed that the main land use changes during these four periods were the following:

1. A total area of 125 km² was brought under cultivation in the post-*Anfal* period (B) and onwards (C, D).
2. A total area of 115 km² became inactive during the post-*Anfal* period (B), but returned to cultivation during the reconstruction (C) and economic growth (D) periods.
3. A total area of 114 km² of previously uncultivated land was converted to cropland in the economic growth period (D).

Overall, these results indicate that, while some of the cropland was affected by the land abandonment caused by *Anfal*, the effects were not permanent. Furthermore, some areas were taken into cultivation just after *Anfal*, which might be an effect of people moving to new places and starting their lives over there. We also see new areas taken into cultivation in period D (2011–2014), during a time when Iraqi Kurdistan was experiencing rapid economic and urban growth. So despite a generally urban trend, and a better economic potential for importing food, new land was claimed for agriculture. These results did not completely support the general narrative about a stalled agricultural and rural development in Iraqi Kurdistan, which raises further questions about both land use and rural–urban dynamics in this area. In order to get a better understanding of the cropland changes since 1984 we decided to further focus on the changes identified in two subdistricts of the Duhok province, Mangesh and Semel, using the same cropland data as Eklund et al. (2015).

Cropland changes in Mangesh and Semel

Duhok province is a mountainous area. Because of the topography there is little suitability for cereal crops and only 17% of the subdistrict area has been cultivated with winter crops during at least one of the four periods (Fig. 9.3). This is hereafter referred to as 'arable land'. The Mangesh subdistrict is located in the centre of the province, north of the province capital, the city of Duhok. In Mangesh, only 3% of the arable land has been cultivated throughout the four periods. Instead, the majority of arable land became actively cultivated in the reconstruction period (C) and the present period (D) (Fig. 9.4). During the post-*Anfal* period, very little land was cultivated and the percentage of cultivated land decreased from 34 to 19% of total arable land. Ten per cent of the arable land was cultivated only during the pre-*Anfal* period, indicating that this land may have been permanently abandoned.

Semel, in contrast to Mangesh, is located in the plains of the Duhok province, south of the mountains. Thus, 70% of Semel subdistrict's area consists of arable land (i.e. land that has been cultivated during at least one of the four periods, Fig. 9.3). In Semel, a majority of the arable land has been cultivated throughout the four periods (approximately 60%). This subdistrict also saw a small reduction in cultivated area during the post-*Anfal* period (B), but an increase in cultivated area in the reconstruction (C) and economic growth (D) periods (Fig. 9.4).

Focusing on Mangesh and Semel we find some interesting results that open up further questions about the rural and agricultural development in Duhok province.

1. There seems to be increased cereal cultivation in mountainous areas, such as Mangesh, from 2000 and onwards, although these areas are not particularly suitable for that type of cultivation. What are the reasons behind this?
2. In both Semel and Mangesh, many areas that were cultivated before *Anfal* were not cultivated in the period immediately following *Anfal*, but cultivated again in the reconstruction period and onwards. This seems to challenge the general narrative that *Anfal* led to long-term land abandonment and a stalled agricultural development. How can this apparent contradiction be explained?
3. Large parts of the Semel subdistrict had areas that had been inactive before *Anfal*, but became active immediately following *Anfal*. What caused this expansion of cultivation just after the *Anfal* campaign?

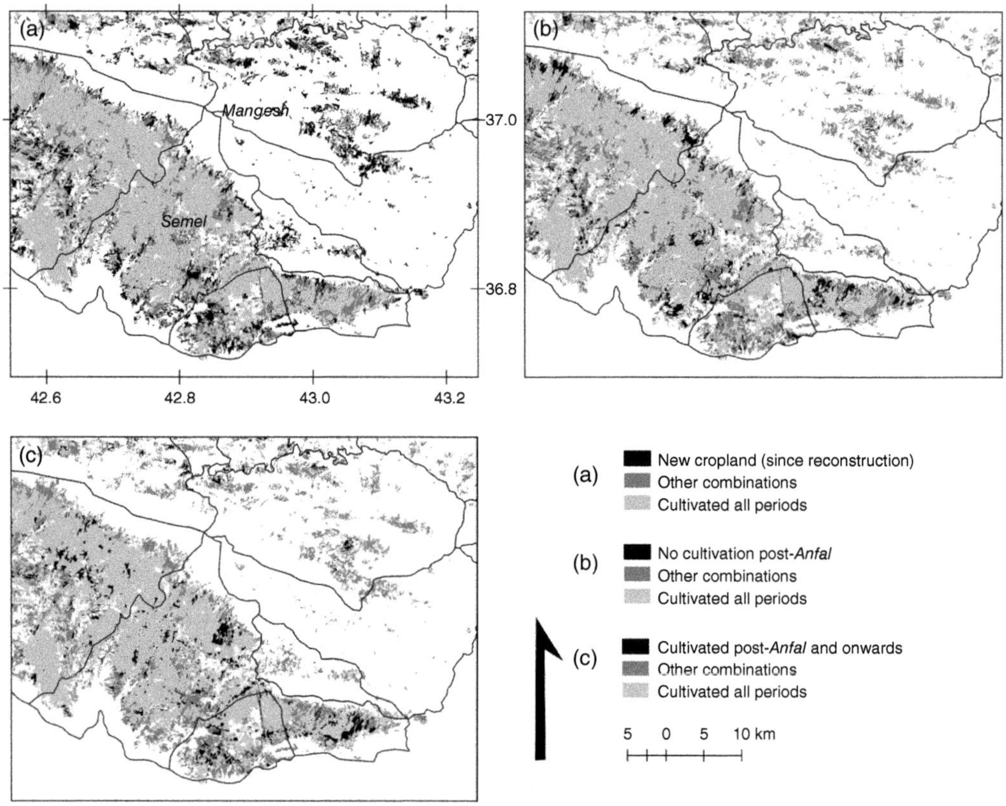

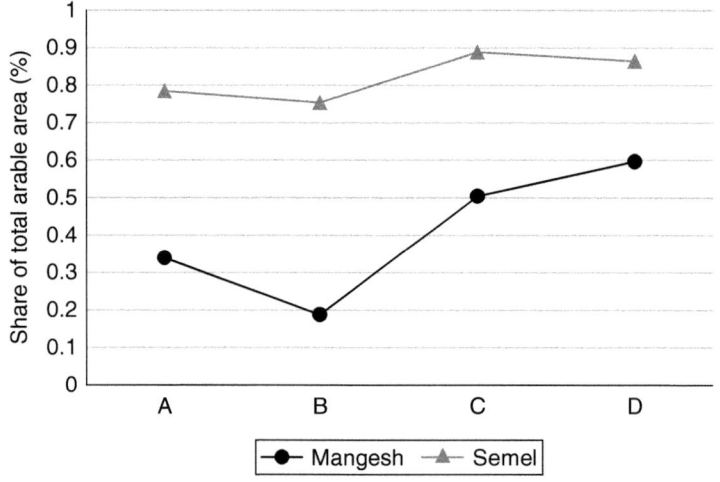

Fig. 9.3. Cropland changes in Mangesh and Semel where (a) highlights areas that have become actively cultivated in the reconstruction (C) or growth (D) periods; (b) highlights areas that were cultivated pre-*Anfal*, and in the reconstruction and growth periods, but not right after *Anfal*; and (c) highlights areas that became active after *Anfal*.

Fig. 9.4. Land under cultivation during the four periods in Mangesh and Semel.

Experiences, Representations and Perceptions: a Qualitative Approach

We argue that the questions posed above cannot be answered by relying on quantitative methods alone, but require a complementary approach, using qualitative methods such as semi-structured narrative interviews, notably with a biographic and family history focus, but also evaluation of written documentation and local publications or unpublished university theses from the Kurdistan Region. Although at this point, coordinated research using qualitative methods to address the abovementioned questions in a systematic way is still in an incipient stage, findings from previous research, carried out in Duhok province by anthropologist Katharina Lange between 2012 and 2015, suggest a direction that these answers could take. At the same time, they also show the methodological challenges presented by this material. These pertain, most of all, to the historic and political situation of the views and perspectives articulated by respondents and interlocutors in Kurdistan. Their responses, articulated in the course of longer conversations, do not give (only) factual accounts. Often, the facts which could be extracted and represented in a form resembling 'quantitative' data (e.g. sizes of land holdings, exact identity of owner, etc.) are seemingly obscured in a richer narrative fabric that evokes, among other things, family relations, political history or personal tastes. Consider, for instance, the following snippet of conversation between the researcher and two respondents from the land-owning stratum in a central Mangesh village:

> Question: I've been told that agriculture has become difficult, [what are your experiences on this]?
>
> Answer (resp. 1): I own two villages [we interpret this as a reference to his family being in authority, and holding property, in these villages], and yet I wait for my salary.
>
> Answer (resp. 2, cousin of resp. 1): His grandfather owns 30 pieces of land, and yet he waits for his salary...
>
> Question: And what happens with the land?...
>
> Answer (resp. 1): Today, the mazari' cultivate it [Arabic: muzāri^c; literally 'planter', here used in the sense of agrarian entrepreneur]. They rent the land on a 50:50 basis [a profit-sharing agreement]. They are rich, cultivate [on a large scale and their situation is] different from a small farmer (fellah) who engages in small-scale agriculture.
>
> Question: And what size of land do you rent to the mazari'?
>
> Answer (resp. 1): I don't know.

The conversation cited above was strongly inflected with criticism of the economic and especially agrarian policies of the KRG, but also with a self-criticism addressed to 'the Kurds', who had 'become lazy' and 'given up' the hard work of agriculture. This echoes a tendency that was evident in many other narrative interviews and conversations. In this and other interviews, accounts of the development of agriculture in Kurdistan were often wrapped up in a language of criticism regarding the political-economic trajectory of the Kurdistan region as a whole, and notably the role of the KRG. The decline of Kurdish agriculture was mainly blamed on failings of the KRG to protect Kurdistan agriculture from cheaper, foreign imports (according to popular wisdom, most products consumed in Kurdistan have been imported from the outside, most of all Turkey and Iran), and, thus, the failure to provide farmers with the opportunity to sell their produce at good prices.

According to local interlocutors, the only crop that could be guaranteed to sell at a profit was wheat: unlike grapes, almonds, chickpeas or coriander, wheat was bought by the Iraqi government at guaranteed prices that even surpassed the global rate. This government policy has certainly been driven at least in part by the considerable increase in population – after all, the population of Iraq has grown from 10 million in 1970 to 32 million in 2007 (United Nations Iraq, 2017).

Let us return once more to the above-cited conversation, held in a village in Mangesh region in spring 2015.

> (Resp.1): Currently, in our area here, if you want to make money, you can only do that with wheat. Only wheat is currently good. [...] Because the government buys the wheat, that's why. The government does not buy the other things [produce], for everything else, [for those

crops] there is only the market. And the market is only in Duhok [the provincial capital], [the produce] does not go to Baghdad, it does not go to Basra...

This remark holds a possible answer to the first question formulated above on the basis of the quantitative data: why increased cereal cultivation can be observed in mountainous areas, such as Mangesh, from 2000 and onwards, although these areas are not particularly suitable for that type of cultivation. The answer would be that, due to economic policies, this crop appears as a more reliably profitable investment to farmers and agricultural entrepreneurs even in a less favourable geographic setting. For farmers, marketing the wheat harvest has not only been more profitable, but also easier, than trying to market other produce.

On the basis of the information cited above, we might also formulate a hypothesis (which still needs to be confirmed or disproved through further research) to the second question – namely, why areas throughout different parts of Duhok province that had been left fallow directly after *Anfal* were brought under cultivation in the reconstruction period and afterwards. At first glance, this seems to belie the larger narrative according to which the decline of Kurdish agriculture is one of the long-term consequences of the *Anfal* campaign.

We assume that the explanation of this seeming contradiction may be a shift in the identity of cultivators, away from small-scale farmers to the larger agricultural entrepreneurs (*mazari'*) referred to above. This explanation would fit both narratives: where prior to the devastations of the 1970s and 1980s, a majority of small-scale farmers cultivated the lands in and around their native villages, they have subsequently given up this livelihood for the reasons named above, became increasingly dependent on state or party salaries and shifted the focus of their lives away from the rural settings to the urban conglomerations, leading – among other things – to intergenerational changes regarding lifestyle choices and skills. In terms of societal perceptions, this means a large-scale shift and, in fact, decline of agriculture. But the answers cited above suggest that the lands left behind do not necessarily lie fallow. Rather, there appears to be investment in agriculture – especially wheat – on the part of large-scale agricultural entrepreneurs who operate at a profit and who seek to cultivate large amounts of land in order to make this investment even more lucrative. This would explain why the remote sensing data show an increase of the actual area under cultivation.

It is most difficult to suggest an answer to the third question – why areas located in Semel that had been inactive agriculturally before *Anfal* were cultivated immediately afterwards. Several interlocutors suggested that this might be related to demographic shifts rooted in the political developments of the region: they recounted that Arabic farmers, many of them from the southern parts of Iraq, had been brought to settle and cultivate land in Semel of which Kurdish farmers had been dispossessed in the 1970s and 1980s. According to this narrative, the new settlers, who were strangers to this part of the country and unfamiliar with the agricultural work needed there, often did not cultivate as much land, or put in as much effort, as they could have. In 1991, they were driven out or left, fearing retribution from local Kurds, and Kurdish cultivators (the original owners as well as, possibly, Kurdish farmers who had been displaced from other parts of the region) began to cultivate these areas again. This renewed shift in ownership, together with the demand for locally produced cereals in the wake of the embargo, resulted in an increase in cultivation. However, this explanation is, at this point, anecdotal and has not yet been corroborated from other than oral sources. Further research is therefore needed to follow up on these tentative suggestions.

Conclusion: an Interdisciplinary Approach

Our initial analysis of changes to agricultural land use in the Mangesh and Semel subdistricts since the 1980s shows a multitude of developments. While the general narrative about the Kurdistan Region describes a deteriorating agricultural situation, we show that both Mangesh and Semel experienced land abandonment in relation to *Anfal*, which was then countered by recultivation during the reconstruction programme (1998–2002). Here we suggest looking further into who uses the land. Our qualitative data point to a shift from small-scale agriculture to agricultural entrepreneurs who cultivate the lands of farmers and share the profits with them. The identified increase in winter crop area in the mountainous areas in Mangesh since 2000 can potentially be explained by the fact that wheat is more profitable than other crops, due to economic policies of the Iraqi government. Thus, crops are possibly chosen not for suitability but rather for profitability. In Semel, we saw a trend of increased cultivation after *Anfal* in areas that were previously not cultivated. A possible explanation for this is the displacement of people during *Anfal*, where farmers resettled in new places. We have through this initial assessment identified some topics that require further investigation to better understand the agricultural development in Duhok between the 1980s and 2010s.

In terms of methodology, this chapter shows that an interdisciplinary approach helps us to gain a more thorough understanding of changes in agricultural land use. The remote sensing approach helps us identify and quantify changes in land use, but it cannot tell us why these changes have happened, or what the implications are. Therefore, we need a qualitative perspective based on interviews and literature reviews to understand how people experienced these changes and why they happened.

However, while interdisciplinary research is useful, it can also be challenging. In the process of writing this chapter we needed to find a common understanding of each other's data and methods, coming from very different disciplinary backgrounds. Differences include understandings of method. How many interviews would, for instance, be needed to reliably provide an explanation for an identified change? Even within qualitative research, there are very different approaches (and indeed, the very terms 'qualitative' and 'quantitative' may be used and understood differently in the humanities/social sciences and the natural sciences). An ethnographic approach based on in-depth narrative interviews, frequently conducted over a longer period of time and/or during several sessions, often seeks out 'gatekeepers' or 'key informants'; persons who, due to their individual biographies and experiences, their social status or their political position, can share particularly relevant and significant information about local contexts. Yet with this time-consuming method, the number of respondents who can be considered is limited. A very different social science approach – not incorporated in the present chapter – would look for a representative sample of a larger number of respondents, typically using much more structured interviews or questionnaire-based surveys. Each of these methodologies (as well as quantitative methodologies relying, for instance, on remote sensing), has its own, specific advantages and limitations; it is a basic requirement for an interdisciplinary collaboration to recognize the legitimacy of each.

Moreover, we found that our writing styles were different: whereas it is one of the objectives of anthropology/ethnography to convey 'local' views through detailed descriptions from the micro-perspective and often literal quotes of interlocutors' words, natural science aims to give abstracted information as concisely and briefly as possible. We met this challenge by adopting an alternating writing technique, where different parts were written by each author, while the conclusion was co-written by both.

In conclusion, we find that interdisciplinary work requires a collaboration guided by openness to different ideas and solutions and, most of all, the mutual recognition of the benefits and limitations of different types of data, generated through different approaches. Based on our own experience, we argue for a complementary, interdisciplinary approach to studies of agricultural change, especially in settings where data from other sources are scarce or not always reliable. Such an approach could bring the strengths of the different methodologies together and thus lead to a more comprehensive understanding of agricultural transformations.

Note

[1] Unfortunately, data restrictions stopped us from including the very interesting period between 2003 and 2010 in the analysis, as the Scan Line Corrector of Landsat 7 failed in 2003 and very few images from the Landsat 4 and 5 missions are available from this period for the specific area.

References

Ahram, A.I. (2015) Development, counterinsurgency, and the destruction of the Iraqi marshes. *International Journal of Middle East Studies* 47(03), 447–466. DOI: 10.1017/S0020743815000495.

Alcantara, C., Kuemmerle, T., Baumann, M., Bragina, E.V., Griffiths, P. *et al*. (2013) Mapping the extent of abandoned farmland in Central and Eastern Europe using MODIS time series satellite data. *Environmental Research Letters* 8(3), 035035.

Baban, S.J. (2012) Achieving sustainable food production and security in Iraqi Kurdistan: challenges and opportunities. In: Heshmati, J., Dilani, A. and Baban, M.J. (eds) *Perspectives on Kurdistan's Economy and Society in Transition: Volume II*. Cambridge Scholars Publishing, Newcastle upon Tyne, UK.

Bartholomé, E. and Belward, A. (2005) GLC2000: a new approach to global land cover mapping from Earth observation data. *International Journal of Remote Sensing* 26(9), 1959–1977.

Barwari, N. (2013) Rebuilding Peace: land and water management in the Kurdistan Region of northern Iraq. In: Unruh, J. and Williams, R.C. (eds) *Land and Post-Conflict Peacebuilding*. Earthscan, London, pp. 363–385.

Baumann, M., Kuemmerle, T., Elbakidze, M., Ozdogan, M., Radeloff, V.C. *et al*. (2011) Patterns and drivers of post-socialist farmland abandonment in Western Ukraine. *Land Use Policy* 28(3), 552–562. DOI: http://dx.doi.org/10.1016/j.landusepol.2010.11.003.

Baumann, M., Radeloff, V.C., Avedian, V. and Kuemmerle, T. (2014) Land-use change in the Caucasus during and after the Nagorno-Karabakh conflict. *Regional Environmental Change* 15(8), 1703–1716.

Bromber, K. and Lange, K. (2016) 'Research with': how to implement a methodological approach and an ethical claim – experiences from an institutional perspective. *Das Forschungsjahr 2015 – Jahresbericht der GWZ*. Geisteswissenschaftliche Zentren, Berlin, pp. 83–97.

Casim, S. (2016) *Die Anfal-Operationen der irakischen Regierung gegen die Kurden*. (Studia Kurdica). Universität Erfurt, Mustafa Barzani Arbeitsstelle für Kurdische Studien, Erfurt, Germany.

DeFries, R., Field, C., Fung, I., Collatz, G. and Bounoua, L. (1999) Combining satellite data and biogeochemical models to estimate global effects of human-induced land cover change on carbon emissions and primary productivity. *Global Biogeochemical Cycles* 13(3), 803–815.

Dziegiel, L. (1981) *Rural Community of Contemporary Iraqi Kurdistan Facing Modernization*. Akademia Rolnicza, Krakow, Poland.

Eklund, L. (2015) No friends but the mountains. Understanding population mobility and land dynamics in Iraqi Kurdistan. Dissertation, Lund University, Lund.

Eklund, L., Persson, A. and Pilesjö, P. (2015) Cropland changes in times of conflict, reconstruction, and economic development in Iraqi Kurdistan. *Ambio* 45(1), 78–88.

Eklund, L., Romankiewicz, C., Brandt, M., Doevenspeck, M. and Samimi, C. (2016) Data and methods in the environment-migration nexus: a scale perspective. *DIE ERDE – Journal of the Geographical Society of Berlin* 147(2), 139–152.

Eklund, L., Abdi, A. and Islar, M. (2017a) From producers to consumers: the challenges and opportunities of agricultural development in Iraqi Kurdistan. *Land* 6, 44.

Eklund, L., Degerald, M., Brandt, M., Prishchepov, A.V. and Pilesjö, P. (2017b) How conflict affects land use: agricultural activity in areas seized by the Islamic State. *Environmental Research Letters* 12(5).

Estel, S., Kuemmerle, T., Alcántara, C., Levers, C., Prishchepov, A. *et al*. (2015). Mapping farmland abandonment and recultivation across Europe using MODIS NDVI time series. *Remote Sensing of Environment* 163, 312–325. http://dx.doi.org/10.1016/j.rse.2015.03.028.

Farouk-Sluglett, M. and Sluglett, P. (1990) *Iraq Since 1958: From Revolution to Dictatorship*. Tauris, London.

Fischer-Tahir, A. (2003) *Wir gaben viele Märtyrer. Widerstand und kollektive Identitätsbildung in Irakisch-Kurdistan*. (Beiträge zur Kurdologie, 7). Unrast, Muenster, Germany.

Fischer-Tahir, A. (2011) Representations of peripheral space in Iraqi Kurdistan: the case of the Qaradagh District Centre. *Études Rurales* 186, 117–130.

Fischer-Tahir, A. (2016) Aufgeschobene Geschichte: der innere Krieg in Irakisch-Kurdistan und seine (Nicht-) Erzählbarkeit. In: Fischer-Tahir, A. and Lange. K. (eds) *Ethnographien des Wandels im Nahen Osten und Nordafrika*. Leipziger Universitätsverlag, Leipzig, pp. 245–268.

Gibson, G.R., Campbell, J.B. and Wynne, R.H. (2012) Three decades of war and food insecurity in Iraq. *Photogrammetric Engineering and Remote Sensing* 78(8), 885–895. DOI: 10.14358/PERS.78.8.895.

Gibson, G.R., Taylor, N.L., Lamo, N.C. and Lackey, J.K. (2016) Effects of recent instability on cultivated area along the Euphrates River in Iraq. *The Professional Geographer*, 1–14. DOI: 10.1080/00330124.2016.1194216.

Jaafar, H.H. and Ahmad, F.A. (2015) Crop yield prediction from remotely sensed vegetation indices and primary productivity in arid and semi-arid lands. *International Journal of Remote Sensing* 36(18), 4570–4589. DOI: 10.1080/01431161.2015.1084434.

Jaafar, H.H. and Woertz, E. (2016) Agriculture as a funding source of ISIS: a GIS and remote sensing analysis. *Food Policy* 64, 14–25. DOI: http://dx.doi.org/10.1016/j.foodpol.2016.09.002.

Jaafar, H.H., Zurayk, R., King, C., Ahmad, F. and Al-Outa, R. (2015) Impact of the Syrian conflict on irrigated agriculture in the Orontes Basin. *International Journal of Water Resources Development* 31(3), 436–449. DOI: 10.1080/07900627.2015.1023892.

Hansen, M., DeFries, R., Townshend, J.R. and Sohlberg, R. (2000) Global land cover classification at 1 km spatial resolution using a classification tree approach. *International Journal of Remote Sensing* 21(6–7), 1331–1364.

Hardi, C. (2011) *Gendered Experiences of Genocide: Anfal survivors in Kurdistan/Iraq*. Ashgate, Farnham, UK.

Hostert, P., Kuemmerle, T., Prishchepov, A., Sieber, A., Lambin, E.F. et al. (2011) Rapid land use change after socio-economic disturbances: the collapse of the Soviet Union versus Chernobyl. *Environmental Research Letters* 6(4), 045201.

HRW (Human Rights Watch) (1993) *Genocide in Iraq – The Anfal Campaign Against the Kurds*. HRW, New York.

Hussein, M., Mohammed, A. and Tahir, R. (2016) Strikes, protests reignite as Kurdistan salary delays worsen. Iraq Oil Report, 20 January 2016. Available at: www.iraqoilreport.com/news/strikes-protests-reignite-kurdistan-salary-delays-worsen-17727 (accessed 12 March 2018).

Ibrahim, F. (1983) *Die kurdische Nationalbewegung im Irak*. Klaus Schwarz Verlag, Berlin.

Kamrava, M. Babar, Z., Woertz, E., Harrigan, J., Bush, R., Salami, H. et al. (2012) Food security and food sovereignty in the Middle East. Summary Report No. 6. Center for International and Regional Studies, Georgetown University, Washington, DC.

Kuemmerle, T., Müller, D., Griffiths, P. and Rusu, M. (2009) Land use change in Southern Romania after the collapse of socialism. *Regional Environmental Change* 9(1), 1–12. DOI: 10.1007/s10113-008-0050-z.

Lange, K. (2016) Verlust/geschäft: Repräsentationen sozialen und wirtschaftlichen Wandels im ländlichen Duhok (Kurdistan-Irak). In: Fischer-Tahir, A. and Lange. K. (eds) *Ethnographien des Wandels im Nahen Osten und Nordafrika*. Leipziger Universitätsverlag, Leipzig, Germany, pp. 203–220.

Leezenberg, M. (2002) Refugee camp or free trade zone? The economy of Iraqi Kurdistan since 1991. In: Mahdi, K.A. (ed.) *Iraq's Economic Predicament* (Exeter Arab and Islamic Studies Series). Ithaca Press, Ithaca, New York, pp. 289–320.

Leezenberg, M. (2003) Economy and society in Iraqi Kurdistan: fragile institutions and enduring trends. In: Dodge, T. (ed.) *Iraq at the Crossroads: State and Society in the Shadow of Regime Change*. Oxford University Press, Oxford, pp. 149–160.

Leezenberg, M. (2005) Iraqi Kurdistan: contours of a post-civil-war society. *Third World Quarterly* 26(4–5), 631–647.

Leezenberg, M. (2006) Urbanization, privatization, and patronage: the political economy of Iraqi Kurdistan. In: Abdul Jabar, F. and Dawood, H. (eds) *The Kurds: Nationalism and Politics*. Saqi Books, London, pp. 130–150.

Leezenberg, M. (2015) Politics, economy and ideology in Iraqi Kurdistan since 2003: enduring trends and novel challenges. *Arab Studies Journal* 23 (1), 154–183.

McDowall, D. (2004 [1996]) *A Modern History of the Kurds*. Third revised and updated edition. I.B. Tauris & Co., London.

Mlodoch, K. (2015) *The Limits of Trauma Discourse. Women Anfal Survivors in Kurdistan-Iraq*. (ZMO-Studien 34). Klaus Schwarz-Verlag, Berlin.

Mubareka, S. and Ehrlich, D. (2010) Identifying and modelling environmental indicators for assessing population vulnerability to conflict using ground and satellite data. *Ecological Indicators* 10, 493–503.

Natali, D. (2010) *The Kurdish Quasi-State. Development and Dependency in Post-Gulf War Iraq*. Syracuse, New York.

Natali, D. (2012) The politics of Kurdish crude. *Middle East Policy* 19(1), 110–118.

Osgood, P., Tahir, R. and Hussein, M. (2016) K.R.G. slashes payroll to stave off budget collapse. *Iraq Oil Report*, 4 February 2016, www.iraqoilreport.com/news/krg-slashes-payroll-stave-off-budget-collapse-17923 (accessed 12 March 2018).

Physicians for Human Rights (1989) *Winds of Death. Iraq's Use of Poison Gas Against its Kurdish Population*. Physicians for Human Rights, Somerville, Massachusetts.

Pittman, K., Hansen, M.C., Becker-Reshef, I., Potapov, P.V. and Justice, C.O. (2010) Estimating global cropland extent with multi-year MODIS data. *Remote Sensing* 2(7). DOI:10.3390/rs2071844.

Sinemillioglu, H. (2011) Wiederaufbau des ländlichen Raumes unter den besonderen Bedingungen der Krise in Kurdistan/Irak. Doctoral dissertation. Technische Universität Dortmund, Dortmund, Germany.

Springborg, R. (1986) Iraqi Infitah: agrarian transformation and growth of the private sector. *The Middle East Journal* 1(40), 33–52.

Stansfield, G. (2003) *Iraqi Kurdistan: Political Development and Emergent Democracy*. Routledge Curzon, London.

Stock, P. and Burton, R.J. (2011) Defining terms for integrated (multi-inter-trans-disciplinary) sustainability research. *Sustainability* 3(8), 1090–1113.

United Nations Development Program (UNDP) (2010) Drought – impact assessment, recovery and mitigation framework and regional project design in Kurdistan Region (KR). Available at: http://reliefweb.int/sites/reliefweb.int/files/resources/B03750804A0EB2EC85257830006B6A97-Full_Report.pdf accessed 12 March 2018).

United Nations Iraq (2017) United Nations Iraq: Country profile. Available at: http://uniraq.org/index.php?option=com_k2&view=item&layout=item&id=941&Itemid=472&lang=en (accessed 26 April 2017).

van Bruinessen, M. (1992) *Agha, Shaikh and State. The Social and Political Structures of Kurdistan*. Zed Books, London.

Wimmer, A. (1997) Stammespolitik und die kurdische Nationalbewegung im Irak. In: Borck, C., Savelsberg, E. and Hajo, S. (eds) *Kurdologie. Ethnizität, Nationalismus, Religion und Politik in Kurdistan*. LIT, Muenster, Germany, pp. 11–43.

Woertz, E. (2013) *Oil for Food: The Global Food Crisis and the Middle East*. Oxford University Press, Oxford, UK.

10 Yemen's Agricultural World: Crisis and Prospects

Max Ajl*

Cornell University, Ithaca, New York, USA

Introduction

By 2017, Yemeni agricultural and food supply systems were collapsing, and the population was on the verge of famine amidst war and the conversion of a sustainable and self-subsistent agriculture to an unsustainable one that cannot feed the country's rapidly growing population. The capital, Sana'a, may be the first city in the world to simply run out of water as the nearby aquifers draw down. Yemen in too many ways is in slow-motion disintegration.

This chapter examines the collapse's history. It focuses on North Yemen, but not myopically so. For Yemen's developmental descent is not the result of purely internal decay – a mismanaged society finally falling to tatters. Indeed, it is much more the fruit of external forces yanking, tugging and ripping apart the country – sometimes with local connivance. Such pulls and pressures remoulded subsistence agriculture surrounding coastal entrepôts into a petroleum capital-incubated social formation that has transformed and grown in human magnitude, but barely developed. Yemen relies on constantly expanding flows of cereals to feed the people it cannot feed on its own. Such flows are subject to extraction both on international and internal markets. Yemen's transition is also the tale of how an ecologically sustainable system, wrenched from its basic stability, now relies on non-renewable sources for water, the lifeblood of human civilization.

This chapter proceeds as follows: it first describes Yemen's pre-1970 agricultural system, before the oil age's regional and global dynamics reworked the country's political ecology. Next, it explains remittances' role in restructuring Yemen, especially its agricultural sector. It then offers a quantitative and qualitative-historical account of Yemen's rising food import dependency – a world systemic process. The chapter traces how remittances levered Yemen into an irrigation system that uses motors to withdraw water from aquifers, and shows how higher wages and the qat economy lubricated this shift. It highlights how diesel subsidies have become a catalyst for the degradation of traditional irrigation, a mechanism of rural social differentiation, and a means for society-wide differentiation and the denial of development. The chapter concludes by discussing additional exacerbating factors, including ongoing external aggression, and comments on policies to cease and eventually reverse the social and ecological de-development of Yemen's agricultural system, and the country itself.

---

* Email: msa95@cornell.edu

Before Oil: Yemeni Agriculture Before 1970

Yemen has long been the most fertile area on the Arabian Peninsula. Before the 1970s, Yemeni agriculture was primarily an affair of smallholders and subsistence farmers, producing primarily sorghum and barley. Most families also kept livestock, often goats, sheep and a cow, mostly for milk production. Women often tended to such animals, which were also one of their sole sources of accumulated capital – a fact that must be kept in mind when one considers the pathology of famine, and adaptation measures, which include selling livestock. Land ownership was primarily private, although people often worked as sharecroppers on land owned by others. Sharecropping was more frequent on more fertile lands. Wage labour was uncommon, mostly occurring in the Tihama – historically a region of relatively high inequality in land ownership (Nugent, 2003). There was next to no state agricultural extension. Yields during this period were – and have remained – relatively low (Table 10.1).

Yemeni farming then relied almost exclusively on a mix of controlled floods and terraces, alongside a system of water rights to regulate the sustainable use of those waters, to intensify production. Yemen has traditionally used flash-flood (*sayl*) and spring (*ghail*) irrigation. In the former, simple dams made of local materials move evanescent floodwaters into a series of channels. Fields sit next to the channels, and those closer to the channels receive water before those further away. This system has predominated in the lowlands – for example, the Tihama. Before the oil age, flash-flood irrigation was 'ecologically stable and efficient', producing no soil salinization, and making large agricultural surpluses (Mundy, 1985, p. 27). In the spring irrigation system, which dominates in the highlands, cisterns collect spring water and direct it to specific terraces. Terraces also receive water through rain – that which falls on them directly, and on abutting ditches and dikes, which direct the clear gold towards the terraces. Such catchment and concentration mechanisms were built in precise ratios to the size of the terraces to ensure adequate hydration for their plantings. Thick fog also envelops the slopes in some sites such as Manakhah in the Haraz mountain massif, protecting the soil from the sun and enhancing soil moisture content by leaving dew. Such terraces relied on rainwater that ran across stony beds using only gravity. That rainwater brought with it regular doses of fertilizer as it gathered elements from the surfaces over which it ran (Varisco, 1983).

According to the Ministry of Agriculture and Fisheries census, as of 1983, even amidst widespread terrace degradation, 75% of the 13,500 km^2 of northern cultivated land was dry-farmed – mostly terraced (Varisco, 1991). The combination of irrigation and terracing meant cultivation could take place in isohyets normally incapable of supporting agriculture (150–300 mm of rain a year). Furthermore:

> Where there is a spring or spring line, terraces will be located in clusters below so that a gravity-flow sequence can irrigate the system as a whole through a common channel network. In these spring-fed irrigation systems the water may descend through channels over a kilometer to the farthest plots. In the higher mountain slopes, where runoff harvesting is practiced, a group of terraces will be arranged so that excess flow will be carried off through the channels. Since farmers will not be present when the flooding occurs, it is important to allow for ample drainage of the flow in order to avoid erosion of the plots. Knowledge of the best location and size of both plots and channels is attained by experience. The principle is to use as much of the fast flow as feasible without jeopardizing the plot.
>
> (Varisco, 1991, p. 168)

Table 10.1. Sorghum and millet production, 1964–1974. (Adapted from: Tutwiler, 1990.)

Year	Area (1000 ha)	Yield (kg/ha)	Production (1000 tonnes)	Economic events
1964–1966	737	837	617	Civil war, 1964–1970
1967–1970	930	720	669	Drought, 1967–1973
1971–1974	909	740	679	Migration, 1974–

These irrigation and terracing systems were the fruit of meticulous knowledge of microclimates, montane contours and the engineering feats needed to farm within those geographies. The water use rights that went along with them were well suited to sustainable use. And they had another signal trait: because they dig not dig very deeply, or were spring-fed, farmers only drew down groundwater in exceptional situations, and at rates which would permit a natural recharge of the aquifers. Use of water in the present did not imperil use of water in the future. Atop these terraces sorghum was the dominant crop. It is resistant to drought, capable of multiple cropping without needing to reconstitute its root structure, and fit for food for humans and fodder for animals (Moore, 2011, p. 42).

The Oil Hothouse: 1973–1990

Yemen had been part of global trading circuits for a millennium at least – an entrepôt for trade from India wending down the African coast, and a sentinel of the Red Sea (Abu-Lughod, 1991, pp. 241, 272). In the 19th century Aden was a colonial British coaling station, forcefully sheared from its hinterland and the rest of Yemen. Later it would be one of the world's most well-trafficked ports (Carapico, 2007, pp. 25–26). But Yemen's rural areas were only lightly touched by the flows and fleets funnelled by Britain through its trading cities. It was only in the late 1960s that massive changes in the world system ripped this sustainable system asunder.

As anti-colonialism coursed and raged through the global south, large segments of Yemen's population not only began to take political action, but in the south especially – the north was never colonized – participated in anti-colonial revolts and wars of national liberation. Amidst the rapid rise in oil prices and widespread regional nationalizations – themselves the fruit of global anti-colonial revolt – the US government connived with the Arab oil principalities to push up rapidly the price of oil (Stork, 1975; Blair, 1976; Oppenheim, 1976; Nitzan and Bichler, 2002). This rejiggering of the proceeds of global oil rents led to a substantial construction boom in Saudi Arabia, which led large swathes of Yemen's male population to abandon agriculture in pursuit of higher wages abroad.

There are only estimates of the number of Yemeni workers who went to Saudi Arabia, particularly from North Yemen. But the consequences for Yemen were clear: out-migration in the 1970s 'created severe labor shortages and worsened already critical skill deficiencies' (Fergany, 1982, p. 762). Much of the adult male population departed North Yemen, leaving behind children, the elderly and women. Those who left were disproportionately highly skilled and highly trained workers, at a time when schooling was reserved for a tiny portion of the population – in the late 1970s just 3% of the economically active population had finished secondary education. Technical departments in the government were left half-empty, as were a fifth of government posts. The lack of qualified personnel hamstrung developmental planning, albeit to an unknown and unknowable degree (Fergany, 1982).

Wages in Yemen grew to parity with those in Saudi Arabia in order to lure workers to remain in place. Even agricultural wages increased explosively. Meanwhile remittances reworked everything about Yemen, weaving it into global commodity flows with jagged and uneven stitches. Its balance of trade worsened markedly: in 1965 imports were four times exports. In 1979 imports were 200 times exports, which were overwhelmingly agricultural: hides, skins, salt, coffee, cotton and biscuits. This imbalanced ratio was both cause and symptom of agricultural disarray. From 1965–1979 the amount of cultivated land decreased by 25%. From 1967–1978 local food production increased by merely 0.3% a year by volume. Amidst population growth – a demographic problem linked to inadequate government support for family planning – per capita food production decreased by 1.6% a year over the same period (Fergany, 1982). Remittances served less to supply that jerryrigged incantation of the economists, 'demand', than to create it: the inflow of remittances spurred large trade deficits due to the atmosphere of free trade prevailing in North Yemen. Remittances reached 40% of North Yemen's gross national product (GNP) and 44% of south Yemen's GNP in 1980 (Fergany, 1982; Okruhlik and Conge, 1997).

Remittances also transformed Yemeni production. Sustainable subsistence agriculture became increasingly remittance-based, marked by local brokerage and import services that emerged to channel and profit from remittances, while at the same time using them to import consumer goods. Internal circular flows broke, replaced by triangular international ones. Saudi oil went to the wealthy industrialized countries, the proceeds of European and Japanese oil consumption went to Saudi Arabia, and remittances derived from this trade passed via Saudi Arabia to Yemen. The US, European states, Australia and Canada heavily supported their own farmers' wheat production when it was otherwise unprofitable. Their petroleum-intense farming technologies used oil to produce cereal surpluses that Yemen, increasingly unable to use its local labour-intense productive matrix to produce cereals, imported (Friedmann, 1982; Patnaik and Patnaik, 2016).

The combination of macro-economic shocks and government promotion of Green Revolution technologies led to the replacement of traditional crops and a transformation of Yemeni diets. From 1970–1980, the area planted with cereals decreased 27% (Table 10.2), at a time of sharp population increases. Barley production plummeted 50% from 1970–1975. Sorghum – the traditional staple crop – peaked in 1975 before declining until the early 1990s. The government introduced highly water-demanding 'improved' Green Revolution semi-dwarf wheat varieties. This led to a 56,000-ha gain for wheat from 1970–1990. Such a supply-led push also had its demand side – 'the standardization of diets on global models', an underexamined process that has often taken the form of a state-led passive revolution in so-called consumer preferences through which wheat replaced traditional arid-land crops such as sorghum, millet and barley in cereal baskets (Mundy and Pelat, 2015, p. 101).

Import dependence has since then further heightened. Yemen imported 75% of its cereal needs in 2015, up from 18% in 1970. That disparity is more marked when one disaggregates cereals into those grown locally like sorghum, barley and millet, and those traded on international commodity markets such as wheat. In 2009, of Yemeni-grown cereals, 55% by weight and 60% by area was sorghum; combined with barley and millet those numbers reached 69 and 80%. But wheat was 74% of Yemeni cereal consumption in 2009, and by 2011, 93% of the national wheat needs were imported (Mundy and Pelat, 2015).

Other mechanisms have also damaged domestic cereal production, undercutting the productive sector where terraces have traditionally done best. The government has failed to support research and investment in terrace and water harvesting systems. The government has subsidized imported grain, too. In 1995, for example, amidst the currency continuously losing value against the US dollar, the government subsidized 81% of the import parity price. The lower prices make production less attractive. This has also been a mechanism of social differentiation, since cereal cultivation has often been the province of lower-income farmers. As Ward notes, 'Apparently the government's cereal policy has not been to increase the incomes of marginal cereal producers, but instead to subsidize the cost of food for the much more visible and vocal constituencies of consumers' (Ward, 2000, p. 388). The long-run consequences of failing to

Table 10.2. Cereal production, 1961–2011, selected years. (From: FAOSTAT.)

	1961	1970	1975	1980	1990	2000	2011
Cultivated area (ha)							
Cereals, of which:	1,199,277	1,082,721	1,261,749	851,000	844,845	619,583	798,619
Sorghum	1,000,000	887,000	1,057,000	644,000	506,770	359,632	460,711
Barley	127,000	111,000	58,295	48,000	52,325	37,075	44,491
Wheat	26,000	42,300	71,655	73,000	97,900	87,334	124,463
Output (tonnes)							
Cereals, of which:	938,400	845,415	1,199,953	865,000	766,873	672,237	816,555
Sorghum	700,000	610,000	921,000	623,000	441,231	375,009	412,031
Barley	141,000	126,000	64,134	47,000	55,003	42,428	30,003
Wheat	34,000	38,500	88,134	73,000	154,937	141,884	232,339

support demand-enhancing measures amongst Yemen's poorest sectors, especially in the countryside, would soon become clear.

The need to import food does not by itself indicate developmental failure. However, alongside such imports, Yemen increasingly failed to produce anything at all of value – other than a trained workforce that it often exported. A hypertrophied tertiary sector dominated national employment. Both that sector and the state channelled oil and geopolitical rents to various non-productive sectors. Such dynamics were neither sustainable nor sufficient to set in train endogenous processes of accumulation. The undercutting of agriculture, the one sector of the Yemeni productive system capable of creating backwards-and-forwards linkages, was a delayed-onset disease that would later harm Yemen's poor severely.

One result of these combined and disastrous policies has been that across Yemen, marginally productive fields are left uncultivated. Elsewhere, farmers farm their fields less intensely, and families cannot maintain terraces. Amidst men's absence, women, already carrying the burden of household labour and social reproduction, must split their time between 'household and field as never before', leading to production shortfalls (Swanson, 1979a, p. 41). Moreover, since Yemeni agriculture is traditionally labour-intense, technology cannot replace labour. As wages rise, the changing production possibilities frontier leads to lands leaving production.

Furthermore, terraces require constant maintenance for the retention structures that hold soil in place for planting to function. As farming families left terraces, the balance between soil formation through sediment loads and rock weathering and erosion disappeared. For the first decade after abandonment, soil erosion in terraces is extremely high. Approximately 60–160 tonnes of soil per acre is lost annually – the same as the ground lowering between 0.4 and 1.6 inches (1.016–4.046 cm) per annum (Vogel, 1987). Once again, tortuously human-made landscape improvements, ever-difficult to value but which have been the basis of sustainable agriculture for millennia, are lost. Once lost, they require stunning amounts of labour to regain and rebuild. Furthermore, the place-based knowledge – the knowledge of the place that is an inseparable part of terraces-as-techniques – is partially lost as terraces collapse and their minders and builders move elsewhere, bear children to whom they do not pass on the knowledge, and then die. Such knowledge is bound to a culture as a holistic life-world of production and consumption. Once lost it is not clear that it can be regained.

One should not understand this as an antiquarian interest in sustainable, low-productivity agriculture. The particular course of 'modernization' created immense vulnerabilities within Yemeni agriculture and society more broadly. The oil age meant a Saudi remittance structure enveloped and transformed Yemen's rural world, filling cracks in the productive structure with the temporary sealant of oil remittances. But that structure's stability rested on an absence of the tremors of geopolitical conflict – on the quiescence and quiet absorption of Yemeni workers, and Yemen itself remaining quiet on the international political stage, an object rather than subject of politics. When this quietude shifted amidst the first Gulf War, conflict took the form of labour expulsion, with consequences I detail below.

The Breaking of Sustainable Irrigation

Another dimension of the prolonged crisis of Yemeni rural society has been the decline in sustainable irrigation and its ties to the growing cultivation of fruits, vegetables and qat, a mildly hallucinogenic amphetamine that needs relatively little labour. Even by 1974, of the 61.4% of gross domestic product (GDP) generated by agriculture, estimates – since official numbers do not exist – were that 50% was from qat cultivation; others suggest that qat was a similar percentage of GDP in 1982 (Varisco, 1986; Colton, 2010). Fruits and vegetables, which need lots of irrigation but also produce more cash value than cereals on a per-hectare basis, also increased rapidly. However, unsustainable techniques – diesel motors relying on abstraction – replaced sustainable terraces (Colton, 2010). The crop area irrigated by groundwater grew from 37,000 ha in 1970 to 400,000 ha in 2012, or one-third of the total cropped area. The quantity of land under fruit and vegetable production rose from 39,000 ha in 1970, 3% of the total cropped area, to 184,000 ha, 14% of the total cropped area. Yemen recorded a 20-fold increase of fruit and vegetable production, from 40,000 tons in 1970 to 800,000 tons in 2010. According to the World Bank, supplemental or full well irrigation accounts for two-thirds by value of crop production (Al-Weshali et al., 2015). The government has also banned fruit and vegetable import, creating a protected market, while it subsidizes cereal import. The effect this incentive structure has on farmers' choices is clear. Given that fruit and vegetable cultivation is based on irrigation from aquifers and grain cultivation traditionally has taken place on sustainable rain-fed terraces, springs or flood-diversions, this has meant a shift from sustainable to unsustainable agronomic techniques. Meanwhile, the law of global value's price system moulded and informed such choices, making it rational for farmers to produce crops valuable on the market, even if such choices would later mean that Yemen would be unable to provide for its own needs and so vulnerable to famine. Qat has acted as this process's hinge. Table 10.3 provides quantitative indicators of production shifts.

Qat increasingly dominates the local market. It negatively affects aggregate and household-level food production and the water system's long-run sustainability. By 2011, the World Bank estimated qat accounted for 25% of GDP and 16% of employment. Qat purchases take up between 18 and 30% of the adult population's estimated expenditures. Most devastatingly, qat absorbs up to 16% of the high-quality potable water, the factor of production most dear in Yemen.[1] Qat production took up around 145,000 ha of arable land in 2000 – 3.5–4% of the total. Furthermore, 80–90% of new wells in the highlands are for qat production. This is only to be expected, given its massive profit margin per cubic metre of water – five times that of grapes (Almas and Scholz, 2006).

Other factors have also accelerated subsistence terrace agriculture's abandonment in favour of qat production. First, men continue to spend about 3 h per day chewing qat, exacerbating labour scarcity. Women work less on farms directly but spend 4 h daily collecting water and fuel – a consequence of state underinvestment in infrastructure capable of using more labour-efficient technologies to accomplish the same tasks, and thus an indicator of the gendered nature of developmental policies. Marketing channels are another mechanism that promotes qat production. A 1997 study showed farmers received 80% of qat's market price, whereas they received between 62 and 72% of maize's and sorghum's (combined in the study's results) market price. For barley, they received between 62 and 53%; wheat, between 40 and 36%, and millet, between 64 and 60% (Hebshi, 2004, p. 10). Middlemen made qat more attractive even on a purely commodity-production basis than various cereal crops, placing aside qat's considerably higher returns per hectare – a function both of remittances, which provide the footloose cash capable of sustaining qat consumption at such an elevated level, and world systemic factors. The relative 'cheapness' of cereals on world commodity markets and local markets is inseparable from their doubly subsidized production in global north granaries – both directly via the state, and indirectly via the use of petroleum-intense inputs whose prices ignore environmental externalities of hydrocarbon incineration or industrialized farming writ large. Both the onset of the oil age in Yemen and Saudi Arabia and their relationship with a shift to increasingly industrialized and hydrocarbon fuelled agriculture in the west have contributed to Yemen's agricultural de-development.

The shattering of sustainable water techniques is not just a tale of terrace deterioration but also of the relative decline of spate irrigation in coastal *wadis* (valleys). In the Tihama Plain, 'improvement' projects in the upstream areas

Table 10.3. Land devoted to commodity crops (1000 ha). (From: Almas and Scholz, 2006.)

	1970	1975	1980	1985	1990	1995	1996	1997	1998
Grapes	10	10	12	13	17	21	21	21	23
Coffee	7	8	8	17	25	27	23	32	32
Qat	8	35	45	56	77	89	91	93	98

reworked traditional diversion structures. They blocked both floodwater and groundwater from flowing downstream. The upstream areas have traditionally been the ones with greater inequality in land ownership. One consequence of such 'development' projects in these spate irrigation zones has been high income from commercial farming upstream, and poverty, malnutrition and even the emptying out of entire villages downstream (van Steenbergen et al., 2010; Yemen Water, 2013). Development in agriculture has thus in this way been an expression of social conflict in agriculture.

On a larger scale, this complex, delicate and unequal system rested on external inputs of capital – massive remittance flows from Saudi Arabia. Those flows, in turn, rested on Yemeni labourers remaining quiet economic cogs in the machine of Saudi production and accumulation, linked to a divided and dependent Yemen. 'Saudi Arabia feared that a united Yemen would be a radical, pro-Soviet state over which the kingdom would exercise little influence' (Katz, 1992, p. 122). When a newly unified Yemen demanded that 'foreign' (United States) forces leave the region, Saudi Arabia correctly interpreted this as a demand that it deal with Iraq on its own, from a position of clear weakness. In retaliation, it essentially expelled the Yemenis for Yemen's muted opposition to the first Gulf War – an economic sanction from a stronger state to a weaker one, but with effects that varied along class lines. Amidst this flood of labour, two dynamics worked in tandem to worsen the situation. First, those spate zones that had been partially spared from dynamics of differentiation were suddenly forced to find room, land and livelihoods for returnees. With less land per farmer, returns decreased. With increased supplies of labour, wages worsened. Furthermore, smallholders have been often forced to sell their land, leading to greater concentration of land in fewer hands in the countryside and of people in the cities. Sand dune encroachment worsened the problem, making some lands totally unviable (Ertürk, 1994). Upstream, the new diversion structures often silted up, converting viable sustainable and relatively egalitarian techniques into mechanisms for short-run enrichment and medium-run spoliation, not merely dividing the pie more unequally but in fact reducing its size in absolute terms. For example, in Wadi Mawr, large upstream landlords have deliberately sabotaged the water diversion project, preventing water from flowing downstream and monopolizing it for their own holdings. In this way, the development project of 'improved' irrigation and diversion systems ended up bringing nothing but de-development and differentiation to the Tihama Plain. Meanwhile, the country needed to find food for the returnees, further straining the import bill.

Oil Discovery: 1980s

The discovery and subsequent export of oil in Yemen in the 1980s did not support the development or sustainability of the agricultural sector through sowing the oil wealth. Instead, it made available cheap diesel that has supported unsustainable irrigation practices, water-intensive agricultural production, and widespread rural social differentiation. This has happened directly, through elite capture of diesel subsidies by exporting diesel, and indirectly, since subsidized diesel has exacerbated social differentiation in the countryside.

The actual diesel that Yemen consumes comes largely but not entirely from outside the country. From 1998 to 2003, as the government subsidized diesel consumption, diesel as a percentage of imports jumped from 6.44% to 14.86% while the diesel-burning machinery imported to use the fuel decreased. Much of the public logically became convinced that the government was smuggling diesel: 'When the government admits that the subsidies on the oil derivatives go to the pockets of smugglers, why doesn't it audit even one of them?' (Phillips, 2005). Such diesel makes its way to neighbouring countries through the Horn of Africa, or the elite simply transfers it before it even enters the country. Various studies suggest that 30–50% of the public money allocated to diesel was captured through elite smuggling (Al-Weshali et al., 2015). Diesel subsidies have also melded with pre-existing techno-institutional shifts towards wells and away from the various techniques for rainwater or spring-fed irrigation.

Such shifts had multiple catalysts and accelerators. First, remittances could finance agricultural investment, and the lack of manpower made labour-substituting techniques particularly attractive. Lack of labour for terraces due to migration was thus a necessary condition for the remittance-fuelled well-digging. Oil, a non-renewable resource, and its exploitation opened the way for a transition from renewable to non-renewable techniques for irrigation amongst Yemeni farmers: 'Economic uncertainty and the devaluation of Yemen's currency during the 1980s also made it "better to have land than money"' (Lichtenthäler, 2000, p. 162; Moore, 2011, p. 44). This rationality made it logical for rural residents to invest in improving their land through digging wells and installing diesel pumps. Furthermore, terraces were a locally managed technology that rested on local expertise and local, difficult-to-commodify knowledge. They did not rely on inputs from which various sectors could profit either by external export or by brokering their import. Finally, they were not useful for state-building along lines of hierarchical patron–client relationships. Wells had none of the benefits associated with terraces and brought with them a bevy of vices (or benefits only for the few). First, state-financed groundwater abstraction was both a channel for patronage and a means to construct a state along particular lines. Wells especially were 'political gifts through which [shaykhs] were coopted into power sharing' (Lichtenthäler, 2000, p. 163). Furthermore, international agencies financed the capital-intense schemes. Their focus was not, per se, supporting patronage-based political ties. It had more to do with an aspect of the technology itself, which also made it suitable for shoring up patron–client ties in ways terraces were not. The contrast was between democratic, centrifugal, local, capital-light, labour-heavy and local knowledge-based technologies as against centralized, centripetal, capital-intense, labour-light and imported technologies. It was less that terraces were somehow preternaturally immune to capitalist investments or justifying project-based loaning and more that they were resistant to them. As a World Bank official put it, overstating the technical dimension but still revealing an important truth, 'The neglect of traditional water control systems can be traced to one key constraint: the lack of easy technical packages that can readily lead to new productivity and attract private or public investment' (Moore, 2011, p. 43).

This, too, provided incentives for the Yemeni government to encourage groundwater abstraction for 'rural development', or a highly unequal process of capitalist change in the countryside. In the process it bartered Yemen's long-run ecological survival for short-run political advantage. It created a range of programmes that would feed this economic sector given its use for patronage-based state construction. Ultimately, 'the government actively distributed the benefits ensuing from groundwater development, and directed a large share of them towards key constituencies, i.e. tribal leaders, large landowners, and the military and economic elites that formed the power basis of the regime, in exchange for their loyalty and support' (Karner, 2014, p. 15).

One mechanism was low-interest loans and credits to ease the purchase of expensive drilling equipment, adding further weight to the preference for capital versus labour-intense hydraulic technology. Another was government subsidies for diesel fuel, which as noted, eventually became a basis for massive rent distribution and capital capture by elites and their clients. Finally, the government did not take any legislative action to allay metastatic growth of high-technology hydraulics. It was not until 2002 that it passed the National Water Law that made mandatory permits for new wells or the deepening of old ones. By 2011, 45,000 to 70,000 wells were active in Yemen. Furthermore, the country contains over 800 private drilling rigs currently running. India, with a population perhaps 50 times the size of Yemen's, has only 100 such devices (Karner, 2014).

In the Tihama, beginning in the late 1970s, as the global law of value reworked Yemeni agriculture, farmers felt pressure to grow more high-value crops. They opted for vegetables and melons, which require well irrigation. Both the state and the development institutions encouraged and subsidized this system. Those with capital to invest did so, contributing to a constantly lowering water table. Because of the capital-intense nature of the new technology, it sharpened class distinctions (Mundy, 1985). Furthermore, large landholders continue to sell groundwater they mine with tube wells to smaller farmers. They install them using state subsidized credit – because 50% of the investment must be self-financed, this is in effect a subsidy to capital (van

Steenbergen *et al.*, 2010). This, too, contributes to rural poverty.

By the late 2000s this combined fuel–water–patronage nexus was gutting state developmental capacity, robbing Yemen's ecological bequest from its past to its future, and sharpening social differentiation. Because of the narrowness of most Yemenis' survival margins, even slight subsidy removal without countervailing programmes could severely damage their budgets and their access to daily bread. Combined with inbuilt corruption arising from the use of diesel as a means for state elites to privatize public wealth from the treasury, the central government increased diesel prices only lightly and hesitantly from 1996 to 2011. Furthermore, as more and more rural families sink wells, the water table sinks. Access to water itself requires more expensive equipment. Water access becomes a function of capital access, ensuring that ecological despoliation, or the lowering of the water table, is experienced along class lines. Since access to the diesel subsidy has then been mediated through capital ownership, the diesel subsidy, even where genuinely used for diesel, has become a mechanism of social differentiation. Well-off farmers benefit far more than do smaller ones from the subsidies – a typical pattern. The richest 25% of households receive 40% of fuel subsidies while the bottom 40% receive just 25%. In rural areas, 29.8% of households have reported diesel use for agriculture, and use 200 l/month to that end, while just 1.7% of those in the poorest third of households reported diesel use for farming, and just 25 l/month (Al-Weshali *et al.*, 2015).

But the diesel question is even more fraught than such indicators suggest. After the 2011 uprising, crisis, civil war and war of external aggression to which Yemen has been subjected, the government temporarily lifted oil subsidies. Although grossly inequitable in quantitative terms, in Yemen as elsewhere removal of diesel subsidies damaged the poorest households qualitatively. Some rural households were unable to irrigate their crops. Others could no longer get crops to market. And a great many others experienced food price inflation, to which increased transport costs – hinging, again, on the breakdown of local provisioning systems and reliance on the market for the basic caloric goods needed for survival – contributed (Al-Weshali *et al.*, 2015).

Furthermore, subsidy removal did not reduce abstraction of water and there is no reason to think that it will. Increases in diesel prices, in the current socio-political-institutional context, have not induced a substitution, or a return, to labour-intense techniques. Rather they have induced a substitution for crops that can continue to secure a profit for their planters even when diesel, and thus water, increases in cost. Since margins on qat are higher than those on fruit, vegetables or other crops grown on well-irrigated land, increased diesel prices mean more farmers are likely to change to qat cultivation. Only in some cases will well irrigation become unviable. As Al-Weshali *et al.* (2015, p. 233) note, 'The danger of increasing the price of water is that farmers will convert to qat production on a large scale because the costs of water for qat are substantially below the value of water... This will trigger groundwater extraction even further.' In the Tihama, small farmers often had to sell off their livestock to purchase diesel at its higher price (Al-Weshali *et al.*, 2015). Livestock is traditionally an emergency reservoir for smallholders and thus a famine buffer. The consequences of razing that buffer would soon come clear.

The Houthis and War

By 2015, questions of competing policy frameworks and how best to reform subsidy systems and their implementation had become moot, because the fighting began to fill the political horizon. On one side have been forces, domestic and to a large extent international, aligned with Abdrabbuh Mansour al-Hadi, who was installed through a US- and Saudi-backed 2012 election in which he was the only candidate. On the other have been large elements of the Yemeni national army, as well as the Houthi movement. The latter evolved out of a Zaydi Shi'a revivalist movement in Sa'ada province, emerging in defence of their identity amidst the spread of Saudi-backed sectarianism. The group adopted anti-Israel and anti-US rhetoric in reaction to the 2003 US invasion of Iraq, spurring the Yemeni government, which was then collaborating on 'counter-terrorism' with the US to kill its leader, Hassan Badreddin al-Houthi. The Houthis, known formally as Ansar Allah, initially allied with the southern Hirak movement. In late March 2015, after lightning advances of the Houthi forces, Hadi fled to Saudi Arabia and an external onslaught began, one enabled, supplied and carried out under the aegis of the US and the UK, the former colonial power (Ryan, 2016; Hill, 2017).

It bears noting that food price inflation and thus a qualitative deterioration in access to food preceded that attack. According to the 2012 Comprehensive Food Security survey, nine out of ten households reported increasing food prices were damaging their ability to secure food. The inflation rate for food was over 20% from 2009–2011. Some items had doubled in price. Household coping strategies included reducing the quantity and quality of their diet and taking on debt. Both strategies would become increasingly difficult to sustain. More than one-third of households had food-related debt; this figure was 45% amongst food insecure households with an already poor diet. Amidst mounting household debt, there was a fear that bad diets would get worse. The menace of food insecurity mapped unevenly over the rural–urban diet, between and amongst governorates, and in different biomes and agro-climactic zones. Reflecting a global pattern, the situation was and is far worse amongst those who grow food than those who merely buy it – nearly four times as many food insecure people lived in the countryside as compared to the city. This was a massive increase over the 2009 figures. Furthermore, rural areas themselves are not homogenous, with patches of extreme hunger and desperate food insecurity spreading even in 2012 (World Food Programme, 2012).

By 2015, a concatenation of causes had led to further damage to Yemen, pushing ever-more regions of the country ever closer to famine. In a report tellingly entitled, 'Summary of the Preliminary Assessment of Damages to the Agricultural Sector in the Provinces of the Republic of Yemen Caused by the inhuman Saudi-USA Air Strikes from 26/3/2015 to 31/10/2015', the Ministry of Agriculture and Irrigation counted nearly 100 poultry farms destroyed, over 40,000 honeybee hives eradicated and over 50,000 head of livestock lost. Air strikes had targeted almost 1000 fields, almost 100 wells and water pumps, and 11 regional fruit and vegetable markets – including 9 in Sa'dah governorate, the Houthi heartland. The air raids had razed or damaged over 3500 greenhouses and a half-dozen headquarters of agricultural associations, overwhelmingly or solely in Sa'dah governorate. All this was recorded between March and October 2015 (Ministry of Agriculture and Irrigation, 2015). Such numbers are merely indicative, since the period for which we have statistics was just 7 months of a war, which by October 2017 had lasted 31 months. Nevertheless, they suggest a pattern of systematic attacks on the agricultural sector. Qualitative survey evidence supports such a conclusion. Around that time, a Gallup poll estimated that 45% of Yemenis had lost their main income source due to the war, and the amount who were finding it 'difficult' or 'very difficult' to get by had increased from 26 to 41% of the polled population (Fakhreddine, 2016). These results say nothing of almost 12,000 dead and 20,000 injured by April 2017 (Legal Center for Rights and Development, 2017) as well as a murderous cholera outbreak.

As Mundy points out, the Saudi military declared the Sa'dah governorate a military target 6 weeks into the war, and there is 'strong evidence that Coalition strategy has aimed to destroy food production and distribution in the areas which Ansarallah (the Houthis) and the General Congress Party control' (Mundy, 2017, p. 15). In April 2017 the World Food Programme reported that, as compared to the 'pre-crisis period', the national average price of wheat flour had risen 29%; red beans, 58.5%; sugar, 28.5%; and vegetable oil, 11.2%. Governorates with active fighting experienced higher inflation. Market disintegration, leading to uneven prices across the country, continued to prevail (World Food Programme, 2017). Cultivated areas and production were 38% lower than during the pre-crisis period. Most fishermen had lost their boats. By March 2017, 70% of the population was food insecure (IPC, 2017).

Remedies

Questions of resolution of crisis must account for internal and external crises, the stressors that cause them, and the root causes of those stressors, all the while attending to the aggravated disarray within Yemeni agricultural and food provisioning systems. If the war has pushed Yemen's poor past the breaking point, especially in the rural world, poverty and hunger were widespread before the war. This was above all the case in the rural sector, where farm households had been the most likely to be blighted by food insecurity of several scales of severity.

But to speak of reconstruction, reorientation of developmental planning, redirection of resources and reassessment of capacities presumes a state capacity to carry out those tasks. Such a capacity can only be the work of a state apparatus working in dynamic consultation with those most vulnerable amongst the Yemeni poor. It follows that resolving Yemeni agricultural crisis begins with stopping the external war. The Gulf States, with the US supplying them logistically and technically such that it is commonly and, I would suggest, correctly understood that it is a US–Saudi war, have imposed that war upon Yemen's people. To speak of state-directed development when the substantial fabric of sovereignty is itself being torn apart by high-explosive weapons are not words that carry much weight.

Upon the war's end, there surely exists the capacity to craft developmental plans capable of contributing to Yemeni endogenous development. The work of Yemeni social and natural scientists provides the basis for such plans. Such intellectual labour can certainly be scaled up provided resources are freed from bourgeoisie rent circuits, particularly through the import licences through which consumption ability of the poorer classes is taxed through price inflation and which circularly provide a structural disincentive for import substitution whether in the agricultural sector or elsewhere. Furthermore, the perceived need for Yemen to armour itself – and such a perception has been borne out amidst the recent US–Saudi aggression – must be addressed through security arrangements such that the stronger states of the region and globe are unable to run pell-mell over the weaker states, or force them to divert scarce capital from developmental spending to cladding for war.

Following from that, one cannot speak of developmental interventions in the agricultural sector before the task of accurately mapping the contours of Yemeni farming, including assaying local land races, is seriously taken up. For example, historically, villages have selected sorghum suitable for their microclimates (Swanson, 1979b). Sorghum remains a dominant crop in Yemen, resistant to the idiosyncrasies and pressures of the Yemeni climate, including drought and aridity, and useful as both food for humans and fodder for herds. The state must support improving sorghum yields through extension work, including surveying endogenous genetic capacities. This has scarcely occurred in Yemen or globally, since sorghum as a crop is not suitable for imported capital-intense techniques or the make-work projects that are often the interventions of the international agencies. It is also a crop of the poor. Yemen must be understood not as a place for the injection of external knowledge, but in the words of Mundy and Pelat, a place for 'Yemen's development of knowledge **of** itself' (Mundy and Pelat, 2015, p. 100, original emphasis).

The government must also act immediately to save or stabilize existing terraces before the remainder of this capital collapses. Once the terraces are gone, there will be no such farming arenas to which to return. Surveying water basins with an eye towards the appropriate medium- or small-scale hydraulic techniques to enhance irrigation capacity sustainably is another project. It is also one that waits for the conclusion of the external aggression – one that has brought literal famine and plague to Yemen, and casts a shadow so dark over Yemen as a country and a civilization that war's end must be the first demand for those concerned with Yemen's future.

Acknowledgements

A very special thanks to Martha Mundy for sharing her knowledge of Yemeni agriculture in general and her comments on a draft of this chapter.

Note

[1] Other estimates put qat's use of total available irrigation water at 30–40% (Weiss, 2015).

References

Abu-Lughod, J.L. (1991) *Before European Hegemony: The World System A.D. 1250–1350*. Oxford University Press, Oxford, UK.

Almas, A.A. and Scholz, M. (2006) Agriculture and water resources crisis in Yemen: need for sustainable agriculture. *Journal of Sustainable Agriculture* 28, 55–75.

Al-Weshali, A., Bamaga, O., Borgia, C., Van Steenbergen, F., Al-Awlaqi, N. and Babaqi, A. (2015) Diesel subsidies and Yemen politics: post-2011 crises and their impact on groundwater use and agriculture. *Water Alternatives* 8(2), 215–236.

Blair, J.R. (1976) *The Control of Oil*. Pantheon Books, New York.

Carapico, S. (2007) *Civil Society in Yemen: The Political Economy of Activism in Modern Arabia*. Cambridge University Press, Cambridge, UK.

Colton, N.A. (2010) Yemen: a collapsed economy. *Middle East Journal* 64, 410–426.

Ertürk, Y. (1994) Implications of labor displacement for production relations in Yemen. *Center for Migration Studies Special Issues* 11 (4), 107–120.

Fakhreddine, J. (2016) Yemenis divided politically, united in misery. Available at: www.gallup.com/poll/188897/yemenis-divided-politically-united-misery.aspx (accessed 18 August 2017).

Fergany, N. (1982) The impact of emigration on national development in the Arab region: the case of the Yemen Arab Republic. *International Migration Review* 16, 757–780.

Friedmann, H. (1982) The political economy of food: the rise and fall of the postwar international food order. *American Journal of Sociology* 88, 248–286.

Hebshi, M.A. (2004) The role of terraces on land and water conservation in Yemen. In: Agriculture Congress 2004: Innovation Towards Modernized Agriculture. Available at: www.yemenwater.org/wp-content/uploads/2013/03/TerraceFarmingSocioEconomics.pdf (accessed 28 October 2017).

Hill, G. (2017) *Yemen Endures: Civil War, Saudi Adventurism and the Future of Arabia*. Oxford University Press, Oxford, UK.

Integrated Food Security Phase Classification (2017). Yemen: projected acute food insecurity situation – March–July 2017. Available at: www.ipcinfo.org/ipcinfo-detail-forms/ipcinfo-map-detail/en/c/522844 (accessed 26 March 2018).

Karner, M. (2014) Water scarcity and human security in Yemen. *Middle East Perspectives Series 2*. Middle East Institute, Singapore.

Katz, M.N. (1992) Yemeni unity and Saudi security. *Middle East Policy* 1(1), 117–135.

Legal Center for Rights and Development (2017) The outcome of 680 days. المركز القانوني للحقوق والتنمية.

Lichtenthäler, G. (2000) Power, politics and patronage: adaptation of water rights among Yemen's northern highland tribes. *Études Rurales* 155/156, 143–166.

Ministry of Agriculture and Irrigation (2015) Summary of the preliminary assessment of damages to the agricultural sector in the provinces of the republic of Yemen caused by the inhuman Saudi–USA air strikes from 26/3/2015 to 31/10/2015. Available at: www.athimar.org/Article-81 (accessed 26 March 2018).

Moore, S. (2011) Parchedness, politics, and power: the state hydraulic in Yemen. *Journal of Political Ecology* 18, 39–50.

Mundy, M. (1985) Agricultural development in the Yemeni Tihama: the past ten years. In: Pridham, B.R. (ed) *Economy, Society, and Culture of Contemporary Yemen*. Croon Helm, London, pp. 22–40.

Mundy, M. (2017) The war on Yemen and its agricultural sector. Presented at the ICAS – Etxalde Colloquium: The future of food and agriculture, 24, 25, 26 April. Available at: http://elikadura21.eus/wp-content/uploads/2017/04/50-Mundy.pdf (accessed 28 October 2017).

Mundy, M. and Pelat, F. (2015) The political economy of agriculture and agricultural policy in Yemen. In: Al-Sarhan, S. and Brehony, N. (eds) *Rebuilding Yemen: Political, Economic and Social Challenges*. Gerlach Press, Berlin, Germany.

Nitzan, J. and Bichler, S. (2002) *The Global Political Economy of Israel*. Pluto Press, London.

Nugent, J.B. (2003) Yemeni agriculture: historical overview, policy lessons and prospects. In: Lofgren, H (ed.) *Food, Agriculture, and Economic Policy in the Middle East and North Africa*. Emerald Group Publishing Limited, Amsterdam, New York, pp. 257–288

Okruhlik, G. and Conge, P. (1997) National autonomy, labor migration and political crisis: Yemen and Saudi Arabia. *Middle East Journal* 51, 554–565.

Oppenheim, V. (1976) The past: we pushed them. *Foreign Policy* 25, 24–57.

Patnaik, U. and Patnaik, P. (2016) *A Theory of Imperialism*. Columbia University Press, New York.
Phillips, S. (2005) Cracks in the Yemeni system. *Middle East Report Online*, July 28. Available at: at: www.merip.org/mero/mero072805 (accessed 28 October 2017).
Ryan, M. (2016) Civilian casualties in Yemen bring charges of U.S. responsibility for Saudi actions. *Washington Post*. Available at: https://www.washingtonpost.com/world/national-security/civilian-casualties-in-yemen-bring-charges-of-us-responsibility-for-saudi-actions/2016/10/03/29a9b606-864d-11e6-ac72-a29979381495_story.html (accessed 28 October 2017).
Stork, J. (1975) *Middle East Oil and the Energy Crisis*. Monthly Review Press, New York.
Swanson, J.C. (1979a) Some consequences of emigration for rural economic development in the Yemen Arab Republic. *Middle East Journal* 33, 34–43.
Swanson, J.C. (1979b) *Emigration and Economic Development: the Case of the Yemen Arab Republic*. Westview Press, Boulder, Colorado.
Tutwiler, R. (1990) Agricultural labor and technological change in the Yemen Arab Republic. In: *Labor and Rainfed Agriculture in West Asia and North Africa*. Springer, Dordrecht, pp. 229–51. https://doi.org/10.1007/978-94-009-0561-0_11.
van Steenbergen, F., Lawrence, P., Mehari Haile, A., Salman, M. and Faures, J.-M. (2010) Guidelines on spate irrigation (No. 65), FAO Irrigation and Drainage Paper. FAO, Rome.
Varisco, D.M. (1983) Sayl and ghayl: the ecology of water allocation in Yemen. *Human Ecology* 11, 365–383.
Varisco, D.M. (1986) On the meaning of chewing: the significance of qāt (*Catha edulis*) in the Yemen Arab Republic. *International Journal of Middle East Studies* 18, 1–13.
Varisco, D.M. (1991) The future of terrace farming in Yemen: a development dilemma. *Agriculture and Human Values* 8, 166–172.
Vogel, H. (1987) Terrace farming in Yemen. *Journal of Soil and Water Conservation* 42, 18–21.
Ward, C. (2000) The political economy of irrigation water pricing in Yemen. In: Dinar, A. (ed.) *The Political Economy of Water Pricing Reforms*. Oxford University Press, New York, pp. 381–394.
Weiss, M.I. (2015) A perfect storm: the causes and consequences of severe water scarcity, institutional breakdown and conflict in Yemen. *Water International* 40, 251–272.
World Food Programme (2012) The state of food security and nutrition in Yemen. World Food Programme. Available at: http://documents.wfp.org/stellent/groups/public/documents/ena/wfp247833.pdf (accessed 28 October 2017).
World Food Programme (2017) Yemen market watch report (No. 12). Available at: https://reliefweb.int/report/yemen/yemen-market-watch-report-issue-no-12-april-2017 (accessed 28 October 2017).
Yemen Water (2013) Yemen Water briefing note 2. Available at: www.yemenwater.org/wp-content/uploads/2013/03/Fact-sheet-2.pdf (accessed 28 October 2017).

11 Farming for Freedom: the Shackled Palestinian Agricultural Sector

Alaa Tartir*

Al-Shabaka: The Palestinian Policy Network, Washington, DC, USA and The Graduate Institute of International and Development Studies (IHEID), Geneva, Switzerland

Introduction

In the occupied West Bank and Gaza Strip, agriculture is commonly perceived by the Palestinian people to be the backbone of the Palestinian society and economy, with farmers widely viewed as the last stronghold of resistance (Sansour and Tartir, 2014). Agriculture is not merely viewed as an ordinary economic sector, but is instead widely perceived to be an act of resistance and an illustration of steadfastness (Dana, 2014a). Farmers preserve and reclaim land, build self-reliance and an economy of resistance, and challenge forced dependency and economic asymmetric containment. In each of these respects farming is also a political act.

Additionally, agriculture and farming are embedded in and well connected to the individual and collective identity of the Palestinians: their history, culture, lifestyle, literature, and overall struggle for freedom and self-determination. In addition to fulfilling the conventional economic and societal role of the agricultural sector, Palestinian farming and agriculture can be seen as domains that challenge oppression and achieve freedom and emancipation.

However, the reality of Palestinian agriculture and farming has proven to be somewhat at odds with the popular narrative and conscience – closer reflection shows that this backbone has been severely distorted and damaged, if not paralysed (Abu Sa'da and Tartir, 2014). The resilience and resistance of the farmers have been hampered, mainly by the continuation of the Israeli occupation and its policies, but also by the unsupportive and rather damaging policies of the Palestinian Authority (PA). In engaging at both points, this chapter argues that Palestinian agriculture and farmers have been shackled by Israeli colonialism, but also by Palestinian neoliberalism.

As the colonial power, Israel continues in its land confiscation and territorial annexation policy: expanding settlements and colonies; nourishing settler violence; stealing land and natural resources; imposing policies of closure, siege and blockade; controlling trade, exports and imports – each being an element within a 'matrix of control' directed towards the colonization of the Palestinians (Farsakh, 2004; Halper, 2010). Each one of these policies is ultimately directed towards the Palestinian agricultural sector, being undertaken with a view to 'destroy[ing] farming as a way of life for Palestinians and thereby weaken[ing] their passionate attachment to their ancestral lands' (Cook, 2016, p. 4).

As the colonization process advanced, Palestinian land was lost and conquered by the

---

* Email: alaa.tartir@graduateinstitute.ch

Israelis, with the remaining lands being fragmented by multiple measures; trees were uprooted and yields destroyed; water was stolen or became unfit for consumption; and the whole agriculture sector was besieged (UNCTAD, 2015). Since 1967, Israeli authorities have uprooted 2.5 million fruit trees (ARIJ, 2015a); 800,000 Palestinian olives trees, equivalent to 33 Central Parks of New York (ARIJ and PMoNE, 2011; Visualizing Palestine, 2013), have also been removed, providing a particularly vivid depiction of the colonial enterprise and its impact on agriculture.

The policies and donor-driven development model of the PA also continue to contribute to the deterioration of the agriculture sector (Abdelnour et al., 2012). Less than 1% of the PA budget is currently allocated to the agricultural sector, which is also completely absent from the organization's state building and development plans. The adoption of the neoliberal approach of joint Palestinian–Israeli agro-industrial zones and qualified economic zones also continues to negatively impact both the needs of Palestinian farmer and the agricultural sector more generally (Tartir, 2012a, 2013; Dana, 2014b). These are just few examples of the PA's pernicious policies and terrible neglect of the agricultural sector.

However, ultimately it is the PA's reluctance to invest in farming that has proven to be the gravest dereliction of duty. The PA's reticence appears to be primarily attributable to the fact that such investment would necessitate direct confrontation with both Israel and the international donor community, a confrontation that the PA is both unable and unwilling to undertake (Tartir, 2016). In addition, the donor community has shown little or no interest in agriculture, evidencing a clear lack of interest in confronting the Israeli occupying authorities and an orientation towards quick-impact projects (Le More, 2008; Taghdisi-Rad, 2010). A total of nearly US$30 billion of international aid has been committed to Palestine during the period 1993–2015. Of this, only a tiny amount has been directly committed to the agricultural sector: between 1994 and 2000 only 1.41% of aid was allocated to agriculture; between 2000 and 2005 the average allocation was 0.74%; and between 2005 and 2009 only 0.9% on average was committed (DeVoir and Tartir, 2009; Tartir et al., 2012; Kurzom, 2017).

In the half-century since 1967, the Palestinian agriculture sector has experienced a pervasive process of de-development that has gradually deprived it of its transformative potential while expanding Israel's territorial dominance and control. The process of de-development is a 'deliberate, systematic and progressive dismemberment of an indigenous economy by a dominant one, where economic – and by extension, societal – potential is not only distorted but denied' (Roy, 2007, p. 33). De-development is therefore a process that forestalls development by 'depriving or ridding the economy of its capacity and potential for rational structural transformation [i.e. natural patterns of growth and development] and preventing the emergence of any self-correcting measures' (Roy, 1995, p. 128). De-development occurs when normal economic relations are impaired or abandoned, preventing any logical or rational arrangement of the economy or its constituent parts, diminishing productive capacity and precluding sustainable growth. Over time, de-development represents nothing less than the denial of economic potential (Roy, 2014). While the implications of this politically constructed process extend beyond agriculture, it is this sector that most clearly conveys its different attributes.

This chapter will initially provide an overview of the Palestinian agricultural sector, which sets out its resources, productivity and contributions. This is then followed by a discussion of the impact of the Israeli occupation, which focuses on the distortion and deterioration of Palestinian agriculture. The chapter also reflects on the PA's neglect of the agricultural sector, along with the pernicious policies that it has enacted, before offering a set of conclusions.

The Palestinian Agricultural Sector: an Overview

In 1967, Palestinian agricultural production was almost identical to Israel's, accounting for, according to some estimates, over half of the Palestinian gross domestic product (GDP). The West Bank exported 80% of the entire vegetable crop it produced, along with 45% of its total fruit production (Hazboun, 1986; PASSIA, 2015).[1] In the aftermath of the 1967 *Naksa*, the agricultural sector's contribution to GDP declined dramatically (UNCTAD, 1990): from nearly 50% of GDP in the 1960s, this share fell to 9.8% in 2000; 5.5% in 2010; and 3.4% in 2015 (UNCTAD, 2015; PCBS, 2017a). Due to the Israeli state's violence and economic policies towards the Occupied Palestinian Territory (OPT), the agricultural labour force fell as a percentage of the total labour force from 46% in 1969 to 27.4% in 1985 (Kahan, 1987); between 1965 and 1989, Palestinian cultivated areas in the West Bank declined by 30% (from 243,500 ha to 170,600 ha). As a result of the 1967 Israeli occupation of what remains of the Palestinian lands and to the construction and expansion of its illegal settlements, the area of field crops and forages evidenced an equally drastic decline, decreasing from 85,000 ha in 1966 to approximately 46,000 ha in 1994 (Al-'Aloul, 1987; Butterfield et al., 2000).

By 1993, food self-reliance (production/consumption) for some products exceeded the 100% level (citrus 213%, olives 190%, vegetables 149%), but declined at a dramatic rate in the years since (Anabtawi, 2016). The average yield per hectare is half that of Jordan and only 43% of Israel, despite the fact that these countries share an almost identical natural environment (UNCTAD, 2015). The agricultural and fishing sector employed 32% of the Palestinian labour force in the early 1980s; however, this percentage declined to 11.8% in 2010 and 8.7% in 2015 (PCBS, 2017a). Between 2000 and 2008, the total cultivated area in the West Bank and Gaza Strip averaged 150,000 ha; however, since 2008 it has dropped by a staggering 40% to 90,000 ha due to the three major Israeli assaults on the Gaza Strip and the illegal settlements expansion, lands confiscation and continuation of the military occupation measures in the West Bank, as well as the PA's neglect of the agriculture sector (Kurzom, 2016; PCBS, 2017a). In addition, the ratio of labour productivity in agriculture to labour productivity of the economy as a whole fell by more than 50% between 1995 and 2011 (World Bank, 2014).

Figures provided by the United Nations Conference on Trade and Development (UNCTAD) indicate that nearly 81% of West Bank cultivated land is committed to the cultivation of low-value and low-yield crops – olives account for 57% of cultivated land while vegetables and fruits account for 19%. The 2011 PCBS agriculture survey reported that there are 7.8 million fruit-bearing olive trees and 1.1 million non-bearing olive trees in the West Bank and Gaza (PCBS, 2011). The olive subsector contributes 15% of total agricultural income and also 'mitigates the impact of unemployment and poverty by providing 3 to 4 million days of seasonal employment per year and by supporting 100,000 Palestinian families' (UNCTAD, 2015, p. 5). However, olive oil production is in decline. It has dropped from an average of 23,000 tonnes per year (2000–2004) to 14,000 tonnes per year

Table 11.1. Palestinian agricultural holdings, average daily wages, and cultivated areas. (Prepared by the author based on PCBS (2011), PA (2014), UNCTAD (2015) and PCBS (2017a).)

Dimension	Statistics
Number of agricultural holdings (2011)	105,238 (85,885 in the West Bank and 19,353 in Gaza)
Type of agricultural holdings (2011)	68% plant holdings; 10% livestock holdings; 22% mixed holdings
Ownership of agricultural holdings (2011)	93% to men; 7% to women
Agricultural sector average daily wage (2015)	US$13 (US$18 in the West Bank and US$6 in Gaza), below the average wage for all economic activities (US$22)
	Of women employed, 82% were unpaid family members
Total cultivated area (2011)	103,490 ha (64% fruit trees; 13% vegetables; 24% field crops)
Type of land (2014)	200,000 ha as rangeland; 62,100 ha for grazing; 9400 ha as forests
Location of land (2015)	62.9% of the arable land is located in Area C; 18.8% in Area B; 18.3% in Area A
Number of livestock (2013)	33,980 cattle; 2058 camels; 730,894 sheep; 215,335 goats; 4,342,910 broilers; 1,154,586 layers; 46,226 beehives
Caught fish and fishermen (2015)	Caught fish totalled 3.2 tonnes; 3617 fishermen; 1261 boats

(2007–2010). As a result of this decline, around half of domestic demand had to be met through imported olive oil in 2009 (Palestinian Ministry of Agriculture, 2010). This is a particularly significant development because the olive harvest 'represents the ultimate kind of resistance by Palestinians: an individual refusal to be moved, and a collective refusal to be ethnically cleansed' (Cook, 2016, p. 3).

Crucially, Palestinian use of water for agriculture was estimated to be one-tenth of Israel's (UNCTAD, 2015); on a per capita basis, Israel's residential water consumption is more than five times that of West Bank Palestinians (Ma'an Development Centre, 2010). Israeli settlers, meanwhile, consume ten times more water than the Palestinians in the occupied West Bank (AIX, 2017; Hass, 2017). The same pattern of water allocation extends to the agriculture sector. Palestinians are allowed to use only 18% of the water resources available in the West Bank, with the remainder used by Israel (ARIJ, 2007, p. 4). Over 70% of communities located entirely or predominantly in Area C^2 are not connected to the water network, meaning that they are forced to rely on tanker water, a resource provided at very high cost, which also extends to the agricultural domain (Abdel Razek-Faoder and Dajani, 2013). The lack and the increasing cost of water have forced many Palestinian farmers to leave the agriculture sector, and instead work as cheap labour within the Israeli economy (Farsakh, 2002; Stop the Wall, 2007; Mansour, 2012).

The deterioration of the Palestinian agricultural sector has clear implications for poverty, inequality and food insecurity. Nearly 30% of Palestinians are caught in a poverty trap, while others experience persistently high levels of unemployment (the unemployment rate in Gaza is around 45%, while youth unemployment surpasses 60% (Wildeman and Tartir, 2016)). Only one in four households in West Bank and Gaza is food secure (UNCTAD, 2014). In Gaza, 57% of households are classified as food insecure, 80% are dependent on humanitarian aid and around 33% have been forced to reduce their daily food intake as a result of the Israeli blockade (ILO, 2014). In 2014, approximately 40% of all Palestinian households reported that they received some form of assistance (84% in Gaza Strip and 17% in West Bank) (FSS and PCBS, 2016). Clean water is a requirement for food security, but approximately 95% of the Gaza regional aquifer's water is unsafe for drinking without treatment. In 2000, the United Nations Relief and Works Agency for Palestine Refugees in the Near East (UNRWA) provided 80,000 refugees in Gaza with food assistance – in 2016, this figure had increased to more than 930,000 refugees (UNRWA, 2016). These bleak indicators derive from a decades-long process of domination and colonization, appearing as the logical conclusion of political decisions and predispositions.

Impact of Israeli Occupation on Palestinian Agriculture

The perpetuation of the Israeli occupation and its colonial project are the central factors that continue to militate against Palestinian development. The occupation actively seeks to curtail Palestinian economic, cultural and political rights (Abunimah, 2014). Palestinian political and economic rights are the key objects to which the occupation is addressed, being clearly directed towards their frustration and even negation. This is why the land, and by extension agriculture and farming, are such core preoccupations for the Israeli occupation (Farsakh, 2004). Land remains at the very core of the conflict, a point which is clearly retained by the policies that successive Israeli governments have pursued since Israel was founded in 1948 (Pappe, 2016).

The Israeli policies of domination, enacted over the course of decades, have created fundamental structural deficiencies and distortions in the Palestinian economy, a feature evidenced in the dramatic erosion of the agricultural base (Calis, 2017). From the Israeli perspective, this damage is not negative but rather translates to direct profits, culminating in clear benefits for the Israeli economy, the occupation and Israeli settlers (Who Profits, 2017a, 2017b). Any attempt to quantify these policies, along with the damage that they inflict, will only reflect part of the reality of de-development. However, despite this limitation, it is still worthwhile to reflect upon quantitative indicators that have been provided by international institutions and scholars that are directly engaged with agricultural questions in the West Bank and Gaza.[3]

In 2011, the Palestinian Ministry of National Economy (PMoNE) and the Applied

Research Institute – Jerusalem (ARIJ) reported that the measurable costs imposed by the Israeli occupation on the Palestinian economy (for 2010) amounted to US$6.897 billion, which represented 84.9% of the total estimated Palestinian GDP. This estimate suggests that, in the absence of the Israeli occupation, the Palestinian economy would be nearly twice its current size. Closer consideration reveals that more than half (56%) of the costs are attributable to the fact that the Palestinians are not able to access their own resources, including land (ARIJ and PMoNE, 2011).[4]

In 2015, ARIJ updated the report, estimating that the overall cost of the Israeli occupation amounted to around US$9.458 billion, a figure representing 75% of Palestinian GDP (ARIJ, 2015b). Direct damages inflicted during the course of the three Israeli military operations in the Gaza Strip during the period 2008–2014, were at least three times the amount of Gaza's GDP. From 2000–2005, the total cost of Israeli incursions during the *Al-Aqsa* (or second) *Intifada* amounted to US$8.4 billion, an amount twice the size of the Palestinian economy (UNCTAD, 2016). Clearly, these costs cannot be reversed or retrieved while the military occupation is ongoing; international aid provides, at best, a partial or qualified amelioration.

The World Bank has also estimated that the potential direct additional output of a number of sectors (in particular agriculture and the exploitation of Dead Sea minerals) would amount to a total annual income of at least US$2.2 billion – equivalent to 23% of Palestinian GDP (2011 figures). This counterfactual estimate suggested that the agricultural sector would benefit by a total of US$704 million, equivalent to 7% of the 2011 GDP (World Bank, 2014). The Office of the Quartet Representative estimated that an increase of nominal GDP (up to US$1 billion) and employment (up to 30,000–55,000 jobs) could be achieved over a 3-year period, despite the continuation of the occupation. Both outcomes, it was suggested, could be achieved through the commitment of an additional 150 million cubic metres (MCM) of water and cumulative capital expenditures of US$190–245 million (Office of the Quartet, 2014, p. 3).

Upon descending from the abstract level of these counterfactual analyses, we see that the reality of Palestinian agriculture is characterized by deterioration and falling productivity, developments that have important implications for the wider economy and society: UNCTAD estimates that the Palestinian agricultural sector is currently operating at perhaps one-quarter of its potential (UNCTAD, 2015). In large part, this unrealized potential is attributable to the access and mobility restrictions that continue to impede the work of Palestinian farmers. The digging of wells is not allowed; security buffer zones are established and Palestinian land is confiscated for Israeli military use; meanwhile, the Separation Wall, checkpoints and other physical barriers impede the ability of the agricultural sector to grow or even function. In Gaza, a large area of agricultural land is inaccessible or out of production as a direct result of its procurement for security purposes; meanwhile, the fishing industry has completely collapsed.

The regime of control and restrictions imposes considerable transaction costs, complicating economic exchange and frustrating economic development. UNCTAD estimates that the 'costs of exporting and importing borne by Palestinian producers are twice as much as those borne by their Israeli counterparts, while procedures for importation require four times the amount of time Israeli importers spend on similar activities' (UNCTAD, 2015, p. i). As a direct consequence of Israeli restrictions on the importation of fertilizers, Palestinian agricultural productivity has declined by between 20 and 33% since enforcement of the restrictions on the importation of fertilizers (UNCTAD, 2015). In 2016, Israel demolished 1023 Palestinian residential and commercial properties in the West Bank (PCBS, 2017b); between 1967 and 2013, 27,000 Palestinian structures were demolished (UNDTAD, 2015, p. 20). In the West Bank, there are around 413 illegal Israeli constructions (including 150 settlements and 119 outposts), of which 48% are built on private Palestinian land (PCBS, 2017b). In 2014, approximately 9333 productive trees were destroyed or vandalized; in January 2015 alone, another 5600 trees were vandalized. At least 10% of the most fertile land of the West Bank has been enclosed behind the separation barrier; meanwhile, only 35% of potentially irrigable land is actually irrigated, coming at an annual cost of 110,000 jobs and 10% of GDP (UNCTAD, 2016, pp. 11–15).

Area C of the West Bank effectively remains out of reach for Palestinians – of the 16,000 arable hectares in the Jordan Valley, Palestinians only farm about 4200. Settlers, meanwhile, farm between 3000 and 5000 ha with the rest being blocked off to Palestinians (AIX, 2017; Hass, 2017). As the consequence of a systematic policy of displacement, the Jordan Valley's Palestinian population has decreased over the period 1967–2009 from 320,000 to 52,000 residents (Abdel Razek-Faoder and Dajani, 2013). While most of the West Bank's aquifer and spring water is located in Area C, Palestinians have not been able to extract their agreed allocation of 138.5 MCM per annum (World Bank, 2014, p. 19). If utilized to its full potential, this resource would irrigate the 32,640 ha of arable land that is notionally available for Palestinian cultivation in Area C, thus increasing Palestinian Area C production by US$1.068 billion (World Bank, 2014, p. 20).

During the course of the 2014 war on Gaza, the Israeli army destroyed major parts of the agricultural infrastructure, a large portion of which has not yet recovered. During that assault, the Israeli army destroyed 1287 ha of tree orchards, 335 licensed water wells, 1754 ha of vegetables, 922 poultry farms, 1532 sheep and cattle farms and 3219 greenhouses (FSS, 2016). In 2011, it was estimated that three-quarters of the Gaza Strip's population were food insecure, a development attributable to Israel's commitment to use food as a means of population control (Zurayk and Gough, 2013). The Israeli 'security buffer zone' usurps about 24% of the total area of Gaza Strip, further constricting the most densely populated area in the world (which has around 5000 capita/km^2) (PCBS, 2017b). In response to rapidly deteriorating living conditions, in which socio-economic conditions have fallen to their lowest point since 1967 (UN, 2015), the UN has suggested that the Strip could become uninhabitable by 2020.

Israeli policies also impede the farming livelihoods of Palestinian Bedouin communities resident in the West Bank and Al-Naqab (Negev). An Al-Shabaka policy brief (Abdelnour et al., 2012, p. 3) suggests that there are 13,000 Palestinian Bedouins resident in the West Bank, with 20 communities just east of Jerusalem accounting for 2300 of this total (IWGIA, 2011); of the 7000 Bedouins resident in Area C, more than 80% are refugees, with 55% being food insecure (OCHA, 2011; 2014). 85% of Bedouins in the West Bank lack connection to electricity and water networks, while two-thirds reported facing settler violence over the course of the preceding 3 years (OCHA, 2014). The struggle for existence of the Palestinian Bedouins against Israel, especially those resident in Area E1 (an area located just east of the Jerusalem 'municipal boundary', on the hills between Ma'aleh Adummim Jewish settlement and Jerusalem), is particularly instructive, resulting in the loss of almost all access to land (Crowe, 2012; Hass, 2016). Many of these residents have demolition orders pending against their homes and face severe security risks, including exposure to land mines on a daily basis, which poses a direct threat to their herding activities and their livelihoods (Heneiti, 2016).

In the Al-Naqab (Negev), similar practices are also being implemented (HRW, 2008; Amara, 2013). The Bedouin village of *Al-Araqib* has reportedly been demolished more than 110 times (MaanNews, 2017), with a view to realizing the Jewish National Fund's plan to plant a 'peace forest' in its place. The violent demolition of the Palestinian Bedouin village of *Umm al-Hiran*, undertaken with a view to enabling the construction of a Jewish village, is another instructive example in this respect (Avis, 2017). The 2013 Prawer Plan envisages the 'relocation' of 40,000 Palestinian Bedouins, with a view to settling 1 million Israeli Jews in the Al-Naqab/Negev.[5]

While the human costs of these actions are considerable at the Palestinian side, they have resulted in substantial profits for Israeli companies and settlers. For the Israeli government, agriculture is an offensive weapon that can be directed towards the denial of Palestinian rights, a denial that in turn implies the inverse (e.g. profits and privileges) for Israel's Jewish citizens. This is clearly reiterated by the fact that a considerable amount of the agricultural export, much of which is grown by Israeli settlers in the occupied Palestinian and Syrian territories (Who Profits, 2014), is exported to Europe. In the process, the illegal Israeli settlements make extensive use of water and other natural resources, clearly reiterating Israel's systematic intention to dry Palestine of its water (Dajani, 2014). In the Jordan Valley, Israel granted almost exclusive use of water to settlements; every year, settlers export around US$285 million worth of agricultural goods to Europe, which is around 28% of the total agricultural goods exported from Israel to Europe (Who Profits, 2014, p. 8); in contrast, Palestinian exports account for a total of around US$19 million.[6] These figures clearly illustrate how the potential profits of Palestinian agriculture are seized and transferred, being transformed into a weapon that helps to perpetuate the colonial condition.

The PA and Agriculture

The Israeli policies and their consequences for Palestinian farming and agriculture have not been challenged, much less reversed, by the Palestinian governing body. For the PA, this imperative is at the bottom of its list of priorities, something that is clearly reiterated in actions and policies (or the lack thereof) and the resources that the PA commits to this issue (Dana, 2014a). The official Palestinian neglect of the besieged agricultural sector has been qualified to some extent by the engagement of an aid-dependent Palestinian civil society. Yet, this remains profoundly problematic; appearing to suggest, by virtue of the fact that it does not openly challenge or contest, a tacit approval of a colonial strategy that seeks to strip the colonized from their productive base and attachment to their land (Sansour and Tartir, 2014).

Since the establishment of the PA in 1994, less than 1% of the PA annual budget has been allocated to the agricultural sector (Abdelnour et al., 2012), with 85% of this total being spent on the salaries of staff within the Ministry of Agriculture (UNCTAD, 2012); in comparison, over the past decade, the security sector has accounted for around 30% of the PA's annual spending (Tartir, 2017). It is particularly instructive to reflect that the budget of some Palestinian non-governmental organizations (NGOs) exceeds the budget set aside for the PA Ministry of Agriculture. The little funding that has reached the agricultural sector has helped in sustaining the status quo and reinforcing the condition of dependency (Mansour, 2012; Saleh, 2012). The consequences that have arisen from specific donor practices (conditionality) and interventions within specific areas (particularly water-related issues) have not been challenged, much less corrected, by the PA (Hanieh, 2016).

These developments serve as further confirmation of the PA's deeply embedded dependence upon external donors (Tartir, 2015). The PA's failure to prioritize agriculture is clearly underlined by the fact that the agricultural sector does not feature prominently within its strategic development plans and state-building programmes. In those instances, where the sector does appear within these plans, there is the suspicion that this is attributable to the need to tick the inclusiveness box rather than the indication of a clear commitment to elevate the sector. In the absence of a clear strategic direction, the PA is neither supporting farmers through subsidies, infrastructure investments or the opening of new markets, nor protecting farmers from the violence of Israeli settlers.

PA policy makers have also frequently yielded to the conceptual and theoretical limitations of the modernization paradigm, thereby coming to view farming and agriculture as being an 'anti-modernity' domain that is fundamentally opposed to the project of modernization that needs to be undertaken. In contrast, the services and real estate sectors have instead been afforded priority, being more clearly aligned with the central prerogatives of the donor-imposed programme of neoliberal economic reform. In all too many respects and instances, the PA leadership has come to view the agriculture sector as a burden, failing to acknowledge, much less engage, the essential contribution that agriculture can make both to the national economy and the wider political struggle (Bisan, 2011; Tartir and Shikaki, 2013).

Instead of empowering farmers, the PA, acting in harmony with its political and economic agenda and complex network of corruption, has instead provided agency to Palestinian crony capitalists and a private sector elite (Tartir, 2012b). In essentially adopting the mantra 'rich individuals, poor nation', the PA unconsciously echoed the practices of the Israeli occupation. In a faithful rendition of the policies pursued by the colonial power, the PA has even forcefully confiscated land from Palestinian farmers, citing 'public use' justifications in the process, with a view to empowering crony capitalists to expand their real estate investments (Dana, 2014b).

As a direct consequence of neoliberal policies, credit facilities have rapidly proliferated. However, and as in many other places around the globe, this development has been of little benefit to the agricultural sector, in large part due to the perception that it is a 'high risk' investment (Mansour, 2012; Abunimah, 2014). Neoliberal policies have also directly cultivated a class system among agricultural producers, spawning an elite of large-scale land-owning agribusiness owners and a rootless and dislocated peasantry willing to undertake wage labour in the service of these agribusinesses (Mansour, 2012). Neoliberal trade policies initiated under the Economic Paris Protocol (the economic annex of the Oslo Accords) have contributed to an influx of cheap food imports, leaving traditional Palestinian producers struggling to compete as a direct consequence (Mansour, 2012).

The PA's close adherence to neoliberal policies has also resulted in the establishment of joint Palestinian–Israeli agro-industrial zones, with the apparent expectation that these export-oriented zones will support Palestinian farmers. There is ample evidence that suggests the exact opposite, with farmers being transformed into cheap labour, ultimately benefiting Israeli companies and Palestinian crony capitalists. Export, as opposed to the fulfilling of urgent local demand, is the underpinning priority informing each of these developments (Tartir, 2012a; Bisan, 2013).

More problematically, these donor-driven and sponsored agro-industrial zones were built on fertile lands (as in the case of Jenin), a development that represents a 'criminal waste of Palestinian agricultural resources' (Cook, 2016). Closer examination reveals that these agro-industrial zones 'are helping to deprive the Palestinian economy of its transformative potential; expand Israel's territorial dominance in the OPT; increase Palestinians' dependency on Israel in both goods and labour markets; and displace small-scale family farming, which has been the sustaining power of the Palestinian people and culture for generations' (Sansour and Tartir, 2014, p. 2).

To summarize, the PA's reluctance to confront the conditions that sustain and embed dependence ultimately contribute to a perpetuation of the colonial condition. In remaining beholden to wider developmental prejudices and its own conceptual limitations, the PA lacks a clear strategy that can be applied to existing challenges and obstacles. Far from addressing the conditions and attributes of de-development, the PA, through its policies, actions and cooperation with the colonizer, ultimately ensures its reinforcement and perpetuation.

Conclusion

The crisis of the Palestinian agricultural sector is not in itself attributable to the absence of a modernized farming sector. Rather, this feature is instead a symptom, which can be traced back to originating political pathologies. The distortions and the fundamental deficiencies that characterize the agricultural sector are driven by the decades-long Israeli colonization of the Palestinian land. This process of colonization rested upon the conquest of Palestinian land that sought to restrict and confine the possibilities of independent Palestinian development (both political and economic). De-development is not an unfortunate or coincidental outcome, but can instead be traced back to a deliberate and focused colonial strategy.

Israel's systemic policies, which targeted Palestinian farming and agriculture, have been sustained and embodied by coercive power and brutal occupation, inflicting substantial economic and social damage. Far from challenging or contesting colonial power, the PA has instead more frequently functioned as a conduit through which it is reproduced. The Palestinian agricultural sector is therefore entrapped between Israeli colonialism and Palestinian neoliberalism, a uniquely pernicious double-bind that frustrates both its contemporary and future development. To continue under these circumstances is to undertake an act of resistance: to farm Palestine is to farm for freedom.

Notes

[1] In this chapter, the scope of analysis is limited to the Palestinian agricultural sector in the occupied West Bank (5840 km^2) and Gaza Strip (360 km^2), with particular focus upon the past two decades. For added context and historical background, see Tamari (1981), Doumani (1995), Sayigh (1979), Nadan (2006), Kurzom (2001) and Samara (2005).

[2] Under the terms established by the Oslo Accords, the West Bank was divided into three areas: Area A, which came under the civilian and security control of the PA (18%); Area B, which came under PA civilian and Israeli military control (21%); and Area C, which came under full Israeli control (61%). According to the World Bank, 'less than 1% of Area C, which is already built up, is designated by the Israeli authorities for Palestinian use; the remainder is heavily restricted or off-limits to Palestinians, with 68% reserved for Israeli settlements, circa 21% for closed military zones and circa 9% for nature reserves' (2014, p. 13).

[3] Hever (2010; 2012; 2016), Bar-Tal and Schnel (2012) and Swirski (2005; 2008) are particularly relevant to this discussion.

[4] Costs imposed by the Israeli occupation on the Palestinian economy were also attributed to restrictions on the value added of irrigation (15.0%), the restrictions on the Jordan Valley agriculture (8.2%), restrictions on the Dead Sea salts and minerals (13.6%) and the uprooting of trees (1.7%).

[5] The amount of land cultivated by the Bedouins in the Negev prior to 1948 was more than three times the amount cultivated by the entire Jewish community (the Yishuv) in all of historical Palestine. Refer to the *Journal of Palestine Studies,* Special Focus on Palestinian Bedouins, published in July 2016 (www.palestine-studies.org/resources/special-focus/palestinian-bedouins).

[6] Most of the pomegranates exported from Israel to Europe are grown in the occupied West Bank, as well as 22% of the almonds, 12.9% of the olives, 5.4% of the nectarines and 3% of the peaches (Who Profits, 2014, p. 11). Additionally, more than 60% of the dates sold in Israel are grown in the occupied part of the Jordan Valley. More than 80% of the dates grown specifically in the Jordan Valley are destined for export (Who Profits, 2014, p. 15).

References

Abdel Razek-Faoder, I. and Dajani, M. (2013) Land dispossession and its impact on agriculture sector and food sovereignty in Palestine: a new perspective on Land Day. *NewPal Paper*. Available at: http://novact.org/wp-content/uploads/2013/10/NEWPal-Land-Day-paper-final.pdf (accessed 13 March 2018).

Abdelnour, S., Tartir, A. and Zurayk, R. (2012) Farming Palestine for freedom. *Al-Shabaka Policy Brief*, July. Available at: http://al-shabaka.org/node/437 (accessed 13 March 2018).

Abu Sa'da, C. and Tartir, A. (2014) From food security to food sovereignty without a state: the case of Palestine. Paper presented at *Developing Agriculture, Cultivating Sovereignty in the Arab Middle-East (1940–2014)* conference, 6–8 November, University of Fribourg, Switzerland.

Abunimah, A. (2014) *The Battle for Justice in Palestine*. Haymarket Books, Chicago, Illinois.

AIX (2017) Improving the Gazan Economy and Utilizing the Economic Potential of the Jordan Valley. Aix Group, Jerusalem. Available at: http://aix-group.org/index.php/2017/01/15/two-further-studies-improving-the-gazan-economy-and-utilizing-the-economic-potential-of-the-jordan-valley (accessed 13 March 2018).

Al-'Aloul, K. (1987) *A Plan Proposal for Afforestation and Land Reclamation in the West Bank*. Arab Thought Forum, Jerusalem.

Amara, A. (2013) The Negev Land question: between denial and recognition. *Journal of Palestine Studies* 42, 4, 27–47.

Anabtawi, R. (2016) Is development under occupation sustainable? Agriculture as a model. MSc thesis. Bethlehem University, Bethlehem, Palestine.

Applied Research Institute – Jerusalem (ARIJ) (2007) A review of the Palestinian agricultural sector. ARIJ, Bethlehem, Palestine. Available at: www.arij.org/files/admin/2007_agricutlure_sector_review_english_lr.pdf (accessed 13 March 2018).

Applied Research Institute – Jerusalem (ARIJ) (2015a) Palestinian agricultural production and marketing between reality and challenges. ARIJ. Available at: https://goo.gl/HsT38I (accessed 13 March 2018).

Applied Research Institute – Jerusalem (ARIJ) (2015b) The economic cost of the Israeli occupation of the occupied Palestinian Territories. ARIJ. Available at: https://goo.gl/zJWD33 (accessed 13 March 2018).

Applied Research Institute – Jerusalem (ARIJ) and Palestinian Ministry of National Economy (PMoNE) (2011) The economic costs of the Israeli occupation for the occupied Palestinian territory. PMoNE and ARIJ. Available at: www.un.org/depts/dpa/qpal/docs/2012Cairo/p2%20jad%20isaac%20e.pdf (accessed 13 March 2018).

Avis. M. (2017) What is evidence anyway? Activism in the era of post-truth. *Open Democracy*, 23 January. Available at: https://www.opendemocracy.net/arab-awakening/maya-avis/what-is-evidence-anyway-activism-in-era-of-post-truth (accessed 13 March 2018).

Bar-Tal, D. and Schnel, I. (ed.) (2012) *The Impacts of Lasting Occupation: Lesson from Israeli Society*. Oxford University Press, Oxford, UK.

Bisan (2011) The myth of development [Wahm Al-Tanmeya]. Bisan Center for Research and Development, Ramallah, Palestine.

Bisan (2013) Critical studies in development in Palestine [Dirasat Naqdiya Fi Waqi' Al-Tanmina Fi Filasteen]. Bisan Center for Research and Development, Ramallah, Palestine.

Butterfield, D., Isaac, J., Kubursi, A. and Spencer, S. (2000) Impacts of water and export market restrictions on Palestinian agriculture. Available at: http://socserv.mcmaster.ca/kubursi/ebooks/water.htm (accessed 13 March 2018).

Calis, I. (2017) Routine and rupture: the everyday workings of abyssal (dis)order in the Palestinian food basket. *American Ethnologist* 44 (1), 65–76.

Cook, J. (2016) Agro-resistance. *The Link*, Americans for Middle East Understanding. September, 49, 4. Available at: www.ameu.org/Current-Issue/Current-Issue/2016-Volume-49/Agro-Resistance.aspx (accessed 13 March 2018).

Crowe, S. (2012) 2,600 Bedouins threatened with displacement as Israeli settlements expand. *Electronic Intifada*, 7 February. Available at: https://electronicintifada.net/content/2600-bedouins-threatened-displacement-israeli-settlements-expand/10903 (accessed 13 March 2018).

Dajani. M. (2014) Drying Palestine: Israel's systemic water war. *Al-Shabaka Policy Brief*, September. Available at: https://al-shabaka.org/briefs/drying-palestine-israels-systemic-water-war (accessed 13 March 2018).

Dana, T. (2014a) A resistance economy: what is it and can it provide an alternative?, *Pal Papers*, Rosa Luxemburg Foundation, November. Available at: www.rosalux.de/fileadmin/rls_uploads/pdfs/sonst_publikationen/A_Resistance_Economy.pdf (accessed 13 March 2018).

Dana, T. (2014b) The Palestinian capitalists that have gone too far. *Al-Shabaka Policy Brief*, January. Available at: https://al-shabaka.org/briefs/palestinian-capitalists-have-gone-too-far (accessed 13 March 2018).

DeVoir, J. and Tartir, A. (2009) Tracking external donor funding to Palestinian non-governmental organisations in the West Bank and Gaza 1999–2008. Palestine Economic Policy Research Institute-MAS and the NGO Development Center (NDC), Ramallah.

Doumani, B. (1995) *Rediscovering Palestine: Merchants and Peasants in Jabal Nablus, 1700–1900*. University of California Press, Berkeley, California.

Farsakh, L. (2002) Palestinian labor flows to the Israeli economy: a finished story? *Journal of Palestine Studies* XXXII(1), 13–27.

Farsakh, L. (2004) The political economy of agrarian change in the West Bank and Gaza Strip. *Robert Schumann Centre for Advanced Studies Working Paper*, European University Institute, Florence, Italy.

Food Security Cluster (FSS) (2016) Agricultural damages and FSS funding March 2016. FSS, 28 March. Available at: http://fscluster.org/state-of-palestine/document/agricultural-damages-and-fss-funding (accessed 13 March 2018).

Food Security Cluster (FSS) and Palestinian Central Bureau of Statistics (PCBS) (2016) Socio-economic and food security survey 2014. Food Security Cluster and Palestinian Central Bureau of Statistics, May. Available at: http://fscluster.org/sites/default/files/documents/sefsec2014_report_all_web.pdf (accessed 13 March 2018).

Halper, J. (2010) *An Israeli in Palestine: Resisting Dispossession, Redeeming Israel*. Pluto Press, London and Ann Arbor, Michigan.

Hanieh, A. (2016) Development as struggle: confronting the reality of power in Palestine. *Journal of Palestine Studies* 45 (4), 32–47.

Hass, A. (2016) Whither the Bedouin once their homes are demolished? *Haaretz*, 19 March. Available at: www.haaretz.com/israel-news/1.709704 (accessed 13 March 2018).

Hass, A. (2017) Israeli and Palestinian experts draw up road map to save Palestinian economy. *Haaretz*, 25 March. Available at: www.haaretz.com/middle-east-news/palestinians/.premium-1.779406 (accessed 13 March 2018).

Hazboun, S. (1986) Continuous destruction of agriculture in the Occupied Territories. *Al-Katib Journal* 97, 45.

Heneiti, A. (2016) Bedouin communities in greater Jerusalem: planning or forced displacement? *Jerusalem Quarterly* 65, 51–85.

Hever, S. (2010) *The Political Economy of Israel's Occupation: Repression Beyond Exploitation*. Pluto Press, London.

Hever, S. (2012) Economic cost of the occupation to Israel. In: Bal-Tar, D. and Schnel, I. (ed.) *The Impacts of Lasting Occupation: Lesson from Israeli Society*. Oxford University Press, Oxford, UK, pp. 326–358.

Hever, S. (2016) How much international aid to Palestinians ends up in the Israeli economy?, *Aid Watch Paper*, Ramallah, Palestine. Available at: http://goo.gl/RnV3f5 (accessed 13 March 2018).

Human Rights Watch (HRW) (2008) Off the map: land and housing rights violations in Israel's unrecognized Bedouin villages. 30 March, Human Rights Watch. Available at: https://www.hrw.org/report/2008/03/30/map/land-and-housing-rights-violations-israels-unrecognized-bedouin-villages (accessed 13 March 2018).

International Labour Organization (ILO) (2014) *Report of the Director General – Appendix: The situation of workers of the occupied Arab territories*. 103rd session of the International Labour Conference, 28 May–12 June, Geneva.

International Work Group for Indigenous Affairs (IWGIA) (2011) The Greater Jerusalem plan threatens Bedouin communities. IWGIA. Available at: www.iwgia.org/regions/middle-east/palestine/886-update-2011-palestine (accessed 13 March 2018).

Kahan, D. (1987) Agriculture and water resources in the West Bank and Gaza (1968–1987). The West Bank Data Base Project, Jerusalem.

Kurzom, G. (2001) Towards alternative self-reliant agricultural development. Development Studies Programme, Birzeit University. Available at: http://sites.birzeit.edu/cds/research/publications/2001/5.pdf (accessed 13 March 2018).

Kurzom, G. (2016) The shocking Palestinian agricultural regression [Al-enhedar Al-Zirai' Al-Filastinin Al-Sademelaayan]. MA'AN Development Center, December. Available at: https://goo.gl/12XaCn (accessed 13 March 2018).

Kurzom, G. (2017) Financing de-development [Tamweel Al-Latanmeya]. MA'AN Development Center, February, Available at: https://goo.gl/oAlH6y (accessed 13 March 2018).

Le More, A. (2008) *International Assistance to the Palestinians after Oslo: Political Guilt, Wasted Money*. Routledge, London.

Ma'an Development Centre (2010) Draining away: the water and sanitation crisis in the Jordan Valley. Ma'an Development Centre, Ramallah. Available at: www.apis.ps/documents/MaanWaterCrisis_JVpdf.pdf (accessed 13 March 2018).

MaanNews (2017) Israeli forces demolish Bedouin village of al-Araqib for 110th time. MaanNews Agency, 9 March. Available at: https://www.maannews.com/Content.aspx?id=775870 (accessed 13 March 2018).

Mansour, A. (2012) Impact of post Oslo aid interventions on the Palestinian agricultural sector. MSc thesis. Bethlehem University, Bethlehem, Palestine.

Nadan, A. (2006) *The Palestinian Peasant Economy under the Mandate: A Story of Colonial Bungling*. Harvard Middle Eastern Monographs 37, Harvard University Press, Cambridge, Massachusetts.

Office for the Coordination of Humanitarian Affairs (OCHA) (2011) Bedouin Relocation: Threat of displacement in the Jerusalem Periphery. September, OCHA. Available at: www.ochaopt.org/documents/ocha_opt_bedouin_FactSheet_October_2011_english.pdf (accessed 13 March 2018).

Office for the Coordination of Humanitarian Affairs (OCHA) (2014) Bedouin communities at risk of forcible transfer. September, United Nations Office for the Coordination of Humanitarian Affairs. Available at: https://www.ochaopt.org/documents/ocha_opt_communities_jerusalem_factsheet_september_2014_english.pdf (accessed 13 March 2018).

Office of the Quartet Representative (2014) Initiative for the Palestinian Economy Agriculture. Office of the Quartet Representative, Jerusalem. Available at: www.quartetrep.org/files/server/agriculture.pdf (accessed 13 March 2018).

Palestinian Academic Society for the Study of International Affairs (PASSIA) (2015) *PASSIA Desk Diary 2015: Economy*. Palestinian Academic Society for the Study of International Affairs, Jerusalem.

Palestinian Authority (PA) (2014) National agriculture sector strategy 2014–2016: 'Resilience and Development'. The Palestinian Authority, Ramallah, Palestine. Available at: www.apis.ps/up/1417423273.pdf (accessed 13 March 2018).

Palestinian Central Bureau of Statistics (PCBS) (2011) Agricultural census report. PCBS, Ramallah.

Palestinian Central Bureau of Statistics (PCBS) (2017a) Numerous agricultural statistics. PCBS. Available at: www.pcbs.gov.ps/site/lang__ar/939/Default.aspx (accessed 13 March 2018).

Palestinian Central Bureau of Statistics (PCBS) (2017b) Press release: on the eve of the 41st anniversary of land day. Palestinian Central Bureau of Statistics. 29 March. Available at: https://goo.gl/viwywc (accessed 13 March 2018).

Palestinian Ministry of Agriculture (2010) Agriculture sector strategy 2011–2013: a shared vision. The Palestinian Authority, Ramallah, Palestine.

Pappe, I. (2016) *The Biggest Prison on Earth: A History of the Occupied Territories*. Oneworld Publication, Oxford, UK.

Roy, S. (1995) *The Gaza Strip: The Political Economy of De-Development*. Institute for Palestine Studies, Washington DC.

Roy, S. (2007) *Failing Peace: Gaza and the Palestinian–Israeli Conflict*. Pluto Press, London.

Roy, S. (2014) Foreword to *The Palestinian People and the Political Economy of De-development: Contesting Colonization, Negating Neoliberalism*. In: Turner. M. and Shweiki, O. (ed.) *Decolonizing Palestinian Political Economy: De-development and Beyond*. Routledge, London, pp. X–XIII.

Saleh, M. (2012) The association between agricultural development and free political will 'Palestine as a model'. MSc thesis. An-Najah University, Nablus, Palestine.

Samara, A. (2005) Development under popular protection [Al-TanmiyaTahat Al-Himaya Al-Sha'abiya]. AlMashraq for Economic and Development Studies, Ramallah, Palestine.

Sansour, V. and Tartir, A. (2014) Palestinian farmers: a last stronghold of resistance, *Al-Shabaka Policy Brief*, July. Available at: http://al-shabaka.org/policy-brief/economic-issues/palestinian-farmers-last-stronghold-resistance (accessed 13 March 2018).

Sayigh, R. (1979) *Palestinians: From Peasants to Revolutionaries: A People's History*. Zed Books, London.

Stop the Wall (2007) Defending Palestinian food sovereignty against occupation and expulsion. Stop the Wall Campaign, Ramallah. Available at: http://stopthewall.org/defending-palestinian-food-sovereignty-against-occupation-and-expulsion-0 (accessed 13 March 2018).

Swirski, S. (2005) *The Price of Occupation*. AdvaCenter, MAPA Publishers, Tel Aviv, Israel.

Swirski, S. (2008) *The Cost of Occupation: The Burden of the Israeli–Palestinian Conflict*. AdvaCenter, Tel Aviv, Israel.

Taghdisi-Rad, S. (2010) *The Political Economy of Aid in Palestine: Relief from Conflict or Development Delayed?* Routledge and LMEI, London and New York.

Tamari. S. (1981) Building other people's homes: the Palestinian peasant's household and work in Israel. *Journal of Palestine Studies* 11(1), 31–66.

Tartir, A. (2012a) Jericho Agro-Industrial Park: a corridor for peace or perpetuation of occupation? *Working Paper*, Bisan Center for Research and Development. Available at: hwww.al-shabaka.org/sites/default/files/Tartir_BCRD_Paper_En.pdf (accessed 13 March 2018).

Tartir, A. (2012b) The private sector and development in occupied Palestinian: who profits? Bisan Center for Research and Development, Ramallah, Palestine.

Tartir, A. (2013) PA industrial zones: cementing statehood or occupation? *Al-Shabaka Commentary*, February. Available at: http://al-shabaka.org/pa-industrial-zones-cementing-statehood-or-occupat (accessed 13 March 2018).

Tartir, A. (2015) Contentious economics in occupied Palestine. In: Gerges, F. (ed.) *Contentious Politics in the Middle East*. Palgrave Macmillan, London, pp. 469–499.

Tartir, A. (2016) Towards a new development doctrine for Palestine [Nahwa Ruya Tanmaweya Filstiniya]. *Journal of Palestine Studies* 105, 61–69.

Tartir, A. (2017) Criminalizing resistance: the cases of Balata and Jenin refugee camps. *Journal of Palestine Studies* 46, 182, 7–22.

Tartir, A. and Shikaki, I. (2013) Development as resistance [Al-Tanmeya ka'ada llimuqawama wa altaharur]. In: Shahin, N. (ed.) *Critical Studies in Development in Palestine*. Bisan Center for Research and Development, Ramallah, Palestine, pp. 13–40.

Tartir, A., Bahour, S. and Abdelnour, S. (2012) Defeating dependency, creating a resistance economy. *Al-Shabaka Policy Brief*, February. Available at: https://al-shabaka.org/briefs/defeating-dependency-creating-resistance-economy (accessed 13 March 2018).

United Nations (UN) (2015) Gaza could become uninhabitable in less than five years due to ongoing 'de-development' – UN report. United Nations, September. Available at: www.un.org/apps/news/story.asp?NewsID=51770#.V4YoYrh97b0 (accessed 13 March 2018).

United Nations Conference on Trade and Development (UNCTAD) (1990) The latest agricultural development in the Occupied Palestinian Land. UNCTAD, Geneva.

United Nations Conference on Trade and Development (UNCTAD) (2012) Report on UNCTAD assistance to the Palestinian people: Developments in the economy of the Occupied Palestinian Territory. UNCTAD, Geneva.

United Nations Conference on Trade and Development (UNCTAD) (2014) Report on UNCTAD assistance to the Palestinian people: developments in the economy of the Occupied Palestinian Territory. UCTAD, Geneva.

United Nations Conference on Trade and Development (UNCTAD) (2015) The besieged Palestinian agricultural sector. UNCTAD, Geneva, Switzerland.

United Nations Conference on Trade and Development (UNCTAD) (2016) Economic costs of the Israeli occupation for the Palestinian people. UNCTAD, Geneva, Switzerland. Available at: https://goo.gl/9Y8uvV (accessed 13 March 2018).

United Nations Relief and Works Agency for Palestine Refugees in the Near East (UNRWA) (2016) Khalifa Bin Zayed Al Nahyan Foundation continues supporting Iftar for Palestine refugees in Gaza. UNRWA, 5 July. Available at: www.unrwa.org/newsroom/press-releases/khalifa-bin-zayed-al-nahyan-foundation-continues-supporting-iftar-palestine (accessed 13 March 2018).

Visualizing Palestine (2013) Uprooted: olive harvest. Infographic, Visualizing Palestine, October. Available at: http://visualizingpalestine.org/visuals/olive-harvest (accessed 13 March 2018).

Who Profits (2014) Made in Israel: agricultural export from Occupied Territories. Who Profits, April. Available at: https://www.whoprofits.org/content/made-israel-agricultural-export-occupied-territories (accessed 13 March 2018).

Who Profits (2017a) Financing land grab: the direct involvement of Israeli banks in the Israeli settlement enterprise. Who Profits, February. Available at: https://whoprofits.org/content/financing-land-grab-direct-involvement-israeli-banks-israeli-settlement-enterprise (accessed 13 March 2018).

Who Profits (2017b) Greenwashing the occupation: The solar energy industry and the Israeli occupation. Who Profits, February. Available at: https://whoprofits.org/content/greenwashing-occupation-solar-energy-industry-and-israeli-occupation (accessed 13 March 2018).

Wildeman, J. and Tartir, A. (2016) A mapping of donors in Gaza 2006–2016: no contact, data chaos and contextually inappropriate programming. A study submitted to Oxfam GB, August.

World Bank (2014) Area C and the future of the Palestinian economy. World Bank, Jerusalem and Washington, DC. Available at: http://documents.worldbank.org/curated/en/137111468329419171/West-Bank-and-Gaza-Area-C-and-the-future-of-the-Palestinian-economy (accessed 13 March 2018).

Zurayk, R. and Gough, A. (2013) *Control Food, Control People: The Struggle for Food Security in Gaza*. Institute for Palestine Studies, Washington, DC.

12 Games Without Frontiers: Development, Crisis and Conflict in the African Agro-Pastoral Belt

Michele Nori[1],* and Edoardo Baldaro[2]

[1]*European University Institute (EUI), Firenze, Italy;* [2]*University of Naples 'L'Orientale', Naples, Italy*

Introduction

Pastoralism and pastoralists are facing important challenges today, due mostly to the reshaping of socio-economic and agro-ecological landscapes of their territories. While in the past herding groups were considered the wealthiest amongst rural people, today the situation is more articulated, and people living on rangelands constitute a large fraction of the world's most vulnerable. Many pastoral populations rank today amongst the poorest and most destitute agricultural peoples in the world and are the most excluded from basic socio-economic services and infrastructure (WHO/UNICEF, 2005; Haughton and Khander, 2009; African Union, 2010). Human development reports indicate that over 50% of the world's most disadvantaged countries are in dryland Africa, where growing inequality and social stratification are reported in most areas, furthering the high socio-economic vulnerability of lower pastoral population strata (Little, 2013). In recent decades even physical security has degraded in portions of these regions, as will be discussed. Why and how are long-term phenomena and short-period dynamics interacting in the pastoral world, degrading pastoralists' security, livelihoods and environment?

Starting from these questions, the aim of this chapter is to explore the main changes linked to these evolutions. Proposing an in-depth analysis of two significant cases from the Horn of Africa and the Sahel, we shed light on the main factors that influence pastoral conditions, strategies and behaviours, and further the structural crisis of their world. In the following sections, we first describe our analytical and theoretical framework. Pastoralism and its territories are introduced, taking into consideration both structural/global trends and local transformations affecting and defining their space. Pastoralists are conceived as rational agents, in their longstanding efforts to cope with and adapt to a shifting context, with a view to defend and evolve their livelihoods and the very existence of their communities. Three main behavioural and strategic patterns to pursue their physical, political and economic security are identified. The second part of the chapter applies this framework to our two cases, with the aim to empirically explore crisis and change in the pastoral world. According to this purpose, we selected two crucial cases where conflict, transnational networks, state weakness and shifting border regimes are reconfiguring the pastoral landscape.

---

* Email: michele.nori@eui.eu

Pastoral Settings

Pastoralism entails extensive livestock-based livelihood systems, where people's economy mostly relies on livestock, whose feeding in turn mostly relies on natural grazing. Pastoral rangelands are characterized by inherently poor soils and extreme climatic conditions; they cover about one-fourth of the global land area, spanning from African drylands to Central Asian steppes, from European mountains to Andean plateaux (Fig. 12.1). In such territories limited and erratic water availability restricts options for alternative land use other than mobile livestock rearing (IFAD, 2010). Such a complex form of natural resource management is an adaptive strategy to a stressful environment, which requires maintaining an ecological balance between pastures, livestock and people. Traditionally pastoral economies are closely integrated with marketing, and nomadic societies are associated with caravan trading and to the extended silk and salt routes. The increasing interface with markets is today pushed by the growing demand for livestock commodities. According to the Republic of Kenya (2012) the term pastoralism refers to both an economic activity and a cultural identity, but the latter does not necessarily imply the former. The range of strategies pastoralists apply to maintain that balance results from and is affected by the larger geo-political system (Nori *et al.*, 2008).

Important portions of pastoral territories form the core of the 'crisis- and conflict-prone' space inside the contemporary pastoral world, including the regions covering Afghanistan and Pakistan through Iraq, northern Yemen, Somalia, Sudan, Libya, southern Algeria and Mali. In this pastoral belt poverty rates and food insecurity rank high, climate hits severely, and violence and insurgency have taken over political dialogue, also because of inappropriate development patterns and state presence. This region overlaps with that identified by Eyal Weizman's *Conflict Shoreline* (2015). In his illustrative work Weizman attests to the close relationship between agro-ecological frontiers and conflict. Using drone strikes as a proxy, he identifies areas where battles are fought in very asymmetric ways, where powerful players have a concern but do not dare to send their armies (which in turn increases the risk for civilian casualties and eventually triggers further resentment and hostility). This takes place along the so-called 'aridity line', areas where the average rainfall is at best 200 mm a year, considered as the minimum for growing cereal crops on a large scale without irrigation; in these areas ecological marginality is at home, and pastoralism often represents the main production and livelihood system.

Trends in Pastoral Areas

A main feature of the agrarian question in recent times is the polarization all over the globe of political and economic interests between areas of high and low economic potential, within countries and amongst regions (McMichael, 1997). Since the 1970s the classical centre–periphery approach, based on the analysis of international political economy, described the structural forces determining the hierarchy shaping international order and affecting north–south relations of domination and dependence (Galtung, 1971; Wallerstein, 1979). Within a country the centre often hosts main urban areas, capital cities and intensive agriculture production, while boundaries and frontiers define state peripheries.

The international setting following the end of the Cold War confirmed this trend, reinforcing the centre–periphery division at different levels, leading to the intensification of spatial differentiations and to the widening of existing inequalities amongst territories with different resource endowments: '[T]he best available set of terms to capture the relationships of the 1990s comes from the centre–periphery approach [...]. "Centre" here implies a globally dominant core of capitalist economies; "periphery" a set of industrially, financially and politically weaker states operating within a set of relationships largely constructed by the centre' (Buzan, 1991, p. 432).

Peripheries can also be presented as 'marginal areas' characterized by fragile resources, vulnerability to climatic events and limited options for production intensification, such as the drylands and mountain ranges covering important portions of the globe. Inhabiting communities face several barriers to participation in 'mainstream' patterns of activities, distant from centres of decision-taking and policy-making and with a structurally lower availability of

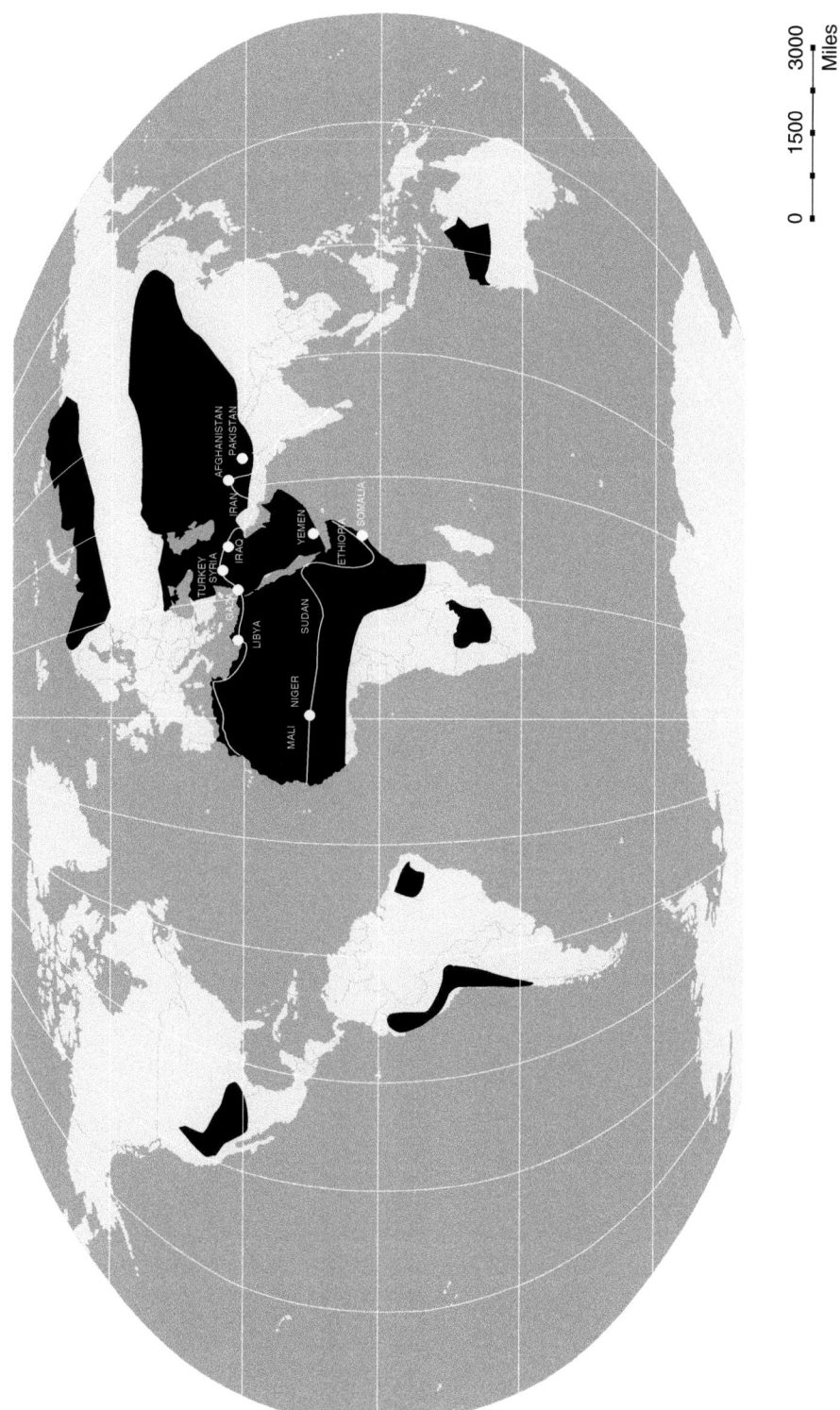

Fig. 12.1. The Conflict Shoreline cutting through pastoral regions around the globe. (Elaborated from WISP (2008) and Weizman (2015).)

institutional services. In these areas poverty and unemployment rank high, with evidence of seasonal and permanent migrations. Moreover, the idea of marginality also takes into account local perceptions and grievances: structural forces and socio-political exclusion push local communities to define themselves as a separate group, strongly opposed to and distinct from the wider national polity (De Haan et al., 2014; Dowd and Drury, 2017; Lind and Luckham, 2017).

Reconfiguring the margins

In political and governance terms, pastoral lands can be considered as areas of 'limited statehood', 'those parts of a country in which central authorities (governments) lack the ability to implement and enforce rules and decisions or in which the legitimate monopoly over the means of violence is lacking' (Risse, 2013, pp. 4–5). Limited statehood does not have a unique cause or the same explanation in every context. Nevertheless, we can remark that in many cases limiting statehood in peripheral areas has been a clear choice made by African state-builders, based on a comparison between potential benefits and effective costs linked to the expansion of ruling power over marginal territories (Herbst, 2000).

In particular, political marginalization of the pastoral belt – not only in Africa – has intensified since the 1980s, as a consequence of shrinking public budgets. Structural Adjustment Programs (SAPs) started reshaping policy and investment agendas, leading to a scaling down of state engagement and sovereignty in these territories, where, together with austerity, an overall feeling of abandonment started creeping in. The shortage of public funding and the disengagement of formal institutions in these areas has challenged the post-colonial paradigm, whereby the central state engaged in providing support to populations residing in remote and low-density areas through infrastructure investment, employment schemes, assistance programmes and subsidies. Maintaining a healthy and loyal population in these regions used in fact to be a strategic choice to secure these territories, where other forms of control and management would have involved much higher transaction costs (Cruz and Repetto, 1992; Riddell, 1992; Konadu-Agyemang, 2000).

On the other hand, state alliances with local communities, clans and tribes also implied tolerating and even promoting to an extent illegality and the informal economic activity pursued by local elites. These activities were conceived as complementary to state engagement and tolerating them was part of the 'socio-political contract' the state had agreed in these areas in exchange for peace, security and loyalty to its structures. Pastoral communities were often part of the equation, as the recent history of different Kurdish, Berber and Bedouin groups attest. This frequently resulted in the development of trafficking, smuggling and other illicit businesses based on the cross-border interfaces through the networks and routes traditionally utilized by pastoralists for grazing and for trading alike. This kind of 'hybrid' governance arrangement can be retraced in the Sahara-Sahel region (Scheele, 2012), in North Africa (Drozdz and Pliez, 2005) and more generally over the whole African continent (Laremont, 2005). The financial and political costs of these contracts were motivated and shared within the Cold War rationale and geo-political settings.

The partial 'retreat' of the central state from peripheral areas following the implementation of SAPs and the end of the Cold War strongly affected local equilibria. As we show in our case studies, both the Horn and the Sahel witnessed a spectacular increase of informal and/or criminal economic activities, while those spaces started opening to the penetration of new international actors. As a matter of fact, transnational entrepreneurs, traffickers, criminals and insurgent groups have reshaped political and economic orders in marginal lands inhabited by pastoralists, contributing both to further isolate these territories from their national 'centres', and to insert them into new trans-border and global dynamics (Harmon, 2014; Dowd and Drury, 2017). In the aftermath of SAP implementation, and out of the Cold War settings, these relationships and related equilibria changed radically, and post-colonial contracts and alliances faded alongside the rationale for sustaining the financial and political costs of such engagements. The 9/11 events in 2001 became another important benchmark in recent history, with relevant consequences for the overall restructuring of the international environment; 9/11 'brought more radical changes to the global periphery, as it was used as a pretext to reconfigure it as a space of (in)security rather than as spaces of underdevelopment and poverty' (Smith, 2009, p. 22).

Marginal territories with their alleged political vacuum and the presumed absence of central authorities were seen as important assets for transnational terrorists. These presumably 'ungoverned spaces' started being perceived as places where terrorists and criminals could find space and resources to plot and organize their activities (Keister, 2014). This 'safe haven myth' is based on the assumption that rural environments, characterized by poverty and bad governance and inhabited by Muslim populations, are the places more likely to become 'black holes' of the international system (Innes, 2008). Many of the regions included in the pastoral belt seem to fit the 'safe haven' definition; this can explain why they have been turned into theatres of the 'long war' against terrorism, bringing us back to Weizman's mapping and reasoning.

Pastoral strategies of change and transformation

The value and the potential of pastoralism are significant in most national economies of dryland sub-Saharan Africa, in terms of employment opportunities, income generation and national food security.[1] Pastoralism furthermore represents the predominant livelihood system for more than 50 million people inhabiting arid and semi-arid drylands of sub-Saharan Africa (De Haan et al., 2014). Official figures, however, only limitedly capture pastoralists' contribution to the economy, and statistics are often unreliable in these regions. In general pastoralists' contributions are poorly accounted for by scientists and disregarded by national and regional policy makers; the end result being that a significant production and livelihood system continues to live on the margins of policy to the detriment of the citizens that depend on it for their survival.

In pastoral areas intense environmental change plays an important role in challenging traditional resource management and institutional governance structures, due to the specific exposure of rangelands to climate variability, and to the pressure of a fast-growing pastoral population.[2] Rangelands are as well increasingly targeted by external investors, and land grabbing for different purposes is a feature characterizing several pastoral regions.[3] Loss and fragmentation of pasturelands amplify the negative effects of droughts, whose present impacts relate as much to these land use changes as they do to climate (Little, 2013). These factors contribute overall to amplify the vulnerability of local livelihoods.

Even though pastoralism represents the most effective way to manage livelihood risk in these environments (COMESA, 2009), increases in rangeland productivity are hardly able to satisfy a fast-growing and demanding population. The rising integration of pastoral economies into market dynamics represents possibly the main mechanism that has supported the growing human population on rangelands in latest decades, through the beneficial terms of trades from exchanging animal proteins with cereal starches. This mechanism has probably come to show its limits; long-standing humanitarian food crises and the food import dependency in parts of these regions might represent an outcome of such dynamics. Economic alternatives through livelihood diversification are increasingly popular measures amongst herding households, through a set of different strategies (Catley et al., 2013).

Pastoral livelihoods reflect and react to changing statehoods, expanding markets and shifting border regimes, and the overall reconfiguration of state and market forces and actors in these regions and beyond – as these provide renewed constraints as well as fresh opportunities. These changes are not without consequences, and contribute in turn to reconfiguring the institutions, relationships and norms that historically governed these societies, affecting their strategies and behaviours and altering internal group cohesion and power relations (Catley et al., 2013).

Although aware that pastoralists cannot be considered as a homogeneous and unitary group most of the time, and acknowledging that more than one strategy can be adopted at one time (Watts, 2015), we propose a classification of the transformations of livelihoods along the pastoral belt – at least in the areas analysed in this work – with a view to develop a further understanding of these dynamics. Three main patterns, at times complementary and intertwined, seem to characterize the rationale driving pastoral strategies and behaviours when facing crisis and conflict: *migrations*, *markets* and *militias*.

1. Emigration and shifting out of pastoralism, which provides the opportunity to support herding households by spreading community members and diversifying the livelihood base, by receiving economic support through remittances as well as by establishing extended social networks.

2. Enhancing market integration of the pastoral economy, through intensification and/or diversification, through developing strategic exchanges and ties with urban settings and regional and global markets, so as to provide important sources for income and employment.

3. Engaging in illicit activities, including trafficking, smuggling and hosting/joining guerrilla/militias, as illegal organizations and networks, together with insurgent movements, have repositioned areas on nations' margins at the core of regional networks and global pathways.

These patterns represent evolutions of strategic behaviours and longstanding efforts of pastoralists to extend, diversify and integrate their economies, in order to cope with and adapt to a shifting context, with a view to defend and evolve their livelihoods and the very existence of their communities.

Exploring Changes and Challenges in African Drylands

Having established and defined the main trends that affect the pastoral world, we turn in the following section to analyse the Horn and the Sahel. These African regions host over 50% of the world's most disadvantaged countries, several of which are characterized by bad governance and failed states. The two empirical cases explored in this chapter could be extended to other pastoral regions. In such a framework we explore the adaptation strategies implemented by pastoralists and the role played by the redefinition of frontiers and border regimes.

Somalia and the Horn of Africa

In order to cope with agro-ecological conditions characterizing the Horn of Africa drylands, Somali pastoral livelihoods traditionally rely upon mobility, information networks and market integration (Swift, 1979). These livelihood strategies typically take place at a regional level without major consideration for borders and frontiers; regional networks crossing and covering Ethiopia, Kenya, Somalia and Djibouti allow for extensive links that facilitate the regional movement of livestock, information and commodities. Recent decades have witnessed an interesting reconfiguration of these territories, where trans-border production, exchange and commercial patterns are importantly contributing to supporting livelihoods development (Nori, 2010).

Somali pastoral communities sharing the same ecological conditions, cultural features and livelihood patterns are scattered amongst the four countries in the Horn of Africa, in what has been defined as the Somali ecosystem; in spite of political divisions the lands inhabited and exploited by Somali pastoralists form in fact a single economic and ecological unit (Lyons, 1994). This 'Somali ecosystem' (Fig. 12.2) is reflective not only of similar ecological conditions but also of a *continuum* that characterizes the manmade networks and relations that make these populations integrated and interdependent. This ecosystem is crossed by and interlinked through corridors, territorial patterns through which complementary movements of livestock, people, food, commodities and finances often take place. Corridors typically develop from the Somali coasts to the Somali-inhabited regions in Djibouti, Ethiopia and Kenya. They serve to interlink the seemingly isolated inner drylands with coastal areas and – through the ports – the international arena.

By allowing continuous exchanges between pastoral products, imported goods and the interrelated flows, these corridors serve the different needs and activities of groups living under different environmental settings (Nori and Majid, 2002). A corridor is constituted by the interaction between a hard and a soft component, a physical and a social infrastructure, which are both critical to ensure its functioning. The institutional setting that governs such infrastructure is critical as well, as it regulates access to and utilization of resources, controls and secures movements and transactions, and provides the enabling environment for such flows and exchanges to happen.

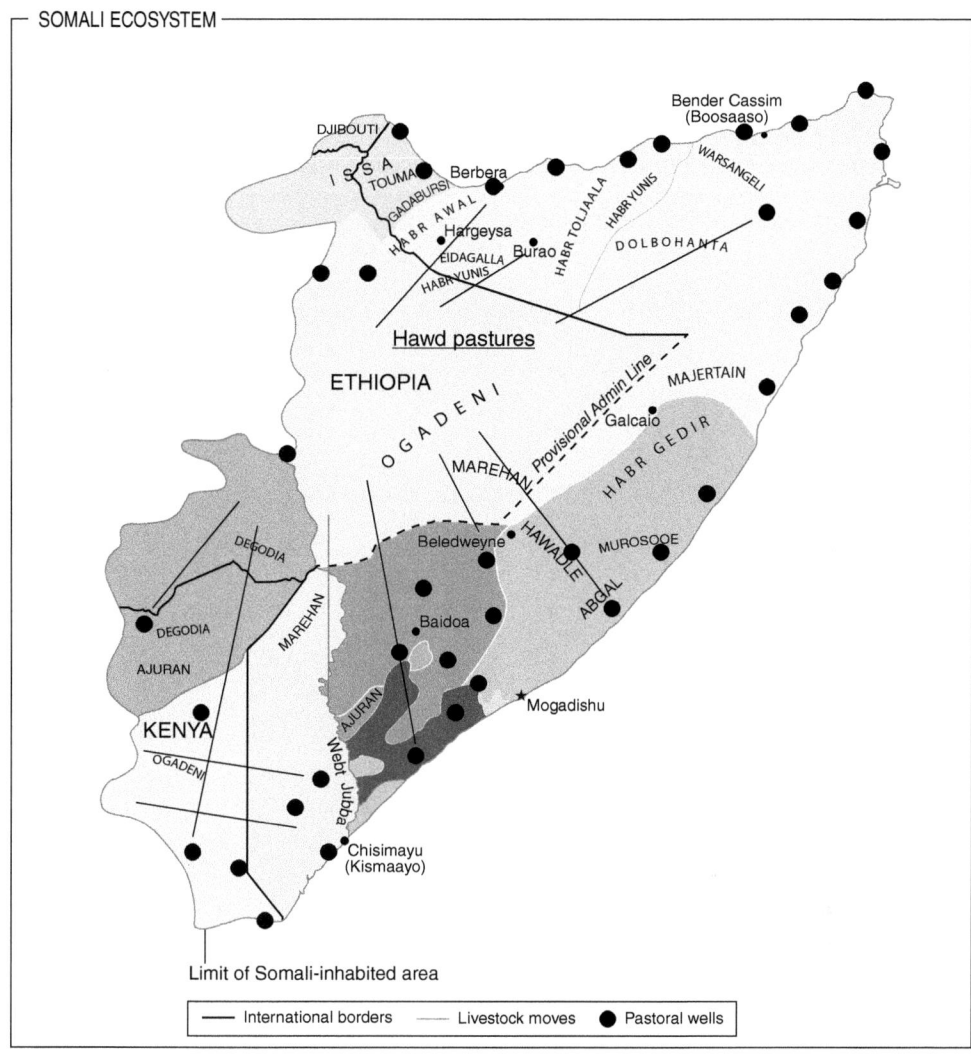

Fig. 12.2. The Somali ecosystem (with indications of mail clans, trade routes and wells). (Elaborated from Nori (2010).)

The failure of Siad Barre's regime (1969–1991) to mediate amongst the interests and power of different groups eventually led to the breakdown of the centralized institutional structure altogether, with the end of the idea of a Somali state. The collapse of the national government in 1991 reflected and triggered important societal changes, as it represented a major driver of change and innovation for the Somali society, opening the way for different development trajectories, with specific paces and patterns in diverse portions of the Somali region (Brons, 2001). As Fairhead and Leach (2005) argue, when the state rolls back, more informal polities roll in and while civil unrest still rules in parts of southern Somalia, in the pastoral north effective systems of local resource management and institutional governance have been established, drawing from customary settings, rearranged and adapted to fit within and serve the current context. Somaliland, the northwestern region of Somalia, provides today an interesting case of institutional blending in the Horn, through a government structure that accounts for traditional and modern governance patterns.

The wide Somali coastal area, the second largest in Africa, has traditionally favoured exchange and cultural crossover with distant and diverse cultures and environments. The large Somali diaspora, and the relevance of international remittance and import–export dynamics, attest to this exposure to the global setting. This is a main reason behind the booming livestock export trade that characterizes the Somali economy. After years of steady growth, in 2015 Somalia exported a record of about 5.5 million heads of livestock to the Arabian peninsula (Fig. 12.3), mostly through its northern ports of Berbera and Bossaso – an astonishing figure for a nation which has not 'enjoyed' any official recognition within the international system since about 25 years. Together with small ruminants export chains to the Arabian peninsula, a range of diverse livestock markets exist today in the region with different degrees of scale and specialization, such as the trade of camels to northern Africa and cattle to Kenya; local markets exist as well for local consumption, herd restocking, or fattening purposes. In northern Somalia the relatively secure environment and the flourishing economies driven by expanded livestock trading have benefited from, as well as triggered, important infrastructural developments in the area, including roads, water points, seaports, animal health and market facilities, mobile phone and radio networks, and financial services. These have come through public funding but also largely through private investments, as the intense land fencing, *berkaad*[4] establishment and range enclosures which currently characterize the Hawd and Sool plateaux demonstrate (Nori, 2010).

Parallel to the booming livestock trade, the stateless Somali society has also witnessed the evolution of the marketing of camel milk, a staple whose commercialization was taboo until late 1980s, and that represents today a main pillar of the Somali pastoral economy (Herren, 1993; Nori, 2010). As a result of the steady growth in population figures and the related expansion of consumption demand from urban dwellers, pastoralists have reconfigured the control and utilization of natural resources in order to cope with change and accommodate opportunities (Platteau, 2000). Nowadays thousands of litres of camel milk produced in Somali rangelands are daily traded through networks that link small kiosks, restaurants and shops in the main cities via long transport routes to the desert hinterlands. In 1998 it was estimated that the amount of camel milk commercialized reached about 8000 l per day during the rainy seasons in Puntland (northeastern Somalia) alone, with certain production areas located as far as 600 km from the terminal markets (Nori, 2010, p. 164).

Production, collection and marketing of camel milk feeds deep into rangelands that cross Somalia, Djibouti, Ethiopia and Kenya. Herds and trucks move through the Somali ecosystem according to rainfall opportunities, grazing resources and market routes in a subregional perspective, through the corridors well nested in the Somali customary institutional setting (Nori,

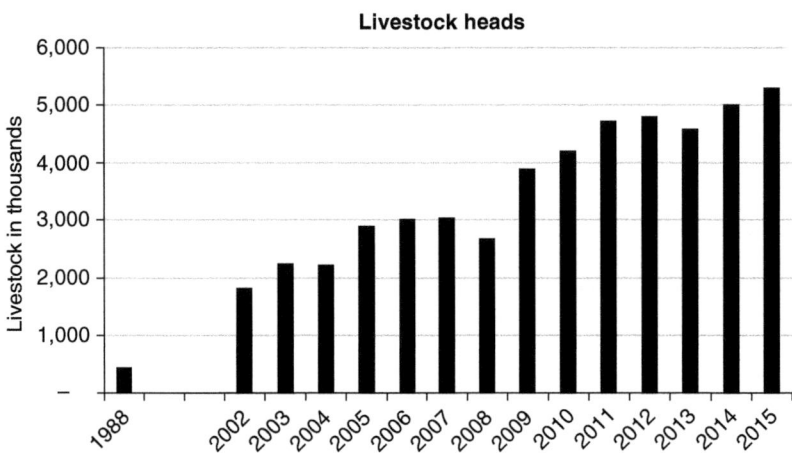

Fig. 12.3. Somali livestock export figures. (Elaborated from FSNAU (2016).)

2010). This social innovation has been brokered by Somali women, who manage and govern camel milk marketing, relying on sophisticated information and credit systems, which enable them to overcome the constraints of large distances, scattered productivity and even drought spells (Nori, 2010). Such trade has eventually also developed to different extents in other parts of the Somali ecosystem (Ethiopian lowlands, northeastern Kenya) as well as amongst other pastoral groups (i.e. the Boran, Gabra, Rendille) (Anderson et al., 2011; Mwaura et al., 2015).

Interestingly, all this is taking place and evolving in a context where the Food and Agriculture Organization of the United Nations (FAO) and the World Bank reported in the early 1990s that there was little potential for commercializing milk production amongst pastoralists, especially those on communal lands (Walshe et al., 1991). In a wider perspective marketing of camel milk represents a further step in the market integration of the Somali pastoral economy, a process characterized by the complementarity between gender-based roles and responsibilities, as men trade what is produced under women's responsibility (i.e. small ruminants), while women commercialize the primary product of clansmen (i.e. camel milk) (Nori, 2010).

The problems affecting Somalia since the collapse of its central state cannot be overlooked – including the clan conflicts that still ravage the southern portion of the country; patterns of unsustainable livelihoods diversification and income generation (through deforestation and charcoal trade, as well as illicit activities related to social banditry, piracy and smuggling that characterize parts of the country); and the presence of insurgent groups and militias, such as *Al-Shabab*, whose activities and the related insecurity pose critical challenges to local livelihoods in certain areas (Maruf, 2016).

A more critical and in-depth look at Somali evolutions enables, however, appreciating the efforts of Somali pastoralists and society as a whole to strengthen their resilience, through risk-taking entrepreneurial attitudes and social innovations. Somalis thus demonstrate that forms of market integration and economic diversification could contribute to enhancing pastoral performance and welfare. These strategies need, though, peaceful and safe conditions, as insecurity significantly raises transaction costs, constraining pastoral livelihoods and thus enhancing their vulnerability (Mahmoud, 2009).

Mali and the Sahelian belt

In the western part of the pastoral belt the Sahel is simultaneously affected by the interconnected effects of deteriorating environmental conditions, changing frontier regimes, the fading presence of the state in its peripheral territories and the impact of different transnational actors. A focus on the situation in Mali helps to clarify the different interests, forces and actors at work in the region.

Pastoralism is one of the dominant economies of the Sahel and is by far the main economy on the fringes of the Sahara (De Haan et al., 2014). With few exceptions, in the Sahara-Sahel pastoralists tend to live in rural, cross-border territories with low population density, usually the most peripheral regions of their countries. The main ethnic groups practising pastoralism are the Berbers, which include the Touareg, the Moors and the Saharawi (in the Maghreb and in the Sahara-Sahel region); the Toubous (living between Chad, Sudan, Libya and Niger); and the Fulani (in an area that extends from Senegal and southern Mali to northern Nigeria and Cameroon) (Fig. 12.4).

With few exceptions, pastoral groups have historically represented an ethnic minority inside Mali, Niger and Chad as much as in Morocco, Algeria, Tunisia and Libya.[5] The specific nature of their economic activities and socio-political identities has frequently fuelled conflict with both central governments and other ethnic groups (De Haan et al., 2014; OECD and CSAO, 2014). On the one hand, trans-border exchanges and ethnic loyalties represented a serious threat for the newly independent Sahelian states, which sought to affirm their full sovereignty over the national territories (Drozdz and Pliez, 2005; Harmon, 2014). On the other hand, access to land and other common natural resources periodically generated tensions between pastoralists with farmers and other sedentary populations (Benjaminsen et al., 2012). Commercial movements and trade routes are also traditional features of pastoral livelihoods, with livestock raised in the Sahelian drylands commercialized in urban areas of western as well as northern Africa. This

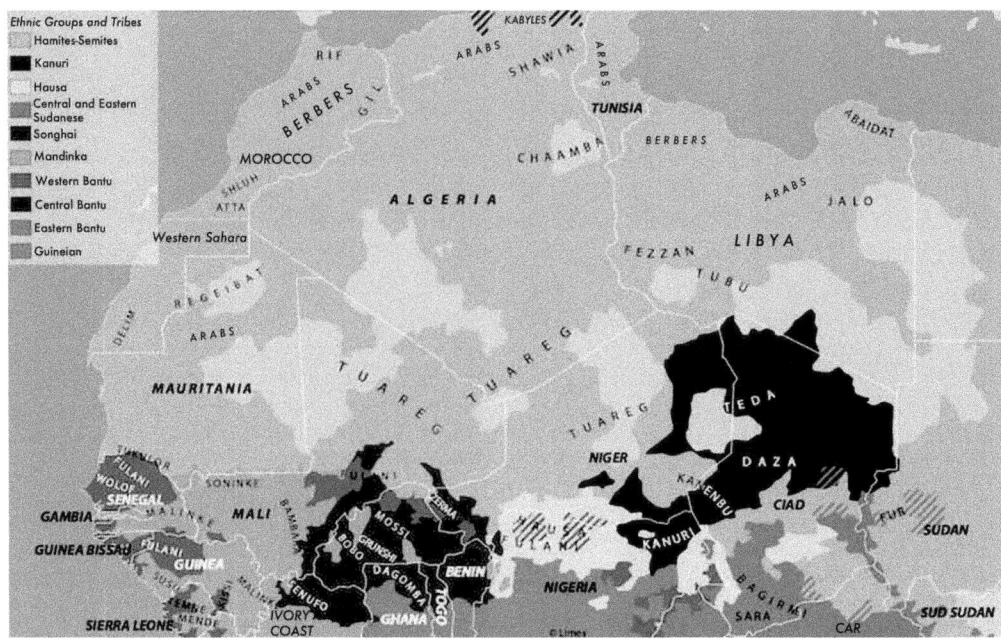

Fig. 12.4. Diffusion of nomad pastoralists in the Sahara-Sahel. (Elaborated from Canali (2012a).)

lucrative business represents a main and increasingly relevant source of income for most pastoral communities in the region (OECD and CSAO, 2014).

In a context already characterized by economic and political tensions, pastoral livelihoods were dramatically altered by droughts that affected the Sahelian belt since the 1970s (Batterbury and Warren, 2001). In this sense, the Touaregs can be considered as a paradigmatic case. Under the combined effect of environmental shocks and political (and military) struggles with the central state – in particular in Mali and Niger – during the 1970s many Touaregs opted to abandon their lands, spreading mainly to the north and west, and forming important diasporas in Algeria and Libya; as a matter of fact, this choice led to a partial crisis of their traditional forms of political and social organization (Lecocq, 2010). In this geo-political setting several Touareg groups have been co-opted since 1980s by the Khadafy regime, as a process where militias issued from pastoral groups throughout the Sahelian belt have been instrumental in inter-state skirmishes and conflicts (International Crisis Group, 2012). Such has been the case for the pastoral groups Toubous in Chad and Zaghawa in Sudan and their role in the regional conflicts in the past decade, including in Darfur. These patterns have not changed much, though today insurgency groups seem better positioned than state armies, as a consistent number of Al Qaeda in the Islamic Maghreb (AQIM) as well as Islamic State (IS) affiliated operators seem today issued from regional pastoral youths (Solomon, 2015). The relevance of these dynamics in the latest revolts and trouble are well acknowledged; the fall of the Khadafy regime in 2011 made modern weaponry available throughout the Sahel, with relevant important domino effects on the geo-politics of the region (Strazzari and Tholens, 2014).

Recent Malian history is characterized by violence and conflicts, in which pastoral communities are playing a central role. In 2012–2013 Mali was affected by a civil war that began as an uprising of the Touaregs, aiming to obtain independence for the lands they inhabited, a region they name Azawad. This has not been an unprecedented event: three previous Touareg rebellions took place from 1962–1964, 1991–1996 and 2007–2009. In the same line, the Fulani population also is recently showing signs of radicalization and a will to defy the central state: deadly inter-ethnic skirmishes have been registered in the Mopti region (where most of the Malian Fulani live), while a Fulani jihadist group made its appearance in 2015 (International Crisis Group, 2016). Open rebellion against the central state has been but one amongst different strategies implemented by pastoralists to adapt to and take advantage of a changing political setting. Once again, the Touaregs can be considered as the most significant example.

Most of the Touareg population lives in northern Mali, a landlocked, mostly desert territory sharing borders with Mauritania, Algeria and Niger (Fig. 12.5). Since the early 1990s the Malian state started to 'retreat' from this area for two main reasons: on the one hand, in accordance with recommendations coming from international financial and development partners – such as the World Bank or the Organisation for Economic Co-operation and Development (OECD) – the Malian government started to implement a programme of decentralization, creating regions and administrative districts in charge of exerting effective ruling power over peripheral areas (Seely, 2001). On the other hand, the peace agreement signed with Touareg leaders at the end of the 1991–1996 rebellion determined the demilitarization of northern regions by Malian security and defence forces (Harmon, 2014).

These forms of disengagement of the central state from northern Mali have had economic implications as well. Table 12.1 reports data reflecting declines in poverty rates at the national level, aggregated for rural and urban environments, and for the cities of Bamako, Gao, Timbuktu and Mopti, as reported by van de Walle (2012). In particular, Gao and Timbuktu are two of the most important cities in northern Mali (even if Kidal is considered as the Touareg 'capital'). Absolute poverty decreased faster and in a more substantial way in Bamako and in all urban environments generally, while in Gao and Timbuktu it even increased between 2001 and 2006.

This mix of structural weakness and policy choices created the conditions for Touareg and other pastoral communities to implement different strategies in order to improve their livelihoods and adapt to the new socio-political context. In the case of northern Mali, price differentials and a network of rural and urban markets historically transformed trans-border trade and exchanges with Algeria in the most important economic activity. This happened for decades with the compliance, and usually even the participation, of local security forces (Harmon, 2014). 'Northern Mali largely lives on informal trade, described as smuggling, which has largely been permitted for all products [...]. Officially the only trade authorized between Algeria and Mali is that in dates, any other trade is defined as smuggling by Algerians. Almost all of the consumption commodities in northern Mali are sourced from Algeria' (Bensassi et al., 2015, p. 18). Moreover, '[W]eekly turnover of informal trade fell from approximately US$2 million in 2011 to US$ 0.74 million in 2014,[6] but continues to play a crucial role in the economies of northern Mali and southern Algeria. Profit margins of 20–30% on informal trade contribute to explaining the relative prosperity of northern Mali' (Bensassi et al., 2015, p. 1).

A similar economic organization has been developed in the Mopti region, along the frontier between Mali and Burkina Faso. In this second case, local administration also participated in the process of transboundary integration, creating a system where some basic social services – such as primary school or public health – are informally shared with cross-border citizens (Söderbaum and Taylor, 2007).

Since the early 2000s transnational criminal groups exploited these informal organizational structures to develop their activities in the Sahel (Fig. 12.6). South American drug cartels in particular transformed the region into one of the main transit hubs for cocaine distributed in the European market (UNODC, 2007). Even if there is a lack of reliable data concerning the illicit drug business, it has been suggested that between 2002 and 2012 cocaine trafficking generated a business of at least US$10 billion in West Africa (Thiolay, 2013; Raineri and Strazzari, 2015).

Participating in transnational trafficking enabled portions of pastoral communities to redefine their economic functions and political power within the region. A new 'entrepreneurial class' has developed across these borders, pursuing both economic business and political agendas. The regional penetration by transnational terrorist groups linked to radical Islamism showed to be particularly appealing for pastoral youth, attracted both by their 'revolutionary'

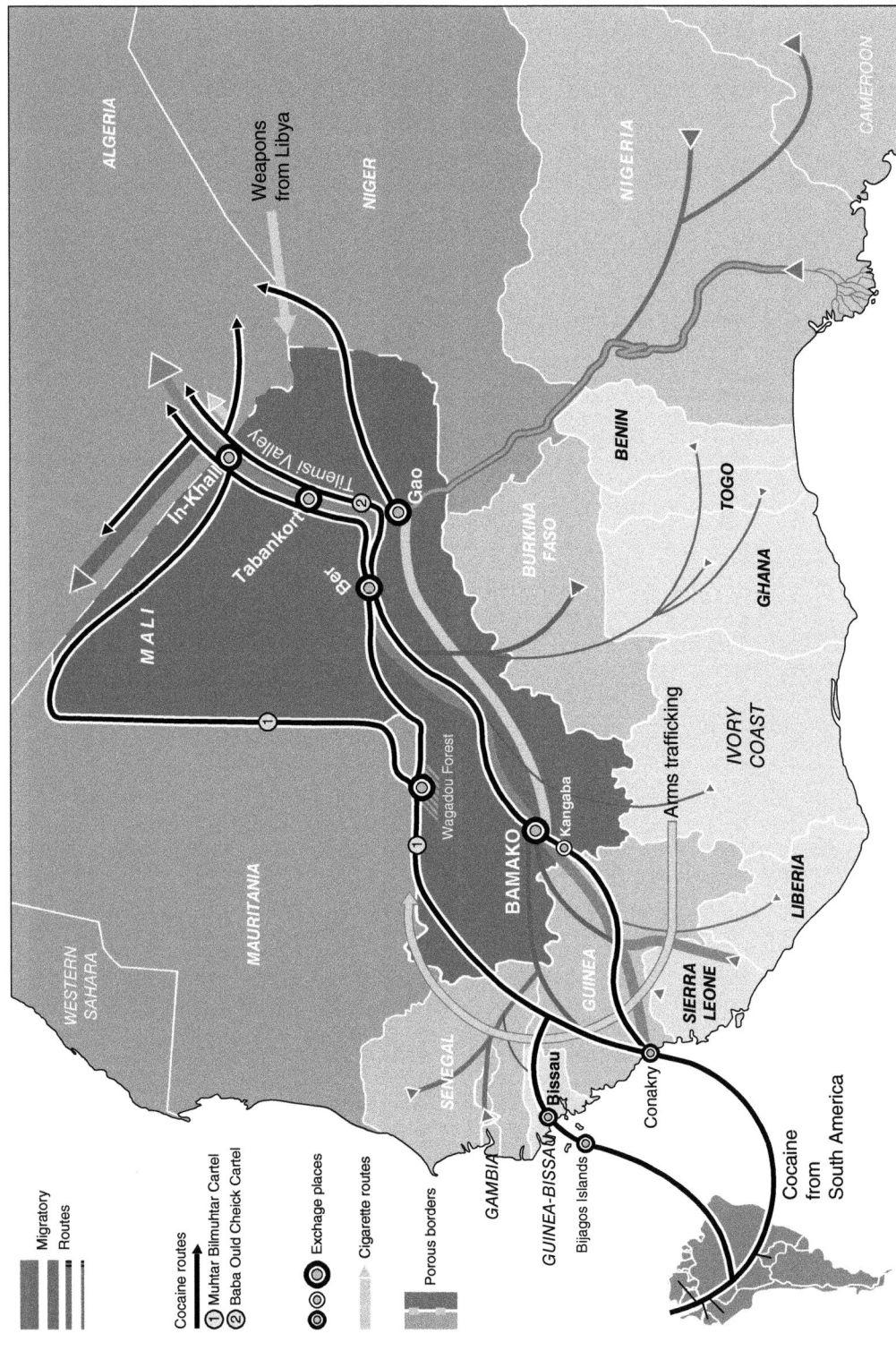

Fig. 12.5. Mali as a crossroads of regional traffic. (Elaborated from Canali (2015).)

message, and by military wages as an alternative livelihood out of pastoral activities (Solomon, 2015). Leaders of these terrorist groups are themselves often issued from pastoral communities, and have weaved networks and gained respect in the region often through illicit cross-border activities; these assets have eventually scaled up when turning from illegal traders to insurgent political leaders.

Lines of division and competing interests have evolved accordingly, along generational, social and political divides. As a matter of fact, pastoral communities must not be considered as a monolithic unit. In the case of Touareg society there is a multilevel political organization based both on 'castes' and on family and clan affiliation (Lecocq, 2005); recent transformations have redefined communities' political and social equilibria. On the one hand, new entrepreneurs started to challenge traditional elites, creating their own political and economic networks. On the other hand, they also started to call for participating in the local system of governance. Especially during the presidency of Amadou Toumani Touré (2002–2012), the Malian government opted for the implementation of a form of indirect rule concerning its peripheral territories, based in the co-optation of local elites and big men through corruption and clientelism. Local businessmen or militia leaders had become 'unavoidable' governing actors, creating at the same time new cleavages and tensions even inside the Touareg community (Briscoe, 2014).

The Malian case shows that the current crisis in the Sahel must be considered as a 'hybrid threat' (Martìn de Pozuelo, 2017) that results from multiple crises affecting local states and the pastoral world. Environmental change, weak governance, fading frontiers, and the redefinition of the economic and political systems are

Table 12.1. Decline in the rate of absolute poverty in Mali. (From: van de Walle (2012).)

	1989	2001	2006
National	72.8	68.3	64.4
Rural	80.6	79.2	79.5
Urban	72.7	37.2	31.8
Bamako	37.2	27.5	11.0
Timbuktu	74.8	54.4	57.9
Gao	66.9	48.2	57.9
Mopti	90.0	78.5	75.2

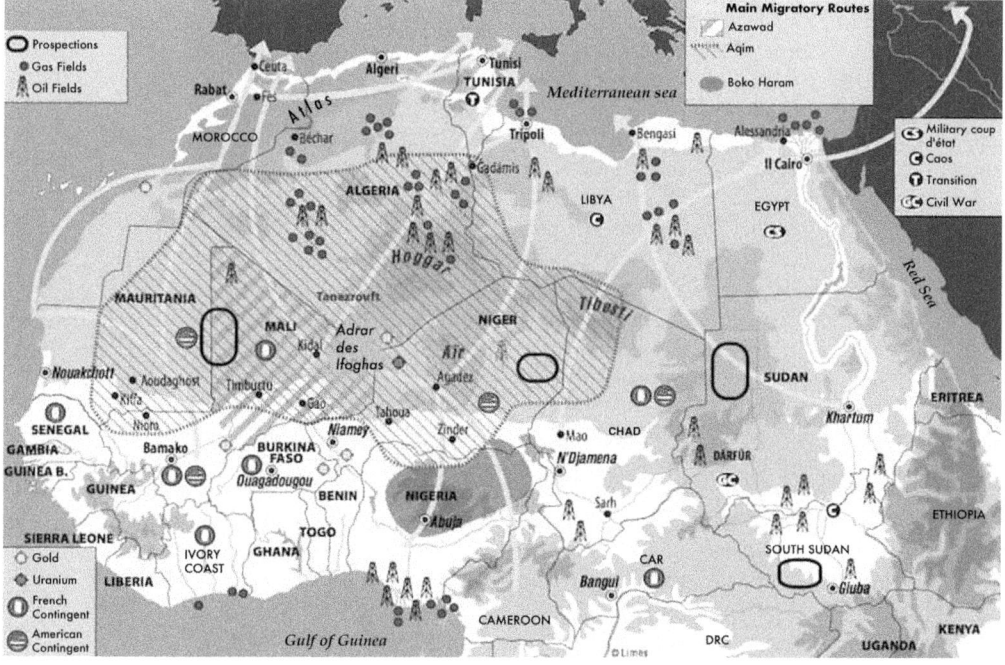

Fig. 12.6. Reinventing pastoral routes in the Sahelian-Saharan belt. (Elaborated from: Canali (2012b).)

creating opportunities and threats that altogether are posing relevant challenges to local pastoral livelihoods as well as to regional security.

Conclusions

Pastoral territories share similar geographical, economic as well as political configurations in their growing incorporation within global chains and networks that combine economic and political dimensions alike. The pastoral belt – spanning from Afghanistan to Somalia to Mauritania – appears today as a new environmental, geopolitical and geo-economic space, defined and modified by the interaction of several international, transnational and local dynamics, forces and actors. On the one hand, environmental changes together with the restructuring of economic agendas contribute to redefining pastoral territories and communities and their place in the global setting. On the other hand, a complex interplay between global trends, regional arrangements and local transformations is determining the rise of new models of social and political governance. As a result, inhabiting communities have become part of a wider network, with implications for the reshaping of pastoral spaces, economies and societies. In order to cope with evolving contexts, customary institutional settings have reorganized along lines that account for modern economic practices.

In the African drylands the diminished state presence in frontier zones has importantly contributed to loosening border regimes, reshaping regional territories and patterns of mobility and exchanges. Crossing borders has become a 'value-adding' activity, thanks to the administrative, economic and political differentials that characterize the different national territories (Meddeb, 2012). The intertwined relationships between the pastoral economy and more recent trans-frontier economic activities materialize in several ways; networks, infrastructure, geographical know-how and socio-political alliances have been reinterpreted to serve new flows, needs and interests. Not only the routes utilized by 'modern' traders, traffickers and smugglers very often re-interpret those forged and traditionally utilized by caravan traders and nomadic pastoralists, but the same trucks and pick-ups that move livestock and milk one way, regularly transport commodities – including humanitarian assistance – on their way back. Regional migratory flows are also embedded in existing networks and routes that traditionally link areas on different sides of frontiers, and often complement and nourish other parallel trades and transactions.

In certain areas ongoing dynamics have evolved from challenging the monopoly of the state to control trans-frontier exchanges, towards challenging its own presence, control and authority. Simultaneously, insurgent movements have found their way through pastoral territories, capitalizing to an extent on the sense of disillusionment, resentment and abandonment local populations feel towards central governments, state structures and international institutions (Nori *et al.*, 2008). The Somalian and the Malian cases show that pastoralists often find themselves sharing, interacting and competing on the same territories with insurgent-jihadist groups such as Al-Shabab and AQIM. While in certain areas this gives rise to tensions and conflicts and insecurity harms the pastoral business, in other areas radicalization might be seen as a strategy pursued by marginalized communities to express their grievance and adapt to a hostile context. These dynamics provide groups that traditionally represented minorities in national settings the chance to take centre stage within regional networks and pathways.

In this evolving framework the pastoralism–security nexus has become visible at different policy levels, often without a clear understanding of the associated threats and challenges. The role and the positioning of international agencies in this process should not be underestimated, as development and humanitarian assistance often represent relevant resources in such contexts (e.g. engaging in strategic dialogue, joining food/cash schemes, sending household members to a refugee camp, etc.), and hold relevant power in assisting the negotiation of local and regional agendas.

In several cases putting an end to transhumance and nomadic herding practices is considered the most viable solution to address security concerns in these regions. As noted by Agence Française de Développement (AFD) in 2013, 'In the eyes of concerned African states as much as in those of the international community, it seems necessary to replace pastoralism, amongst the lasting solutions to tackle the insecurity problems affecting the Sahel'.[7] Other more illuminated perspectives attest to the relevance of putting an end to pastoralists' marginalization and fully supporting herding practices and to engaging and integrating pastoral communities in the management and governance patterns of regional territories as the way to ensure a sustainable, peaceful and inclusive development (Annex 1). In these proactive perspectives, pastoral communities are conceived as the most relevant and powerful allies to maintain safe and productive rangelands, as herding represents the best way to safely occupy and secure vast, remote territories, where the costs of any other form of producing as well as of controlling, monitoring and patrolling would be significantly higher.

Annex 1 Reconfiguration of the Political-Institutional Landscape

African Union, 2010 – Policy Framework for Pastoralism in Africa:

> The development challenges of pastoral areas in Africa are multi-dimensional and complex; poverty, environmental degradation, marked rainfall variability, human and animal diseases, conflicts and civil strife must be dealt with simultaneously. Inappropriate development policies, ineffective institutional settings, unfair market relationships and increased pressure on pastoral ecosystems add to these challenges, and place many pastoralists in a situation of worsening vulnerability.

Déclaration de N'Djaména, 2013:

> The participants consider that the future of Sahelo-Sahelian areas cannot be conceived without pastoralism and its irreplaceable roles for economic, social, environmental and territorial development; (...) In the Sahelo-Sahel region, where security is seriously threatened, the relationship between herd mobility and security operates in both directions. As pastoralism and trade are main safe and peaceful activities in these areas, they represent an essential line of defense against insecurity through the occupation of space; [...] invite the States of the region and development partners to place pastoralism at the heart of the strategies of stabilization and development in the short, medium and long term of the Sahelo-Sahelian areas (...)

Déclaration de Nouakchott, 2013:

> [...] Unanimously, we affirm that pastoralism must be placed at the heart of the strategies and policies of stabilization, of sustainable development and of agricultural development at the national and regional levels, by integrating issues of sustainable management, equitable resource sharing, political inclusion, security, markets connections, health, education and gender. [...] Together we declare our commitment to [...] accelerate the political inclusion of pastoral communities through: a) systematic consideration of pastoralism in development policies, plans and programs; b) recognition of the legitimacy of traditional pastoral institutions; c) the inclusion of pastoralists in the processes of participation, consultation and decision-making put in place by the decentralized institutions.

Farmers' Forum, 2016:

> International Cooperation agencies[8] (...) should reinforce the institutional capacities and governance of pastoralist organisations and extensive livestock breeders to influence policy processes at local, national and regional levels. (...) should support the independent engagement of pastoralist organisations in policy dialogues at local, national, regional and global level through adapted legislations for pastoralists and extensive livestock breeders and the creation and reinforcement of enabling platforms for policy making with governments and regional institutions. (...) should continue to implement its policy on improving access to land and tenure security with specific attention to the security and tenure of pastoralist communal land and the governance of natural resources. Particular attention should be devoted to cross-border movement, mobility and conflict in these areas.

Notes

[1] Apart from their contribution to GDP, pastoral areas are the major source of animal proteins to the people inhabiting the region.

[2] The Middle East and North Africa (MENA), Sahel and the Horn of Africa are considered the regions undergoing the most intense demographic growth and amongst those most exposed to climate change impacts, and in most African countries the pastoral population is often growing faster than national averages (IFPRI, 2011; IPCC, 2014; UNCCD, 2014).

³ Reference is made to the work of the International Land Coalition (n.d.), specifically the Rangelands Observatory project (https://landportal.info/library/resources/rangelands-observatory).
⁴ Small local water reservoirs.
⁵ In both Mali and Niger, Touaregs and Fulanis together represent approximately 15% of the local population (CIA, 2017). The situation is somewhat different in Mauritania, where Moors are a net majority; and Chad, where pastoral groups are more embedded in state structures.
⁶ This fall can be explained considering the impact of the 2012 conflict and its consequences, in particular on border control; since 2013 the French military mission Barkhane, the Algerian army and American drones directly participate to the monitoring of northern Mali's frontiers.
⁷ Authors' translation of the original text: 'Aux yeux des Etats africains concernés et de la Communauté internationale, il apparaît en particulier fondamental de replacer l'élevage pastoral, parmi les solutions durables au problème d'insécurité du Sahel'.
⁸ Specific reference is made here to the UN International Fund for Agriculture Development.

References

African Union (2010) *Policy Framework for Pastoralism in Africa: Securing, Protecting and Improving the Lives, Livelihoods and Rights of Pastoralist Communities*. African Union, Department of Rural Economy and Agriculture, African Union, Addis Ababa, Ethiopia.
Agence Française de Développement (AFD) (2013) Développement et sécurité des espaces Saharo-Sahéliens: l'atout de l'élevage pastoral. Agence Française de Développement. Colloque de N'Djamena: Éléments de contribution aux débat. Ndjamena, Chad.
Anderson, D.M., Elliott, H., Kochore, H.H. and Lochery, E. (2011) *Camel Milk, Capital, and Gender: The Changing Dynamics of Pastoralist Dairy Markets in Kenya*. The British Institute In Eastern Africa, London.
Batterbury, S. and Warren, A. (2001) The African Sahel 25 years after the great drought: assessing progress and moving towards new agendas and approaches. *Global Environmental Change* 11(1), 1–8.
Benjaminsen, T.A., Alinon, K., Buhaug, H. and Buseth, J.T. (2012) Does climate change drive land-use conflicts in the Sahel? *Journal of Peace Research* 49(1), 97–111.
Bensassi, S., Brockmeyer, A., Pellerin, M. and Raballand G. (2015) *Algeria-Mali Trade: The Normality of Informality*. World Bank Group, Washington, DC.
Briscoe, I. (2014) *Crime after Jihad: Armed Groups, the State and Illicit Business in Post-Conflict Mali*. Conflict Research Unit, the Clingendael Institute, The Hague, The Netherlands.
Brons, M.H. (2001) *Society, Security, Sovereignty and the State in Somalia: From Statelessness to Statelessness?* International Books, Utrecht, The Netherlands.
Buzan, B. (1991) New patterns of global security in the twenty-first century. *International Affairs* 67(3), 431–451.
Canali, L. (2012a) Popoli Sahariani e Saheliani. *Limes – Rivista Italiana di Geopolitica* 5/12.
Canali, L. (2012b) Sabbie Mobili. *Limes – Rivista Italiana di Geopolitica* 5/12.
Canali, L. (2015) Il Mali al Centro dei Traffici. *Limes – Rivista Italiana di Geopolitica* 3/15.
Catley, A., Lind, J. and Scoones, I. (2013) *Pastoralism and Development in Africa. Dynamic Chance at the Margins*. Routledge, New York.
Central Intelligence Agency (CIA) (2017) The world factbook. Available at: https://www.cia.gov/library/publications/the-world-factbook/fields/2259.html (accessed 22 June 2017).
Common Market for Eastern and Southern Africa (COMESA) (2009) Income diversification among pastoralists: lessons for policy makers. *COMESA Economic Diversification Pastoralists Policy Brief 3*. COMESA, Lusaka.
Cruz, W. and Repetto, R. (1992) *The Environmental Effects of Stabilization and Structural Adjustment Programs: The Philippines Case*. World Resources Institute, Washington, DC.
Déclaration de N'Djaména (2013) Elevage pastoral: une contribution durable au développement et à la sécurité des espaces Saharo-Sahéliens. N'Djaména, 29 May 2013, Colloque Régional et Conférence Ministérielle, 27–29 May 2013.
Déclaration de Nouakchott sur le Pastoralisme (2013) Mobilisons ensemble un effort ambitieux pour un pastoralisme sans frontières. 29 October 2013.
De Haan, C., Dubern E., Garancher B. and Quintero, C. (2014) *Pastoralism Development in the Sahel: A Road to Stability?* World Bank Global Center on Conflict, Security, and Development, Nairobi, Kenya.

Dowd, C. and Drury, A. (2017) Marginalisation, insurgency and civilian insecurity: Boko Haram and the Lord's Resistance Army. *Peacebuilding* 5(2), 136–152.

Drozdz, M. and Pliez, O. (2005) Entre Libye et Soudan: la fermeture d'une piste transsaharienne. *Autrepart* 4, 63–80.

Farmers' Forum (2016) International statement, Rome, 13 February 2016, Special Session of the Farmers' Forum with Pastoralists and Livestock Breeders, Jointly organized by IFAD and VSF, IFAD, Rome.

Fairhead J. and Leach M. (2005) The centrality of the social African farming. *IDS Bulletin* 36(2), 86–90.

Food Security and Nutrition Analysis Unit (FSNAU) (2016) Market data update. Food Security and Nutrition Analysis Unit – Somalia. Available at: www.fsnau.org/downloads/Market-Data-Update-February-2016.pdf (accessed 13 March 2018).

Galtung, J. (1971) A structural theory of imperialism. *Journal of Peace Research* 8(2), 81–117.

Harmon, S.A. (2014) *Terror and Insurgency in the Sahara-Sahel Region: Corruption, Contraband, Jihad and the Mali War of 2012–2013*. Ashgate, Farnham, UK.

Haughton, J. and Khander, S.R. (2009) *Handbook on Poverty and Inequality*. World Bank, Washington DC.

Herbst, J. (2000) *States and Power in Africa: Comparative Lessons in Authority and Control*. Princeton University Press, Princeton, New Jersey.

Herren, U.J. (1993) *Cash from Camel Milk: The Impact of Commercial Milk Sales – Southern Somalia*. EPOS, Uppsala University.

Innes, M.A. (2008) Deconstructing political orthodoxies on insurgent and terrorist sanctuaries. *Studies in Conflict and Terrorism* 31(3), 251–267.

International Crisis Group (2012) *Mali: Avoiding Escalation*. Africa Report no. 189. ICG, Brussels.

International Crisis Group (2016) *Central Mali: An Uprising in the Making?* Africa Report no. 238. ICG, Brussels.

International Food Policy Research Institute (IFPRI) (2011) *Food Security and Economic Development in the Middle East and North Africa, Current State and Future Perspectives*. International Food Policy Research Institute, Washington, DC.

International Fund for Agricultural Development (IFAD) (2010) Livestock Position Paper 2010. Livestock planning, challenges and strategies for livestock development in IFAD. IFAD, Rome

International Land Coalition (ILC) (n.d.) Rangelands Observatory. International Land Coalition, Rome. Available at: https://landportal.info/library/resources/rangelands-observatory (accessed 13 March 2018).

International Panel on Climate Change (IPCC) (2014) Fifth assessment report. IPCC. Available at: www.ipcc.ch/report/ar5 (accessed 13 March 2018).

Keister, J. (2014) The illusion of chaos: why ungoverned spaces aren't ungoverned, and why that matters. *Cato Institute Policy Analysis* no.766.

Konadu-Agyemang, K. (2000) The best of times and the worst of times: structural adjustment programs and uneven development in Africa: the case of Ghana. *The Professional Geographer* 52(3), 469–483.

Laremont, R.R. (2005) *Borders, Nationalism, and the African State*. Lynne Rienner Publishers, Boulder, Colorado.

Lecocq, B. (2005) The Bellah question: slave emancipation, race, and social categories in late twentieth-century northern Mali. *Canadian Journal of African Studies/La Revue canadienne des études africaines* 39(1), 42–68.

Lecocq, B. (2010) *Disputed Desert: Decolonisation, Competing Nationalism, and Tuareg Rebellions in Northern Mali*. Brill, Leiden, The Netherlands.

Lind, J. and Luckham, R. (2017) Introduction: security in the vernacular and peacebuilding at the margins; rethinking violence reduction. *Peacebuilding* 5(2), 89–98.

Little, P.D. (2013) Reflections on the future of pastoralism in the Horn of Africa. In: Catley A., Lind J. and Scoones I. (eds) *Pastoralism and Development in Africa. Dynamic Chance at the Margins*. Routledge, New York.

Lyons, T. (1994) Crises on multiple levels: Somalia and the Horn of Africa. In: Samatar, A.I. (ed.) *The Somali Challenge. From Catastrophe to Renewal?* Lynne Rienner Publishers, Boulder, Colorado.

Mahmoud, H.A. (2009) Conflicts and pastoral livelihoods in the Kenya-Ethiopia-Somalia borderlands. In: Goldsmith, P. (ed.) *Fighting for Inclusion: Conflicts Among Pastoralists in East Africa and the Horn*. Development Management Policy Forum, Nairobi.

Martìn de Pozuelo, E. (4 March 2017) La Amenaza Hìbrida. La Inestabilidad Norteafricana Pone en Guardia al Flanco Sur Europeo, *La Vanguardia*. Available at: www.lavanguardia.com/internacional/20170304/42527034121/norte-africa-amenaza-europa-yihadismo-crimen-organizado.html (accessed 29 March 2018).

Maruf, H. (2016) Al-Shabab seizes somali herders' livestock. *Voice of Africa* 26 December 2016. Available at: https://www.voanews.com/a/somalia-al-shabab-farmers-livestock/3652155.html (accessed 13 March 2018).

McMichael, P. (1997) Rethinking globalization: the agrarian question revisited. *Review of International Political Economy* 4(4), 630–662.

Meddeb, H. (2012) La course à el khobza aux frontières de l'Etat. PhD thesis, CERI-Sciences-Po, Paris.

Mwaura, M.W., Wasonga, O.V., Elhadi, Y.A.M. and Ngugi, R.N. (2015) Economic contribution of the camel milk trade in Isiolo Town, Kenya. IIED Country Report. IIED, London. Available at: http://pubs.iied.org/10123IIED (accessed 13 March 2018).

Nori, M. (2010) Milking drylands: gender networks, pastoral markets and food security in stateless Somalia. PhD dissertation thesis, CERES Wageningen University, Lambert Academic Publishing.

Nori, M. and Majid, N. (2002) Enhancing Somali cross-border networks. Save the Children, UK, project proposal. Unpublished.

Nori, M., Taylor M. and Sensi A. (2008) Browsing on fences: pastoral land rights, livelihoods and adaptation to climate change. IIED Drylands Series #148, London. Available at: www.iied.org/pubs/display.php?o=12543IIED (accessed 13 March 2018).

Organisation for Economic Co-operation and Development (OECD) and Cahiers de l'Afrique de l'Ouest (CSAO) (2014) *Un Atlas Du Sahara-Sahel: Géographie, Economie et Insécurité*. OECD, CSAO, Paris, France.

Platteau, J.P. (2000) *Institutions, Social Norms, and Economic Development*. Harwood Academic Publishers, Amsterdam.

Raineri, L. and Strazzari, F. (2015) State, secession, and jihad: the micropolitical economy of conflict in northern Mali. *African Security* 8(4), 249–271.

Republic of Kenya (2012) Sessional Paper No. 8 of 2012, on National Policy for the Sustainable Development of Northern Kenya and other Arid Lands, *Releasing Our Full Potential*, Ministry of State for Development of Northern Kenya and Other Arid Lands, Republic of Kenya, Nairobi.

Riddell, J.B. (1992) Things fall apart again: structural adjustment programmes in sub-Saharan Africa. *Journal of Modern African Studies* 30(01), 53–68.

Risse, T. (2013) Governance in areas of limited statehood: introduction and overview. In: Risse, T. (ed.) *Governance Without a State? Policies and Politics in Areas of Limited Statehood*. Columbia University Press, New York, pp. 1–35.

Scheele, J. (2012) *Smugglers and Saints of the Sahara: Regional Connectivity in the Twentieth Century*. Cambridge University Press, Cambridge, UK.

Seely, J.C. (2001) A political analysis of decentralisation: coopting the Tuareg threat in Mali. *Journal of Modern African Studies* 39(3), 499–524.

Smith, M.S. (2009) *Securing Africa: Post-9/11 Discourses on Terrorism*. Ashgate, London.

Söderbaum, F. and Taylor, I. (2007) *Micro-Regionalism in West Africa: Evidence from Two Case Studies*. Nordiska Afrikainstitutet, Uppsala, Sweden.

Solomon, H. (2015) *Terrorism and Counter-Terrorism in Africa. Fighting Insurgency from Al Shabaab, Ansar Dine and Boko Haram*. Palgrave Macmillan, London.

Strazzari, F. and Tholens, S. (2014) Tesco for terrorists' reconsidered: arms and conflict dynamics in Libya and in the Sahara-Sahel region. *European Journal on Criminal Policy and Research* 20(3), 343–360.

Swift, J.J. (1979) *The Development of Livestock Trade in a Pastoral Economy: The Somali Case*. Cambridge University Press, Cambridge, UK.

Thiolay, B. (21 March 2013) Mali: La Guerre de La Cocaïne. *L'Express*. Available at: https://www.lexpress.fr/actualite/monde/afrique/mali-la-guerre-de-la-cocaine_1233028.html (accessed 29 March 2018).

United Nations Convention to Combat Desertification (UNCCD) (2014) Desertification: the invisible frontline. UNCCD, Bonn, Germany.

United Nations Office on Drugs and Crime (UNODC) (2007) *Cocaine Trafficking in West Africa: The Threat to Stability and Development*. UNODC, Vienna, Austria.

van de Walle, N. (2012) Foreign aid in dangerous places. The donors and Mali's democracy. *UNU-Wider Working Paper* 61.

Wallerstein, I. (1979) *The Capitalist World – Economy*. Cambridge University Press, Cambridge, UK.

Walshe, M.J., Grindle, J., Nell, A. and Bachman, M. (1991) Dairy development in sub-Saharan Africa – a study of issues and options. Technical paper no. 135, *Africa Technical Development Series*. World Bank, Washington, DC.

Watts, M.J. (2015) Now and then: the origins of political ecology and the rebirth of adaptation as a form of thought. In: Perreault, T., Bridge, G. and McCarthy, J. (eds) *The Routledge Handbook of Political Ecology*. Routledge, Abingdon, UK.

Weizman, E. (2015) *The Conflict Shoreline: Colonialism as Climate Change in the Negev Desert*. Steidl Verlag, Göttingen, Germany.

World Health Organization (WHO) and United Nations Children's Fund (UNICEF) (2005) *Global Immunization Vision and Strategy 2006–2015*. HO/IVB/05.05. 2005.

World Initiative for Sustainable Pastoralism (WISP) (2008) Learning from the delivery of social services to pastoralists. IUCN, Nairobi.

13 Border Change and Conflict in Central Asia: the Case of Agro-Pastoral Communities in Cross-Border Areas of the Ferghana Valley

Asel Murzakulova[1],* and Irène Mestre[2]
[1]*Mountain Societies Research Institute, University of Central Asia, Bishkek, Kyrgyzstan;* [2]*University of Jean Moulin-Lyon 3, Research Unit UMR 5600 Environnement Ville Société, Lyon, France*

Introduction

Conflicts between communities in the Ferghana Valley are often seen as having the potential to destabilize the entire Central Asia region (Reeves, 2005). In this chapter we investigate the conflict dynamics at play in the Ferghana Valley through the lens of the institutional changes over the past century that have affected the agro-pastoral systems. The conflict dynamics in border communities of Tajikistan and Kyrgyzstan are characterized by varying intensity of incidents associated with natural resources. The most acute conflict takes place between the communities of the Vorukh enclave and neighbouring village of Khoji-A'lo, Tajikistan, and the communities of Ak-Say, Kyrgyzstan. In other transborder areas of Kyrgyzstan such as Kara-Bak, Samarkandek and Ak-Tatyr rural municipalities, tensions occur only seasonally when infrastructure, such as roads leading to the pasture and irrigation canals, is used in common with neighbouring villages of Tajikistan (Matveeva, 2017).

The Ferghana Valley is one of the most highly populated areas of Central Asia and largely relies on the use of natural resources to carry out agro-pastoral activities. Since ancient times it has been a hub between different communities because of its location at the crossroad of communication and trade flows. The nature and shape of the borders crossing it have shown great variability and flexibility over time, and today the Ferghana Valley stretches across eastern Uzbekistan, southern Kyrgyzstan and northern Tajikistan (Fig. 13.1).

Although the territories and communities of the Ferghana Valley have different pre-Soviet histories and different ecological attributes, they share the experience of being part of the Soviet Union and going through the post-Soviet transformation, albeit with different outcomes. This area is also characterized by the large share of the population involved in agro-pastoralism, a mix of extensive livestock keeping with seasonal migrations to pastures and cultivation. Irrigation water and pastures, as well as all the infrastructure which allow their use, thus play a crucial role in the region.

In pre-Soviet times, even before the countries existed as such, the Ferghana Valley functioned as a nexus for common use of natural

---

* Email: asel.murzakulova@ucentralasia.org

Fig. 13.1. Elevation and political map of the Fergana Valley of Central Asia. (From: Mountain Society Research Institute of the University of Central Asia.[1])

resources and exchange of agricultural production. These connexions were maintained during the Soviet period, as the borders between the Soviet Socialist Republics of Uzbekistan, Tajikistan and Kyrgyzstan that cross this area were considered as internal. They had no impact on people's lives and inhabitants could cross easily and share infrastructure. Several government commissions were appointed to address the matter of border demarcation under the Soviet Union. Decisions on the border tended to reflect the view that ethnicity and nationality should match, and that view was in turn reinforced by the border's delimitation. Several enclaves were created, including Vorukh, which we investigate in this chapter.

After 1991, as the Soviet states became independent, the nature of the border changed and different border regimes were implemented. Uncertainty grew as the conditions to cross the borders transformed quickly. The delimitation process started again in a completely different context and resulted in the creation of even more enclaves. The strengthening of the border regimes also created a number of pene-enclaves. A pene-enclave is a territory that is accessible by wheel transportation only by crossing another state, and/or and which relies on infrastructure administrated by another country. Despite the delimitation process, borders are still undefined: Tajikistan and Kyrgyzstan share 978 km of borders, of which roughly 396 km remained unmarked in 2016 (Kabar, 2016). The areas where the border is conflictual are usually densely populated and inhabited by citizens from Tajikistan, Kyrgyzstan and Uzbekistan. In some situations, the border makes little sense, for example, when it separated a house from its kitchen garden (Reeves, 2014a).

In the Ferghana Valley until 2005, there were still landmines on the land adjacent to the Uzbek–Kyrgyz border. In 2010 the border itself was closed to citizens and goods from Kyrgyzstan after violent clashes in southern Kyrgyzstan. In contrast, the Kyrgyz–Tajik border was open and low-militarized until recently, even during the civil war in Tajikistan that lasted from 1992 until 1997. However, since 2010, both sides have restricted crossing of the Kyrgyz–Tajik border. According to the Kyrgyz border services there was an active growth of the conflict between border guards, from 19 incidents in 2012 (Kabar, 2012) to 32 incidents in 2014 (IA 24, 2015).

This chapter examines conflicts over water and pastures between agro-pastoral communities in the Ferghana Valley, which lies along the Kyrgyz–Tajik border. This chapter provides insight into the transformative environment of the border regimes and natural resource management, since the time of Soviet collectivization up to the reforms of the modern independent states. This chapter argues that cross-border communities have a set of specific vulnerabilities that come from the transformation of border regimes and an institutional shift from state-centric management of natural resources to governance by the users. This chapter sheds light on the little studied aspect of conflicts in post-socialist societies at the stage of forming a new class structure, from collective farm employees to owners of small-scale farms in the newly formed states.

Building on a case study, this chapter aims at investigating how the reconfiguration in terms of border regime and management of agricultural resources of agro-pastoral socio-ecological systems have affected transborder communities. We focus on a case study of interactions between the communities of Vorukh and Khoji-A'lo with the neighbouring communities of Ak-Say. Vorukh is a rural municipality of the Soghd province of Tajikistan formed by several villages: Tozhikon, Bedak and Vorukh. Khoji-A'lo is part of Chorku rural municipality, Tajikistan. Ak-Say is a rural municipality of the Batken province of Kyrgyzstan and is also formed of several villages: Kapchygay, Ak-Say, Bakay, Uch-Dobo and Kok-Tash. Settlements of the communities of Ak-Say are located in a territory that is a pene-enclave, as they depend on infrastructure belonging to Tajikistan for their water supply. Inhabitants of Ak-Say also largely rely on infrastructure crossing through Tajikistan such as the road to reach summer pastures and irrigation and drinking water. We investigate the tensions' dynamics occurring in their territory in the past century. The numerous transformations and issues that local inhabitants went through are crucial to understand the conflictual situation. The selected case highlights the consequences of the multiple reconfigurations of agro-pastoral communities and their present issues.

Conflict in Transborder Areas of the Ferghana Valley: Disentangling New Borders and Inherited Collaborative Use of Natural Resources

A large variety of understandings of conflict are used in research to analyse the issue of agriculture and conflict (see Chapter 1, Bahn and Zurayk, this volume). They are usually viewed through the prism of resources: conflict over land, water, etc. Another widespread approach is the typology of conflicts according to participating actors, traditionally divided into civil, interstate, extra-systemic and non-state (Sarkees and Wayman, 2010). In this chapter, we use a definition of conflict for describing tensions and clashes between communities that share interlinked territory and natural resources under the mandate of different states. Thus, our definition of conflict cannot be identified as interstate because the main stakeholders are the farmers who use local natural resources, although stakeholders with responsibilities at a higher level can also get involved. Along with this, we refer to the tensions related to linkages between access to natural resources, infrastructure, and border regime, as shown by Bichsel (2005, 2009a, 2009b, 2013, 2014) regarding irrigation water and by Reeves (2005, 2011, 2014a, 2014b, 2016) regarding state and identity.

Exploratory research in border communities of Tajikistan and Kyrgyzstan showed the importance of natural resources management and practices on the conflicts in border areas (Murzakulova and Mestre, 2016). After the collapse of the Soviet Union, borders became less permeable to the common use of irrigation water and livestock pastures. Transhumance from Tajikistan to the mountains of Kyrgyzstan had been a long-standing practice in the Soviet period and most rivers take their sources in mountainous areas of Kyrgyzstan before reaching neighbouring countries. Reeves (2005) highlights the paradox between international and local discourses on conflicts in this area. The international discourse tends to underline the potential for violent conflicts in the Ferghana Valley because 'it does not fit into normative accounts of the "proper" relationship between territory, ethnicity and citizenship' (Reeves, 2005, p. 67), while local inhabitants see the absence of demarcated borders as an asset offering flexibility and an opportunity to share a common resource. However, such discourses are changing under the influence of strong messaging on statehood, nationalism and ethnicity.

Methodology and Focal Area

Field data for this chapter were collected from 2015 until 2017 in two neighbouring rural

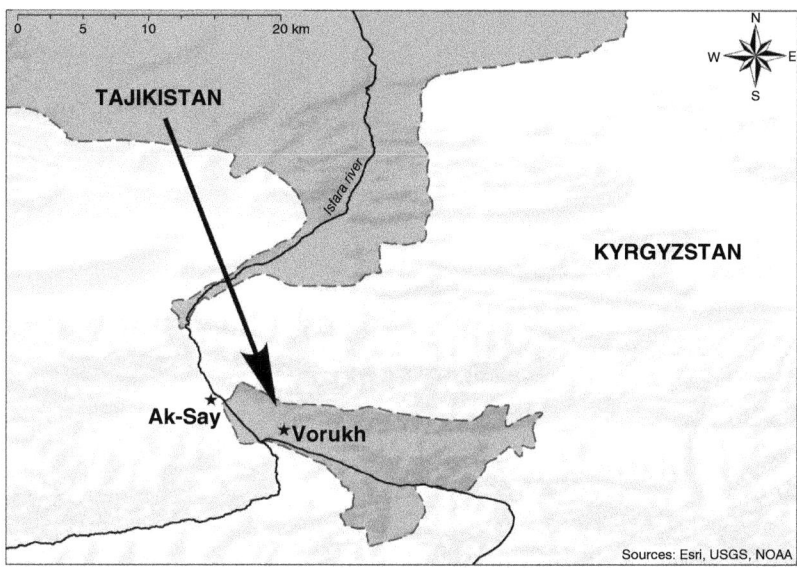

Fig. 13.2. Map of Vorukh and Ak-Say. (From: Mountain Society Research Institute of the University of Central Asia.[1])

municipalities (*zhamoat* in Tajik, *ajyl ajmak* in Kyrgyz), Vorukh and Ak-Say (Fig. 13.2). The total area of both rural municipalities covers around 600 km² and counts 39,000 inhabitants. We base our work on an embedded single case study (Yin, 2013), using mixed methods for data collection such as participant observation, in-depth interviews and critical analysis of academic research, archives and grey literature. The selection of the local respondents from the municipalities was based on criteria related to their social status (farmers, businessmen, migrants, officials, etc.), and economic assets. In total, 68 interviews were conducted with local respondents and nine with respondents working in non-governmental organizations, international organizations and governmental agencies.

The Soviet Period in the Ferghana Valley: Centralized Political System, Settlement of Semi-Nomadic Populations and Intensification of Livestock Breeding and Agriculture

Since ancient times, agriculture as practised within the territory of modern Kyrgyzstan and Tajikistan has been dominated by agro-pastoralism. Its two pillars are irrigation water, to cultivate crops for human consumption and winter fodder for the animals; and pastures used by seasonal altitudinal and horizontal migrations of herders with multispecies flocks. There are variations in agro-pastoral practices, as communities in Tajikistan are more oriented towards cultivation and communities of Kyrgyzstan towards livestock keeping. Moreover, agriculture holds not only economic, but also social and political value.

Tajikistan and Kyrgyzstan share the experience of being Soviet Socialist Republics and of the following transformations. Despite very strict frames and little flexibility, the Soviet period was also characterized by the variety of legal forms of agricultural production. Households could cultivate small private plots and keep small numbers of animals. People worked in large state farms or collective farms. In line with decisions made centrally, the sizes of farms and their number fluctuated. The integration of both countries in the larger USSR drastically impacted the agricultural sector, management of natural resources, and the way of life of the inhabitants. These changes did not occur in a uniform way throughout the Soviet period, but were rather the result of various policies. In the early Soviet republics, the objective in agriculture was to achieve collectivization of land and livestock by creating numerous small-scale collective farms. After the

1950s, they were aggregated to create larger units. If in 1928 one collective farm counted on average 13 households, in 1981 this figure had risen to 489 (Central Department for Statistics, 1988). The Soviet Union made important efforts to develop infrastructure such as irrigation canals or reservoirs, and roads to pastures in Tajikistan and Kyrgyzstan. This infrastructure oriented to pasture and irrigation water, as well as to the transport of people and goods, was designed to create a network between the Soviet republics and to enhance connections. This was especially the case in the Ferghana Valley between Tajikistan and Kyrgyzstan.

In Vorukh and Ak-Say, these Soviet policies translated into drastic changes in land use as well as in the processes of conflict resolution and the legal system. Traditionally, communities of semi-nomadic livestock keepers and settled farmers lived in the territory of Ak-Say and Vorukh. This territory was part of the Kokand Khanate (18th–19th centuries) and the Russian Empire until the October Revolution in 1917. These communities were represented by various tribes and clans, which, in the process of Soviet national policy, began to be identified as nations. Kyrgyz were engaged mainly in livestock keeping and migrated in search of pastures for their flocks while Tajik were mainly engaged in cultivation. With the establishment of Soviet power between 1920 and 1930, the compulsory settling of semi-nomads began, since the Soviet vision of progress regarded the semi-nomadic way of life as a vestige of the past and backwardness. Like in other parts of the Ferghana Valley, it faced fierce resistance from local residents, which is known in history as the Basmachi movement (Chokushov, 1968). Soviet power forced the local population to create collective farms, the development of which should have produced a class of peasant workers, although local residents were not familiar with the project of global construction of socialism. The assets of the collective farms consisted of forcibly alienated property, land and livestock of local residents, and this process is known in history as a policy of collectivization (Khalid, 2007, pp. 76–80). Soviet authorities carried out measures to achieve specialization in livestock keeping for the newly created collective farms formed by former semi-nomadic communities. During that period, Vorukh was not an enclave because the border had only an administrative existence. Communication, trade and cultural ties were common. At that time, the dominance of a national state on which group identities were built did not yet exist. However, the Soviet Union pursued an active national policy, according to which ethnic identity was considered as the basis for the formation of Soviet nations (Abashin and Bushkov, 2004; Abashin, 2007). Thus, with the development of the two Soviet national republics, Kyrgyzstan and Tajikistan, enclaves were created within their territories.

The Soviet period is characterized by different stages of building interactions between Ak-Say and Vorukh. From 1920–1950 several collective farms were created: 'Kyzyl-Ay', which is now used as a pasture in the Keravshin mountain area; 'Kommunism' (transformed into the settlement of Kapchygay, part of the Ak-Say municipality); 'Kirov' (now Kok-Tash); 'Ozgorush' (now divided among Bedak and Vorukh); 'Shvernik' and 'Pravda' (now Vorukh). Members of the collective farms could not move freely because of the registration at the place of residence (*propiska*) introduced during the Soviet period, which tied the place of residence to the place of work. The authorities thus strictly regulated migration, since without this registration it was impossible to gain access to health care and education or to find housing.

All types of land and pasture were under the supervision of the state. To avoid even greater resistance, the pastoralist tribes were assigned to collective farms, whose members were predominantly of the same tribe (in our case, the Uru-Avat and Kypchaks). Even though the nomadic way of life was banned by the government, the traditional knowledge of regulated grazing was adapted and partly modernized by the Soviet state, by the institutionalization of the seasonal use of pastures and the introduction of motorized transport for the transfer of livestock to distant high-altitude pastures. Private ownership was limited to 20 sheep or goats per household (Oshskoye oblastnoye upravleniye sel'skogo khozyaystva, 1959), while collective flocks counted 8000–15,000 sheep and goats (Karl-Marks atyndagy kolkhoz, 1959). All the production from collective farms was sent to the government, and members of collective farms received a share from the sale of collective farm products usually not in money terms, but as consumption goods.

In the 1950s and 1960s, the Soviet authorities decided to consolidate the collective farms and to create state farms through merging of collective farms. The most obvious difference between collective and state farms was that farmers, as members of state farms, received fixed salaries from the government and had to implement a plan to produce agricultural products. In the territory of the case study, the '100 Letiia V.I. Lenina' state farm owned 40,000 sheep, 20,000 goats and 1000 yaks (V.I. Lenin 100 zhyldygy atyndagy sovkhoz, 1961). The territory of this state farm was recognized as a territory of Kyrgyzstan. The only collective farm that would not be transformed into a state farm was 'Pravda' in the territory of Vorukh. During this period, large irrigation projects including the construction of the Tortkul irrigation reservoir were implemented in the Ferghana Valley. The opportunity to cultivate new lands became a new prospect for the state farm of '100 Letiia V. I. Lenina' and the collective farm 'Pravda'. This led to the extension of the cultivated territories, which thus became adjacent to the Tajik settlement of Khoji-A'lo.

In 1970–1971, a conflict took place between the communities of Khoji-A'lo and Kok-Tash (part of Ak-Say) over the right to cultivated land. This land was under the administration of the central Forest Agency in Moscow. Two opposed practices of determining the right to land collided. Nomadic and semi-nomadic communities widely apply the traditional right to land, which is not fixed by legal documents or property signs such as fencing, for example (Kozhonaliev, 1963). As the Soviet government viewed nomads as a group carrying an archaic way of life, the traditional right to land was not considered as a source of law, especially as all land in the USSR was the exclusive property of the states of the Soviet Socialist Republics. The Soviet central authorities applied the principle 'the one who cultivates the land owns it' as a basis to solve disputes. Thus, the Kyrgyz communities were not in a favourable position, because they could not support their right to the land by cultivation practice, since they were more involved in livestock keeping (see also, Bichsel, 2009a). In turn, Tajik communities viewed Kyrgyz settlements as migrants coming from the mountains. Disputes between communities that occurred during the period of 1970–1971 could not be resolved at the local level, and police were brought in to maintain order by force (Bushkov, 1990).

In 1975, a new clash erupted between the communities in the new settlements of Tozhikon (part of Vorukh) and Bakay (part of Ak-Say). Tozhikon started cultivating plots that inhabitants of Bakay considered as part of their territory. To resolve that conflict, in 1975 the central government in Moscow proposed an agreement between the two parties: inhabitants of Tozhikon would cultivate this land and in return, inhabitants of Ak-Say would gain a right to irrigation water from the Mekhnatobod-Ak-Say canal. However, this agreement was not implemented, and, instead, a pumping station was built for Ak-Say with a government subsidy for the electricity supply (Bichsel, 2009a, p. 29). The Soviet infrastructure played the role of *de facto* border; however, this was complicated by the long-term land use arrangements (and leases) that collective farms could conclude to make use of land or infrastructure on the other side of the border (Reeves, 2014a).

Later, in 1989, the conflict over the right to cultivate and build houses on conflicted territories escalated, and resulted in the removal of 40 Kyrgyz families living in Vorukh to neighbouring Ak-Say for permanent residence.[2]

The confrontations of 1970–1989 revealed the conflict potential of the agrarian policies pursued by the authorities. At the same time, the issue of determining borders between the communities began to take the form of a long-term dispute, since no party had foreseen that the Soviet Union would soon cease to exist and that internal borders would suddenly become international.

The Post-Soviet Period in the Ferghana Valley: Multiplication of Household Farms and Transformation in Land and Water Management

In comparison with the deep upheaval due to the implementation of the Soviet system in the agrarian and political domains, the end of the Soviet Union and measures related to implementation of a market economy were a shock to rural communities as changes happened in a short period of time. Agrarian reforms mainly contributed to shape rural landscapes in two ways. First, the dismantling of state and collective farms resulted in the appearance of a multitude of household farms (Statistic Agency under the President of Tajikistan, 2015; National Statistic Committee of the Kyrgyz Republic, 2016) and the creation of rural municipalities, responsible for the social care and management of part of the agricultural and industrial lands as well as residential areas. Second, institutions managing natural resources – such as irrigation water and pastures – were reformed.

Independent Tajikistan is characterized by an institutional variety of farms, between family-managed (household plots and family or individual *dekhan*) and collective (*dekhan* and agricultural enterprises). The latter are the successors of Soviet collective farms and state farms. Although the successors of the collective farms (nowadays called 'collective *dekhan* farms') and state farms still manage important shares of arable land, household farms control around 65% of the arable land. Each of the 65,000 rural households in the country manages a kitchen garden of around 0.3 ha. Statistics provided by the government do not allow us to highlight the share of land used by each farm category, but according to estimates household farms cultivate 65% of the total land of the country (Lerman, 2012).

At independence, rural inhabitants in Kyrgyzstan and Tajikistan were severely hit by the dramatic decrease of social support from the government in those times of already widespread poverty. Despite the decline of the share of agriculture in the gross domestic product (GDP) (Fig. 13.3), farming still represents a crucial subsistence and income-generating activity for rural households. Around a quarter of rural inhabitants in both countries were reported to be living below the poverty line in 2013 (World Bank, 2017).[3]

Both countries have maintained large rural populations – 73% in Tajikistan and 64% in Kyrgyzstan in 2013 – with a high proportion of agro-pastoralists (World Bank, 2017). Since 1997, substantial economic growth has been observed due to agricultural activities. Despite this growth, Tajikistan remains the poorest country in Central Asia (World Bank, 2017). Both countries identify increased production of livestock and agricultural cultivation in their national strategies as central to stimulating economic growth (see, e.g. the Programme of Pasture Development of the Republic of Tajikistan (Republic of Tajikistan, 2015) and the National Sustainable Development Strategy for the Kyrgyz Republic 2013–2017 (Kyrgyz Republic, 2013)).

Labour migration has grown quickly since the end of the Soviet period in rural areas as an alternative strategy to the shrinkage of work opportunities. According to the World Bank, in 2013 Tajikistan and Kyrgyzstan were among the most remittance-dependent countries in the world (World Bank, 2017). According to local

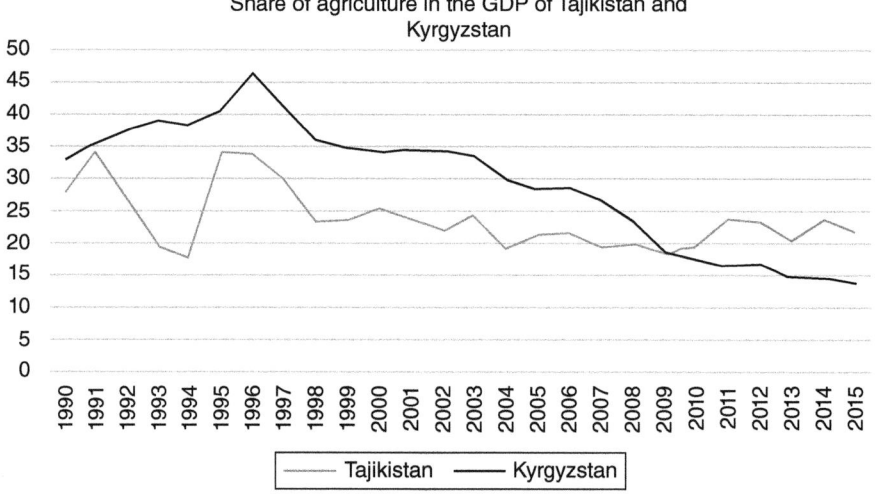

Fig. 13.3. Share of agriculture (including fisheries and forestry) in the GDP of Tajikistan and Kyrgyzstan. (From: FAO (2017).)

authorities, whose representatives were interviewed in the framework of our field research in 2015–2016, 30% of the inhabitants of Ak-Say participate in the permanent or seasonal labour migration; while in Vorukh, 85% of the employable population work abroad seasonally and 15–20% of them on a permanent basis. Labour migration makes the system of natural resource management less sustainable by driving away the labour force and active social leaders (Murzakulova and Mestre, 2016). The lack of leadership at the local level is an important element in the creation of new community-based management institutions replacing the Soviet centralized management system.

In 1994, Kyrgyzstan was the first of the Central Asian countries to introduce Water Users Associations (WUAs) through donor-funded projects, though their legal framework was not adopted until 2004. The Water Code was adopted one year later, introducing the new concept of watershed management. In 2006, Tajikistan passed the Law on Water Users Associations, which in many aspects is similar to the legal code of Kyrgyzstan. In both countries, non-commercial, non-governmental WUA institutions became responsible for the operation and maintenance of irrigation systems inherited from the state farms and collective farms of the Soviet era. According to the regulations, WUAs are also responsible for the implementation of measures against land degradation. WUA work is based on a member-based decision-making body, which elects a board to serve as its executive. Members can be individuals, commercial organizations or farms registered with various legal statuses. However, in many cases users do not actually hold decision-making power (Abdullaev et al., 2010).

In Tajikistan, a legal framework leaning towards integrated water resource management (IWRM) was adopted in parallel with the reform of the agricultural sector, creating the Ministry of Water and Energy and the Agency of Irrigation and Melioration in 2012. The aim of this reform was to clearly divide policy-making responsibilities from operational and management responsibilities (Rahimova, 2016). WUAs are under the authority of State Water Institutions, but they face difficulties working independently and are dominated by local authorities (Sehring, 2007).

The transfer of irrigation management to the local level was implemented in order to internalize the ecological, economic and social externalities of irrigation (Herrfahrdt et al., 2006). In the specific post-Soviet context, it also aimed at countering the decay of the infrastructure. In Kyrgyzstan and Tajikistan, as in many other countries, this has in practice translated to implementation of large reforms in the irrigation water sector within a relatively short period of time.

As with the management of irrigation water, Kyrgyzstan emphasized the decentralization of pasture management. Pasture User Unions (PUUs) were allocated the responsibility for pasture management at the rural municipality level in 2009. The pasture users do not rent these pastures but pay a pasture user fee. However, pastures under the State Forest Fund are still managed through local Forest Management Units (FMUs). These pastures represent around 30% of the total area of the pastures (Shukurov et al., 2006). Although theoretically FMUs were in charge of reforesting these areas, most of these areas appeared not to be suitable and were instead leased to graze livestock. They are commonly used in combination with pastures under the PUU administration.

In Tajikistan, since 2013 the law authorizes the creation of PUUs at the district level. Unlike in Kyrgyzstan, Tajik users have little role in decision-making. In Tajikistan, the introduction of the PUUs and of the Commission on Pastures at district level did add a layer of complexity and uncertainty to the allocation of pasture rent and the monitoring of pasture use (Wilkes, 2014). Moreover, four management models exist in the country, through the PUUs, State Forest Fund, State Land Reserve and State Pasture Reserve. Pasture allocation mechanisms can also vary from formal to less formal (Wilkes, 2014).

In summary, the institutional framework for the management of irrigation water and pastures is now scattered among different organizations, each with its own level of dependence to the central government and its own regulations on resource allocation. Moreover, community-based institutions are themselves producing a hybrid Soviet management system framed in a wider inherited institutional landscape (Isaeva and Shigaeva, 2017).

Conflicts and Implications of Natural Resource Management Policy for Vorukh and Ak-Say

The end of the Soviet Union and the subsequent institutional turbulence severely affected the territories of Vorukh and Ak-Say. One of the main consequences of this transformation was that they became border villages of the Republic of Tajikistan and of the Kyrgyz Republic. In the territory of Ak-Say, all families received their land plots, which they could use at their own discretion: to build a house or to cultivate. In Vorukh, the collective farm 'Pravda' continued to operate until 2005, when the authorities decided to dismantle it into Vorukh cooperative; after 2015 it was transformed into an Association of the *dehkan* farms. It should be noted that despite the preservation of the collective farm, its content has radically changed. Members of the collective farm could suddenly choose their production,[4] freely move, and go abroad for labour migration. However, for Vorukh, outward migration did not relieve the tension related to the shortage of land, as it was impossible to build houses on the land. The housing issue is acute for Vorukh, with a population of 33,000 people in an area of 33,084 ha (Vorukh Zhamoat, 2017). In comparison, Ak-Say has a population of 7616 people in an area of 31,027 ha (National Statistic Committee of the Kyrgyz Republic, 2017).

The flexible nature of the border was historically an asset in the joint use of infrastructure such as roads and markets. However, it began to change radically after independence. For example, in the past, the road to Vorukh was used by both Tajiks and Kyrgyz to reach markets where both currencies were used, serving as platforms for exchange between communities from both sides of the border. In this context, the construction of the new road after independence changed the nature of the border as 'citizens of neighbouring states increasingly use different roads, different routes, and different minibuses to get to different markets using different currency' (Reeves, 2014a, p. 254). This caused dissatisfaction among Vorukh's residents, as it hindered their movement to mainland Tajikistan (Reeves, 2014a, p. 254).

During the post-independence transition period, conflicts between communities were quite low and had a seasonal character as they occurred during the irrigation period and during the transhumance. These seasonal confrontations took place along the perimeter of the borders in the Ferghana Valley and were known as '*ketmen wars*'.[5]

After 2000, both countries launched policies to strengthen their borders, which resulted in setting up border checkpoints on public roads. Procedures to cross bordering areas were reinforced and increased the tensions in the communities. During that period, the two countries implemented symmetrical responses by blocking roads and cutting access to water to protest against the detention of citizens by the border guards. The Kyrgyz side proposed to the Tajik government a memorandum on renting pastures to foreign citizens in order to mitigate the tensions because of the vague legal framework of the transborder transhumance routes from Tajikistan to the pastures of Kyrgyzstan (Rural Development Fund, 2010). The Tajik government refused the memorandum since agreeing would have been a tacit acceptance of the border delimitation suggested by the Kyrgyz side. It led to the break of the traditional practice of sharing pastures between Ak-Say and Vorukh.

Aside from the interests of governments in terms of border delimitation, the joint use of pastures has become an issue as the institutions inside Kyrgyzstan pursue different goals. The management of pastures in Kyrgyzstan is scattered among local FMUs and PUUs, whose interests may be contradictory. For example, the FMU is interested in renting their pastures to shepherds from Tajikistan because they pay more than citizens of Kyrgyzstan. The PUU of Ak-Say is also interested in the use of forestry pastures under the responsibility of the FMU, but access to them is limited by the absence of a legal framework to support cooperation with them. Internal conflicts between the FMU and the PUU creates an opposition between PUU and Tajik communities over the joint use of pasture.

The conflict over the access to the pastures reached a peak in the summer of 2014, when Kyrgyzstan attempted to build an alternative road to summer pastures bypassing the Vorukh enclave (Fig. 13.4). Inhabitants of the Vorukh enclave were strongly opposed to this new road as the decrease of the use of the road would

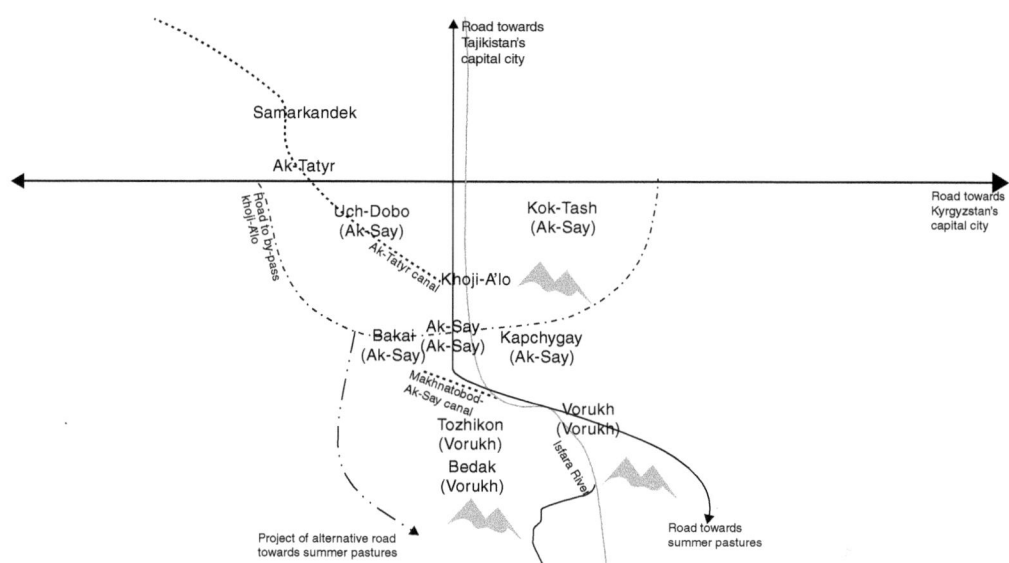

Fig. 13.4. Main patterns of infrastructure and location of settlements described in this case study. The construction of the alternative road towards the pasture meant to bypass Vorukh was interrupted by violent clashes from the side of Tajikistan border guards. Borders are deliberately not shown, to highlight the complexity of infrastructure and access to natural resources for communities.

weaken their interactions with the communities of Kyrgyzstan and leave them in a deep seclusion. The conflict escalated to the point that Tajik border guard detachments fired mortars. The border was then closed for several months, and the tensions held even when cross-border trade was interrupted. After the re-opening of the border, previous links between communities could not be restored. Traders now prefer to trade only in the markets of their communities, young people communicate only within their communities and residents try to avoid staying in the neighbouring community. The border patrols have become even more armed, and the authorities have become convinced that independent infrastructure – roads, irrigation canals, etc. – and separate resource use are necessary to avoid conflicts.

The most acute issues in the domain of irrigation management are related to the Ak-Tatyr canal (Fig. 13.4). The Ak-Tatyr canal is 19 km long. It starts on the territory of Ak-Say (part of Ak-Say), crosses the territory of Tajikistan through Khoji-A'lo for 4 km, then re-enters the territory of Kyrgyzstan, through Uch-Dobo (part of Ak-Say) and Ak-Tatyr and Samarkandedk rural municipalities. The situation related to the canal is a clear example of how poor management of natural resources can trigger conflicts between transborder communities. Formally, the management of the canal is scattered between different institutions. Four kilometres are under the responsibility of the Batken Water Department. However, this oversight is not legitimate in the eyes of inhabitants of Khoji-A'lo, who claim that if the infrastructure is located in their country, it should be owned by their country and administrated accordingly. Because of this dispute, the Batken Water Department has almost no access to the canal to carry out repairs on the portion of the canal located in Tajikistan.

Management of the Ak-Tatyr canal also raises tensions between Kyrgyz institutions. On-farm parts of the canal are under the stewardship of the WUA Tortkul' Tolkunu, which was created in 2007 by the rural municipality of Samarkandek, which is located at the end of the canal (Fig. 13.4). This initiative was not supported by the neighbouring rural municipalities of Ak-Say and Ak-Tatyr, as the creation of the WUA raised the question of the water allocation among rural municipalities. Although they are part of the same country, their interests differ. Ak-Say is located upstream and does not rely on water from the canal, and thus, its municipal council is not actively involved in the conflict resolution pushed by Samarkandek. The Ak-Say municipal council instead prioritizes the negotiations over the building of the new road bypassing Vorukh. As Ak-Tatyr is located in the middle of the canal length, it is interested in maximizing the water use of its own inhabitants rather than sharing with other rural municipalities, and does not take part in the process initiated by Samarkandek to clarify and reconsider the water allocation. Samarkandek, being at the end of the canal and having uncertain access to irrigation water, is the most interested and active stakeholder in the work of the WUA to resolve water access issues.

Tensions grow in summer, when access to water is an especially acute problem. Samarkandek and Ak-Tatyr rural municipalities of Kyrgyzstan experience lack of water and blame Khoji-A'lo for pouring water from the canal into the river. Thus, the poor management of water negatively impacts the inter-community relationships and leads to the ethnicization of the conflict over access to water. In this context, a new type of migration in transborder areas emerges: households forming an ethnic minority, in this case Kyrgyz living in Khoji-A'lo and Vorukh and Tajik living in Ak-Say, migrate to communities where their ethnic group is the majority. This trend rearranges social networks by enhancing their ethnicization and limits inter-community cohesion, which is replaced by cohesion inside ethnic groups.

Discussion and Conclusions

Agro-pastoral systems in the Ferghana Valley underwent two forceful shocks, collectivization of agriculture and political independence, which affected every aspect of their functioning and caused various adaptation responses from communities. First, Soviet agrarian policies led to a drastic reorganization of the way of life of semi-nomadic and sedentary communities of the Ferghana Valley. Later, the shift from internal to international borders between Tajikistan and Kyrgyzstan contributed to the dismantling of a system in which communities from Soghd and Batken used to collaborate despite their differences in population and stream position. The enclaves and surrounding territories have become more vulnerable to the new border regime and the natural resource management systems that have been introduced in both countries since the mid-1990s. Disputes exacerbate questions around the ownership of arable land, irrigation water, pastures and their related infrastructure. In turn, these tensions weaken the ancient practices of common use of natural resources. Moreover, this common use of natural resources has begun to be understood as one of the main causes of inter-community conflicts.

In the countries of the Ferghana Valley, the long-lasting agricultural crisis following the end of the Soviet Union was meant to be solved by the implementation of reforms in the field of natural resource management. However, communities from both sides of the border face difficulties in creating cooperative relationships for natural resource management, especially as the management is scattered inside the countries themselves. The dissymmetry and uncertainty between the institutions from Kyrgyzstan and Tajikistan strengthen the potential for conflicts. Institutions for water and pasture management on both sides of the border do not have the same competences. The legal resolution of one question, for example, the grazing of livestock of Tajikistan in Kyrgyzstan, requires the involvement of many stakeholders from both sides of the border. At the same time, despite the trend towards decentralization, there is an opposing process to centralize the competences to deal with institutions located in neighbouring countries. Moreover, the rapid adoption of agrarian reforms, their frequent amendments and their uneven implementation hinders collaboration. Institutions have little information about the tasks allocated on the other side of the border, and institutions themselves cannot clearly define their role.

This complex and unstable institutional setting also reflects the situation among institutions inside each of the countries. The diversity of their legal frameworks, functions and aims leads to variation in their interests in transborder use. While some stakeholders try to avoid joint use of natural resources, others try to maintain it (formally or informally) because it represents a source of income or because they rely on resources or services provided by institutions from the other country. This results in a lack of coordination.

Since the governments of both countries are reluctant to make concessions, the issue of borders will stand on the agenda for a long time. In this context, conflict dynamics between the agro-pastoral communities of Ak-Say and Vorukh may be regarded as a status quo situation.

Acknowledgements

The authors want to thank Michele Nori as well as the editors for their valuable comments and editing on previous versions of this chapter. The authors are grateful to the Department for International Development (DFID)-funded project 'Improving stability and better natural resource management in Kyrgyzstan and Tajikistan', which supported the authors' field research in the Ferghana Valley. The views expressed are the authors' own and do not represent the views of any organization.

Notes

[1] The maps presented as Figs 13.1 and 13.2 were derived from the Openstreetmap database. Some borders in the area are still in the process of being demarcated by intergovernmental commissions. Accordingly, the Mountain Societies Research Institute (MSRI) takes no responsibility for errors, omissions or positional accuracy, and makes no warranties, express or implied, accompanying the products.
[2] Data from in-depth interview materials, Ak-Say, June 2017.
[3] Specifically, 23.6% of rural inhabitants in Tajikistan and 32.6% in Kyrgyzstan fall under the countries' respective national poverty lines (World Bank, 2017).
[4] State purchasing was preserved but only for cotton, which was not cultivated in the area of Vorukh or Ak-Say.
[5] *Ketmen* is an agricultural tool for manual weeding of the land.
[6] The transliteration was done according to the BGN/PCGN system, except for the nouns that already exist in English.

References

Abashin, S.N. (2007) Natsionalizmy v Sredney Azii: V poiskakh identichnosti [Nationalisms in Central Asia: In search of identity]. Aleteyya, Saint Petersburg, Russia.
Abashin S.N. and Bushkov V.I. (2004) Ferganskaya Dolina. Etnosi, etnicheskie prochesi, etnicheskie konflikty [Ferghana Valley. Ethnicity, Ethnic processes, Ethnic conflicts] Russian Academy of Sciences Institute of Ethnology and Anthropology named after N.N. Miklouho-Maclay, Moscow.
Abdullaev, I., Kazbekov, J., Manthritilake, H. and Jumaboev, K. (2010) Water user groups in Central Asia: emerging form of collective action in irrigation water management. *Water Resources Management* 24(5): 1029–1043. DOI: 10.1007/s11269-009-9484-4.
Bichsel, C. (2005) In search of harmony: repairing infrastructure and social relations in the Ferghana valley. *Central Asian Survey* 24(1): 53–66. DOI: 10.1080/02634930500050008.
Bichsel, C. (2009a) *Conflict Transformation in Central Asia: Irrigation Disputes in the Ferghana Valley*. Central Asian Studies. Routledge, Abingdon, UK.
Bichsel, C. (2009b) It's about more water. Natural resource conflicts in Central Asia. In: Péclard, D. (ed.) *Environmental Peacebuilding: Managing Natural Resource Conflicts in a Changing World*. Swisspeace, Bern, Switzerland, pp. 32–40.
Bichsel, C. (2013) Dangerous divisions: peace-building in the borderlands of post-Soviet Central Asia. In: *Violence on the Margins: States, Conflict, and Borderlands*. Palgrave Macmillan, New York, pp. 145–165.
Bichsel, C. (2014) The transformation of Tajikistan. The sources of statehood. *Europe – Asia Studies* 66(6), 1017–1018. DOI: 10.1080/09668136.2014.924757.

Bushkov, V.I. (1990) On some aspects of interethnic relations in the Tajik SSR. Interethnic relations in the USSR. Series A. Document #9. Studies on applied and urgent ethnology. Institute of Ethnography of the USSR Academy of Sciences. Available at: http://static.iea.ras.ru/neotlozhka/9-Bushkov.pdf (accessed 15 March 2018).

Central Department for Statistics (1988) Kolhozy SSSR (Kratkiy statisticheskiy sbornik) [Collective farms of USSR (Short Statistic Report)], Moscow.

Chokushov B. (1968) Klassovaya bor'ba i uprochneniye sovetskoy vlasti v Kirgizskikh Ayylakh. Ministerstvo Narodnogo Obrazovaniya Kirgizskoy SSR. [Class struggle and strengthening of the Soviet power in the Kyrgyz villages (1918–1924)] Ministry of Education Kyrgyz SSR. Kyrgyz State University.

Food and Agriculture Organization of the United Nations (FAO) (2017) FAOSTAT. Available at: http://faostat3.fao.org (accessed 22 January 2017).

Herrfahrdt, E., Kipping, M., Pickardt, T., Polak, M., Rohrer, C. and Wolff, C. (2006) *Water Governance in the Kyrgyz Agricultural Sector: On Its Way to Integrated Water Resource Management?* German Development Institute, Bonn, Germany.

IA 24 (2015) V Kyrgyzstane v 2014 godu zafiksirovano 40 prigranichnykh konfliktov [Kyrgyzstan has registered 40 border conflicts in 2014]: Informational Agency 24. Bishkek 20 April. Available at https://24.kg/parlament/11074_v_kyirgyizstane_v_2014_godu_zafiksirovano_40_prigranichnyih_konfliktov (accessed 25 June 2017).

Isaeva, A. and Shigaeva, J. (2017) Soviet legacy in the operation of pasture governance institutions in present-day Kyrgyzstan. *Journal of Alpine Research* 105-1. Available at: http://rga.revues.org/3631 (accessed 23 June 2017).

Kabar (2012) S nachala 2012 goda na kyrgyzsko-tadzhikskoi granitse proizoshlo 19 intsidentov [Since the beginning of 2012, 19 incidents occurred on the Kyrgyz-Tajik border]: Press Agency Kabar. Bishkek, 22 November. Available at: http://old.kabar.kg/regions/full/44397 (accessed 25 June 2017).

Kabar (2016) Kyrgyzstan i Tadzhikistan soglasovali boleye 500 km gosudarstvennoy granitsy [Kyrgyzstan and Tajikistan agreed on more than 500 km of governmental border]: Press Agency Kabar. Bishkek, 16th January. Available at: www.kabar.kg/rus/SMIoKG/full/100975 (accessed 4 March 2016).

Karl-Marks atyndagy kolkhoz (1959). Godovoy otchet, Batken rayonduk arkhiv, Fond 15/1/1 [Annual report, Batken district archive, Fond 15/1/1].

Khalid, A. (2007) *Islam after Communism: Religion and Politics in Central Asia*. University of California Press, Berkeley, California.

Kozhonaliev, S. (1963) Sud i ugolovnoye obychnoye pravo kirgizov do oktyabristskoy revolyutsii. [Court and criminal customary law of the Kyrgyz before the October Revolution] Akademia Nauk Kirgizskoi SSR Otdel Filosofii I Prava [Academy of Sciences of the Kyrgyz SSR. Department of Philosophy and Law]. Frunze.

Kyrgyz Republic (2013) National sustainable development strategy for the Kyrgyz Republic for the period of 2013–2017 (2013). Available at: http://faolex.fao.org/docs/pdf/kyr143374E.pdf (accessed 15 March 2018).

Lerman, Z. (2012) Agrarian reform of the Republic of Tajikistan: farm reform and restructuring cooperative development. Available at: www.fao.org/docrep/017/aq331e/aq331e.pdf (accessed 22 January 2017).

Matveeva, A. (2017) Divided we fall… or rise? Tajikistan-Kyrgyzstan border dilemma. *Cambridge Journal of Eurasian Studies* 1, 1–20. DOI: 10.22261/94D4RC.

Murzakulova, A. and Mestre I. (2016) Dynamics in the management of natural resources in the border communities of Kyrgyzstan and Tajikistan. Research Report. MSRI UCA. Available at: www.ucentralasia.org/Resources/Item/1148 (accessed 12 June 2017). DOI: 10.13140/RG.2.1.1862.3601.

National Statistic Committee of the Kyrgyz Republic (2016) Sel'skoye khozyaystvo Kyrgyzskoy Respubliki 2011–2015 [Agriculture in the Kyrgyz Republic 2011–2015]. Bishkek.

National Statistic Committee of the Kyrgyz Republic (2017) Census of the population for 2016. Available at: www.stat.kg/ru/statistics/naselenie (accessed 1 June 2017).

Oshskoye oblastnoye upravleniye sel'skogo khozyaystva (1959) Document N°1584, Batkenskiy rayonnyy arkhiv, Fond 15/1/1 [Batken district archive, Fond 15/1/1].

Rahimova, S. (2016) Reforma Vodnogo Sektora V Tadzhikistane: Proyekt 'Upravleniye vodnymi resursami v Tadzhikistane' [Water sector reform in Tajikistan: Project 'Management of water resources in Tajikistan']. Conference presentation, Bishkek.

Reeves, M. (2005) Locating danger: Konfliktologiia and the search for fixity in the Ferghana valley borderlands. *Central Asian Survey* 24(1), 67–81. DOI: 10.1080/02634930500050057.

Reeves, M. (2011) Fixing the border: on the affective life of the state in southern Kyrgyzstan. *Environment and Planning, D, Society and Space* 29(5), 905–923. DOI: 10.1068/d18610.

Reeves, M. (2014a) *Border Work: Spatial Lives of the State in Rural Central Asia*. Cornell University Press, Ithaca, New York.

Reeves, M. (2014b) Roads of hope and dislocation: infrastructure and the remaking of territory at a Central Asian border. *Ab Imperio* 2, 235.

Reeves, M. (2016) Infrastructural hope: anticipating 'independent roads' and territorial integrity in southern Kyrgyzstan. *Ethnos* 82(4), 711–737. DOI: 10.1080/00141844.2015.1119176.

Republic of Tajikistan (2015) Programmy razvitja pastbiŝ v Respubliki Tažikistana na 2016–2020 gody [Programme for pasture development in the Republic of Tajikistan for 2016–2020]. Republic of Tajikistan.

Rural Development Fund (2010) Mechanisms of management of natural resources with the involvement of communities: analytical report on the conflict in Ak-Say and Ak-Tatyrsk village districts of the Batken region of the Kyrgyz Republic. Available at: http://pdf.usaid.gov/pdf_docs/pnaea247.pdf (accessed 15 March 2018).

Sarkees, M. and Wayman, F. (2010) *Resort to War: A Data Guide To Inter-State, Extra-State, Intra-State, And Non-State Wars, 1816–2007*. CQ Press, Washington, DC.

Sehring, J. (2007) Irrigation reform in Kyrgyzstan and Tajikistan. *Irrigation and Drainage Systems* 21(3–4), 277–290. DOI: 10.1007/s10795-007-9036-0.

Shukurov, E., Makeev, T. and Koshoev, M. (eds) (2006) Kyrgyzstan, environmental and natural resources for sustainable development. State Agency on Environment Protection and Forestry under the Government of the Kyrgyz Republic and United Nations Development Programme in the Kyrgyz Republic, Bishkek.

Statistic Agency under the President of Tajikistan (2015) Tajikistan in figures: 2015. 160pp. Statistic Agency under the President of Tajikistan.

Wilkes, A. (2014) Institutional analysis on pasture management in Tajikistan. Final report. Unique, GIZ, Freiburg, Germany.

V.I. Lenin 100 zhyldygy atyndagy sovkhoz (1961) Godovoy otchet, Batken rayonduk arkhiv, Fond 15/1/1 [Annual report, Batken district archive, Fund 15/1/1].

Vorukh Zhamoat (2017) Spravka o naselenii i territorii Zhamoata Vorukh [Official note about population and territory of the Vorukh Zhaomat].

World Bank (2017) World Bank country data, poverty and equity database. Available at: http://databank.worldbank.org (accessed 20 January 2017)

Yin, R.K. (2013) *Case Study Research: Design and Methods*, 5th edn. SAGE, Thousand Oaks, California.

14 Conflict and Resistance in Southern Punjab: a Political Ecology of the 2010 Floods in Pakistan

Ali Nobil Ahmad*

Leibniz-Zentrum Moderner Orient, Berlin, Germany

Introduction

Classical historical materialist studies of rural South Asia addressed the relationship between agricultural modernization and social change – the 'agrarian question' – in a rich body of work dating to the 1970s concerned with transitions to capitalism in subsistence economies, modes of production and peasant class struggles (e.g. Gough et al., 1973; Byres and Mukhia, 1985). In line with more recent scholarship, this chapter approaches the agrarian question in environmental terms, building upon the work of authors such as Nancy Peluso, Michael Watts and others associated with the 'ecological agrarian question' (AQ7, within the typology outlined by Bahn and Zurayk, Chapter 1, this volume). Whereas the basic premise of AQ7 relates conflict to the 'biophysical contradictions' of capitalism, my own approach encompasses the wider process of natural resource development through science, technology and infrastructure. Irrigation in particular mediates and underpins logics of accumulation and profiteering in 'hydraulic' societies such as rural Punjab, Pakistan, where class frictions are predicated upon who benefits from a given hydro-social order through access to land, water, capital and technology. In addition to the socio-economic inequality resulting from unequal resource access, irrigation regimes are bound up with uneven geographies of vulnerability. At moments of environmental crisis such as floods, these spatial dimensions come abruptly into play, shaping the way in which class struggles are experienced and articulated.

In addition to agriculture and socio-natural relations, this chapter considers the role of culture, language, ethnicity and region – in short, identity – in shaping patterns of conflict. Much of Pakistan's political history is rooted in antagonism between the central state and its ethnically diverse, economically marginalized geographical peripheries. At moments of environmental distress, this contradiction, sometimes referred to as the 'national question', has tended to flare into political conflict because regional ethnonationalist movements have historically experienced environmental disasters as opportunities to mobilize opposition to centralized state power. Most famously, in 1971 Pakistan lost over half of its population and much of its territory to the movement for the establishment of Bangladesh, which gained considerable momentum after a massive cyclone that exposed the state's inability to provide an adequate rescue and rehabilitation response. Following the catastrophic 2010 floods in Southern Punjab, the phenomenon of indigenous Saraiki (ethno-linguistic)

---

* Email: alinobil@googlemail.com

nationalism played a prominent role in protests against the government following its inadequate disaster response. Accordingly, I consider the importance of Saraiki politics within the context of its longer-term role in opposing dominant irrigation regimes, exploring the question of ethno-nationalism and culture as an important aspect of the agricultural ecological question.

The chapter begins with a brief sketch of the historical background to natural resource development in Pakistan, focusing on the evolution of state and society in relation to irrigation infrastructure in the province of Punjab since the colonial era. Then, after a brief note on the methodology and data used in this study, an empirical discussion of the 2010 floods and its aftermath follows. Discussion is divided into two sections that relate, respectively, to the causes and consequences of the 2010 floods in the short and longer term. I conclude with some broader remarks on what might be extrapolated from Pakistan's experience of the ecological agrarian question.

The 2010 Floods in Context

The aspiration to control water – to direct its course for economic, political and ideological purposes – is by no means an exclusively modern impulse. It was not until the latter decades of the 19th century, however, that the contours of a distinctly modern colonial hydrology came into view, marking a qualitative historical break with the continuities displayed under prior phases of colonial rule by the East India Company. Between 1860 and 1920, the construction of public works dramatically increased in pace, scope and form, part of a far-reaching transformation in colonial capitalism tied to developments in science and new orthodoxies in the field of economics, which increasingly directed policy towards a new emphasis on infrastructure and technology as the key to successful rule – materially and ideologically. Irrigation policy became a key plank of Britain's 'civilizing mission', particularly in Punjab's 'canal colonies', where radical transformation of its western landscape through the building of perennial canals and permanent headworks opened up 14 million acres (5.67 million ha) of previously arid land to cultivation, a development credited with having turned the province into a 'bread basket'. Extensive military recruitment and land grants to hereditary castes helped socially engineer a fiercely loyal class of agrarian magnates, cementing an allegiance between the colonial state and province of Punjab, with fateful consequences. South Asia's agricultural modernization or 'transition' was further complicated by the establishment of an indigenous lower level irrigation bureaucracy that thrived on the basis of its power to confer or deny patronage, favours and water privileges within rural society (Ali, 1987; Gilmartin, 1994; D'Souza, 2006).

During the 1950s and 1960s, industrialization in Pakistan, hailed by many economists as a model developing economy, was subsidized by agricultural modernization continuous with colonial regimes of hydrology. The so-called Green Revolution saw the building of barrages in Sind and Southern Punjab provinces (including Taunsa Barrage, the focus of this chapter) planned in the colonial period. The symbolism of barrage opening ceremonies, like their colonial predecessors, fantasized about transforming Sind's 'deserts into gardens', ending 'backwardness' through irrigation (Haines, 2011, pp. 187–188). Replete with nationalist rhetoric, officials portrayed irrigation and flood management infrastructure as harbingers of progress and civilization, in some respects outdoing the British in their modernist will to master the natural environment and wide-eyed orientation towards growth.

Pakistan's technocratic, almost fetishistic predisposition towards scientific technology parallels neighbouring India's fixation with dams as 'modern temples', to cite Nehru's famous metaphor. However, the vigour with which Pakistan's governing elites have pursued the 'hydraulic mission' (Molle et al., 2009) has, according to various estimates, made it the most irrigated large country on earth. Over 80% of its agricultural area is cultivated by canal water and, to a lesser extent, tube wells (tanks and other harvesting devices are prevalent in Baluchistan province). Other countries in South Asia still have less than half their agricultural area covered by irrigation. Moreover, whilst some of these other countries, including post-colonial India, have made strides towards redistribution of rural landholdings, ownership patterns in Pakistan have remained highly skewed, as is reflected in the continued political significance of landed power elites referred to somewhat misleadingly as 'feudal' or in Urdu, *zamindars, jagirdars, waderas* and various other clan and kinship markers associated with rural influence depending on locality (e.g. the *Chaudhrys* of Punjab's Gujrat district, *Sardars* in rural Baluchistan, *Khans* in Khyber Pakhtunkhwa's Swat). The development of electoral democracy has facilitated deepening integration of political and economic power so that the heirs of local power brokers established in the colonial period are arguably even more influential in contemporary rural settings. As members of the national and provincial assemblies, they exert considerable influence over local irrigation bureaucracies as well as other state institutions within local domains, not least the police, who are notorious for acting in their interests (Human Rights Watch, 2016).

In recent decades, an 'iron triangle' linking Pakistan's 'hydrocracy' (hydraulic bureaucracy) to politicians and private companies responsible for the construction and maintenance of public works has evolved. Indeed, the increasingly important role of development banks in financing water infrastructure in the global south means the geometry of this nexus of interests might in fact be classified as a 'rectangle' (Molle *et al.*, 2009). This is not to mention the role of other parties predisposed towards supporting megaprojects – consultant engineering firms, the military, geopolitical allies and regional 'friends' such as China and Gulf monarchies with an interest in developing Pakistan's commercial agriculture as a solution to their own water woes. The vast scale of this complex of interests accounts for why water managers have 'kept a sharp eye on the benefits [managers] could extract from the Indus without regard for the hazards so integral to living in river basins' (Mustafa and Wrathall, 2011, pp. 127–128). Over-spending and irregularities in dam and barrage construction and remodelling are endemic, with allegations of corruption involving state agencies and private companies regularly surfacing. Megaprojects are aggressively pursued irrespective of their impact on local ecosystems. Affected communities regularly report loss of livelihoods and engineering failures that exacerbate flooding and lead to other environmental problems (Ahmad, 2015b).

This dysfunctional political economy of irrigation is viewed by Pakistan's peripheral regional populations through the prism of ethnicity rather than class alone. Punjab being the political–economic core of Pakistan's central state, its middle and upper social classes are perceived as alien rulers buttressed by the treachery of local landed and political interests within the regions themselves. As such, class contradictions are articulated as local, ethno-national resistance to centralized water governance. This tension between the centre and provinces is exemplified by the impasse centred on the famous (or rather infamous) Kalabagh dam, the building of which has been blocked since the 1980s by objections from other provinces due to its likely negative environmental impact for all but Punjab (Ahmad, 2015b).

In Southern Punjab, where the breach studied in this chapter took place, a renewed centrist, technocratic push to boost commercial agriculture and power generation since the 1990s has infused the politics of infrastructure-building with fresh intra-provincial tensions. Political mobilization around ecological issues has been led by a vocal cohort of activists and scholars concerned by the social and environmental impacts of large-scale irrigation on Saraiki-speaking populations who reside in affected districts such as Muzaffargarh and Dera Ghazi Khan. The Chashma Canal Right Bank Project (CRBIP) (the final phase of which was financed by the Asian Development Bank and German government in 2004); the remodelling of Taunsa Barrage in 2005–2007 (financed by the World Bank at a cost of more than 11 billion Pakistani Rupees, or approximately £96 million) and more recently, ambitious plans to build a number of coal, atomic and thermal power plants linked with Chinese finance as part of the Pak-China Economic Corridor (CPEC) (Ahmad, 2016b) have each been mired in controversy, financial irregularities and fierce contestation by communities adversely affected.

It is against the backdrop of political mobilizations against these megaprojects that protests against the government's handling of the 2010 floods must be interpreted. Dubbed by officials, commentators, journalists, policy makers and indeed many scholars as the most devastating 'natural' disaster in Pakistan's history, the floods were characterized internationally as a meteorological event and have been linked with climate change (Gray, 2010). A historical sketch of interventions in the riverbed suggests that far from being a random aberration or consequential upon unprecedented rainfall, the floods constitute the latest episode in the above-discussed history of 'normal' capitalist resource development of perennial irrigation along the Indus. As Mustafa and Wrathall have pointed out, the 2010 floods were 'a bi-product of national decisions about water development and irrigation, integrally linked to the social geography of the basin'. As such, they cannot be understood 'in isolation from the routine river management of the basin' (2011, p. 127).

Methodology and Data

In disciplinary terms, the methods used in this chapter include oral history based on several dozen testimonies of affected persons and activists, analysed textually through the lens of sociological theories, as well as ethnographic observation. Historical archival data and secondary material such as press reports have provided important background. However, the bulk of my primary data are based on a large archive of video and audio recordings in Muzaffargarh and Dera Ghazi Khan in the weeks and months after the 2010 floods, captured with the dual purpose of research and making a documentary (Ahmad, 2015a).[1] In the diagram below (Fig. 14.1), which shows areas affected by the 2010 floods, the principle field is the dark zone bisected by the Indus and Chenab rivers adjacent to Taunsa Barrage. The area has been profoundly shaped by the building and expansion of hydrological infrastructure around the Indus in the post-colonial period, and as such provides an excellent case study for the role of post-colonial water

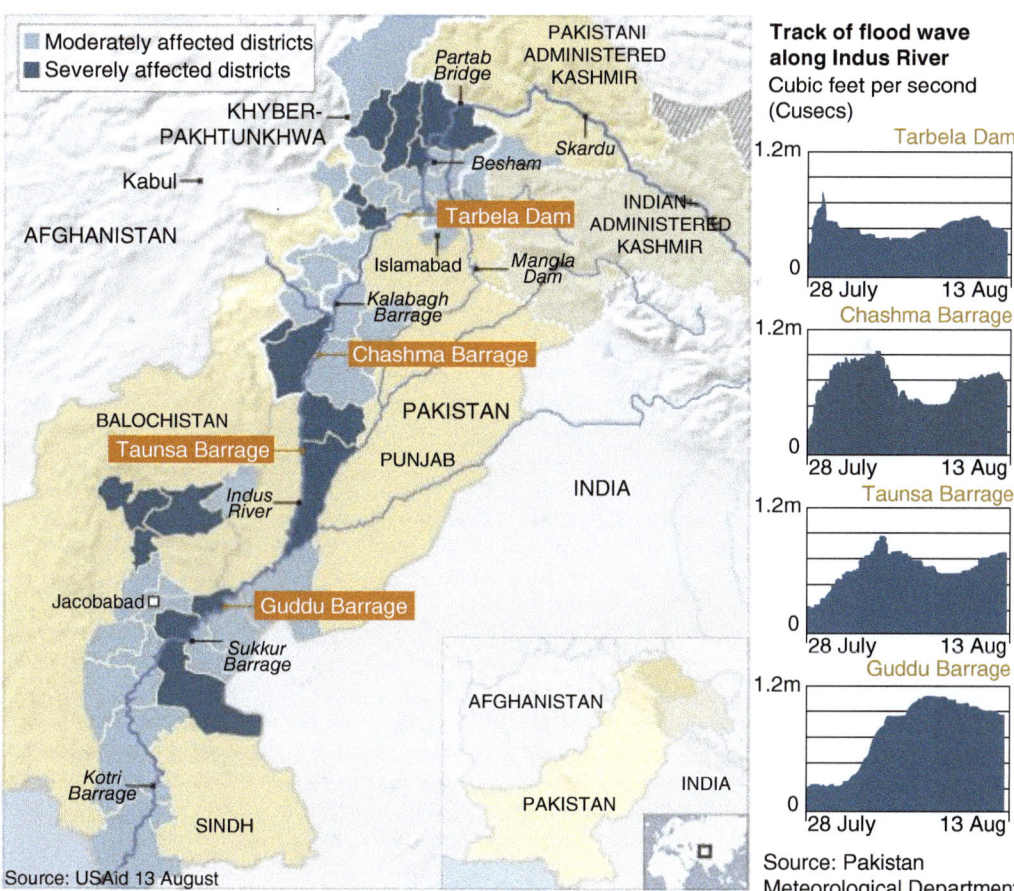

Fig. 14.1. Areas affected by the 2010 floods. (From: www.bbc.com/news/world-south-asia-10986220.)

governance as a factor within agricultural modernization.

Unlike Britain's original canal colonies to the north, Southern Punjab's transition from pastoralism to commercial agriculture came largely in the post-colonial period, with canal irrigation driving its increasingly export market-oriented cultivation of wheat, cotton and mango orchards. Large concentrations of sugar and textile mills in Muzaffargarh have driven an increasingly capital-, technology- and water-intensive regime of production; sugar cane cultivation has expanded considerably since the 1980s. Landholdings are highly unequal in comparison with the rest of Punjab; human development indicators and high poverty levels suggest the benefits of growth have been extremely uneven. Studies of the impact of Southern Punjab's most important irrigation schemes since the 1970s reveal impressive increases in land brought under cultivation, but growing concentration in ownership and poor delivery of social services to the landless and smallholding majority (Naqvi, 2013). All this suggests Punjab's southern-most districts are more akin to rural Sindh, Baluchistan and Khyber Pukhtunkhwa than its relatively prosperous central and northerly zones. Like all Pakistan's backward zones, it has been ruled and exploited by distant metropolitan centres with the collaboration of local functionaries; its marginality and sense of grievance are thus rooted in place and ethno-linguistic identity (Ahmad, 2015b).

Muzaffargarh and Dera Ghazi Khan lie in the heartland of a region populated in considerable measure by a peasantry that identifies as being Saraiki – an identity derived from a language distinct from the Punjabi spoken by generations of military officers and Urdu speaking *mohajirs* (migrants) from India around the time of Partition who have settled in their midst. Both districts and other parts of the Saraiki belt are part of what might be described as an internal frontier filled by generations of colonial and post-colonial settler migrants from the north. These more recent waves of arrivals have further marginalized Saraiki 'sons of the soil', whose subordination to rural elites installed by the British has acquired an increasingly visible ethno-linguistic dimension. As a direct response to the earlier discussed series of infrastructural projects in Southern Punjab since the 1990s, lower income, landless and smallholding Saraiki-speaking populations in both districts have begun to mobilize around ethnicity in recent years, advancing demands for access to resources, environmental protection, cultural recognition and greater regional autonomy through largely non-violent protest and identity politics. If in some respects, their resistance bears similarities to the environmentalism of indigenous peoples around the world, it must be borne in mind that their identity has emerged relatively recently, and is not accorded any kind of official recognition by the Pakistani state; the distinctiveness of their language, known as 'Multani' before the 1960s, tends to be disputed by the increasingly urbanized Pakistani political establishment, which regards Saraiki as a dialect of Punjabi (Shackle, 1977).

An impressive manifestation of grassroots political organization in Southern Punjab's environmental politics from below is the *Sath* or 'People's Tribunal', an innovative sort of public gathering in which local speakers enact participatory forms of justice and judgement about events of concern such as the 2010 floods. Deliberations take place in Saraiki, itself a fact of considerable political significance. *Saths* have allowed, in the words of one activist involved in their revival, ordinary villagers to pass judgement on those involved in projects they deem to have been detrimental to their livelihoods in their own language, on their own terms (Farooq, 2013). For sociologists and oral historians, they provide an important means of giving proper weight to the testimonies of local communities and lay experts in studies of environmental change. As data presented below show, despite their village settings, testimonies reveal the imbrication of local events and experiences within global and national political economics. What follows, then, is not so much a local history as a glimpse of the nation-state and globalization from the perspective of a marginalized region and its people.

The ethnography consists primarily of observation of the aftermath of flooding that resulted from the breaching of an embankment at Taunsa Barrage, located 39 km south of Taunsa Sharif on the Indus River. This fateful occurrence led to some of the worst flooding in August 2010, displacing 1.5 million people and reportedly causing the deaths of an unknown numbers of persons in Muzaffargarh district (many unreported deaths are likely to have occurred). Research was conducted with the support of a non-governmental civil society organization, Hirrak Development Centre, funded by Action Aid, which sought to raise awareness and politicize the poor through cultural and political mobilizations, whilst engaging in social work and rehabilitation after the floods. Anthropological notes were made during six separate trips to flood-affected Southern Punjab spread over September 2010 to July 2011, each of which lasted several days. Subsequent follow-ups to Multan and Muzaffargarh City in September 2014, 2015 and 2016 have yielded evidence of the longer-term implications of the floods, allowing the study of their implications to be placed within the longer-term trajectory of economic development and political change (Ahmad, 2014, 2015a, 2015b, 2016a, 2016b).

Flood Causality and the Politics of Blame

In the early hours of 3 August 2010, several days after heavy monsoon rains caused the severest flooding in northwest Pakistan for 80 years, the river Indus burst through an embankment on its left side in Southern Punjab. Upstream of Taunsa Barrage, floodwaters surged eastwards across densely populated areas, engulfing villages but also the city of Kot Adu, where I arrived to conduct research and investigative news reporting some weeks after the floodwaters had receded. Around 400,000 acres (16,187 ha) of Muzaffargarh was inundated; 1.5 million people were displaced – close to half of the district's population. Official estimates claimed 68 people died (Punjab Government, 2010). My own inquiries suggest the number who perished by drowning was in fact likely to be much higher since the bodies of many of those washed away were never found and likely never reported (Ahmad, 2016a).

Within Muzaffargarh, cities and towns were given advanced warnings by officialdom of the flood, and in some cases evacuated despite not being encroached upon by floodwaters. In rural areas, the first to be affected, the opposite was often true: floodwaters arrived without any warning; no provision was made for evacuation and transport. A vast eastward exodus of rural dwellers adjacent to the Indus towards semi-rural towns and the district capital, Muzaffargarh city, underscored the fundamental inequality between town and country. The agricultural sector, which absorbs most of Pakistan's labour, suffered massive damage to crops soon to be harvested. Farmers received some compensation. However, the majority of rural residents affected by the 2010 floods across Pakistan were landless (Budhani and Gazdar, 2011). With little to return to other than situations of extreme exploitation at the hands of landlords, an unknown number of tenants living and working as agricultural labour never returned to their villages of origin following displacement, effectively migrating to cities and semi-rural towns (Ahmad, 2016a). Those who returned traumatized from camps for internally displaced persons to rebuild their homes did so amidst a profound sense of crisis. Dwelling in tents (if they were lucky), they queued for medical provisions, food aid and small entitlements of government compensation in the post-monsoon climate of humidity and mosquitoes.

Within a short space of time after the floods, allegations that 'feudal' families with powerful connections to the incumbent provincial Muslim League government under Nawaz Sharif (PML-N) engineered the breach to protect cash crops began to surface at the margins of mainstream media, mostly within Pakistan (Ahmed, 2010). Despite their plausibility, these excited little international media attention and did not alter the dominant narrative of the floods as a meteorological event. My own enquiries in the weeks and months after the floods confirm that senior figures in the engineering bureaucracy repeatedly dismissed the concerns of local residents in the lead up to the arrival of the flood crest as the ignorant ranting of 'non-technical' people. Insisting on the need to 'save the barrage', they oversaw a controversial breach of the heavily populated eastern banks of the Indus, and at the same time appear to have been responsible for sparing drainage catchments to the west. The decision to protect 'pondage' areas, *whose very purpose is to absorb floodwaters in emergencies*, became the focus of particular controversy, since it was widely believed, and evident for all to see, that they were being used to grow cash crops. These remarkably lush cotton fields, located where the pond areas were supposed to be, were, according to many local affected persons and political representatives, saved at the behest of the PML-N-backed Hinjra and Khosa families.

Arguments about landed power directing floodwaters away from their crops are persuasive. However, criticism of *waderas* (landed elites) needs to be viewed in the context of an obvious blame game in which accusations and perceptions based on local political struggles were shaping accounts of what caused the floods. In the weeks and months after the floods, many of the poor affected or displaced by the floods blamed landed power and the political elite for their plight, and spoke of a conspiracy that involved the enactment of a deliberate plan, complete with bomb explosions detonated to ensure their ruination. A man I interviewed at an aid distribution point set up in the home of another influential and powerful landlord [Mian Sahab], for instance, had no doubt that the Hinjrais were to blame, when I interviewed him:

> The water ruined us, our houses were levelled. Mian Sahab is helping us in the situation. The Hinjrais have not helped us in this situation, nobody has except Mian Sahab.
> *What caused this?*
> The Hinjrais blew up the *Bund* so it broke. Sultan Haider and Afzal did that. They placed the bomb there. They did that on our side and ruined us.
> *How did you know that?*
> We were informed by eye witnesses.
> *What are they [the Hinjrais] going to get from it?*
> They saved their own farms. Mian Sahab is the only one helping us.

The factual accuracy of this lurid but not necessarily outlandish set of claims is hard to verify. Its real importance lies in two noteworthy aspects of the above-cited indictment. The first is a perception of class conspiracy of rich against poor that is clearly based on a relationship which predates and outlasts the moment of disaster. It suggests the floods – an ecological crisis – in fact ignited the classical 'agrarian question', triggering class frictions based on prior relations of production and land ownership patterns. Elements of this scenario are powerfully reminiscent of the loud 'boom' black residents of poor neighbourhoods claim to have heard when the Levees famously broke in New Orleans, as captured in Spike Lee's epic 2006 documentary about Hurricane Katrina.[2] Both are instances of communities experiencing traumatic and disorienting environmental events interpreted through the prism of long-standing class oppressions rooted in the unequal distribution of resource access, above all inequalities in land and property ownership.

The other point to note in the above-cited passage is that this man felt it necessary to repeatedly praise his powerful and affluent patron (Mian Sahab) in front of the camera, underlining the limited reliability of the often highly partisan testimony that singularly blames individuals and families within the so-called feudal class. Whatever local accuracies and elements of truth in such accusations, it reflects an all too common popular tendency to blame Pakistan's ills on an ill-defined notion of 'feudalism' personified in morally corrupt landlords (Zaidi, 2015). Whilst I found little reason to doubt the distortive and indeed sordid role of landed elites during the floods, many of the broader issues of irrigation and longer-term development policies cannot be pinned on the corruption and greed of individuals, nor indeed on the class system alone. Just how incomplete such personalized finger-pointing is as an account of the causes of the disaster can be gleaned from the glaringly obvious fact that 25,000 acres (10,117 ha) of ponding would not necessarily have prevented the destruction, which inundated some 400,000 acres (161,874 ha) of the district as a whole.

A more convincing, ecologically holistic analysis has been hinted at by local educators, activists and scholars, whose somewhat more structurally oriented and less locally partisan writings and statements about irrigation projects before the 2010 floods have appeared in a number of public platforms. Pointing to the tendency of man-made water management systems to fail, for instance, Mushtaq Gaadi (2010), wrote soon after the floods that the underlying cause of flooding is the rising riverbed caused by upstream sediment deposition at Taunsa Barrage. This much is hinted at in those sections of technical engineering reports comprehensible to lay readers, and clearly evident in the decision to embark upon a massive rehabilitation project funded by the World Bank in 2005–2008. This remodelling, as Gaadi points out, was supposed to involve raising embankments, but 'criminal negligence in this regard resulted in no such measures being undertaken' (2010).

Such claims were corroborated by opposition politicians, and a number of construction workers I spoke with added that embankments were breached precisely where pitching work to raise its height had been done as part of the remodelling. Equally suggestive was a claim first aired to myself and colleagues on our first visit to flood-stricken areas during early September 2010, when a whistle-blower from the irrigation department approached us through mutual contacts to draw attention to an aspect of the remodelling work he felt had been especially responsible for the extent of the flooding, namely the failure to remove cofferdams – temporary embankments built to allow dewatering (the removal of water) in and around dams and barrages so that building and remodelling can take place. This 'failure', submerged and thus hidden by the river water itself, was a means of cost-cutting, facilitated by the lack of oversight by the irrigation department, engineering companies and others that commissioned and supervised the work.

The actual factors that caused the floods almost certainly included faulty barrage repair work, itself a product of the dysfunctional political economy of construction involving irrigation bureaucrats, local subcontractors, international financial institutions, consultants and engineering companies. The implications of this cleavage between perception and reality, to which we now turn, underline a need to disaggregate causality from consequence; a need, that is to analytically separate the environmental triggers of the disaster from its social and political fallout.

Consequences of the Floods

Neither the role of landed elites nor the dysfunctional construction economy received much attention in any of the flood inquiry reports commissioned by the government (Punjab Government, 2010). The general air of secrecy around hearings upon which they were based, released after extensive delays and barely publicized to minimize press interest, merely added to the sense of conspiracy and cover up surrounding the causes of the 2010 floods. With little or no reference to the context and impact of the long-standing and dangerous circumstances identified by activists before 2010, the Flood Commission's legacy as a body appointed to conduct an exercise in distraction and exoneration has been cemented.

This abdication of responsibility or, in the lexicon of technocratic management speak, leadership, presented an important opportunity for local political workers, civil society activists and intellectuals to articulate criticisms of officialdom and call into question the overall strategy of resource development within the region. Many had already been working in the development sector for non-governmental organizations (NGOs) before the injection of national and international humanitarian aid that followed the 2010 floods strengthened their position. If in many parts of Pakistan including Southern Punjab, cynicism towards NGOs for their apparent opportunism is rife, the work of at least two organizations I encountered appeared to be awakening and politicizing the poor through authentic cultural and political mobilizations for environmental justice. Hirrak's critiques of mega-development, 'feudal' (landed) elites and poor governance were made in a distinctly Saraiki idiom, giving momentum to calls for a Saraiki province that echo the other ethno-linguistic nationalisms, past and present, that have (and will continue) to haunt Pakistan's Punjabi elite and political–military establishment.

With a cohort of some half a dozen activists, journalists and academics as well as several students I attended (and filmed) a *Sath* that took place near Taunsa Barrage some 3 months after the floods during the winter of 2010. Attended by some 1500 men, women and children, the choice of venue was in itself significant; the insistence of local activists on linking their protests to the infrastructure of irrigation reflected a conscious attempt to connect the floods with historical water management. Speakers included charismatic local spokespeople, intellectuals, academics and activists as well as ordinary villagers and concerned individuals with knowledge of the factors that triggered the floods. All bore witness, recounting their experiences and expressing views about what had come to pass, emphasizing time and again the political–economic context of the floods. In addition to the *Saths*, demonstrations, 'Long Marches', *dhurnas* [sit ins] and hunger strikes outside the offices of international financial institutions in Islamabad were also organized; so too protest poetry readings, which underline the role of language, culture, aesthetic and religious responses to disaster.

Saraiki nationalism's rebellious peasant spirit, which drew upon the memory of Maoism, Gandhi's anti-colonial legacy and other radical, Sufi folk traditions of the Punjab, posits an unusually sophisticated critique of capitalist development. This distinguishes it from some of the more prominent anti-systemic movements with rural origins in Pakistan, not least the Taliban, famously engaged in an ongoing insurgency against the Pakistani military since at least 2004. Like other ethno-nationalist movements in Pakistan's history, it presents a cultural, political and ideological challenge to Punjab's domination of the central state, demanding a more equitable redistribution of resources and less centralized approach to development. Its emphasis on cultural activism, non-violent protest and engagement with mainstream political parties is notable. Stopping short of separatism, it embraces many eco-socialist ideas and practices, including the involvement of women (Ahmad, 2017). As such, it can be considered an authentic 'environmentalism of the poor' – a struggle for justice among those who depend on the local environment for their livelihoods, as opposed to bourgeois environmentalism, widespread in advanced industrial economies, which is concerned with 'greening' the planet in rather more abstract ways. As is often the case with many environmental movements of the poor, place, identity and culture are of central importance to their struggle (Anguelovski and Alier, 2014).

Indeed, Saraiki nationalism's avowal of traditional aesthetics and eco-moorings point to an organic embeddedness in the physical environment reminiscent of indigenous movements around the world. Imaginative strategies adopted by Saraiki activists in the months and weeks after the floods include aesthetic and cultural practices and performances that underscore the role of vernacular language. Resistance poetry readings were interspersed with political diatribes against the government, and had a distinctly political function. As with East Pakistan's Bengali language movement of the 1960s, demands for cultural recognition were tightly wedded to calls for socio-economic change. The organization of people's tribunals, meanwhile, served to forge a sense of legitimacy surrounding long-term grievances of the poor, some of which were only indirectly related to the devastation triggered by the floods. Two of the activists I spoke with said *Saths* were inspired partly by South African land struggles, news of which reached local civil society through internationally mobile discourses, practices and languages of resistance (Farooq, 2013); most were bolstered by international donor funding. International socialist reference points and their rights-based language suggest the protest movement analysed here, like many movements classifiable under 'environmentalism of the poor', is in fact 'glocal' rather than narrowly local (Anguelovski and Alier, 2014). Its direct confrontation of the World Bank for wastefully and unaccountably financing costly and damaging barrage remodelling projects indicates a multi-scalar framework of vision that encompasses both local and national actors on the one hand (police, landed elites, irrigation officials), and on the other, powerful institutions of governance that operate at the international level.

Despite gathering steam for a brief period after the floods, criticism of the governing authorities and accompanying demands for a Saraiki province did not translate into electoral gains for Saraiki political parties in the 2013 election. The cumulative political impact of denunciations of politicians, the administration and large landowners with ties to the state (itself dominated by Punjabi speakers) was to undermine political parties of both Left and Right, empowering a new generation of populist political workers and politicians who broke the traditional two-party stronghold. Jamshed Dasti, unusual for his working class background and vocal criticism of the landlord class, gained much prominence as a Robin Hood figure, his political capital rising steeply in the aftermath of the floods thanks to frequent public appearances where his fiery rhetoric and foul language against the 'feudals' won much support.

With the defeat of the left wing People's Party during the 2013 elections, an uncompromisingly growth-led, technocratic approach to development and infrastructure-building has gathered further momentum. Incumbent Prime Minister Nawaz Sharif's pro-business 'Noon' wing of the Muslim League, now at the helm of national as well as Punjab's provisional government, has outlined an aggressive growth strategy emulating India, Turkey and China, tying itself tightly to the latter's fortunes. China is financing infrastructural mega-development in Pakistan as part of its purportedly 'game-changing' Pak-China economic corridor (CPEC). The building of roads, coal and atomic power plants has already begun in earnest, defining the parameters of Pakistan's official energy needs without reference to the 2015 Paris Agreement.

Since 2015–2016, both the districts studied in this chapter have been subjected to intensive energy regimes. Plans to build controversial water, atomic and coal power plants suggest the current regime of internationally financed dam, canal and barrage building, which thrives off outsourcing construction to local companies will if anything become more entrenched. Infrastructure development now absorbs an ever greater proportion of jobless locals, and has become an even more important aspect of Southern Punjab's economy. Conflict between the state and local populations due to be displaced by a raft of new megaprojects is ongoing, as many new projects are being fiercely resisted. Fear of pollution and environmental degradation is a major concern in Kot Adu in particular (see Ahmad, 2016b, in which I report on the People's Tribunal pictured in Fig. 14.2). Protests led by Saraiki environmental activists have thus far warded off many of the most audacious of these projects and associated plans to acquire land for their realization. However, the mysterious and chilling murder of Hirrak's director Zaffar Lund in the summer of 2016 (Ahmad, 2017) has dealt the Saraiki struggle for environmental justice a crippling and intimidating blow. It came soon after his coordination of a major protest against the current wave of proposals to extend energy infrastructure in and around Kot Adu tehsil (subdistrict), which threatens to displace populations affected and resettled by the 2010 floods.

Conclusion

This chapter has sought to analyse and foreground environmental dimensions of the 'agrarian question' in contemporary Pakistan, where irrigation has been fundamental to the modernization of rural subsistence economies since the

Fig. 14.2. People's tribunal in Mauza Spira, 2016: rural dwellers meet to plan opposition to the building of coal plants near their habitats. (Photo Credit: Ali Nobil Ahmad.)

colonial period. It has done so by focusing on the way in which the politics of irrigation – its dysfunctional political economy, damage to rural livelihoods and ecosystems – exacerbates class contradictions and shapes patterns of conflict at moments of crisis and by exploring the causes of, and organized local responses to the 2010 flood disaster in Southern Punjab and its aftermath within two districts.

At a glance, the case confirms what we might expect: environmental contradictions in capitalism ignite class struggles, with important political consequences. However, we must be careful to avoid environmental determinism here. Any claim that 'nature' is causing or displacing the agrarian question would be erroneous for four reasons. First, rural Punjab has been a hydraulic society, in which nature and society are inextricably linked, since the early decades of the 20th century; any attempt to accord nature causal agency in abstraction from the state and rural society would be misguided. Second, rainfall itself can no longer be viewed as a purely meteorological event in the era of climate change. Third, patterns of inundation and resulting injury were dictated by human factors. Fourth, although the timing of mobilization by the peasantry was dictated by the immediate symptoms of an ostensibly ecological trigger, the articulation of political protest and resistance was by no means 'green' (environmental), in any pure sense. Popular anger was directed at landowning elites and the bureaucracy in the immediate aftermath of the floods; only later were environmental dimensions highlighted by civil society activists who sought to direct and organize chaotic, undirected rebellious energies into longer-term campaigns against the provincial government and state's centralized, growth-led approach to energy policy, water governance and infrastructural development. In other words, old-fashioned class contradictions associated with the classical 'agrarian question' retain much significance.

Furthermore, 'the ecological agrarian question' is not simply a matter of adding environmental concerns to the study of class struggles. The whole question of place, language and cultural identity shapes the way in which red–green politics is articulated. Protest campaigns took a distinctly Saraiki (regional and linguistic) idiom. The ecological agrarian question (AQ7, in Bahn and Zurayk's terminology) thus intersected with the 'national question' in ways that reflect Pakistan's history and political economy. Saraiki protest movements are reminiscent of the language movement that led to east Bengal's secession and the resulting establishment of Bangladeshi independence in 1971. For the Pakistani state, which is facing similar antagonistic responses from local populations in Baluchistan, Sindh, Gilgit and, indeed, Khyber Pukhtunkhwa, Saraiki political activism poses awkward questions. As processes of modernization and climate change combine to reinforce its internally uneven geography of development, the capacity of Pakistan's federal structures of governance to mediate discontent will be severely tested by ethno-nationalist demands for a larger share of the wealth generated by local resources, coupled with resistance to the polluting effects of power generation infrastructure. The nation-state's failure to accommodate diversity, meaningful devolution and fairer resource distribution – acute at moments of environmental disaster – remains its principal contradiction. The floods of 2010 must be viewed within the context of the long and violent history of confrontations between the nationalist (Punjabi) centre and ethno-linguistic peripheries subjected to internal colonization of their land, and extractive big development projects experienced as inimical to local interests. To this extent, Saraiki ethno-nationalism in Southern Punjab is imbued with a locally embedded, eco-socialist environmentalism of the poor. Its non-violent, creative and culturally vibrant aspects, together with its mainstream leanings, make it worthy of consideration for policy makers seeking progressive allies in an era during which Islamists and the military have dominated perceptions and realities of Pakistani politics.

Acknowledgements

Research for this chapter was supported by travel grants from the Leibniz-Zentrum Moderner Orient in Berlin, and discussions among The Politics of Resources Group, a cluster of researchers at the centre working on environmental change in Muslim majority societies across Asia, Africa and the Middle East (for a discussion of its thematic interests see Lange et al., 2016). It was also supported with funds from the Faculty Initiative Fund (2015–2016) of the Lahore University of Management Sciences. Parts of it were presented at conferences at the National University of Singapore in 2014 and at the College of Europe in Bucharest (Ahmad, 2015b). It has benefited greatly from comments and suggestions by the editors of this book and numerous participants at the conference, 'Crisis and Conflict in the Agrarian World: An Evolving Dialectic', held on 1–3 March 2017 at Sciences Po in Paris. Errors of fact or opinion are my own.

Notes

[1] The selection of these districts was guided by my own location in Lahore, the capital of Punjab, where I was living and teaching at a university at the time; Asad Farooq, a colleague with long-standing connections to activists, connected me with local civil society and generously shared his knowledge of the Saraiki Wasaib. Errors of fact and opinion are my own.

[2] Testimonies suggest the black residents believe white neighbourhoods were being saved by the authorities.

References

Ahmad, A.N. (2014) Urban marginality in Pakistan's smaller cities: rethinking disaster through the 2010 floods. Paper presented at 'The Quotidian Anthropocene', Asia Research Institute, National University of Singapore, 15–18 October.

Ahmad, A.N. (2015a) *Waseb* [Nation], documentary film, 22 minutes.

Ahmad, A.N. (2015b) Flood-causality and post-colonial states: the 2010 disaster in Pakistan's Southern Punjab. Paper presented at 'Floods, State, Dams and Dykes in Modern Times: Ecological and Socio-Economic Transformations of the Rural World', New Europe College, Bucharest, Romania.

Ahmad, A.N. (2016a) Urban marginality in Pakistan's smaller cities: rethinking disaster through the 2010 floods. ZMO Working Paper 15. Available at: www.zmo.de/publikationen/WorkingPapers/ahmad_2016.pdf (accessed 30 April 2017).

Ahmad, A.N. (2016b) The vagaries of a new coal regime, published in *The News*, 20 March. Available at: http://tns.thenews.com.pk/vagaries-new-coal-regime/#.V2q1lk0kqfA (accessed 15 March 2018).

Ahmad, A.N. (2017) The mysterious murder of an NGO worker in Pakistan. *The Guardian*, 3 March. Available at: https://www.theguardian.com/world/2017/mar/03/the-mysterious-murder-of-an-ngo-worker-in-pakistan-zafar-lund (accessed 15 March 2018).

Ahmed, I. (2010) Pakistan floodwaters subside as a tide of allegation rises. *Christian Science Monitor*. Available at: www.csmonitor.com/World/Asia-South-Central/2010/0910/Pakistan-floodwaters-subside-as-a-tide-of-allegation-rises (accessed 15 March 2018).

Ali, I. (1987) Malign growth? Agricultural colonization and the roots of backwardness in the Punjab. *Past and Present* 114(1), 110–132.

Anguelovski, I. and Alier, J. (2014) Environmentalism of the poor revisited: territory and place in disconnected glocal struggles. *Ecological Economics* 102, 167–176.

Budhani, A. and Gazdar, H. (2011) Land rights and the Indus flood, 2010–11. Oxfam Research Report, June 2011. Available at: www.researchcollective.org/Documents/Land_Rights_and_The_Indus_Flood_2010_2011.pdf (accessed 30 April 2017).

Byres, T.J. and Mukhia, H. (1985) *Feudalism and Non-European Societies*. Frank Cass, London.

D'Souza, R. (2006) Water in British India. *History Compass* 4(4), 621–628.

Farooq, A. (2013) On People's Law Tribunals (saths) and water struggles in south Punjab. *Naked Punch*. Available at: www.nakedpunch.com/articles/159 (accessed 30 April 2017).

Gaadi, M. (2010) Engineering failures. *The Dawn*, 16 August. Available at: www.dawn.com/news/844334/engineering-failures (accessed 30 April 2017).

Gilmartin, D. (1994) Scientific empire and imperial science: colonialism and irrigation technology in the Indus basin. *The Journal of Asian Studies* 53(4), 1127–1149.

Gough, K., Sharma, H.P. and Sharman, H.P. (1973) *Imperialism and Revolution in South Asia*. Monthly Review Press, New York.

Gray, L. (2010) Pakistan floods: climate change experts say global warming could be the cause. *Telegraph*, 10 August. Available at: www.telegraph.co.uk/news/worldnews/asia/pakistan/7937269/Pakistan-floods-Climate-change-experts-say-global-warming-could-be-the-cause.html (accessed 31 May 2017).

Haines, D. (2011) Concrete 'progress': irrigation, development and modernity in mid-twentieth century Sind. *Modern Asian Studies* 45(1), 179–200.

Human Rights Watch (2016) This crooked system: police abuse and reform in Pakistan. Available at: https://www.hrw.org/report/2016/09/25/crooked-system/police-abuse-and-reform-pakistan (accessed 31 May 2017).

Lange, K., Ahmad, A.N., Dağyeli, J., Evren, E., Schukalla, P., Schumacher, J. and Serels, S. (2016) (Re)valuing natural resources in the Middle East, Africa and Asia. *ZMO Programmatic Texts* 11. Available at: http://d-nb.info/1103598287/34 (accessed 30 April 2017).

Molle, F., Mollinga, P.P. and Wester, P. (2009) Hydraulic bureaucracies and the hydraulic mission: flows of water, flows of power. *Water Alternatives* 2(3), 328–349.

Mustafa, D. and Wrathall, D. (2011) Indus basin floods of 2010: souring of a Faustian bargain? *Water Alternatives* 4(1), 72–85.

Naqvi, S.A. (2013) *Indus Waters and Social Change*. Oxford University Press, Karachi, India.

Punjab Government (2010) Report of the judicial flood inquiry tribunal. Available at: https://www.punjab.gov.pk/system/files/Ch7.pdf (accessed 31 May 2017).

Shackle, C. (1977) Siraiki: a language movement in Pakistan. *Modern Asian Studies* 11(3), 379–403.

Zaidi, A. (2015) *Issues in Pakistan's Political Economy*, 3rd edn. Oxford University Press, Karachi, India.

15 India: Rural Roots of Naxalite–Maoist Insurgency

Archana Prasad*
Centre for Informal Sector and Labour Studies, Jawaharlal Nehru University, New Delhi, India

Agrarian Capitalism and Conflicts

This chapter explores the relationship between the classical and contemporary agrarian questions and the rise of Naxalite–Maoist insurgency in rural India. It situates the origins and the character of such insurgency within the debates on the nature of agrarian capitalism in India and the evolving resistance to it. In this sense this chapter will be located in the contemporary history of agrarian transformations in constitutionally designated 'tribal areas' of central India, which are popularly known as the Red Corridor or the areas of operation of the Maoists. In this chapter, the term 'Maoists' is used for political activists who are associated with or identify themselves with the Communist Party of India (Maoist). It is not used for local Adivasi people, who may participate in struggles but do not form the cadre of the Communist Party of India (Maoist). The term 'Maoism' is used for the ideology propagated by the Communist Party of India (Maoist) and not necessarily akin to forms of Maoism elsewhere in the world.

The contemporary history of 'tribal areas' shows four stages of agrarian transformations that have brought about adverse integration of Adivasis (an indigenous term for tribal people) into the capitalist commodity and labour markets.

The first stage is that of the displacement of Adivasis from fertile landholdings into the forested areas and their fringes. This period between the late 18th and late 19th centuries laid the basis for the proletarianization of Adivasis and their virtual dependence on agricultural labour and forests for their survival. The nationalization of both minerals and forest lands by the colonial government played an important part in reducing the Adivasis from farmers to wage labourers, even though they may have held small landholdings. Important historical examples show us that the protection of traditional rights on land was intimately linked with the process of dispossession and the settlement of Adivasis into forest lands. For example, the Indian Constitution formulated the Schedule V and Schedule VI in the Adivasi-dominated areas primarily to protect the land rights of Adivasi/tribal people. But most of these tribal-dominated regions were in low-productive and forested regions. Hence, while these schedules protected Adivasi lands, they also ensured that the lands of the working classes outside the scheduled areas were open for purchase at ridiculously low prices as evident in the expansion of arable land immediately after independence (Prasad, 2003;2011).The second phase can be traced from the period of the late19th to the mid-20th centuries, when the accent was

---

* Email: archie.prasad11@gmail.com

largely on setting up plantations and on expanding the base of mining and other industries into forested areas. The direct contradiction between nature and capital created a series of conflicts and displaced the livelihoods of many Adivasis and petty producers like small-scale local miners and crafts-based industries. This once again showed that the crisis in nature and displacement of livelihoods had taken a new pattern. Most of the Adivasis were semi-proletariat (i.e. though they owned land a large part of their survival still depended on labour) and could not meet their needs only through agriculture. Hence the interface between forest-based and casual wage labour and agriculture became important to understand the changes within Adivasi societies. This semi-proletarianization was also structured by the fact that laws had been passed to ensure the protection of land rights within 'tribal areas'. But since a large portion of these lands were within the forests, very minimal land rights and titles were recorded (Prasad, 2004). The third phase of capitalism was the post-Green Revolution period of the mid-late 20th century where the rise of new social classes and the patterns of uneven development led to an aggressive articulation of Adivasi identities. But the more interesting development of this period related to the class-based mobilization of Adivasi peasants known as the 'Naxalbari' or the first phase of the Naxal insurgency. This insurgency confined itself to west Bengal in the struggle for redistribution of lands. The fourth phase of agrarian capitalism in Adivasi areas started with neoliberal policies in the late 20th century that facilitated the corporate harnessing of nature in the Adivasi areas. The failure of Adivasi politics to counter this rapidly growing penury and the absence of the state both contributed to the growth of Maoist insurgency (Prasad, 2010a).

In this context this chapter studies the roots of the Maoist insurgency and compares the different types of Naxal–Maoist actions in the 1970s and the contemporary period. It argues that the character of the insurgency and the failure of the contemporary 'Maoists' to expand their operations and bring about social transformation is largely linked to their misreading of capitalism in general and the agrarian question in particular. This particular failure results from an ideological underpinning which states that Indian agrarian society is a feudal, rather than a capitalist one. Such a perspective therefore contends that the militaristic overthrow of the state would help to establish socialism by bypassing the capitalist stage in transition. Conforming to this understanding the Maoist strategy is linked to its inability to understand the development of capitalist relations in agriculture. As the classic mode of production debate showed, the development of capitalism within rural societies created its own forms of non-economic coercive methods of value extraction (Patnaik, 1990). Since then the nature of the contemporary agrarian question has undergone a considerable change (Moyo et al., 2015), but the Maoist movement has not taken cognizance of the transforming relationship between labour and capital in changing agrarian landscapes. In the light of this analysis this chapter shows that the Naxal movement (1969–1971), which led to the formation of the Communist Party of India (Marxist–Leninist), was acutely aware of the classical agrarian question, even though its interpretation of the question is debatable. However, its current avatar in form of the Maoist movement has misinterpreted and indeed ignored both the classical and contemporary agrarian questions, because they are too pre-occupied with maintaining their guerrilla might to 'overthrow the state' (Prasad, 2010a).

In the light of this analysis, the first part of the chapter describes the growth of the Naxal movement in the 1970s. The second part of the chapter describes the agrarian transformations in Adivasi societies and the changing nature of the conflict in order to foreground contemporary Maoism. In this context the third part of the chapter looks at the character of contemporary Maoism. Finally, the last part documents the impact of 'Maoism' on rural society in Maoist-dominated districts.

The Rural Roots and Origins of the Naxalite Movement

The origins of the Naxalite movement can be traced to the problem of social transformation and class-based mobilization of the Adivasi peasants in west Bengal. Two strands of the communist movement interpreted the nature of the agrarian structure differently and therefore developed contrasting strategies for transformation. Before getting into the details of the Naxal movement it must be noted that such a political mobilization was largely a result of the revolts that occurred in the post-World War II period and were carried on well into the 1950s and 1960s. The most famous of these revolts were the Telengana, Tebhaga and the Warli struggle.[1] However, one of the oldest communist-led tribal organizations was the GanamuktiParishad in the northeastern state of Tripura. Though the struggles were largely a result of the organization of landless agricultural workers, many of them led to the specific organization of 'tribal people' and also created alternative, open-ended identities (Prasad, 2014). This was especially the case with Godavari Parulekar's Warli struggle and the continuous organization of tribal people but the lesser known movements of Tripura's GanamuktiParishad and Kerala's Panna Valayar revolt. Of these the development of the Warli and the Tripura organizations in the post-independence period are especially interesting as they help to analyse the relationship between class and community structures. This relationship was mediated by the understanding that the organization of tribal people would help in building a common understanding and strengthening the alliance between peasants and workers (Prasad, 2014).

The communist work in the tribal regions has already led to the formation of multiple political identities that have been shaped by their own local context. However, there were some similarities in the ways and in the methods of organization and mobilization that led to the articulation of an 'Adivasi' identity that was both open-ended in character and also embedded in class relations and consciousness. This overlap between community and class consciousness was an organized attempt and not the natural outcome of the process of differentiation within designated tribal and Adivasi people. For example, Godavari Parulekar's narrative *Adivasi Revolt: The Story of the Warli Peasants in Struggle* traces the existence of the Adivasi identity to the process of accumulation that was borne out of the penetration of colonial and national capitalism. Hence the whole idea of the 'Adivasi' was the result of a certain dispossession and was thus constituted in opposition to the concentration of capital and accumulation of surplus by the ruling classes represented by large landowners and traders (Parulekar, 1975). At the same time such an identity was also articulated vis-à-vis a peasant landholder with whom the 'Adivasis' had a historical contradiction. The nature of this dispossession was a temporary one as successful land struggles in these regions have not only led to land occupation by tribals but also protected the landholdings of small peasants against the large landlords. This meant that the articulation of the 'Adivasi' identity in Thane and the 'Upjati' identity in Tripura was not inimical to the idea of a larger class unity amongst all dispossessed people. It also aided the process of class formation of 'Adivasis' as both peasants and agricultural workers through a concerted sectional organization. But since the state-led processes of primitive accumulation continued, the conditions for the continuous reproduction of the 'Adivasi' identity also remained. In this sense the Adivasi consciousness articulated within and embedded in the communist movement was democratically constituted and transformative in character. It was also an open-ended identity that was to be organized in a way that it entered into strategic alliances with other class-based organizations to overcome the problem of dispossession. But this consciousness was essentially a non-peasant rural worker consciousness that aspired to combat their historical dispossession. However the problem of explaining the complex relationship between class and the 'Adivasi' identity has remained within communist-led agrarian movements in the Adivasi regions since the post-Nehruvian period (Prasad, 2016a). The agrarian question and resistance in Adivasi India has to be seen in this context.

The Agrarian Question in Adivasi India

Though the first phase of the Naxal movement was based on the question of land and land redistribution, this was specific to the Bengal context. The situation of Adivasis in central and eastern India, where contemporary Maoism is entrenched, has been different since many of the Adivasis have been driven into and on the fringes of forest lands. Hence, for them, the agrarian question is not merely about land redistribution, but also about their claims on nature, which forms the very basis of their existence. In this sense the agrarian question in Adivasi areas goes beyond the question of land rights and has to be also studied in terms of the struggle for legitimate rights in forested areas (Shah, 2013; Prasad, 2016b).

A short contemporary history of agrarian change in these areas can be traced to the post-Nehruvian Green Revolution era, which was marked by the adverse integration of Adivasis into the labour market. The end of the Nehruvian era (1947–1964) has seen a rise of social conflict and protests in the tribal areas, particularly of central and eastern India. These protests were largely a result of the opposition to state capitalism that was characterized by the monopoly control over natural resources in pre-reform decades and was largely symbolized by colonial land and forest laws.[2] While these laws created private property in land, they nationalized forests, minerals and water resources, which were commonly termed as common property resources (Prasad, 2010b). However, the nature of this nationalization was neither socialistic nor welfare oriented in character as it was based on a system of public sector-oriented land acquisition on the one hand and the servicing of industrial fuel needs (through a system of plantation forestry and mineral extraction) on the other. A large part of this acquisition was in non-scheduled areas and was confined to basic raw materials and heavy industries. This process led to enhanced inequity and uneven development that adversely impacted the development of Adivasi vis-à-vis non-Adivasi areas. It also led to the penury and relative marginalization of the tribal people, thus making them the source of cheap labour and raw materials (Prasad, 2010b).

Though state capitalism led to the adverse integration of Adivasis into agrarian capitalism, it also bore some part of the social cost of this integration. However, this situation has changed with the structural changes promoted by the economic reforms. With the opening up of the natural resource sector to corporate penetration, the dispossession and displacement of the tribal people has only increased. The push factors for these changes have largely come from neoliberal state policies that have attempted to ensure the operation of the free market regime in the resource-rich areas. At the same time the welfare functions of the state declined, thus leaving the tribal people bereft of any minimal support structure to cope with the loss. In this situation the question of rights over land, forests and other resources and the opposition to big corporate projects merged at the centre within the debate. This has been accompanied by two associated political factors: the changing nature of the Indian state and the inability of primordial 'Adivasi' politics to deal with the neoliberal challenge. While the state withdrew completely from the tribal areas, it has also become more and more repressive to safeguard corporate interests. At the same time the weakness of both class- and community-based politics has also created the space for the 'Maoist' mobilization in these areas (Prasad, 2010a).

There were at least three main drivers of agrarian change from the early 1990s onwards. First was the penetration of foreign capital and aid into natural resource management programmes and a dismantling of the discourse of centralized control over natural resources (Malhotra, 1997). Second, the critique of state capitalism was an important ideological instrument for International Monetary Fund (IMF)–World Bank-inspired economic reforms by the government since the early 1990s and provided the basis for the market to be established as an alternative to the state as it would provide a level playing field in terms of trade in forest produce and manufacturing of forest-based products. It was believed that the tribal people enjoy a comparative advantage because they are closer to the source of raw material and can offer cheap labour opportunities. Another aspect of the opening up of markets concerned the trade in non-timber forest produce (Marshall et al., 2006). Third, the intent of the policy to open up forests fed into the larger policy perspective that governs the neoliberal natural resource management regime. This understanding is based on the principle that environment and environmental services are to be valued both in terms of their usufruct value and the opportunity costs that such practices incurred. Thus it was believed that such costs needed to be compensated if the environment was either degraded or polluted. This commodification of the environment was evident in the National Environment Policy, 2006, which laid down the framework for the reordering of environmental management. It is interesting to note that the policy blames environmental degradation almost entirely on population growth and institutional failures to settle rights. It also states that it will follow the Public Trust Doctrine where the state is 'not an absolute owner, but merely a trustee of all natural resources, which are by nature meant for public use and enjoyment, subject to reasonable conditions, necessary to protect the interests of a large number of people' (Ministry of Environment and Forests, 2006).

However, within this framework, too, there was some sign of resistance as signified by the enactment of the Scheduled Tribes and Other Traditional Forest Dwellers (Recognition of Forest Rights) Act, 2006, which attempted to correct the 'historical injustices' to which the scheduled tribes had been subjected ever since the colonial times. The Act not only revised the deadline for the settlement of rights from 1980 to December 2005, but also enumerated 13 types of rights over land and produce. It widened the scope of the rights to include people who were displaced and people living in national parks and sanctuaries. But above all, it attempted to democratize the rights settlement process by making the panchayats (village councils) responsible for the initiation of the process (Anon, 2005). It also made stringent conditions for the diversion of forest lands that were settled under the Act. Some of these radical and important provisions were not the initiative of the state, but the result of the recommendations of the joint parliamentary committee. This committee opened up the process of legislation to public debate and negotiated changes by providing a platform to tribal rights organizations, democratic movements and some political formations like the left (Prasad, July–August 2006).But on closer examination it is evident that the neoliberal state has been able to make some fundamental changes in the Act by inserting some limiting provisions in the fine print. It has thus been able to use its state power to protect the interests of the forest bureaucracy, the conventional environmentalists and industry (Prasad, 30 December 2006–12 January 2007).

In this context the agrarian question in Adivasi India has been transformed significantly, from one that looked at the state as the principal enemy, to one that is now directly developing into a confrontation between the Adivasi working classes and corporations who have now been given rights to acquire lands in these resource-rich regions. This has led to the changes in the ways in which Adivasis have been integrated into the labour markets of both their regions and nationally. Such a change is reflected in material basis of the areas which the Maoists are now dominating.

Agrarian changes in states with Maoist insurgency

Four states have been centrally affected by the contemporary insurgency, and these are Chhattisgarh (the Bastar Division having the largest Maoist base), Jharkhand, Odisha and Madhya Pradesh. In this section I discuss some of the important changes within Adivasi societies as a result of broader policy changes.

The first change is the growing landlessness amongst the Adivasis within the past decade, which shows the increase in rural inequality within the region. The decadal changes in the land ownership patterns of these four states in three different periods – between 1999–2000 and 2010–2011 (a decade that is temporally comparable with the census data enumeration in 2001 and 2011)– are displayed in Table 15.1.

Table15.1 shows that the decadal increase in landlessness amongst the scheduled tribes has been the highest in Madhya Pradesh in the period between 2000 and 2010. As shown in the table, the increase in landlessness is lower than the all India average in all states except Jharkhand, the percentage of marginal holdings below 1 ha has registered a significant rise in all the four states in the period between 2004 and 2010 after the formation of the new states of Jharkhand and Chhattisgarh. This clearly indicates that medium size landholdings are getting fragmented but that the loss of land amongst the Adivasis may not be absolute in its character. This means that those with larger landholdings are losing a significant part of their land but not all their land to infrastructural projects and industry. But such landholders cannot be classed as 'landless'. Chhattisgarh is especially significant in this regard since there seems to be an unusual increase in medium Adivasi landholders, a phenomenon that has possibly arisen out of the Chhattisgarh government's contract farming initiative where Adivasi peasants are directly

Table 15.1. Percentage changes in access to total cultivated land by scheduled tribes, 2000–2010. (Calculated from National Sample Survey Organization, 2001, p. 49; 2006, p. 70; 2012, p. 74)

	Change: 2000 to 2010						Change: 2004–2005 to 2010					
	0	0.01–0.40	0.41–1.00	1.01–2.00	2.01–4.00	Above 4.00	0	0.01–0.40	0.41–1.00	1.01–2.00	2.01–4.00	Above 4.00
Chhattisgarh[a]							−0.8	14.7	6.7	−3.3	−15.8	−1.5
Jharkhand[a]							8.2	−4.9	3.2	−7.7	0.9	0.3
Madhya Pradesh	23.1	−3.5	−13.2	−4.6	−0.2	−1.4	17.9	1.1	−10.9	−3.6	−1.8	−2.5
Odisha	4.7	−0.2	−5	−0.3	1	−0.2	1.5	8.8	−7.5	−1.4	−0.6	−0.7
All India	5.1	−2.2	−1.6	4.6	1.1	0	3.6	1.1	−1.7	5.4	−1.2	−0.1

[a] Specific data for the newly formed states of Jharkhand and Chhattisgarh were not collected in 1999–2000. The Chhattisgarh figures have been included within Madhya Pradesh.

linked to corporate houses. This rise in marginal and medium landholdings at the same time indicates a fundamental change within the class structure of the Chhattisgarh Adivasis and can explain the spurt in urban growth rates of Adivasis in the state. The secular rise in marginal landholdings has to be seen as a part of the larger proletarization of the tribal people. It is even more interesting to note that the rate of decline of large and medium landholdings within scheduled tribes is considerably less than that of small and marginal holdings. At an all India level, the picture emerges in a more complex form. The rate of decline of large landholdings is much slower than marginal and submarginal holdings. This indicates that the tribal people with larger landholdings are able to retain their ownership, whereas the marginal farmers were becoming dispossessed, increasing the inequities between the landholders and the landless Adivasi workers (Prasad, 2010b).

The importance of the enactment and implementation of the Forest Rights Act has to be considered in this context and perspective. At the time of its enactment the advocates of tribal rights anticipated that this Act could be an antidote to both displacement and dispossession. But its implementation, when compared with the diversion of forest lands for other projects, serves as a grim reminder of the reality. According to the Comptroller and Auditor General of India (CAG) Report on the Implementation of the Compensatory Afforestation scheme in India, Chhattisgarh, Jharkhand, Madhya Pradesh and Odisha account for about 51% of the diversion of forest lands for corporate projects. If Maharashtra, Andhra Pradesh and Rajasthan are added to this list then these seven states account for about 70% of the land diverted for non-forestry purposes. However, this fact is also accompanied by the lack of recognition of land rights under the Forest Rights Act (Prasad, January 2014). Of the four states with high Adivasi populations, Jharkhand, Chhattisgarh and Madhya Pradesh have a poor record in the settlement of claims under the Forest Rights Act. Chhattisgarh and Madhya Pradesh also have the highest rate of diversion of forest lands for non-forestry purposes. Most of this diversion is for the purposes of private mining projects, which have a big impact in the displacement of tribal livelihoods. This is clearly seen in the decadal changes in land ownership as shown previously. In fact in Madhya Pradesh landlessness has increased by 23.1% in the decade from 2000 to 2011, and in Chhattisgarh by 8.2% between 2005 and 2011. This clearly indicates that the class position of the Adivasi as a rural worker rather than as a peasant has been further reinforced ever since the post-economic reform period. But today, most Adivasis are unable to find gainful employment opportunities in agriculture. Such a conclusion is only reinforced by the Census data of 2011, which shows a steep rise in non-agricultural and marginal agricultural labour amongst the Adivasis and a steep fall in the percentage of farmers and cultivators (Prasad, January 2014). This conclusion is especially significant because of the displacement caused by diversion of lands, which has led to large scale migration and unrest in the countryside. In this context it is pertinent to ask the question of whether the Maoist insurgency has grown as a result of and had an impact on this intensifying rural crisis.

Contemporary Maoist Insurgency and the Rural Crisis

Contemporary history of Maoist operations

The roots of the Maoist insurgency in central India, particularly in Chhattisgarh where they command a 'liberated zone,' can be traced to the early 1980s. Unlike the Naxalbari movement of Bengal, the Communist Part of India (CPI) (Maoist) grew in central India as a result of the outflux of its cadres from the neighbouring state of Andhra Pradesh, where democratic political forces put up a resistance to their violent politics and the police action intensified. Hence the early leadership of the Maoists in the area was not 'local' in character. In Chhattisgarh the expansion of the 'Maoists' was a result of the formation of the People's War Group (PWG) in Andhra Pradesh in the 1980s (Sudhakar, January 2006, p. 3). This group also started to function in the Dandakaranya region in the districts of Bastarthat bordered the Telengana and the area of Andhra where the PWG was strong (Prasad, 2010a).

The 'Maoists' consolidated their struggles through an increase in the frequency of sporadic raids that were aimed at weakening the state system and parliamentary parties of the region from the mid-1980s until 2000. In the mid-1990s they formed two mass organizations: the Dandakaranya:AdivasiKisanMazdoorSangathana (DAKMS) and the KrantikariMahila Adivasi Sangathana (KMAS). Under the aegis of these organizations the Maoists made several demands that would give the Adivasis control over their land, water and forest resources (Prasad, 2010a). For the first 5 years they conducted a long strike during the *tendupatta* (tendu leaf) season and in 1999 managed to get the people Rs115 per 100 bundles for their labour from the contractors. Simultaneously the contractors were also made to pay a development tax that went up to Rs10,000 per year in some cases (Shankar, 2006, p. 38). Similar struggles were carried out for other non-nationalized forest produce like tamarind and bamboos. In the case of nationalized produce the argument of the Maoists was not to remove the middlemen from the system but to force the denationalization of forest produce, so that they could control the operations of the contractors and other private players. Maoists argued that if put under pressure, the traders would give a better market price to the produce collectors. Further, the DAKMS and KAMS also demanded that if the government was unable to pay a proper price to the collectors then it should ensure that the contractors and private traders pick the produce at the proper rate. The selling of produce should be done from the place most convenient to tribal collectors and the full market price should be given to the collectors for their produce (Shankar, 2006, p. 48). This shows that Maoist mass organizations were not against the intrusion of the market that strengthened the exploitative classes within the Adivasi society. Nevertheless, these activities of the organization won them some public support in the absence of the state, which was considered as the primary exploiter of the Adivasi people. Hence the 'Maoists' appeared to get some support in the initial years.

However, the developments after the earlier striking years showed that the partial struggles by the Maoists were used to create the ground work for the guerrilla zone that formed the basic unit of a 'liberated' region. In 'Maoist' jargon, the area under their control has been broadly identified into two zones: the raiding zone (or the Chhappamar zone) and the guerrilla zone. The guerrilla zone is a fortified area that has been termed as a 'liberated area' where the Maoists run a parallel government called Janatasarkars. The raiding zone on the other hand is a peripheral buffer zone from where surpluses are extracted to maintain the guerrilla zone. It is worth noting that the maintenance of Janatasarkars or the guerrilla 'liberated' zones as isolated islands within the larger political landscape requires the expansion of the raiding zone, which is done through several militias and committees. Such support comes through the system of taxation by the Maoists as well as the harnessing of 'voluntary labour'. It is therefore not surprising that the establishment of one guerrilla zone in Dandakaranya had taken around 25 years. The first step in the development of the guerrilla zone is the formation of mass fronts that carry out agitations to gain popular support. These mass organizations act as 'cover organizations' for the growth of 'Maoist' guerrilla activity. They have a strict code of conduct where undercover Maoist leaders do not disclose their real identities. In Dandakaranya, the formation of people's committees had begun and taken place by 1994 and this was followed by the making of 'revolutionary people's committees' in these regions. The objective of such committees was to organize the villagers into collective productive activities that would support the work of the guerrilla army and the militia. By 1997–1998 most of the regions under their influence had such committees (Prasad, 2010a). The formation of these committees was the first step towards setting up the conflict between the state and the Maoists. This strategy was based on a particular understanding of the agrarian question.

Impact of Maoist insurgency on agrarian relations and change

The CPI (Maoist) believed that the overthrow of the semi-feudal Indian state was the only way to bring about radical transformation in agrarian relations, as we have seen earlier. Such an understanding was fundamentally different from the ideological analysis of the earlier Naxal movement, which was based on a concrete analysis of the state and the need to bring about an agrarian transition through a process of social transformation. This has not been the case with the Maoists, whose ultimate objective is not to bring about an effective and egalitarian agrarian transformation, but to bring about the 'ultimate resolution' (i.e. their version of socialist revolution) by bypassing the capitalist transition. This distinguishes Bengal's Naxalbari movement from the current phase of 'Maoist' insurgency and has grave implications for the Adivasi people.

As we have seen in the course of this chapter, the Maoists' strategy does not acknowledge the changing character of the rural context, especially through the penetration of capital and labour markets. The adverse integration of Adivasis into the capitalist labour market has gone virtually unchecked despite the presence of the Maoists. This is reflected in different ways in the life of the Adivasi people. First, the agenda of land reforms was not taken on board by the Maoists and hence they failed to influence the effective and just implementation of the Forest Rights Act. In the latest report the largest rates of rejection for title claims on forest rights is in Chhattisgarh and Jharkhand. At the same time the unabated growth and pressure of mining giants has forced the government to start military operations to protect the leases and build infrastructure for big business. It is therefore not surprising that military pickets have come up all over the roads that provide access to mining-rich areas like Raoghat in Kanker and Bastar. By the same measure all Adivasis who dare to oppose the mines have been termed as 'Maoists'. A case illustrating this is seen from the eastern state of Odisha where the Adivasis and other communities resisted the Vedanta project through a broad-based class alliance of diverse political forces. Though the Indian Supreme Court gave the protestors some relief, the government continued to state the protestors were 'Maoists' as they were opposing the mining corporates. Thus, the branding of all protestors as 'Maoists' has been used by the state to curb and repress resistance (Padel and Das, 2010).

In continuation with the previous observation, the second point to note is that the Adivasis of the region are caught in a violent war between the government and the Maoists. The pressure by the Maoists on the villagers for boycotting of governmental flagship schemes (like the employment guarantee and public distribution scheme) has led to a displacement of livelihoods. At the same time the lack of state commitment to resolve the conflict and recognize the legitimate rights of the Adivasis has led to a greater proletarianization of Adivasi people. Given this reality a recent fact-finding report in the liberated zone showed that the level of poverty of the Adivasis was only increasing. An example of this is seen in the recent visit to some interior villages in the region. During its visit, the study team tried to ascertain whether the villagers were receiving the benefits of the schemes run by the state government for areas affected by leftwing extremism, and the general conditions of survival in remote areas. It ascertained information about the main livelihood strategies, namely, agriculture, collection of *tendupatta*, the public distribution system (for which the state is famous), and the work generated through the Mahatma Gandhi National Rural Employment Guarantee Act (MgNREGA). Almost all the blocks visited revealed one common feature: the villages near the camps of the Indian state security forces or where these camps were located had better facilities than the ones that were in the remote areas. This is largely because the government is using development as an incentive to get villagers to cooperate with security agencies. Further, the result of the conflict was that the villagers were not even being paid for their legitimate work. The lack of work in the MgNREGA is accompanied by machine-driven construction of roads in the entire area. However, the lack of MgNREGA work is not only a function of the state government's inability to provide work and make regular payments. The pressure from the Maoists to stop this form of work also creates fear in undertaking it. In one case from 2016, the Maoists detained villagers for up to 12 days as punishment for cooperating with the district administration, including demanding MgNREGA work (Prasad, 25 June 2016).

Hence, developmental work has suffered because of the ongoing conflict between the Maoists and the state. The villagers are beaten up if they cooperate with state schemes for their own survival, and face the ire of the security forces if they refuse to cooperate. In both cases, the victim is the ordinary Adivasi whose survival needs are compromised in the current situation, and whose lands are under threat of being taken away, especially in mining areas such as Raoghat of Kanker district. Further, though the Chhattisgarh government states that it has addressed 100% of the claims made under the Forest Rights Act coming to a total of 8.5 lakh (850,000 INR) claims, with 3.47 lakh (347,000) claims accepted and 5.07 (507,000) lakh claims rejected, there were many educated youth in the area who had not found any employment and were at home cultivating (Prasad, 25 June 2016).

The third important aspect of the evaluation of the Maoist strategy concerns the methods of repression. The two main methods are forced surrenders and fake encounters. A recent report entitled 'Caught in an Irresponsible War: both the State and Maoists Responsible' shows that a majority of the surrenders in the eastern and central Maoist dominant regions are fake (Prasad, 25 June 2016). Another report by the All India Peoples Front highlights that more than 70% of the surrenders are forced and without any basis (All India Peoples Front, 2016). Several media reports have also shown that the surrendered people lost their livelihoods and government compensation because their 'surrenders' were not considered genuine. Hence in Bastar alone, less than 3% of the surrenders were considered for compensation in 2016 (Ghose, 2017). This effectively means that 97% of the people lost not only their livelihoods, but also the compensation on which they were meant to rebuild lives. But these actions of the state further reinforce Maoist violence, thus driving the Maoists to keep up their confrontation at all cost.

Supporters of the Maoists have tended to compare them to other revolutionary guerrilla movements. They believe that Maoist violence has the popular backing of the people and will therefore be successful in bringing a socially just transformation of Adivasi societies. But this argument is not tenable because it is not based on the concrete analysis of the Maoist practice today. This chapter has shown that the failure of the Maoists to grasp the rapidly changing material reality of Adivasi India has informed their overall flawed strategy.

As the chapter has shown, the conflict between the Maoists and the Indian state reinforce the tactics of each other, rather than tackling the basic questions that have arisen from providing resolution to the problems confronted by the Adivasis. In the Maoists' perception, the capitalist transition can be bypassed through these tactics. But the material realities of Adivasi life show that the Adivasis, who are largely dependent on selling their labour power, are being rapidly integrated into mainstream neoliberal capitalism. The problems arising out of such integration have to be both understood and addressed by the Maoists. In order to do this, the Maoists must take into account the diverse aspects of rural life and their uneven impact on Adivasi consciousness. Hence, despite having rural roots, the contemporary Maoist insurgency is limited by its own perspective and strategy. The lack of a proper interpretation of the contemporary agrarian question leads to the failure of both the theory and practice of Maoist politics in India.

Notes

[1] The Telengana, Tebhaga and Warli struggles were some of the biggest communist-led peasant and Adivasi struggles in the decades of the 1940s and 1950s. They were anti-slavery struggles against the nexus of the landlords and the states and led to occupation of lands by the tiller of the land in many regions. They also laid the foundations for the demands and implementation of land reforms in west Bengal and other regions in subsequent years.

[2] The National Forest Policy 1952 stressed this point when it stated that one of the main objectives of forest conservation in India should be to meet the needs of industry by promotion of plantation forestry. In Madhya Pradesh alone, 600 acres (243 ha) of plantation forest replaced natural forest for this purpose (Government of India, 1952).

References

All India Peoples Front (2016) Bastar: where the constitution stands suspended. All India Peoples Front, Delhi.

Anon. (2005) *Forest and Tribals* (Special Issue). Seminar, August.
Ghose, D. (2017) Chhattisgarh 'Naxal surrenders': screening cleared just 3 per cent in 2016. *Indian Express*, 29 January.
Government of India (1952) *National Forest Policy*. Government of India, Delhi.
Malhotra, K. (1997) Joint management of forest lands in west Bengal: a case of Jamboni district. In: Malhotra, K. (ed.) *Policy to Commissioning of Joint Forest Management*. Inter-India Publications, Delhi.
Marshall, E., Schreckenberg, K. and Newton, A.C. (2006) Commercialization of non-timber forest products: factors influencing success: lessons learned from Mexico and Bolivia and policy implications for decision-makers (No. 23). UNEP/Earthprint.
Ministry of Environment and Forests (2006) *National Environment Policy*. Ministry of Environment and Forests, Government of India, Delhi.
Moyo, S., Jha, P. and Yeros, P. (2015) The agrarian question in the 21st century. *Economic and Political Weekly* 50 (37), 36–41.
National Sample Survey Organisation (2001) Report of employment and unemployment situation by social group, 1999–2000. Report 469. Ministry of Statistics and Programme Implementation, Government of India, Delhi.
National Sample Survey Organisation (2006) Report of employment and unemployment situation by social groups, 2004–05. Report 516. Ministry of Statistics and Programme Implementation, Government of India, Delhi.
National Sample Survey Organisation (2012) Report of employment and unemployment survey by social groups, 2009–10. Report 543. Ministry of Statistics and Programme Implementation, Government of India, Delhi.
Padel, F. and Das, S. (2010) *Out of this Earth: East India Adivasis and the Alumunia Cartel*. Orient Blackswan, Hyderabad, India.
Parulekar, G. (1975) *Adivasis Revolt: The Story of the Warli Peasants in Struggle*. National Book Agency, Calcutta, India.
Patnaik, U. (1990) *Agriculture and Accumulation: 'The Mode of Production' Debate in India*. Oxford University Press, Delhi.
Prasad, A. (2003, 2nd edn 2011) *Against Ecological Romanticism: Verrier Elwin and the Making of an Anti-Modern Tribal Identity*. Three Essays Collective, Delhi.
Prasad, A. (2004) *Environmentalism and the Left: Contemporary Debates and Future Agendas*. Leftword Books, Delhi.
Prasad, A. (July–August 2006) Conservation and tribal development. *Social Scientist* 34(7/8).
Prasad, A. (30 December 2006–12 January 2007) *Survival at Stake*. The Hindu, Delhi.
Prasad, A. (2010a) The political economy of 'Maoist violence' in Chhattisgarh. *Social Scientist* 38(3/4), 3–24.
Prasad, A. (2010b) Neo-liberalism, tribal survival and agrarian distress: the experience of two decades of economic reforms. In: Alternative Survey Group (ed.) *Two Decades of Neo-Liberalism: The Alternative Economic Survey*. Daanish Books, Delhi, pp. 111–138.
Prasad, A. (2014) Class, community and identity: politics of the 'Adivasi' in contemporary India. In: Alternative Economic Survey Group (ed.) *Marxism with and Beyond Marx*. Routledge, Delhi.
Prasad, A. (January 2014) Structural changes in tribal societies: evidence from 'four least developed states'. Planning Commission, Government of India, Delhi.
Prasad, A. (2016a) Adivasis and the trajectories of political mobilisation in contemporary India. In: Radhakrishna, M. (ed.) *First Citizens: Studies on Adivasis, Tribals and Indigenous People in India*. Oxford University Press, Delhi.
Prasad, A. (2016b)Adivasi women, agrarian change and the forms of labour in neo-liberal India. *Agrarian South: Journal of Political Economy* 5(1), 20–49.
Prasad, A. (25 June 2016) Adivasis and the anatomy of a conflict zone. *Economic and Political Weekly* 26, 12–15.
Shah, A. (2013) The agrarian question in Maoist guerrilla zone: land, labour and capital in forests and hills of Jharkhand, India. *Journal of Agrarian Change* 13(3), 424–450.
Shankar, P. (2006) *Yeh Desh Hamara Hai: Dandkaranya Ke Karantikari Andolan Ka Itihas*. New Vista, Delhi.
Sudhakar, A. (January 2006) Twenty five years of glorious struggle: an epic of people's transformation. *Peoples March* 7(1), 3.

16 Agrarian Transition, Adaptation and Contained Conflict in Cambodia and Vietnam since the 1990s

Christophe Gironde*
IHEID, Geneva, Switzerland

Introduction[†]

Cambodia between 1967 and the late 1990s, and Vietnam from 1945 to the mid-1980s, suffered decades of mass destruction, economic stagnation and downturn caused by wars and agrarian collectivism, which culminated with atrocities and dramatic starvation under the Khmer-Rouge regime. Although the two countries were a battlefield of the Cold War, these events undeniably had deep-rooted causes in long-standing agricultural stagnation and peasants' struggle for land.[1]

Since then, both countries have witnessed rapid agrarian change, driven by an average 4% annual growth rate in crop output (World Bank, 2015) and economic diversification. The process has had a positive outcome when measured using economic growth and living standard indicators, which partly explains why rural areas in both countries have not recorded any major troubles or episodes of what historian Milton Osborne called 'the turbulent presence of revolt and rebellion, and on occasion revolution' (2010, pp. 280–281).

Yet, both countries have been subject to rural discontent and protests. However, I argue, these protests were generally sporadic, that is, did not last nor spread territorially, did not fundamentally change the pace and model of development, and did not substantially modify state–business peasant power relations. I argue also that these protests have been marginal insofar as peasants' predominant reaction has been to adapt to rather than oppose the agrarian transition. This chapter aims at analysing why this was so. What types, nature, and magnitude of hardships and crises have Cambodian and Vietnamese peasants experienced over the past 20–30 years? Which conflicts occurred, in contexts that appear similar to previous hardships? If outright conflicts have not occurred, why is this so? What constitutes peasant action or reaction if not revolts, rebellions and revolutions?

Although Cambodia's and Vietnam's agrarian societies share rice farming and the French colonial episode as key axioms to their respective conflicts, several differences provide possible explanations for why and how agricultural crisis did (not) lead to conflict, the forms of conflict experienced and with what results. First, Khmer[2] (Cambodian) villages were historically not a strong social institution (Chandler, 1996; Brickell and Springer, 2016) unlike in northern Vietnam (Papin and Tessier, 2002). Second, uneven levels of educational and human development turned

---

* Email: christophe.gironde@graduateinstitute.ch
[†] I am grateful to Jean-Christophe Diepart, Christian Culas, Suon Seng, Emmanuel Pannier, Frédéric Fortunel and Peter Larsen for insightful talks.

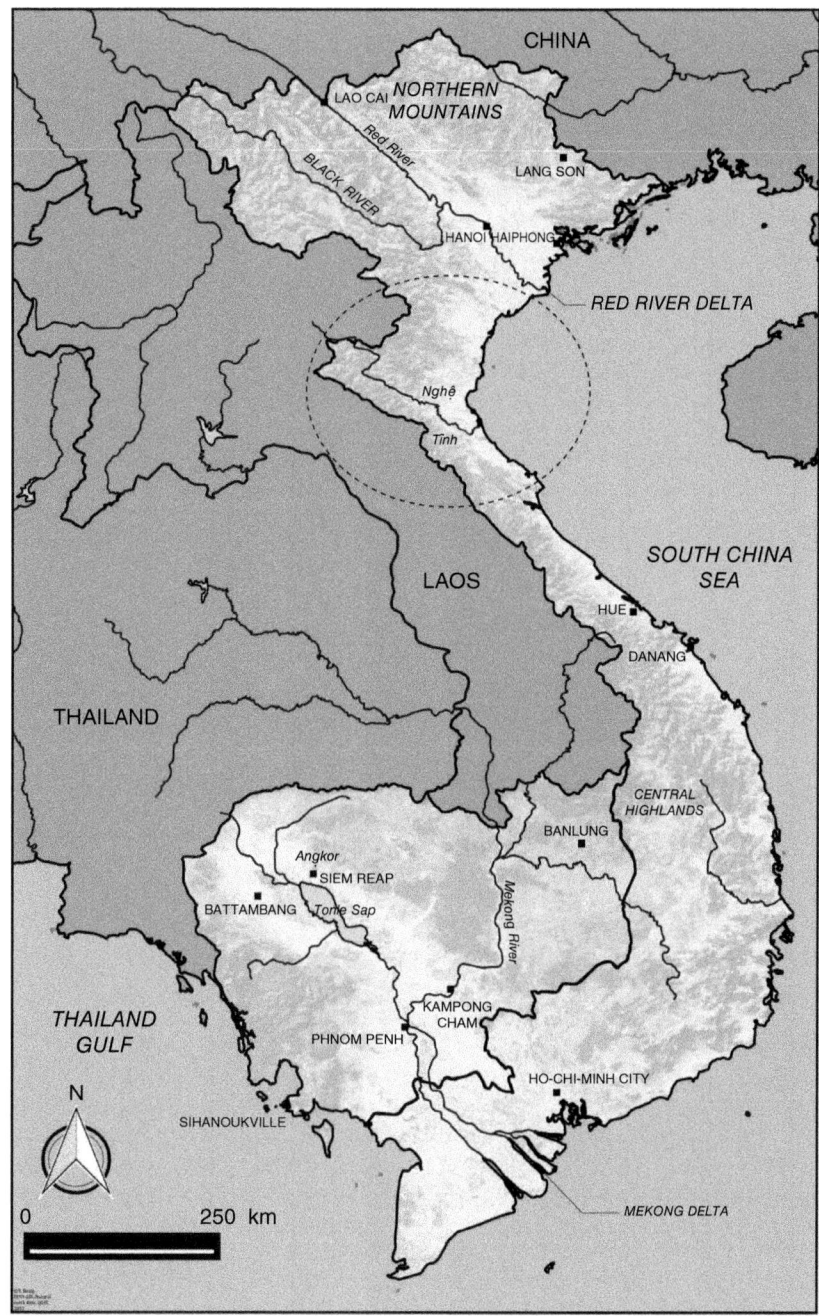

Fig. 16.1. Cambodia and Vietnam: relief and main cities. (Production: Téphanie Sieng.)

into uneven impacts of revolutionary ideas at the time of independence. For instance, one can wonder if Cambodian peasantry had a weaker 'revolutionary potential' (Moore, 1993) than its Vietnamese counterpart and/or lacked the 'middle peasant' that Wolf (1969) deemed crucial.

The chapter is organized as follows. The first section distinguishes the crisis–conflict dialectic under three historical regimes, namely pre-colonial, colonial and communist revolutionary. The second section addresses the concepts and paradigms of resistance and conflict, and presents the analytical framework used to explain their dynamics. The third section examines a series of contemporary crisis–reaction–conflict nexuses, which I deem significant and constitutive of the current era of agrarian transition. The fourth section analyses the reasons for peasants' reactions over the contemporary period, finding them to be adapting rather than opposing in nature; and shows that through intensification in labour and capital and through indebtedness, peasants can be temporarily resilient to the agrarian transition into which they are forced and to which they also aspire. The conclusion reformulates the terms of the dialectic, renamed from crisis–conflict to transition–adaptation.

Revolt, Revolution and Resistance

Stagnation and peasants' agitation in the 18th and 19th centuries

Cambodian agriculture in the 18th and 19th centuries was dominated by rice production, mostly rain-fed but also employing irrigation technology (Nesbitt, 1997). As important as rice farming, people relied on abundant resources from forests and lakes (Chandler, 1996). In cases of insufficient harvests, communities moved to other regions, even crossing country borders (Osborne, 2010). Fights between Thai and Vietnamese forces across Cambodian territory, which adversely impacted agriculture, marked the period. Rural populations mostly reacted through displacement and escape into the forests, also to avoid tax (Chandler, 1996).

Vietnamese agriculture in the same period was comparatively advanced and more intensive, particularly in the north, with two to three yearly crops of irrigated rice and the development of secondary food crops such as maize, beans, sweet potatoes and peas (Nguyên Thanh Nha, 1970). Vietnamese agrarian society was marked by structural inequality between landlords and ordinary peasant families who did not have enough land. This inequality was regulated by the periodic allocation of communal lands to those in need, particularly the landless, widows, orphans and soldiers. Yet, powerful families (notables) and public officers (mandarins) used communal land to privilege their relatives and progressively developed accumulation strategies through land sales (Sansom, 1970). The tension around land largely determined peasants' movements, which, once sporadic, became systematic during the 18th century and particularly after 1732, when the state delegated more autonomy for land management to the communes and their notables; and notables, in turn, pursued an unprecedented increase in land grabs (Woodside, 1971). Ordinary people also suffered an increase in taxes, mandatory chores and military conscription (Nguyên Thanh Nha, 1970). In desperation, many peasants abandoned their villages (Nguyên Xuân Linh, 1981). Bands of wandering peasants and their leaders scoured the countryside and citadels, abolished taxes and distributed land to the rebellious landless. The most famous revolt was that of the Tay Son, which became a dynasty that ruled for 40 years (1771–1802). However, these rebellions did not create inroads for substantial and longer-term changes for the peasantry (Nguyên Thanh Nha, 1970).

Crisis, revolts and (a) revolution under the French (1884–1954)

In Cambodia, the beginning of the French Protectorate was marked by numerous attacks against the French military between 1885 and 1916. These engaged rural populations, but were not strictly peasants' movements in that both peasants' interests and also hierarchical relations between the king, Norodom I, and the French representative were at stake (Devillers et al., 1971). Colonial military operations ruined peasants, and some abandoned their land, hid in the wild or escaped to Siam (modern-day Thailand) (Devillers et al., 1971). Populations from highland forested areas, however, fought against the expansion of colonial land concessions. This occurred in both Cambodia and central Vietnam, where the French established rubber plantations. Peasants' movements intensified as chores and forced recruitment for road infrastructure increased (Forest, 1981). These manifested through brigandage, cattle theft, looting of houses belonging to the Chinese and other vengeful acts followed by escape into the forest. Protests grew, reaching up to 100,000 people across Cambodia in 1916 (Osborne, 1978), including 30,000 who marched to Phnom Penh to bring their claims to King Norodom. Attacks against the French ceased once the French stopped repression: they lacked forces, as Cambodian battalions were sent to France to fight the Germans (Devillers et al., 1971). And yet, rebellions could be of great magnitude, but 'sans lendemain' (Forest, 2008, p. 50). Peasants' movements in Cambodia were hugely influenced by Buddhism (Brocheux and Hémery, 1995), and peasants often reacted by gathering around Buddhist monasteries in search of the social cohesion that local authorities had dissolved.

In Vietnam, agriculture and peasants' socio-economic conditions did not improve during the French colonial period (1884–1945). Except for the export-oriented plantations, mostly rubber and rice in Cochinchina, France paid little attention to agriculture (Fourniau, 1989) until the late 1930s. The colonial administration increased tax rates and strengthened tax enforcement. Many peasants had to borrow from landlords, and were driven into indebtedness and landlessness. Unsurprisingly, most revolts were related to tax issues (Brocheux, 2011). Colonization contributed to disabling traditional redistribution and reciprocity mechanisms within communities; inequality increased (Ngo Vinh Long, 1973), which exacerbated tensions. The French presence contributed also to the emergence of revolutionary ideas: modern critical thought on individualism, particularly amongst youth, questioned intergenerational and gender relations, as well as hierarchy (Brocheux and Hémery, 1995). Vietnamese societal cohesion was thus shaken in its private sphere.

In Vietnam the rebellion–repression dialectic ultimately pushed peasants and workers to revolution shaped by nationalist, anti-imperialist, and communist ideas and organizations (the Viêt Minh), as exemplified by the 1930–1931 factory workers strikes and many peasant uprisings in north-central Nghê Anh and Ha Tinh provinces. These events, called the Nghê Tinh, were against the French, landlords and mandarins (Marr, 1981). With the Nghê Tinh movement, the newly created *Parti Communiste Indochinois* (PCI) organized peasants and provided them weapons; it preluded the 1945 revolution. Land reform became core to the PCI, which gave priority to the peasantry as a strategy to take power (Brocheux, 2011). The years 1944–1945 saw bad harvests, interruption of rice trade by war and 1 million people dead from hunger; the potential and conditions for revolution were met. Once the Japanese withdrew[3] and before the French returned, the vacuum of power gave the PCI and the Viêt Minh underground resistance the '*occasion favorable*' (Brocheux, 2003, pp. 134–142; Hy Van Luong, 2010).

In comparison, nationalism came late to Cambodia, and communist ideas did not develop before 1940. Protests before the 1940s were not of a political nature (Forest, 1993). The absence of revolutionary thought and force comparable to the PCI can be explained by the lower level of human development, with few educated individuals, and a political space largely occupied by Buddhism (Devillers et al., 1971). Later, in the 1970s, the Khmer-Rouge mobilized rural people against an urban class, which had distanced itself from traditional values and behaviours (relations to forest spirits, community moral ethic, etc.). Their rhetoric emphasized the distinction between highland and lowland populations, and manipulated the youth to oppose the elderly and family hierarchical order. In the meantime, the situation had deteriorated by the end of the French Protectorate and further worsened under King Norodom Sihanouk (1953–1970), as illustrated by many prohibitions affecting peasants, including going barefoot, wearing shorts, and driving their pushcarts and cyclos in certain streets of Phnom Penh (Ponchaud, 1998). References to tradition made the Khmer-Rouge revolution quite different from the Vietnamese: while the former drew on the past, with the project to construct large-scale cooperatives and irrigation networks that resonated with ancient Angkor gigantism, the Vietnamese revolution was oriented more towards '*l'homme socialiste nouveau*'[4] (Lê Thanh Khôi, 1978, pp. 99–124).

Hidden resistance and killing fields under communist regimes (1954–1979)

In north Vietnam, the majority of peasants who received land through the 1953–1956 reform were not enthusiastic about the establishment of cooperatives in 1958–1961, as the land was effectively taken back. They rapidly became disillusioned. Resistance to collectivization was not one of the types described by Osborne, as Vietnamese peasants opted for hidden resistance within cooperatives: non-compliance with management instructions, botched work, misappropriation of cooperatives' inputs and tools, harvest thefts, delivery of rice mixed with sand, etc. These were concomitant to collectivization (Nguyên Duc Nhuân, 1992). Crucially, local authorities and cooperatives' management staff used the same strategy, feigning compliance with upper level instructions, but leaving peasants room for manoeuvre, for example, to farm larger areas as their own. Local authorities also allowed arrangements among cooperatives and exchange among communes so that they could achieve production plans (Gironde, 2008). This general laxity was also linked to the American war in Vietnam. The Democratic Republic of Vietnam (DRVN) government needed massive popular engagement in the war against the US, and consequently softened its control on the cooperatives (Beresford, 1988).

After the American war ended in 1975, food availability deteriorated as the government attempted to scale up cooperatives from commune to district unit in north Vietnam (Nguyên Duc Nhuân, 1992) and to establish cooperatives in the south. In the Mekong delta in particular, peasants' resistance was strong and agricultural production was rapidly disrupted, leading to a country-wide food shortage, as rice production from the Mekong delta was crucial to national food security. The situation forced the Communist Party to renounce the socialist model, to endorse the arrangements that had developed locally, and to rehabilitate the household economy and embark on market-oriented reforms (Gironde, 2008).

In contrast to the Vietnamese case, where collectivization remained largely 'on paper' and where the government refused to use force to implement its policies (Fforde, 1989), the Khmer-Rouge collectivization project was put into practice with force and systematic violence. The Khmer-Rouge revolution was actually not peasants' action, rather the Khmer-Rouge apparatus forced peasants to execute its orders. The Khmer-Rouge left no room for resistance, and populations could at best hide their identities or escape abroad and into the wild (Vickery, 1984; Kiernan, 2005).

Analytical Framework and Methodology

Agrarian crises under pre-colonial, colonial and communist revolutionary periods were characterized firstly by the stagnation or decline of food supply. This is no longer the case: the robust agricultural output growth and diversification calls for a reconceptualization of agricultural crisis. For the contemporary period, crisis is conceptualized as a process of disruption, specifically the disruption of rules of access to land and the various exclusions from land encountered by peasants (Hall et al., 2011); the inability of most smallholders to meet their needs through farming alone and the consequent obligation to diversify their activities and places of residence; and the fraying of 'village-community' (Rigg et al., 2012) as smallholders' interests are increasingly different and diverge.

In studying resistance, two main distortions must be overcome: its romanticization, and the view that those who protest are unified in their interests and strategies as they are *peasants*. Romanticization overinterprets any act of resistance as a refusal of the system (Abu-Lughod, 1990). Conversely, a lack of resistance or passive attitude is seen as evidence of collaboration and/or consent (Iglesia, 2010). To what extent were Vietnamese peasants forced or free to join cooperatives in 1958–1961? Did they resist, consent or surrender? In fact, they entered the cooperative 'one-foot-in, one-foot-out' (Vô Nhân Tri, 1990, p. 15), a strategic mix of apparent compliance with resistance and non-compliance whenever possible (Pelzer White, 1986). Reactions can indeed be more complex, ambiguous and seemingly contradictory, sometime 'having nothing to do with compliance or resistance' (MacLean, 2013, p. 212)[5] or being 'dialectical' (Bottomley, 2009, p. 276). The second distortion regards what we call 'peasant movements' and how we interpret them. In the context of agrarian transition characterized by the reduced importance of farming and the growing importance of non-farming activities as well as of activities outside villages and rural areas, movements may not always be of peasant nature, or only resistance with the objective to maintain peasantness. Thus, concepts of peasant and resistance need to be reconceptualized (Moyo and Yeros, 2005).

Recent accounts of agrarian change, particularly land grabbing over the past 10 years, have led to a reconceptualization of resistance that considers a broader range of 'reactions from below' including reactions 'against' land grabs as well as demands for integration into land grab-related businesses (Hall *et al.*, 2015, p. 472). Empirical research shows that people' reactions are actually highly contingent on the context in which land grabs occur and that reactions evolve over time (Baird, 2017; Gironde *et al.*, 2015). This chapter aims at moving beyond the resistance–compliance nexus and a focus on socio-political resistance to identify what Ploeg (2010, p. 16) calls 'resistance of the third kind' pertaining to the reorganization of economic activities.

For the contemporary period, I examine a series of crises and conflicts (or non-conflicts). These cases do not treat countries in their entirety or in all contexts, but are significant examples of the current dialectic between crisis induced by contemporary agrarian change and a range of heterogeneous reactions, contained conflicts and adaptation efforts. The analysis below builds on field research of others; my own field research, conducted since 1996 in Vietnam and since 2010 in Cambodia; and discussions conducted to inform this chapter with experts who have relevant academic and/or practical experience in the two countries.

Building on the historical recall of peasant revolt, revolution and resistance, I use several criteria to assess the varied reactions of the peasantry: what people claim; who and how many engage or participate in reactions, and against whom; if and how community authorities support or oppose resisters; the duration of the reactions; the geographic coverage and spread of reactions; and the scale-up of reactions (Swift, 2015), or the process whereby upper level government addresses reactions and ultimately changes policy, as in the cases of China and Vietnam where peasant resistance was crucial to the decision by central government to put an end to collectivization (Kelliher, 1993; Kerkvliet, 1995).

Agrarian Transition and Contained Conflict

All rural areas and population groups of Cambodia and Vietnam have engaged in the rapid and profound process of agrarian change, albeit unevenly and, for some, adversely. Generally, peasants have found more opportunities to adapt to agrarian change in the lowlands, and conflicts there have been limited and of different nature than in the highlands. The highlands have experienced more acute crisis and conflicts. Yet, agrarian change has been imposed on indigenous populations without major resistance or significant violence, unlike what happened in the Philippines and Indonesia for instance (Singh *et al.*, 2003).

Adaptation and new types of conflict in the lowlands

Cambodian agricultural output has grown regularly over the past 20 years, thus one cannot immediately speak of crisis (World Bank, 2015). However, the overall productivity of agriculture has not improved much due to farmers' inability to pay for improved varieties and inputs and the absence of government assistance to disseminate better technologies and support smallholders. Thus, the conditions for agricultural intensification comparable to that in Vietnam are unmet. To the contrary, market price trends, local monopolies and power relations along value chains have lowered peasants' incomes. Additionally, farmers have faced increasing population density, especially around the Tonle Sap Lake, where many people have settled for security reasons since the time of the civil war.

In the face of these pressures and concerns farmers' main reactions have been to withdraw, partially or totally, from agriculture, as men migrate in search of non-farm jobs in industrial and urban areas (Diepart, 2015). Farmers often have no choice but to borrow increasing amounts of money (Bylander, 2015); the process is encouraged by the lucrative micro-finance industry (Kimty, 2017). Migration and credit have become crucial to sustain livelihoods. However, as rice, cassava and maize prices fell in recent years, many farmers were unable to repay the banks and fell into a vicious cycle of obtaining new loans to repay the ones coming due, borrowing from individual money lenders who are more flexible but costlier than the banks. Agricultural price decreases have actually triggered protests in the lowlands: spontaneous demonstrations, roadblocks and marches to Phnom Penh have become more frequent in recent years.

Vietnamese lowlands have developed more remarkably: paddy yields have doubled since the mid-1980s; the household economy has diversified; rural areas have been incorporated into broader markets and business networks; absolute poverty has declined sharply; and material living conditions have improved (Gironde, 2018). If Vietnam's lowland peasants have been relatively silent as compared to pre-colonial and colonial times, there have been nevertheless cases of open conflict (Fig. 16.2). The progress and enrichment of the well-off is remarkable, and indeed it is enrichment, not poverty, which caused the first spectacular peasants' movements of the Đổi Mới[6] era: In 1997 several thousand people petitioned and demonstrated during several weeks in Thai Binh province in the Red River delta, molesting local officials and even travelling to demonstrate in Hanoi. The main reason for protests was the corruption and illegal enrichment of local cadres (Nguyên Van Suu, 2007). Corruption around land and tax was widespread and publicly known, but the Thai Binh events were unique at that period (Thayer, 2003). One can assume that Thai Binh was simply a case of overabuse, and it is striking that there was no other Thai Binh in the region. In other places where countless cases of families claiming land rights against local authority were recorded (Nguyên Van Suu, 2007), people did not go beyond official land claims procedures, what Kerkvliet (2014) calls 'rightful resistance'.

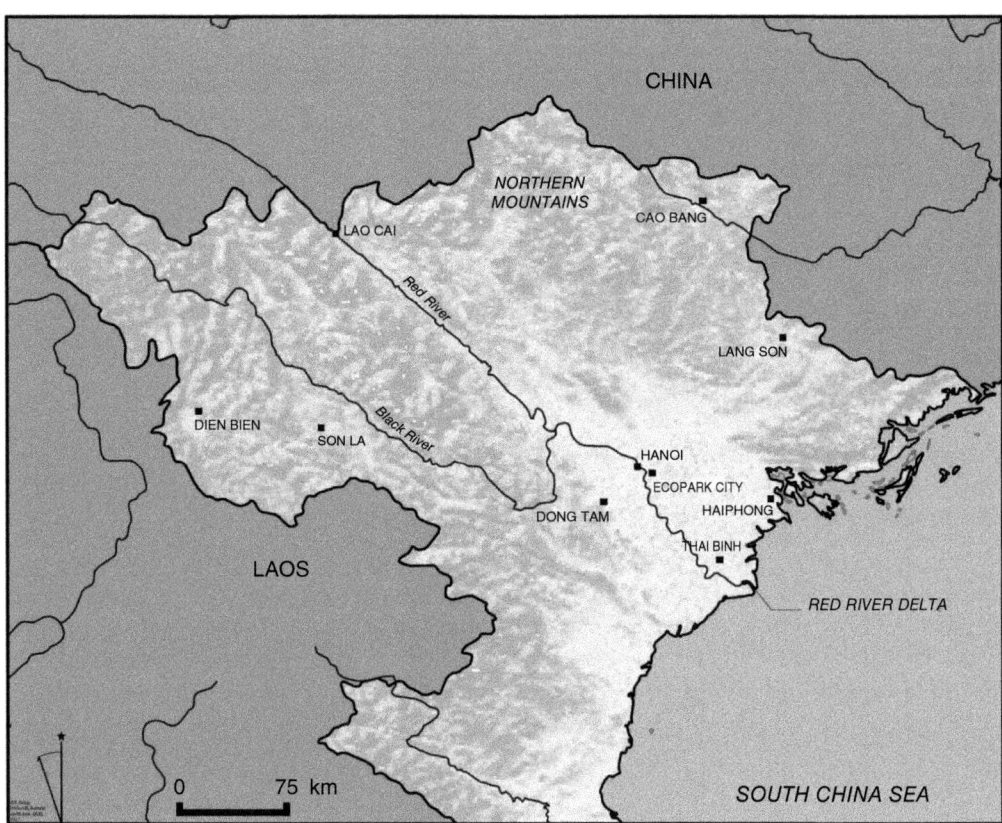

Fig. 16.2. Main places of conflict in northern Vietnam. (Production: Téphanie Sieng.)

Another remarkable conflict occurred a decade later outside Hanoi, where the government attempted to evict peasants from their farmland to build a satellite city called EcoPark.[7] In parallel to filing compensation claims, farmers staged mass protests in 2006 and again between 2009 and 2012. The core of the conflict was the amount of compensation to be paid to dispossessed farmers. Beyond the dispute and its outcome (EcoPark was built), it is noteworthy that the compensation was attractive compared to the income farmers could generate from those lands (Nguyên Van Suu, 2014), and that many used the compensation to develop non-farm activities that provide higher income, including well-off entrepreneurial families aiming at developing artisan–industrial zones (Fanchette, 2015).[8] Those who triggered the EcoPark protests may be peasants, but the protests were not of a peasant nature: most protesters were struggling to pursue economic diversification in an environment that was transforming from agricultural/rural to peri-urban, where agriculture is considered an activity of the past.

Similarly, events that occurred in Đồng Tâm commune 40 km from Hanoi in April 2017 may be considered a post-agrarian conflict that did not last and was limited to the locality where urban/industrial projects were undertaken. Residents protested the revocation of construction land, where they had lived for years and built houses. The residents held hostage 38 civil servants including police officers for several days, blocked road access, and threatened to set fire to public buildings in case security forces would intervene. Protesters physically expressed their grievances, but did not engage in much violence: hostages were reportedly well-treated and released once central government officials were sent from Hanoi to discuss with the protesters.[9]

Land rush, conflicts and resilience on the highlands

Highland areas of Cambodia and Vietnam have experienced more tensions and more profound and sometimes long-lasting conflicts in response to government policies that prompted a land rush for the development of industrial crops and triggered massive migration from the lowlands. The agrarian transition here has undermined indigenous people's livelihoods, as public authorities challenged traditional swidden agriculture as well as free access to natural resources from the wild (forests, waters). In parallel, indigenous people suffered the pressure of newcomers, who were made legitimate to colonize their lands and natural resources (Fortunel and Gironde, 2011). Thus, the magnitude of hardships for native populations differs significantly from the lowlands.

In Cambodia, the development of the concession economy since the 1990s has transformed into extensive and uncontrolled appropriation of land, including land used by small-scale farmers (Diepart and Sem, 2016). The process has intensified since the adoption of the Economic Land Concessions (ELC) policy in 2005.[10] The total area granted to outside individuals and companies – 2.2 million ha in 2013 – represents approximately 20% of national arable land (Peeters, 2015); most of that was in the highlands. In many cases ELCs expanded outside concession areas (Diepart, 2015). In some cases, they led to extreme dispossession whereby populations lost all their land and assets as they were displaced (Gironde et al., 2014). Overall, the concession economy has reduced small-scale farmers' land assets and free access to natural resources including grazing land, forest and water.

There is broad consensus that the concession economy has not benefited indigenous populations much and that ELCs have negatively impacted their income, size of available cultivable land and livestock holdings (Jiao et al., 2015). Others argue that ELCs create few jobs and that working conditions 'today are not fundamentally different from colonial time indentured labour' (Slocomb, 2007, p. 106). Structurally, highlanders have entered a latent crisis, as former food crop-based livelihoods cannot provide for their increasing needs and because they can hardly catch new opportunities. In Ratanakiri, for instance, profitable crops such as rubber remain beyond the reach of the majority of Jarai and Tampuan indigenous groups, who lack financial capital and know-how; they also lack access to regular salaried jobs, which benefit almost exclusively Khmer in-migrants (Gironde and Senties Portilla, 2015). Peasants increasingly rely on borrowing at usurious interest rates (up to 30% over 6 months for US$1000 in Ratanakiri in 2016)[11] and have become chronically vulnerable to over-indebtedness that ultimately results in desperate sales of land.

The most common image of resistance to ELCs is that of face-to-face confrontation between villagers brandishing axes and knives (Bottomley, 2009) and military and police forces 'siding with company owners and provincial and district authorities' (Neef et al., 2013, p. 2).[12] Other open reactions have included angry demonstrations against company staff and confiscation of their vehicles, road blockades, and cutting and uprooting trees (Neef and Touch, 2012; Baird, 2017). Most of those protests remained sporadic and did not last, spread or connect to similar ones in other locations. Communities' confidence to protest depends, inter alia, on the new occupants: interviewees in Ratanakiri stated that they 'can resist' Chinese and Vietnamese companies, but that in the case of Khmer holders, 'you don't play against big men with big guns'.[13] Communal authorities, who had little or no ability to oppose the award of land grants by national and provincial authorities, typically did not support peasants' on-the-ground protests. The fear of reprisals was particularly strong in the case of land grabs by military (Schneider, 2011). A frequent stance of communal authorities has been that claims must be addressed at the national level and cannot be their initiative (Schneider, 2011; Schoenberger, 2017). There are many assertions that communal authority representatives were rewarded for compliance, and although this is rarely documented (see, for instance, Baird, 2017, p. 12), it is highly plausible.[14] However, the role of local authorities has been more complex in many instances, as found in Ratanakiri: communal authorities complied with upper level authority and were rewarded accordingly, but they simultaneously warned their communities about forthcoming territorial expansion of ELCs and encouraged them to occupy lands by planting trees or farming continuously. Elsewhere, the populations occupied lands themselves. This strategy worked well in cases where ELCs did not develop plantations as initially planned, which left villagers some time and space. In other cases, ELCs planted rubber trees rapidly after signature of land leases and villagers could not respond.

If 'overt forms of resistance' are most frequently reported, scholars have also drawn attention to 'covert actions' and the combination of both (Schneider, 2011), for example refusing to sign land deals, farming the lands about to be lost (Schneider, 2011), or resisting relocation by sleeping in the fields (Neef and Touch, 2012). Many of those acts were desperate and unsustainable. Protesters often met police force, including destruction of the houses of those who refused to leave (Touch and Neef, 2015) and dispersal of community patrols protecting forests from ELCs (Nghin and Verkoren, 2015). Community leaders were jailed and police roadblocks prevented community members from attending their meetings (Schneider, 2011).

Overall, many of the communities who undertook independent legal action were rejected by Cambodia's administrative and judicial systems. With the support of non-governmental organization (NGOs), and sometimes of foreign donors and international organizations, as well as media coverage, some procedures were able to follow their course (Global Witness, 2014) and reach upper level authorities for dialogue, negotiation and ultimately compensation (Cismas and Paramita, 2015). NGOs helped communities to organize, for instance for forestry management (Fox et al., 2008). However, in most cases, those procedures have not stopped companies clearing and planting land (Norman, 2015). The Cambodian government has responded to peasants' land claims and denunciations of ELCs through a moratorium on ELCs in 2012; an announcement that ELC licences would be withdrawn in cases where companies did not develop production; and with Order 01, which stipulated that the lands people claimed to have lost because of ELCs should be measured and returned. Order 01 was primarily triggered by electoral imperatives (Biddulph and Williams, 2016) and is highly controversial (Milne, 2013; Beban et al., 2017). It has helped to calm many localities and was partially effective in providing land titles, as observed in Ratanakiri in summer 2012 when it was implemented. Time played against those who claimed the land, as when the state initially recognized peasants' rights in line with Order 01, but then 'erase[d] people's claims' when the ELC 'exerted its influence to undo the survey' (Schoenberger, 2017, p. 874). In cases where land inside the concession was surveyed, it did not systematically translate into the provision of land titles; thus, the State ended being 'unresponsive' to peasants (Schoenberger, 2017, p. 886).

It is impossible to quantify what Cambodian peasants have saved (land) and/or gained (compensation) except on a case-by-case basis; and opinions on the merits of Order 01 differ (Nghin and Verkoren, 2015; Swift, 2015). Ultimately, protests have not led to major change in the implementation of ELCs. They have at best enabled peasants to save some land for some time, avoid further encroachment or obtain some compensation for dispossession. Furthermore, in most cases where peasants managed to win back some land, it is because companies had not fully developed the entire amount of land they had leased because of a lack of capital (machinery) or declining commodity prices and lower profit prospects. In sum, peasants' 'victories' were possible in part due to the limitations or failures of large-scale companies.

In Vietnam, it is in the Central Highlands that tensions were the strongest. One of the largest protests saw several thousand indigenous people gather simultaneously in several urban centres in the Central Highlands in 2001 and again in 2004, calling for the restitution of their ancestral land (Fortunel, 2005). These demonstrations agitated analysts who were eager to find cracks in Đổi Mới,[15] but did not threaten the Vietnamese government and have not led to substantial change in the territorial expansion of commercial agriculture. Ethnic groups that were then the majority have since become a demographic minority, or approximately 15% of the total population of the Central Highlands. They have been incorporated into agro-industry value chains (coffee, pepper, cashew, rubber) via contract farming or salaried labour.

Livelihoods in Vietnam's northern mountains have undergone a similar process of agrarian change and crisis. State policies resulted in greater inequalities with regards to access to land, natural resources and new economic opportunities (Castella and Dang, 2002). The most vulnerable groups were forced to return to ancient but no longer sustainable practices such as slash-and-burn agriculture (Castella and Dang, 2002). The not-so-poor group adopted sedentary agricultural lifestyles. Yet, with few exceptions, there has been no open peasant protest in this process. One noteworthy protest began in Diên Biên in April 2011 and reportedly involved 5000–7000 people (BBC, 2011), mostly from the minority Hmong ethnic group. This protest was first interpreted as Protestant Hmong people demanding more religious freedom. Claims for better land rights and more autonomy were also reported. Another factor must be considered: a Vietnamese government project to develop rubber plantations by leasing land to companies for large-scale production would affect lands currently used by the Hmong, who do not hold land title and are at risk of eviction. Although what occurred was spectacular, as several officials were taken hostage, events such as Diên Biên belong, in my view, to the category of sporadic conflicts that the government has no difficulty to stop rapidly.

Ethnic minorities are well aware of their historical, political weakness vis-à-vis the state (Turner and Michaud, 2009). Their strategy when targeted by government projects in the name of 'development' or 'poverty reduction' has been mostly to refuse to cooperate or to delay, as in the case of Son La hydropower project. The dam under construction since 2005 entails the displacement of 100,000 people, the largest displacement in contemporary Vietnam. In the first place, people did not sign the papers that stipulated their acceptance to move, or signed but did not move (Dao, 2016). A second approach was to plant trees or build houses in order to extract higher compensation for land loss. The ultimate act was last-minute refusal to move, again with the objective to secure higher and immediate compensation. Nevertheless, increasing pressure to move forced some families to dismantle their houses and pack housing materials and their belongings. After displacement, people pursued more open and 'confrontational' resistance (Dao, 2016) to obtain better and timely compensation as well as infrastructure (irrigation schemes, roads, schools) at new locations. In some cases, resettled people returned to their ancient villages to farm; some communities dug trenches to block road access. Resettled people were successful in that they received the promised compensation and part of what they asked additionally (Dao, 2016). In some cases, the engagement of local authorities (village party-secretary and People's Committee headman) who wrote letters and met upper level government and project management representatives, was crucial for the populations to secure compensation. However, in general, the attitude of local authorities has been to accommodate (Thomas, 2003). Actions of this type only served to delay development projects and the unavoidable adjustment to the economic system promoted by the government.

Finally, one must not neglect intimate conflicts that develop within communities and places of salaried work, migration and resettlement. Hall *et al.* (2015) call these 'poor-on-poor conflict'. They occur in Cambodia and Vietnam, between migrant workers and native villagers in places where industrial factories have developed (Fanchette, 2015); between resettled people and inhabitants of the host villages (Dao, 2016); and between indigenous populations and in-migrants on Cambodian pioneer fronts. They occur also amongst indigenous peoples, as in Ratanakiri where better-off households sometimes opportunistically profit from the disorder created by government programmes and land grabs to encroach on the lands of their peers (Gironde *et al.*, 2014). Poor-on-poor conflicts include conflicts found within households. Preliminary results of current research in Ratanakiri indicate that children may contest the authority of their parents and refuse to contribute to the household economy, that is, not farm with them. Youth know their parents will be unable to provide land as inheritance, so may pursue activities such as wage labour and income that may be kept, partly or totally, for themselves rather than contribute to the household's common resources. Conflicts develop also around the use of resources, with divergent priorities and practices between men and women. Finally, tensions between in-laws were also reported. After marriage, husband and wife traditionally live for several years at the wife's parents' house and contribute to in-laws' livelihoods by farming her parents' land. Tensions and negotiations between the parties occur as husbands put conditions, such as being provided a motorcycle, to join their wife's parents' household.

Analysis

Four dialectics of agrarian crisis and conflict (Table 16.1) since the 18th century can be distinguished. Pre-colonial time was marked by agricultural stagnation, recurrent abusive preemption of peasants' resources (land) and output (tax) by local elites, but also compensatory mechanisms (allocation of communal land to the poor). Peasants reacted mainly by escaping and sporadic revolts.

The French colonial system aggravated these dynamics: agricultural output did not significantly improve; colonial practice further unbalanced power relations between the majority rural population and the elite; and earlier compensatory mechanisms were dismantled. The French also contributed to the germination of revolutionary ideas and the organization of protests. Rural social cleavage led to revolutions that combined nationalist and social demands.

Revolutionary experiences failed to increase agricultural production, and indeed fell short. It was disastrous in the case of the Khmer-Rouge, who persecuted and massacred both the urban and the rural population. In Vietnam, collectivization was not fully implemented and remained 'on paper' (Fforde, 1989). In Vietnam, it is not only peasants' various resistances but also the connivance and resistance of local authorities that was crucial. Furthermore, the need to mobilize peasants for the war against the US was crucial to central government behaviour.

Since then, with agrarian transition well underway, rural angst has remained latent except in remarkable cases. These cases are remarkable because so few, and not comparable

Table 16.1. Dialectics of agrarian crisis and conflict.

Period	Agriculture	Main reactions
Pre-colonial	Stagnation	Escapes and sporadic revolts
Colonial	Stagnation	Organized open resistance
Communist revolutionary	Decline (relative in Vietnam, absolute in Cambodia)	Hidden daily resistance and forced compliance
Transition to the market	Growth and diversification, and economic diversification	Economic engagement, sporadic open protests

to those of the pre-colonial and colonial periods. Rural populations' main reaction has consisted of efforts to engage in intensive commercial agriculture and non-farming occupations at the cost of high debt and risk of over-indebtedness.

Several factors may explain why conflicts have been marginal. First, there is general satisfaction, not discontent, with development-related changes in livelihoods and ways of life. Rural populations have not faced output crisis, but rather limitations to the development and diversification of their economic activities. Inequality has increased, but even the worse-off have improved material living conditions. Except in cases of land dispossession, most rural poor nowadays suffer relative rather than absolute poverty. It appears that discontent is not of 'such a deep and desperate nature that men and women have seen no alternative to revolt' (Osborne, 2010, p. 281).

Second, beyond material betterment, values and norms have changed. Change, opportunity and improvement count more than deprivation, inequality or redistribution. In Vietnam, there is little regret or nostalgia for cooperatives and production plans. In Cambodia, the memory of violence under the Khmer-Rouge and even the Sihanouk regime plays a role in popular hesitation over and renouncement of confrontation. People generally believe that there are opportunities and that if one makes an effort, invests or takes a risk, one can be successful. Rising inequality is acknowledged, but is not core to these narratives. The idea that the development and betterment of some may contribute to the stagnation and difficulties of others is not common. Yet, not all is acceptable: Thai Binh protests expressed popular disgust for those who enriched themselves through no talent other than abusing their power. The most common understanding of hardship and crisis is rather termed 'failed development' (Cismas and Paramita, 2015, p. 262), which reflects the acceptance and resignation of people having low self-esteem with regard to their formal educational achievement and experience outside farming and village life. This attitude can be particularly prevalent amongst ethnic minorities, for instance, Jarai indigenous people who compare themselves to Khmer migrants settling in and around their villages.[16]

Peasants' mobilization is also impeded by a real or perceived lack of viability of traditional agriculture: people see that returning to former cropping systems or swidden agriculture is impossible as land is coveted and must be occupied, farmed or planted. Resources previously obtained from the forest have considerably diminished and forest product sales can no longer meet their needs. NGOs that promote alternative livelihoods are often misaligned with the aspirations of the populations they aim to support, as illustrated by the reaction of local populations when they thought our research team was an NGO coming to advise them not to plant rubber but to continue traditional rice-based farming.[17]

The potential for peasants' mobilization has also fallen as households become less homogenous: they have increasingly different resources, interests and strategies. Solidarity and reciprocity within villages fade as individuals develop activities outside, build partnerships with outsider actors, integrate into broader markets and move away for years in search for jobs (and sometimes never return). The fraying of community (Rigg et al., 2012) manifests in the case of conflict. In Cambodia, divergent interests have aggravated the weakness of collective organizational capacity of Khmer villages, which were never cohesive units that could deal collectively with officials (Brickell and Springer, 2016). Analysts have noted that communities have not always been sufficiently unified, and identified 'serious internal problems and internal conflicts' (Pen and Chea, 2015, p. 29) encountered by communities when they engaged in resource-claiming procedures (Bottomley, 2009). Furthermore, traditional community leaders have lost their legitimacy and *raison d'être* in the eyes of many, who argue that the elderly could not oppose or endorsed provincial-level decisions to lease land to outsiders, and even 'sold the land'. Traditional leaders have seen their power eroded by official leaders (commune heads) and political parties, which have financial resources to play villager groups off against each other. Organized groups and villages are further subject to the patronage system (Öjendal and Sedara, 2006), in which people pay gratitude to local officers and beg for favours rather than contesting or confronting them.

Next, public authority has been crucial in containing rural protests, including through the use of force, negotiation and compromise, and legitimation. In both countries, the state has strengthened its territorial, administrative and political control over rural populations, directly through administrative development and public policies (poverty reduction programmes), and indirectly through the incorporation of individuals and communities into regulatory frameworks (for access to land, schools and health services) and financial institutions. Populations can play Scott's 'everyday resistance' (1985), but must also consider their everyday engagement with the state. In the case of Vietnam, the party-state has kept tight control on society, including through mass organizations, which hampers potential protests and rapidly represses open ones. The government has, however, shown its capacity to discuss with protesters when the latter were virulent, as in the case of Đồng Tâm. It has also subtly played 'participation' to filter and silence opposing views, and to increase the legitimacy of its interventions (Culas and Nguyên Van Suu, 2010).

In comparison to Vietnam, the Cambodian state apparatus exerts less formal control. However, it has not been less powerful to impose its economic concession model. To date, peasants' claims are overall radically rejected. Local authorities mostly discouraged communities from making land claims, yet hidden resistance such as encouraging or tolerating occupation of land that was leased to ELCs should not be overlooked. In other cases where communities showed more determination, public authorities have deployed security forces, as well as played for time by objecting to claims on the basis they were not addressed to the appropriate authority rather than rejecting them outright; promised to intervene; and asked communities to wait. The government has also played a dialectical policy of repression and 'avoidance' (Schoenberger, 2017, p. 886), which hampered or delayed peasants' reactions, not ELCs' development. Finally, such policy and practice serve to legitimize land grabs (Beban et al., 2017).

The support provided by civil society organizations to communities has proven effective whether for land claims, legal contests or land management; however, such support has been limited in scope because most NGOs have few resources. Intimidation of NGOs has been constant, and repression has significantly increased since 2015, including prohibition of their activities, tighter conditions for foreign NGOs and even closure. Researchers have recently experienced the same, with requirements to obtain research permits and a ban on research on land-related topics. The perspectives for advocacy politics (Schneider, 2011) are not good, as Western donors' influence on the Hun Sen regime is declining (Biddulph and Williams, 2016). Rural communities' claims have received uneven support from the international community. Repeated, laconic calls for good governance and the communication and promotion of the Voluntary Guidelines for Responsible Investment prepared by the Food and Agriculture Organization of the United Nations (FAO), have had no significant impact on land grabs. The Voluntary Guidelines are simply not implemented. Furthermore, the FAO's recent classification of rubber plantations as 'forest' perpetuates a colonial conception of land as a natural resource to be turned into productive assets.

Conclusion

This chapter adopted an historical perspective starting from pre-colonial times to detail how agricultural stagnation and abusive practices by public authorities have contributed to the generation of peasants' revolts, rebellions, and ultimately revolutions in Cambodia and Vietnam.

Over the contemporary period, the crisis–conflict dialectic has become far more complex and could be renamed 'transition–adaptation'. This is not to say there is no longer a crisis, but it is of a different nature, consisting of a relative decline in the importance of farming in rural livelihoods, and growing 'powers of exclusion' (Hall et al., 2011) of peasants from land and natural resources. This is best illustrated with ELCs in Cambodia and peri-urbanization and industrial zone projects in Vietnam. While peasants' struggles persist, outright conflicts have not been the predominant reaction of rural populations.

With regards to the strong will to engage in new opportunities despite costs and risks, Popkin's 'rational peasant' (1979) has gained momentum on Scott's 'moral' one (1976). The determination to seize opportunities has taken over political consciousness of inequality and marginalization/exclusion, even among the most vulnerable populations. Reactions are sometimes seemingly contradictory, when some peasants sell land that they later lack, or stop growing rice that they must then purchase; in other cases, wait-and-see attitudes can be disconcerting to analysts. These hesitations, dilemmas and refusal to choose are attempts to adapt to opportunities and constraints in an increasingly fast-changing and uncertain economic environment.

The dialectic has not only changed, but has also become more complex, as peasants are no longer only peasants, having new horizons beyond home-village and communities of origin, and new aspirations. Complexity comes also from the growing presence of outsiders such as businesses and in-migrants, new actors such as civil society organizations and the new ways the state exerts its power.

For now, rural populations have turned away from revolutionary ideas; revolt, rebellion and revolution (Osborne, 2010) are bad memories of failures, atrocities and wartime. However, as many of those who are adversely incorporated into the agrarian transition have exhausted all coping mechanisms and may rapidly lose ground, it cannot be ruled out that sporadic protests will intensify, spread and harden. Resilience is not unlimited.

Acknowledgements

Research in Cambodia was supported by the Swiss Network for International Studies of Geneva (2011–2013), and the Swiss National Science Foundation and Swiss Development Cooperation R4D programme (Demeter project, 2015–2020).

Notes

[1] Vietnam's revolution and wars have been the subject of innumerable scholarly works. Recalling the historical analysis of peasants' inclination to communist or counter-revolutionary ideas goes beyond the scope of this chapter; refer to Paige 'Social theory and peasant revolution in Vietnam and Guatemala' (1983), which recalls main contributions such as Wolf's *Peasant Wars of the Twentieth Century* (1978) and Migdal's *Peasants, Politics and Revolution* (1974).

[2] The term Khmer refers to the majority ethno-linguistic group in Cambodia, as distinguished from other 'indigenous peoples'. Similarly, in Vietnam, the Kinh majority group is distinguished from 'ethnic minorities'. The distinction also refers to power relations, specifically the political and economic dominance of the majority over the others, a key structural feature of Southeast Asian societies.

[3] After Germany defeated France in 1940, the colonial administration of French Indochina passed to the pro-German Vichy French government, which ceded control of Hanoi and Saigon to Japan. Japan then extended its control over the whole of French Indochina, until August 1945 when defeated by the Allies.

[4] Socialist new man.

[5] MacLean (2013) detailed behaviours 'that cannot be neatly categorized as self-interested compliance, on the one hand, and everyday forms of resistance, on the other' in the case of Vietnam.

[6] *Đổi Mới* is the market-oriented economic reforms initiated in Vietnam in 1986.

[7] The highest number of protesters reported in 2011 was 2000, facing 3000 security forces (RFA, 2012).

[8] I found the same in Hưng Yên province in 2015 and 2016 among individuals who had received compensation for the loss of their land in cases of industrial zone installation (Tân Dan commune) and highway construction (Minh Châu commune). The majority expressed satisfaction as they could develop other activities with the compensation.

[9] Sentences were pronounced over ten local officers for power abuses in August 2017 (Viêt Nam News, 2017).

[10] Economic Land Concession (sub-Decree 146 dated 27 December 2005) refers to a mechanism to grant private state land through a specific contract to a concessionaire to use for agricultural and industrial-agricultural exploitation (RGC, 2005).

[11] Based on interviews in Ratanakiri, April 2016.

[12] There are countless records of this kind in media (see Cambodian Daily) and civil society organizations' reports (see NGO Forum, LICADHO, ADHOC). Those reactions are also documented by numerous videos available on the internet showing villagers and security forces face-to-face and burning houses.

[13] Based on interview in Ratanakiri, 2013 (author's translation).

[14] The authority representatives–population distinction is too simplistic, as the populations best endowed in land and financial capital could also favour the implementation of ELCs and other investors, from whom they could expect opportunities with regards to new crops.
[15] Đổi Mới designates both the economic reform process and its success in terms of economic growth.
[16] Based on interviews in Ratanakiri, 2013 and 2016.
[17] Based on interviews in Loum Choar, Ratanakiri, July 2010.

References

Abu-Lughod, L. (1990) The romance of resistance. Tracing transformation of power through Bedouin women, *American Ethnologist* 17(1), 41–55.

Baird, I. (2017) Resistance and contingent contestations to large-scale land concessions in southern Laos and Northeastern Cambodia. *Land* 6(1). DOI: 10.3390/land6010016.

BBC (2011) Vietnam 'seals ethnic Hmong protest site'. Available at: www.bbc.com/news/world-asia-pacific-13306362 (accessed 5 September 2017).

Beban, A., So, S. and Un, K. (2017) From force to legitimation: rethinking land grabs in Cambodia. *Development and Change* 48, 590–612. DOI: 10.1111/dech.12301.

Beresford, M. (1988). *Vietnam: Politics, Economics and Society*. Pinter Publishers, London.

Biddulph, R. and Williams, S. (2016) From chicken wing receipts to students in military uniforms: land titling and property in post-conflict Cambodia. In: Brickell, K. and Springer, S. (eds) *The Handbook of Contemporary Cambodia*. Routledge, New York, pp. 169–178.

Bottomley, R. (2009) Contested forests: an analysis of the Highlander response to logging, Ratanakiri Province, Northeast Cambodia. In: Bourdier, F. (ed.) *Development and Dominion: Indigenous Peoples of Cambodia, Vietnam and Laos*. White Lotus Press, Bangkok, pp. 275–295.

Brickell, K. and Springer, S. (2016) *The Handbook of Contemporary Cambodia*. Routledge, New York.

Brocheux, P. (2003) *Ho Chi Minh – Du révolutionnaire à l'icône*. Payot, Paris.

Brocheux, P. (2011). *Histoire du Vietnam Contemporain. La nation résiliente*. Fayard, Paris.

Brocheux, P. and Hémery, D. (1995) *Indochine, la colonisation ambigüe (1858–1954)*. La Découverte, Paris.

Castella, J.C. and Dang Dinh Quang (2002) *Doi Moi in the Mountains: Land Use Changes and Farmers' Livelihood Strategies in Bac Kan Province, Viet Nam*. IRD Éditions, Paris.

Bylander, M. (2015) Credit as coping: rethinking microcredit in the Cambodian context. *Oxford Development Studies* 43(4), 533–553. DOI: 10.1080/13600818.2015.1064880.

Chandler, D.P. (1996) *A History of Cambodia*, 2nd edn. Westview Press, Boulder, Colorado.

Cismas, I. and Paramita, P. (2015) Large-scale land acquisitions in Cambodia: where do (human rights) law and practice meet? In: Gironde, C., Golay, C. and Messerli, P. (eds) *Large-Scale Land Acquisitions: Focus on South-East Asia*. International Development Policy series 6. Graduate Institute Publications, Geneva; Brill-Nijhoff, Boston, Massachusetts, pp. 249–272.

Culas, C. and Nguyên Van Suu (2010) *Norms and Practices in Contemporary Rural Vietnam – Social Interaction Between Authorities and People*. IRASEC, Bangkok.

Dao, N. (2016) Political responses to dam-induced resettlement in Northern Uplands Vietnam. *Journal of Agrarian Change* 16, 291–317.

Devillers, P., Fischer, G. and Cayrac-Blanchard, F. (1971) *L'Asie du Sud-Est*, Tome II. Sirey, Paris.

Diepart, J.-C. (ed.) (2015) *Learning for Resilience. Insights from Cambodia's rural communities*. The Learning Institute, Phnom Penh.

Diepart, J.-C., and Sem, T. (2016) Fragmented territories: incomplete enclosures and agrarian change on the agricultural frontier of Samlaut district, north-west Cambodia. *Journal of Agrarian Change*. DOI: 10.1111/joac.12155.

Fanchette, S. (2015) Gestion foncière métropolitaine et confrontations entre société villageoise et Etat/province (delta du Fleuve Rouge), *Hérodote* 157, 187–198.

Fforde, A. (1989) *The Agrarian Question in North Vietnam, 1974–1979 – A Study of Cooperator Resistance to State Policy*. M.E. Sharpe, New York.

Forest, A. (1981) Les manifestations de 1916 au Cambodge. In: Brocheux, P. (ed.) *Histoire de l'Asie du Sud-est – révoltes, réformes, révolutions*. Presses Universitaires de Lille, Lille, France, pp. 63–82.

Forest, A. (1993) *Le Cambodge et la colonisation française. Histoire d'une colonisation sans heurts (1897–1920)*. L'Harmattan, Paris.

Forest, A. (2008) Pour comprendre l'histoire contemporaine du Cambodge. In: Forest, A. (ed.) *Cambodge contemporain*. IRASEC, Bangkok; Les Indes Savantes, Paris, pp. 17–140.
Fortunel, F. (2005) L'amertume du café dans les plateaux du Centre Viêt Nam, les structures productives d'État et les autochtones. In: De Koninck, R., Durand, F. and Fortunel, F. (eds) *Agriculture, environnement et sociétés sur les hautes terres du Viêt Nam*. IRASEC-Arkuiris, Bangkok and Toulouse, France, pp. 163–168.
Fortunel, F. and Gironde, C. (2011) Transitions agraires et recompositions sociales en Asie du Sud-Est. In: Guibert, J. (ed.) *Dynamiques des espaces ruraux dans le monde*. Armand Colin, Paris, pp. 215–235.
Fourniau, C. (1989) *Annam-Tonkin 1885–1896 – Lettrés et paysans face à la conquête coloniale*. L'Harmattan, Paris.
Fox, J.M., McMahon, D., Poffenberger, M. and Vogler, J. (2008) *Land for my Grandchildren: Land Use and Tenure Change in Ratanakiri: 1998–2007*. Community Forestry International (CFI) and the East West Center, Honolulu, Hawaii.
Gironde, C. (2008) Grandes réformes et petits arrangements dans les campagnes vietnamiennes. *Autrepart* 48, 113–128.
Gironde, C. (2018) Le monde rural. In: de Tréglodé, B. (ed..) *Histoire du Viet Nam de la Colonisation à nos Jours*. Editions de la Sorbonne, Paris, pp. 141–159.
Gironde, C. and Senties Portilla, G. (2015) From lagging behind to losing ground: cambodian and laotian household economy and large-scale land acquisitions. In: Gironde, C. *et al.* (eds) *Large-Scale Land Acquisitions: Focus on South-East Asia*. International Development Policy series 6. Graduate Institute Publications, Geneva; Brill-Nijhoff, Boston, Massachusetts, pp. 172–204.
Gironde, C., Golay, C., Messerli, P., Peeters, A. and Schoenwenger, O. (2014) Large-scale land acquisitions in Southeast Asia: rural transformations between global agendas and peoples' right to food. Swiss Network for International Studies, Geneva. Available at: www.snis.ch/system/files/gironde_working_paper_lsla_southeast_asia_17.08.2014_0.pdf (accessed 19 March 2018).
Gironde, C., Golay, C. and Messerli, P. (2015) *Large-Scale Land Acquisitions: Focus on South-East Asia*, International Development Policy series No. 6, Graduate Institute Publications, Geneva; Brill-Nijhoff, Boston, Massachusetts.
Global Witness (2014) Cambodian communities submit complaint to World Bank over bankrolling of Vietnamese rubber giant behind land grabs and human rights abuse. Available at: https://www.globalwitness.org/fr/archive/cambodian-communities-submit-complaint-world-bank-over-bankrolling-vietnamese-rubber-giant (accessed 5 September 2017).
Hall, D., Hirsch, P. and Li, T. (2011) *Powers of Exclusion: Land Dilemmas in Southeast Asia*. National University of Singapore Press, Singapore.
Hall, R., Edelman, M., Borras, S.M., Scoones, I., White, B. *et al.* (2015) Resistance, acquiescence or incorporation? An introduction to land grabbing and political reactions 'from below'. *Journal of Peasant Studies* 42(3–4), 467–488.
Hy Van Luong (2010) *Tradition, Revolution, and Market Economy in a North Vietnamese Village, 1925–2006*. Hawai'i University Press, Honolulu, Hawaii.
Iglesia, A.C. (2010) Passive resistance. Notes for a more complete understanding of the resistance practices of the rural population during the Franco dictatorship. *Amnis* [Online] 9. DOI: 10.4000/amnis.265.
Jiao, X., Smith-Hall, C. and Theilade, I. (2015) Rural household incomes and land grabbing in Cambodia. *Land Use Policy* 48, 317–328.
Kelliher, D. (1993) *Peasant Power in China: The Era of Rural Reform*. Yale University Press, New Haven, Connecticut.
Kerkvliet, B.J.T. (1995) Village-state relations in Vietnam: the effects of everyday politics on decollectivization. *Journal of Asian Studies* 54(2), 396.
Kerkvliet, B.J.T. (2014) Protests over land in Vietnam: rightful resistance and more. *Journal of Vietnamese Studies* 9, 19–54.
Kiernan, B. (2005) *The Pol Pot Regime: Race, Power, and Genocide in Cambodia Under the Khmer Rouge, 1975–79*, 2nd edn. Silkworm Books, Chiang Mai, Thailand.
Kimty Seng (2017) Rethinking the effects of microcredit on household welfare in Cambodia. *Journal of Development Studies*. DOI: 10.1080/00220388.2017.1299139.
Lê Thanh Khôi (1978). *Socialisme et développement au Vietnam*. Presses Universitaires de France, Paris.
MacLean, K. (2013) *The Government of Mistrust: Illegibility and Bureaucratic Power in Socialist Vietnam*. The University of Wisconsin Press, Madison, Wisconsin.
Marr, D. (1981) *Vietnamese Tradition on Trial 1920–1945*. University of California Press, Berkeley, California.

Milne, S. (2013) Under the leopard's skin: land commodification and the dilemmas of Indigenous communal title in upland Cambodia. *Asia Pacific Viewpoint* 54(3), 323–339.

Moore, Jr, B. (1993) *[First published 1966] Social Origins of Dictatorship and Democracy: Lord and Peasant in the Making of the Modern World*. Beacon Press, Boston, Massachusetts.

Moyo, S. and Yeros, P. (2005) *Reclaiming the Land – The Resurgence of Rural Movements in Africa, Asia and Latin America*. Zed Books, New York.

Neef, A. and Touch, S. (2012) Land grabbing in Cambodia: mechanisms, narratives, resistance. Global Land Grabbing Conference II at Cornell University. Ithaca, 17–19 October.

Neef, A., Touch, S. and Chiengthong, J. (2013) The politics and ethics of land concessions in rural Cambodia. *Journal of Agricultural Environment Ethics* 26, 1085–1103.

Nesbitt, H.J. (ed.) (1997) *Rice Production in Cambodia*. International Rice Research Institute, Manila, Philippines.

Nghin, C. and Verkoren, W. (2015) Understanding power in hybrid political orders: applying stakeholder analysis to land conflicts in Cambodia. *Journal of Peacebuilding and Development* 10(1), 25–39.

Ngo Vinh Long (1973) *Before the Revolution – The Vietnamese Peasants Under the French*. Columbia University Press, New York.

Nguyên Duc Nhuân (1992) Le district rural vietnamien ou l'Etat en campagne. In: Matras-Guin, J. and Tailard, C. (eds) *Habitations et Habitat d'Asie du Sud-Est Continentale – Pratiques et représentations de l'espace*. L'Harmattan, Paris, pp. 343–376.

Nguyên Thanh Nha (1970) *Tableau économique du Vietnam aux XVIIème et XVIIIème siècles*. Editions Cujas, Paris.

Nguyên Xuân Linh (1981) Panorama des mouvements paysans vietnamiens. In: Brocheux, P. (ed.) *Histoire de l'Asie du Sud-est – révoltes, réformes, révolutions*. Presses Universitaires de Lille, Lille, pp. 83–110.

Nguyên Van Suu (2007) Contending views and conflicts over land in Vietnam's Red River Delta. *Journal of Southeast Asian Studies* 38(2), 309–334.

Nguyên Van Suu (2014) Conflits fonciers entre l'état et les paysans: l'anthropologue confronté au terrain. In: De Terssac, G., An Quoc, T. and Cattla, M. (eds) *Viêt-Nam En Transitions*. ENS editions, Lyon, pp. 91–106.

Norman, G. (2015) Cambodian communities follow different routes to justice over Socfin rubber project. Available at: https://news.mongabay.com/2015/10/cambodian-communities-follow-different-routes-to-justice-over-socfin-rubber-project (accessed 5 September 2017).

Öjendal, J. and Sedara, K. (2006) Korob, Kaud, Klach: in search of agency in rural Cambodia. *Journal of South East Asian Studies* 31(3), 507–526.

Osborne, M. (1978) Peasant politics in Cambodia: the 1916 affair. *Modern Asian Studies* 12(2), 217–243.

Osborne, M. (2010) *Southeast Asia – An Introductory History*. Allen & Unwin, Crows Nest, Australia.

Paige, J.M. (1983) Social theory and peasant revolution in Vietnam and Guatemala. *Theory and Society* 12(6), 699–736.

Papin, P. and Tessier, O. (2002) *Le village en questions*. Trung Tram Khoa Hoc Xa Hoi va Nhan Van Quoc Gia, Hanoi.

Peeters, A. (2015) Disclosing recent territory-lift and rural development contributions of Cambodian large-scale land acquisitions. International Conference on Global Land Grabbing, Chiang Mai, 6–8 April.

Pelzer White, C. (1986). Everyday resistance, socialist revolution and rural development: the Vietnamese case. *Journal of Peasant Studies* 13(2), 49–63.

Pen, R. and Chea, P. (2015) Large-scale land grabbing in Cambodia: failure of international and national policies to secure the indigenous peoples' rights to access land and resources. World Bank Conference. Washington DC, 23–27 March 2015. Available at: https://europa.eu/capacity4dev/eu-working-group-land-issues/document/large-scale-land-grabbing-cambodia-failure-international-and-national-policies-secure-indig (accessed 19 March 2018).

Ploeg, J.D. van der (2010) The peasantries of the twenty-first century: the commoditization debate revisited. *Journal of Peasant Studies* 37(1), 1–30.

Ponchaud, F. (1998) *Cambodge année zéro*. Editions Kailash, Paris.

Popkin, S. (1979) *The Rational Peasant: The Political Economy of Rural Society in Vietnam*. University of California Press, Berkeley, California.

RFA (2012) Vietnam: mass security clampdown in land seizure. Available at: www.eurasiareview.com/25042012-vietnam-mass-security-clampdown-in-land-seizure (accessed 5 September 2017).

Rigg, J., Salamanca, A. and Parnwell, M. (2012) Joining the dots of agrarian change in Asia: a 25 year view from Thailand. *World Development* 40(7), 1469–1481.

RGC (Royal Government of Cambodia) (2005) *Sub-Decree No. 146 ANK/BK on Economic Land Concessions*. 27 December 2005.

Sansom, R. (1970) *The Economics of Insurgency in the Mekong Delta of Vietnam*. MIT Press, Cambridge, Massachusetts.

Schoenberger, L. (2017) Struggling against excuses: winning back land in Cambodia. *Journal of Peasant Studies* 44(4), 870–890.

Schneider, A. (2011) What shall we do without our land? Land grabs and resistance in Rural Cambodia. International Conference on Global Land Grabbing, Chiang Mai, 6–8 April. Available at: www.iss.nl/fileadmin/ASSETS/iss/Documents/Conference_papers/LDPI/49_Alison_Schneider.pdf (accessed 19 March 2018)

Scott, J.C. (1976) *The Moral Economy of the Peasant: Rebellion and Subsistence in Southeast Asia*. Yale University Press, New Haven, Connecticut.

Scott, J.C. (1985) *Weapons of the Weak. Everyday Forms of Peasant* Resistance. Yale University Press, New Haven, Connecticut.

Singh, D., Smith, A.L. and Siow Yue, C. (2003) *Southeast Asian Affairs 2002*. Institute of Southeast Asian Studies, Singapore.

Slocomb, M. (2007) *Colons and Coolies: The Development of Cambodia's Rubber Plantations*. White Lotus Press, Bangkok.

Swift, P. (2015) Transnationalization of resistance to economic land concessions in Cambodia. International Conference on Global Land Grabbing, Chiang Mai, 6–8 April. Available at: https://www.iss.nl/fileadmin/ASSETS/iss/Research_and_projects/Research_networks/LDPI/CMCP_10-_Swift.pdf (accessed 19 March 2018).

Thayer, C. (2003) Political developments in Vietnam: the rise and demise of le Kha Phieu, 1997–2001. In: Drummond, L. and Thomas, M. (eds) *Consuming Urban Culture in Contemporary Vietnam*. Routledge Curzon, London, pp. 21–34.

Thomas, M. (2003) Spatiality and political change in urban Vietnam. In: Drummond, L. and Thomas, M. *Consuming Urban Culture in Contemporary Vietnam*. Routledge Curzon, London, pp. 170—188.

Touch, S. and Neef, A. (2015) *Land grabbing, conflict and agrarian-environmental transformations: perspectives from East and Southeast Asia*. International academic conference at Chiang Mai University, Chiang Mai, 5–6 June.

Turner, S. and Michaud, J. (2009) 'Weapons of the Week': selective resistance and agency among the Hmong in northern Vietnam. In: Caouette, D. and Turner, S. (eds) *Agrarian Angst and Rural Resistance in Contemporary Southeast Asia*. Routledge, New York, pp. 45–60.

Vickery, M. (1984) *Cambodia 1975–1982*. Silkworm Books, Bangkok.

Viêt Nam News (2017) Court gives sentences in Đồng Tâm land case. Available at: http://vietnamnews.vn/society/381752/court-gives-sentences-in-dong-tam-land-case.html#Qem0AyDq60Vk7er0.99 (accessed 5 September 2017).

Vô Nhân Tri (1990) *Vietnam's Economic Policy since 1975*. Institute of Southeast Asian Studies, Singapore.

Wolf, E. (1969) *Peasant Wars of the Twentieth Century*. Harper and Row, New York.

Woodside, A. (1971) *Vietnam and the Chinese Model: A Comparative Study of Vietnamese and Chinese Government in the First Half of the Nineteenth Century*. Harvard University Press, Cambridge, Massachusetts.

World Bank (2015) Cambodian agriculture in transition: opportunities and risks. Available at: www.worldbank.org/en/news/feature/2015/08/19/cambodian-agriculture-in-transition-opportunities-and-risks (accessed 19 March 2018).

17 Beyond Displacement by Armed Conflict: the Relationship Between Environmental, Economic and Armed Displacement in Colombia

Carolina Castro Osorio[1,]* and Edinso Culma[2]

[1]*Secretary of Culture, Bogotá, Colombia;* [2]*National Center for Historical Memory, Bogotá, Colombia*

Introduction – Conflict and Forced Displacement in Colombia

In the mid-1960s the National Liberation Army (ELN) and the Revolutionary Armed Forces of Colombia (FARC) emerged in Colombia. The emergence of these leftist guerrilla groups was a response to the political control of the National Front, a power-sharing agreement between the two main parties (the Liberals and Conservatives) from 1958 until 1974, the unequal distribution of rural land and the high levels of rural poverty. From the mid-1960s until today, Colombia has been subject to an armed conflict intensified with the consolidation of paramilitary armed groups (right wing, counter-subversive illegal groups) in the 1990s and the association between illegal groups with the coca economy. In 2016, the state and the guerrillas of the FARC signed a peace agreement and the process of negotiation with the ELN guerrillas began. These two agreements would bring an end to the armed conflict in the country.

In the context of conflict, the country experienced high levels of internal displacement. The victims of forced displacement due to the internal armed conflict have been recognized by the state through laws 387 of 1997 and 1448 of 2011. Under this framework, from 1985 until 1 January 2017, the National Information Network had registered 7,521,661 people as victims of forced displacement in the country. Although it has had an official shift register only since 1985, as we explained the armed conflict in Colombia dates back to the 1960s. Since 1985 the displaced families abandoned or lost more than 6 million ha of land. The country has 114 million ha of land and only 45 million can be used for agricultural purposes. The rest of the land is protected land in national parks and is part of communal properties of indigenous people. This means that the peasants abandoned or lost 13% of exploitable land.

The level of displacement not only speaks to the magnitude of the humanitarian tragedy generated by the dynamics of the armed confrontation between the state, the guerrillas and the paramilitary groups. In some cases, it also speaks to state actions and private actions that promote or counteract the extractive industries in the country, particularly in the areas or territories of the agricultural frontier, prompting civilian displacement.

Forced displacement due to armed conflict has interacted with diverse interests tied to the

*Email: tantalia16@gmail.com, ccastroo@unal.edu.co

exploitation of nature. By analysing the case studies of oil exploitation and palm farming, this article argues that the displacement of populations cannot be understood only as a result of the various dimensions of armed conflict, but also in relation to changes in the territory due to these two types of exploitation. Forced displacement due to armed conflict overlaps with and reinforces displacement due to economic (i.e. mining and oil development projects) and environmental reasons (e.g. the presence of agro-industry such as palm oil cultivation and the use of glyphosate spraying to control coca crops). At the same time, the types of displacement intertwine with each other and have various levels of incidence in each territory depending on the groups (armed and unarmed) that are operating in each region.

The high level of attention in rural studies in Colombia given to land tenancy and its distribution (Fajardo, 2002; Kalmanovitz and López, 2006; Reyes, 2009; Restrepo and Bernal, 2014) circumscribes our understanding of conflict as the struggle for land. Little has been said about rural space beyond understanding it only as land and its distribution and uses. It is also important to recognize socio-environmental and economic drivers of tensions and conflicts. The simplified vision that we point out has led to limited academic perspectives and public policy interventions. Some of the limitations of this vision include the following three points. First, inequality as the main driving force of rural conflicts hides other factors of conflict in rural spaces such as access to collective goods like water, roads and the quality of ecosystems. Second, studies on the impact of armed conflict on rural spaces concentrate their attention on forced displacement due to violence and the seizure of land, overlooking other economic and environmental factors that contribute to displacement, or understanding these factors as a consequence of conflict. Through two case studies, we demonstrate that, although the presence of agro-industry plantations does not necessarily result from processes of appropriation and the dispossession of land, they do incur environmental dispossession. Third, the policy of land restitution for the victims of conflict centres on the restitution of property rights, and does not include broader studies that would allow for the anticipation of factors that could impede the presence of restituted farmers on their land. Law 1448 (2011), better known as the Victims and Land Restitution Law, 'dictates measures for the provision of assistance and restitution to the victims of the internal armed conflict'. The main purpose of the law is to return stolen and abandoned land to internally displaced people and provide reparations to victims of human rights violations.[1]

This chapter contributes to a better understanding of forced displacement, with the objective of guaranteeing that processes subsequent to the restitution of lands allow for the development of the projects that the victims decide. This chapter also promotes an analysis so that public policies that encourage the development of certain economic activities in the countryside must take into consideration the inherent tension between human rights and public interest to which the development projects give rise (Twomey, 2014, p. 16).

Several studies (Duncan, 2006; CNRR *et al.*, 2010; National Center for Historical Memory (CNMH using the Spanish acronym), 2015) have highlighted the following characteristics of forced displacement in Colombia. First, the threats, massacres and general condition of violence of the armed confrontation between the army, guerrilla groups and the paramilitary have partially caused the national exodus. Second, there are also actors with economic interests who have taken advantage of the general condition of violence to encourage displacement and the seizing of land of farmers in various areas of the country. The latter characteristic was studied in depth by CNMH in its report, 'La tierra en disputa', which focused on an analysis of the phenomena of displacement, abandonment and the seizing of land in the northern part of the country. The deployment of paramilitaries that began in the mid-1990s is a central factor that characterizes the phenomenon of violence in this territory. Academic literature agrees that this deployment did not exclusively originate as an anti-subversive struggle, but also as a desire for political and economic control of territory by the regional elite. In this area of the country, there was an alliance among economic groups, regional politicians and the paramilitary that encouraged displacement and took advantage of the situation of abandonment of lands in order to take control. As a consequence of this type of displacement, we see what Gutiérrez has called 'active paramilitary dispossession', which refers to 'the planned and conscious act of taking away land from somebody with the purpose of redistributing the property rights among other actors' (Gutiérrez, 2014, p. 45).

Displacement has been a means to guarantee territorial control, carry out an anti-subversive struggle, and control military and drug-trafficking corridors. It was also used to recuperate land that was given to small farmers under the agrarian reform.[2] Regarding this last motive, the CNRR affirms that it is evident that the municipalities with higher levels of displacement were the places where more land was given to small peasants in the framework of agrarian reform processes. Land distributed among small peasants in the 1970s and 1980s was again appropriated by large landowners (CNRR et al., 2010, p. 493).

Understanding displacement, abandonment and the seizure of land in the northern part of the country shows that economic interests and agro-industrial projects take advantage of the general situation of violence to impose new developments and uses of rural land. In this case, the displacement is not economic, but due to armed conflict, which is in part motivated by economic interests that require the appropriation of more land. Displacement occurs through illicit mechanisms of alliances between businessmen with paramilitary groups, using illegal measures such as forced sales, the falsification of property titles, and through more diffuse mechanisms of buying land at low costs and buying mortgages that the peasant could not pay, taking advantage of the vulnerability of displaced people. '[I]n Colombia, the displacement of the population cannot be explained exclusively as a consequence or "collateral effect" of war and the logics of confrontation between armed actors (displacing to take away the offensive capacity of the "enemy"). The population has also been expelled due to legal and illegal political and economic motivations that are intertwined and coexist with the armed conflict' (CNMH, 2015, p. 131).

In addition to displacement due to economic reasons, we can see that, although economic actors have not contributed directly to the conditions of displacement, they took advantage of the conditions of conflict and the emptying of territories to encourage new farming undertakings and to buy large areas of land at a low cost. Quantitative analyses show that after years of high levels of abandonment, buying and selling transactions are reactivated 3 or 4 years after the peak of conflict (CNMH et al., 2016).

In sum, economic actors, and especially agribusiness actors, have been linked to displacement either directly, as allies or promoters of armed groups that encourage the displacement of farmers; or indirectly, taking advantage of the conditions of abandonment for the massive purchasing of land, and thus limiting the opportunity for the return of the population.

We argue that it is not enough to recognize that actors with economic interests in rural areas are partly responsible for the loss of farmers' land. There are also economic impacts of agrobusiness and the sector that exploits the land that cause the displacement of farmers, without high levels of violence due to armed conflict as a necessary condition. We insist that in addition to forced displacement due to the armed conflict, there are changes in the life conditions of a territory that lead to forced displacement for economic and environmental reasons. It is precisely these other types of displacement that can be identified when analysing the second wave of abandonment centred in the southwest region of the country.[3] We will explore a typology of causes of displacement. We then consider the case of the presence of oil in the department of Putumayo[4] and continue with the case of palm farming in various areas of the country.

Typology of Causes of Displacement

The literature and international consensus (ONU, 1998; World Bank, 2017) are still hesitant in the classification and characterization of displacements that differ from those provoked by conflict or natural disasters. As we will show below, this hesitancy has consequences that states must take into account, especially in countries that, like Colombia, are implementing processes of restitution and are facing the conditions of a recent peace agreement.[5]

International agreements, especially the principles of the United Nations (UN) that guide interventions in internal forced displacement, emphasize the context of violence that defines displacement.[6] However, there is another type of forced migration caused by economic and environmental phenomena.[7] The academic literature shows a demand for recognizing that not only does violence oblige households to change and abandon their territory; forced displacement due to environmental changes and socio-environmental conflicts[8] is increasingly common, as is displacement caused by large-scale development projects. In this sense, Van Hear proposes that forced migration should include 'individuals or communities obligated or induced to move, when they would have preferred to stay in their place of residence; the coercion that induces the forced migration can be direct, open, and focused, or indirect, covered, and diffuse' (Van Hear, 1998, p. 10). As we can see in this definition, the causes of displacement are not included; it only mentions the forced and undesired condition of displacement. It is in this broad definition that we can establish three categories of forced displacement: (i) due to conflict; (ii) due to development projects and economic changes; and (iii) due to environmental reasons (Fig. 17.1).

Only a part of this demand for recognition of displacement due to other phenomena is already included in the UN principles of internal forced displacement. Paragraph c of principle 6, point 2, explicitly includes the topic of displacement due to development projects: 'The prohibition of arbitrary displacement includes displacement: . . . (c) in cases of large-scale development projects that are not justified by a superior or fundamental public interest' (ONU, 1998, p. 5). The main driving forces of displacement due to development projects are: projects for the 'provision of water (reservoirs, dams, or for irrigation); the development of physical and transportation infrastructure (roads, canals, etc.), the provision of energy (mining and hydrocarbon extraction, energy plants), agricultural expansion, parks and nature reserves, and plans for the redistribution of the population' (Robinson, 2003, p. 11). In the UN declaration, the displacement for extractive activities is incorporated, in this case reference is made to the need to change place of residence for a group of people because the land they occupy will be the location of extractive activities. The UN principles do not include human settlements contiguous to operating areas, which will see their living conditions modified by the presence of exploitation.

In the case of economic displacement we find that changes in the use of land have decreased working opportunities in various regions due to a decreased demand for labour in comparison to the demand for traditional agricultural exploitation. The presence of large farms transforms local economies and alters the levels of prices. For example, the presence of a complex of oil exploitation in a remote region of the country causes increased circulation of money and therefore the overall increase of prices of consumer goods, which ends up affecting family

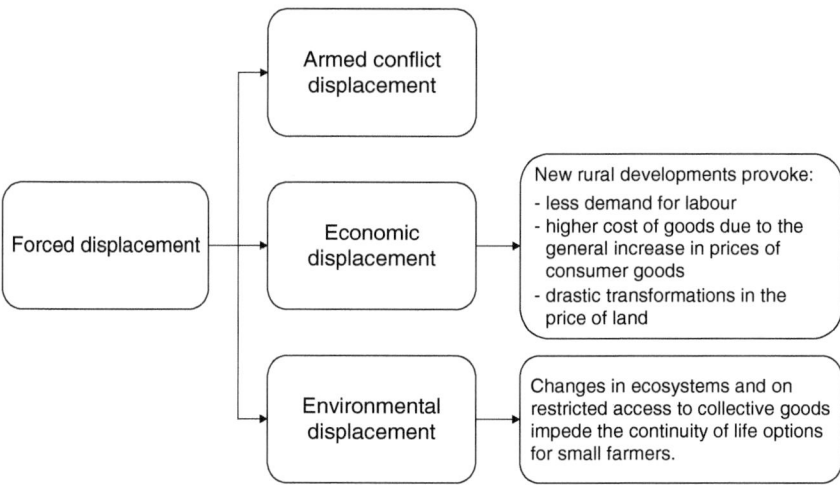

Fig. 17.1. Categories of displacement and their consequences.

economies not related to exploitation. Finally, it has also been shown that changes in the use of land are related to the drastic alteration of the price of land, which impedes small farming communities' access to land.

Environmental displacement 'currently constitutes one of the dominant conditions of forced migration within countries' (Terminski, 2012, p. 7). While it is a more recent phenomenon, since 1985 the environmental programme of the United Nations (UNEP) has defined environmental refugees as 'those people who are forced to abandon their traditional habitats, temporarily or permanently, due to intense environmental transformation (natural and/or caused by human action) that endangers their existence and/or seriously affects their quality of life'. Most case studies about this kind of displacement have focused on so-called 'natural' disasters or those caused by climate change (Gemenne et al., 2013), and to a lesser degree, on ecological transformations caused by agro-industrial practices that jeopardize small farming subsistence. De Moor and Cliquet identify three categories of this kind of displacement: (i) environmental degradation due to climate change and the loss of biodiversity; (ii) sudden disasters, including natural and technological disasters; and (iii) intentional environmental destruction (De Moor and Cliquet, n.d., p. 3). For the case of Colombia, it is pertinent to analyse the last category in which transformations of ecosystems due to new uses of the land lead to a loss in the quality of life of rural inhabitants. In summary, when we refer to environmental displacement, we will focus our attention on the changes in ecosystems and on restricted access to collective goods that impedes the continuity of life options for small farmers.

A comprehensive understanding of forced displacement must include an analysis of the three mentioned categories and their interconnections, and must recognize how the development of public policies implemented until now in Colombia, within the context of the armed conflict, supports other phenomena that cause forced displacement. In any event, maintaining and clarifying the differences between these types of displacement must be a priority, both for the development of adequate policies and for accurately determining who is responsible in each case.

The cases presented below are partly a result of the implementation of three public policies nationwide. The first is the 2010 mining and energy policy, a directive which sought to capitalize on high international prices of energy resources. The objective of this policy was to ensure that the energy sector grew to a 60% share of gross domestic product (GDP). The second was the fight against drug-trafficking and in particular the control of crops through the spraying of glyphosate. The last one, the biofuel policy of 2008, had the objective of positioning the country as an exporter of palm oil.

The Case of Oil and Counter-Narcotics in the Department of Putumayo

In the case of the department of Putumayo, there have been 236,056 victims of forced displacement over the past 30 years. Many of the people and families from Putumayo (indigenous, Afro-descendant and small farmers), who have been forcibly displaced from their territories and who are recognized as such in the Registry of Victims of the Colombian State, have moved not only to survive the risks and the dangers of the confrontation between the 'actors' of the armed conflict, but also as a way to confront the damages done by the oil industry and from the aerial spraying of glyphosate by state programmes to eradicate illegal coca crops. This is because the municipalities of this department, where the war-like events of the armed conflict have been concentrated from the 1990s until 2015, are the same places where oil exploitation is concentrated, where coca is grown, and where the glyphosate spraying occurred during the same period: Puerto Asís, San Miguel, Valle del Guamuez, Orito, Puerto Caicedo, Villagarzón, Puerto Guzmán, Mocoa and Leguízamo.

The oil industry and the glyphosate spraying have destroyed or caused deterioration in the minimum conditions that allow for the continuity of many communities in the rural territories of Putumayo. The contamination of water sources, land and air with the residue of crude oil and glyphosate have reduced food security, human health and the farming economies of the population in this department.

In October 2002, a year after starting aerial spraying with glyphosate in Putumayo under Plan Colombia,[9] the Office of the Ombudsman had registered '... 318 complaints about the impact on 6,076 families and 5,034.25 hectares ... in the municipalities of Puerto Asís, Orito, and Valle del Guamuez alone'. The complaints were about damage caused to land with small farming crops of yucca, banana, corn, fruit trees and pasture. The glyphosate damaged plots where peasants had already eradicated 70–100% of coca crops in fulfilment of the voluntary eradication pact that the national government had signed as part of Plan Colombia (Defensoría del Pueblo, 2002).

These impacts on human health caused by the glyphosate spraying during the mentioned period were not insignificant in the three municipalities named above. The Subdirection of Public Health of the department of Putumayo identified that 85% of people surveyed about this issue in Puerto Asís, Orito and Valle del Guamuez (4883 of 5929) showed deterioration of their health after the aerial spraying with glyphosate. Reported symptoms included respiratory, gastrointestinal, skin and psychological problems, fever, general discomfort and dizziness (Defensoría del Pueblo, 2002).

According to a report by the National Center for Historical Memory, until 2015, there had been no complete official inventory of the environmental liabilities caused by the oil activity that has been ongoing for 50 years in Putumayo (CNMH et al., 2015, p. 177). Despite this lack of information, there are some records that point to the magnitude and gravity of the damage caused by the oil industry in Putumayo territory:

> ....the oldest record of hydrocarbon spills systematized by [the environmental authority] Corpoamazonia is from 2012. We have the following information about it: In total there were 121 spills, of which 87 were due to illegal actions (attacks with explosives, hydrocarbon leaks due to manipulation and illegal perforation of infrastructure), and 34 were either operation accidents or were due to the deterioration of infrastructure...
> (CNMH et al., 2015, p. 178).

There are scattered reports about oil spill events. As an illustration of these cases, the Human Rights Network of San Miguel (2013) of the region presented the case of oil spill in the municipality of San Miguel, Putumayo in 2013. In this event:

> ... the community manifests that the industrial wastewater drainage from the Colón storage container of ECOPETROL S.A. has generated an environmental problem that affects the rural communities (...) In the area of the wastewater drainage canal, a storage container overflowed, showing signs of contamination with crude from the wetlands (next to the school of the rural community El Espinal). It appears that the pipe broke due to mechanical damage and lack of maintenance by the company Ecopetrol.

> ...This year alone there were four oil spills that flowed into the San Miguel River... The local population cannot access drinking water or water for cooking or bathing, and they suffer from headaches, stinging eyes, respiratory and epidemiological problems due to the foul smells (sic)...
> (Human Rights Network of San Miguel, 2013)

The oil industry and the glyphosate spraying have not only destroyed and deteriorated the material, economic and environmental conditions that allow these communities to remain in rural territories of Putumayo; they have also become phenomena that feed political violence and the armed conflict in this department, which in turn has translated into the accumulation of victims of forced displacement, as defined by Colombian legislation. This is due in large part to the fact that law enforcement by the Colombian state in this department owes its presence to the existence of oil exploration and exploitation by private companies, and to illegal coca crops whose income is capitalized upon by illegal armed groups (the FARC and paramilitaries). This became evident starting in 2000, when Plan Colombia substantially reformed the military and expanded the repressive potential of law enforcement of the Colombian state, creating military units capable of making incursions into territory historically dominated by the FARC, maintaining their presence and implementing in those territories programmes of forced eradication of coca crops, and providing protection to the oil industry (CNMH et al., 2015, pp. 142–151). The increase in oil exploration and exploitation in Putumayo is directly related to the increase in clashes between the national army and the FARC guerrillas. These clashes not only cause fear in the population, but also destroy homes, collective goods and roads, causing displacement.

Community recognition of the impacts of oil exploration and exploitation has led to the organization of processes of resistance. Some communities and social organizations that are opposed to the presence and expansion of the oil industry and the programme of forced eradication through aerial spraying with glyphosate were criminalized and persecuted by law enforcement. This stigmatization and persecution materialize in the criminalization and violent repression of protest, the selective assassination of social leaders, damage to civilian goods and forced displacement.

> For the families and communities that inhabit lands where oil is produced and transported, this type of state presence has not only meant the deterioration of their citizen security, but also the vulnerability of their basic social and political rights. The Inter-Church Justice and Peace Commission has documented and disseminated various denouncements that social organizations in Putumayo have carried out. One of them was carried out on June 15, 2011 by the Departmental Committee of Social, Farmer, Indigenous, and Afro-Descendant Organizations of Putumayo. According to this denouncement, on May 25 of that year, members of the national army threatened the population of the village of Miraflores (*inspección* of Guadualito, *corregimiento* of El Tigre, municipality of Valle del Guamuez) with forced displacement for opposing the construction of a road and oil pipeline.
>
> (CNMH *et al.*, 2015, p. 150)

In addition to environmental events and armed confrontations related to the presence of oil exploitation, other problems related to access to land are identified, such as the impossibility of awarding land that, despite having been long occupied, cannot be delivered by the state as the property of small farmers because they are within the range affected by oil operations designated by law (within 5 km of the wells). Likewise, land prices have increased in the areas of exploitation, impeding or limiting the purchase of land by small farmers and indigenous communities due to the potential demand of oil companies (CNMH *et al.*, 2015).

The Case of Palm Oil Crops[10]

Various studies have pointed out the direct relationship between the expansion of palm oil crops and the abandonment and seizure of land belonging to the small farming population.[11] Related to the predominant characterization centred in the northern area of the country, one finds that 'in municipalities like María la Baja and Ovejas, the business expectations around agrofuels – biodiesel from palm oil – and carbon deposits – in the form of teak plantations – articulated with a massive project of counter-agrarian reform dressed in Green' (Camargo and Ojeda, 2017, p. 61). But this is not a generalized tendency for all palm oil crops, as the CNMH identifies in the book, *Tierras y conflictos rurales*. Through geographic tracking and case studies in the areas of expansion of this type of crop, it was determined that not all crops were planted through violent processes of the expulsion of populations or by taking advantage of the generalized conditions of armed violence in a region (CNMH, 2016). What can be generalized is the fact that palm plantations have led to the seizure of collective goods, which eventually causes forced displacement.

Just like with oil exploitation, palm oil production in the country has been encouraged by state development policies at the national level, which do not warn about the consequences it could have for specific local contexts. Although there were antecedents of these policies in the 1950s, the largest state push for this kind of exploitation happened after 2008, with the promulgation of the biofuel policy, which had various objectives, including energy security, environmental protection and the positioning of the country as a palm oil exporter.[12] In only 3 years, there was an 80% growth in the area of the country that was set aside for this crop, growing from 237,000 ha in 2008 to 427,000 ha. Although Latin America is not one of the largest producers of palm oil, Colombia dedicates the most space (24%) to this type of crop among all the countries in the region, and is the fourth largest producer globally (CNMH, 2016).

In addition to the changes in economic conditions caused by the planting of palm crops,[13] here we highlight the changes in ecosystemic conditions (Goebertus, 2008; López, 2010; García and Calderón, 2012; Inter-Church Commission, 2007). This crop is most productive in the most biologically diverse areas of the world, called hotspots, due to the conditions and quality of the habitat (Inter-Church Commission, 2007). Some of the impacts of palm oil production in these areas may include the devastation of existing vegetation; the disappearance of a favourable habitat for certain plant and animal species; deforestation;[14] the destruction of the ecosystem in which farming communities had supported themselves through activities such as hunting, fishing, wood for construction and medicinal plants (Inter-Church Commission, 2007); the transformation of the water system, specifically the drying out of flood zones and transformations to the subsoil because of the need to maintain dry land for crops; the contamination of water due to the residue of exploitation and the use of pesticides, particularly because a significant part of these sources supplies community aqueducts, which, when contaminated, jeopardize access to water and inhabitants' health; and the presence of new infestations due to the decrease in biological control based on biological diversity and due to the use of pesticides. These are the principal environmental impacts of the predominance of monoculture, which also provokes the displacement of agricultural and fishing practices, as well as subsistence practices, which are crucial for food security and small-scale commercial farming.

The following testimony exemplifies the way food security is jeopardized. Speaking about the quality of water and its effect on fishing in Caño Campano in the village of Buenos Aires, in the municipality of El Peñón, Bolívar a peasant said:

> This was a large fish hatchery, around 1985, we got 2,000, 3,000, 4,000, 5,000 fish – that was how catfish fishing was – you don't get *bocachicos* anymore, no! But there were – what happened is that now they died due to the contamination of the water. There was a little catfish but [today] the water brings it bad. It can't survive anywhere, the weakest one dies.
> (quoted in Soler and León, 2009, p. 14)

Changes in the quality of nature are not the only cause of displacement. Practices by palm companies may strip local farmers of access to basic resources, especially access to and control of water sources. For example,

> Well, both the farmer and the rancher need water. You know they need it more. The palm oil corporations don't need water for themselves. They aren't going to have cattle, or pigs, or plants for subsistence. What are they doing? What did they do? In the 600 hectares that were damaged, they immediately covered up the wells and they dried out the branches of the river. Then they came here and started to throw out the land that had piled up and went about covering the wells. That is the first thing they do, cover all the wells in order to plant palms...That is what we are afraid of, that if the government wants to give us this land in 3 years, we won't find branches of the river anymore, or water sources, because they will have dried up. Then, how will we provide for our cattle? How will we provide for pigs without water? How are we going to irrigate?
> (quoted in Soler and León, 2009, p. 15)

There is a constant dispute between big and small farmers over water and conflict over its use for food crops and palm crops. An example of this is inequality in access to and administration of irrigation districts. In the area of Montes de María, 'in María La Baja, the irrigation district that exists in this municipality, which was once used by farmers for their food crops (mostly rice), and currently is destined for more than 5,000 hectares out of 11,873 for the cultivation of palm oil, carried out by the Association of Palm Growers of the Asopalma District' (CNMH, 2014). Palm farming has grown since the transformation of current ecosystems, exacerbating demand for water resources: 'the principal embankment that we are going to see is here in the Caimán stream. These are all palm growers. All these wetlands were dried up to put palm here. They put embankments, a wall so that the Elvira stream is accumulating all this water that passes through. And this water doesn't pass through sediment anymore; it doesn't have any way out' (quoted in Soler and León, 2009, pp. 15–16).

Regarding the increase in infestations and the change in the quality of the land, there were isolated accounts of displaced farmers from Cucal and Cascajalito about the loss of productivity of fruit trees and the appearance of foul-smelling fruit on some plants, among other problems.

Finally, there are the transformations of territory by which some farmers have been 'closed in'. 'The predominance that the crop begins to have in one area encourages the confinement of farmers who continue growing traditional crops and hinders the development of their life ways and food security' (CNMH, 2016, p. 468).

Displacement and Dispossession

The cases described above call for a better understanding of conflicts in rural places. As mentioned earlier, the attention centred on the concentration of land in Colombia (1% of landowners have 42% of private rural property) and the studies on violence based on this perspective do not recognize other conflicts such as socio-environmental ones. In the particular socio-political context of Colombia, it is necessary to recognize the problem of dispossession in a broader sense. The National Commission for Reparations and Reconciliation (CNRR) had already proposed an exercise of conceptualizing dispossession as including material goods as well as territory and collective goods. For the CNRR, 'with dispossession – unlike abandonment – there is a manifest intention of stealing, expropriation, disfranchisement, misappropriation, of a good or a right. It can also be associated with the disfranchisement and dispossession of the use and enjoyment of personal property and real estate, social spaces and community environments, culture, politics, the economy, and nature' (CNRR, 2009, p. 25).

The second argument of this chapter is related to what we previously described, and suggests that economic and environmental impacts of rural development generate dispossession, without necessarily passing through a previous stage of abandonment; and that the dispossession of collective public goods is an issue that will require special attention to guarantee a successful post-agreement period. Figure 17.2 presents a modification of the CNMH outline.

By warning about the seizure of places and territories, beyond the seizure of the land, we can complement our understanding of rural conflict with the conflict that arises by having different perspectives on values and the use of nature, which is understood as environmental conflict.

Conclusion

Recognizing other types of displacement, and in general conflicts other than that exclusively associated with armed conflict, will be fundamental for guaranteeing the success of the peace accords in Colombia and in general the diminution of violence. Bearing in mind that part of the armed conflict originates from an unresolved agrarian problem, integral rural development is recognized as central in the post-conflict context, such that it is included as the first mentioned section in the accords. The first of the six points of the 'final agreement for the ending of the conflict and the construction of a stable and lasting peace', celebrated between the state and the FARC guerrillas, is integrated rural development. This chapter emphasizes land allocation processes for farmers and support to ensure that

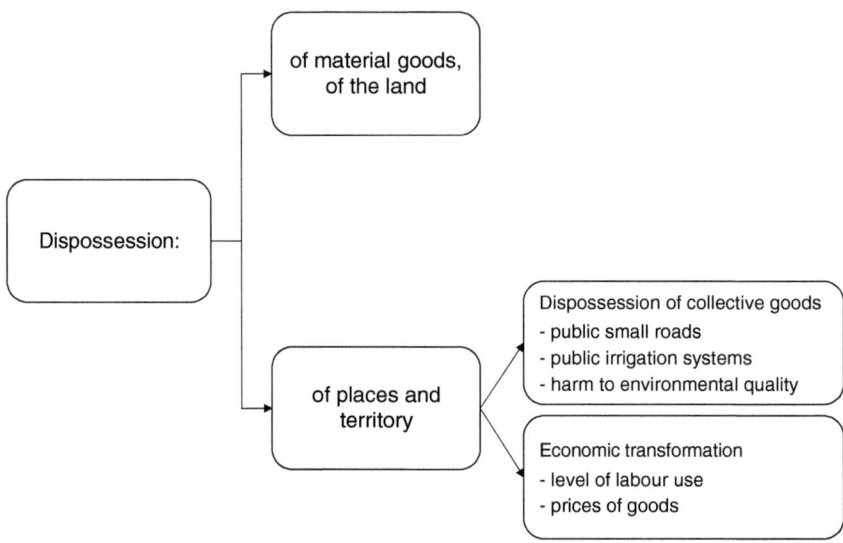

Fig. 17.2. Types of dispossession. (Authors' elaboration of CNMH (2016), p. 470.)

small farmers have opportunities for productivity and improving their living conditions. However, in this chapter and in other texts of rural public policy, little is said about the conflicts between different and opposite visions of rural development that may lead to processes of displacement, to revictimization, to new violent conflicts and to the deterioration of the life conditions of rural residents. If we do not take into account the conflicts about diverse uses of land, it is probable that the violence in the rural zones will persist.

Therefore, it is essential to highlight the challenges related to the factors that cause environmental and economic displacement presented in this article. The first challenge is related to the threats and assassinations of social leaders that have recently occurred in the country. In 2016, during the peace negotiations, it is estimated that more than 70 leaders of social and small farmer organizations were assassinated. Some of these homicides were carried out by re-armed anti-subversive groups: old paramilitary groups that are called criminal gangs today, with interests in land or that are financed by economic elites who seek to guarantee the expansion of their crops or control larger expanses of land. Some of them are also directly trying to avoid the return of displaced small farmers and the processes of restitution, such that today there is an illegal armed group known as the 'anti-restitution army'. Other armed groups have specifically targeted leaders who have participated in movements to defend and protect territories and water sources. There are still not enough legal investigations to affirm that their struggles were the factors that caused their persecution, but at least nine of the assassinations appear to be directly related to the kind of conflicts mentioned in this chapter (see Appendix 1).

The development projects that conflict with the livelihood options of small farmers, demonstrated in this chapter, have been encouraged by state policies such as the energy and mining policy and the biofuel policy. The analysis put forward in this article leads us to consider more carefully the negative consequences of this kind of national strategies. There have been small advances by including in public discussion the need to compare the economic benefits of mining and oil development with the environmental consequences and include them in national accounts.[15] Although ecosystem analyses of the loss of environmental quality are very important, we insist in this chapter that we must approach this issue from a broader perspective of socio-ecosystem balance, which allows us to recognize in a comprehensive way the negative consequences of mining for rural communities and their agricultural and everyday practices.

A last challenge relates to the incorporation of this kind of analysis in regions where the restitution processes are happening. The policy of restitution has focused on the restitution of property rights and measures of reparation, but it does not include broader studies that allow us to anticipate factors that could impede returned small farmers from staying on their land.[16] The interests in land that drive some development policies can work against certain kinds of crops, certain ways of using collective goods and the very social organization of displaced farmers. 'In scenarios of post-conflict, special attention must be given to disputes over the use of resources and the loss of opportunities for taking advantage of these resources. It is necessary to build collective consensuses about the development goals of each territory' (CNMH, 2016, p. 412).

Acknowledgements

This article is part of a 3-year research project on land and social organization by the National Center of Historical Memory. The research was coordinated by Rocío Londoño Botero. While the ideas presented in this article are specifically the authors', they are in part the result of a collective research project.

Notes

[1] Law 1448 (2011), better known as the Victims and Land Restitution Law, 'dictates measures for the provision of assistance and restitution to the victims of the internal armed conflict'. The main purpose of the law is to return stolen and abandoned land to internally displaced people and provide reparations to victims of human rights violations.

² The presence of the National Association of Farmer Users (ANUC) in this area of the country was strong. This association represented the ideals of the farmer movement in the 1960s and 1970s that struggled to improve the distribution of land, demanded the expropriation of uncultivated land of large properties for its subsequent return to farmers, and requested property for rent on land with crops. In addition to the management of the entities of the Agrarian Reform of the State for the assignment of land, during the 1970s, this movement carried out various peaceful takeovers of large ranches, which the movement called 'the recovery of land'. After the takeover of land, a state institution called the National Institute for Agrarian Reform (INCORA) negotiated with large landowners for the donation or purchase of land for its redistribution to small farmers. Between 1962 and 1986, the Colombian state distributed more than 700,000 ha in the entire country. Of these, 54% were bought by the state from large landowners, 7.5% were expropriations and 36.2 % were cessions (CNMH, Londoño et al., 2016, p. 150).

³ According to historical analysis of the reports of abandoned and seized lands, the periods with the highest numbers are 1999–2002 and 2005–2008. Most reports from the first period came from municipalities in the northern part of the country, and those of the second period came from municipalities in the southwest (CNMH, Londoño et al., 2016, pp. 362–367).

⁴ The department of Putumayo is located in southwest Colombia. It has a population of 322,681: 148,711 live in the municipal centres and 173,970 live in the rest of the territory (SIDIH-OCHA Colombia, Department profile, 2009). It has an indigenous population of 51,700. Given these geographic conditions, the department's territory of 24,885 km² is divided into three subregions: upper Putumayo (the Andes mountain range), middle Putumayo (Amazon foothills) and lower Putumayo (Amazon basin). The subregions have four, three, and six municipalities respectively, for a total of 13 (Culma Vargas, 2010).

⁵ In 2016, the state and the FARC guerrilla signed the peace agreement. The FARC was the largest guerrilla group, and had the greatest territorial control in the country. For that reason, this historic moment is identified as the end of the armed conflict in the country. There is still a second group of guerrillas called ELN, with whom peace talks have begun.

⁶ Included in the category of internal forced displacement are those 'people or groups of people who have been forced or obligated to escape or flee from their home of habitual residence, particularly as a result of or in order to avoid armed conflict, situations of generalized violence, violations of human rights, or natural catastrophes or those provoked by humans, and who have not crossed an internationally recognized state border' (ONU, 1998).

⁷ See selected bibliography on displacement caused by development projects at: http://forcedmigrationguide.pbworks.com/w/page/7448135/Selected%20Bibliography%20on%20Displacement%20Caused%20by%20Development%20Projects.

⁸ As part of his struggle to achieve greater recognition of social and environmental conflicts, Joao Martinez Alier, in the framework of the European project of Environmental Justice Organizations, has developed the interactive project, 'Environmental Justice Atlas' (https://ejatlas.org). More than 300 of the conflicts described on this platform are from Latin America and 122 correspond to cases in Colombia. The described conflicts fit into the following categories: mineral ores and building extractions, waste management, biomass and land conflict, fossil fuels and climate justice, water management, infrastructure and built environment, tourism recreation, biodiversity conservation conflicts and industrial and utilities conflicts.

⁹ The Plan Colombia was launched in 2000 by the governments of Colombia and the United States to fight drug mafias and leftist guerrillas. One component of the programme was designed to destroy coca and poppy crops cultivated to produce cocaine through spraying concentrated glyphosate from aeroplanes.

¹⁰ This section is a reworking of CNMH, 2016, 455–472.

¹¹ Existing studies have corroborated a direct relation between displacement due to the armed conflict and the presence of palm crops in the following cases: Montes de María, Bajo atrato chocoano, the department of Cesar, Magdalena (banana area) and Nariño (Goebertus, 2008; Maughan, 2011; Verdad abierta, 2013).

¹² This last objective is called into question, as 'the production of biofuel is not currently competitive despite receiving fiscal incentives. The production of palm oil has a modest impact on local employment, and its main impact as a supply chain is for fertilizers produced outside of the region' (Benavides, 2010, p. 25).

¹³ This type of crop requires less wage-earning labour and fewer workers with specific technical abilities who are not found in these regions. Regarding less employed labour, it is estimated that 'while 1.5 workers per hectare are necessary for banana farming, you only need one worker for every 10 hectares in palm oil farming' (Goebertus, 2008, p. 167).

¹⁴ Not all palm plantations develop through deforestation. Preliminary studies on the transformation of land use through satellite images show that most palm crops in the northern part of the country were implemented in areas previously used for cattle (López, 2010).

15 The accountability office of the republic thoroughly analyses this issue and calls on the state to carry out a macroeconomic analysis, including a calculation of the loss of environmental services by kind of resource exploitation. See Contraloría General de la República (2012).

16 For example, there are cases in the Antioquia region where peasants who have returned to their lands have been unable to undertake agricultural projects that would guarantee them economic sustainability because of the presence of electrical micropower that have changed the water regime in the region and have limited the access to water.

References

Benavides, J. (2010) The economic development of Orinoquia, Executive Summary, Presidential debates. CAF-Fedesarrollo. Bogotá, Colombia.

Camargo, A. and Ojeda, D. (2017) Ambivalent desires: state formation and dispossession in the face of climate crisis. *Political Geography* 60, 57–65.

Centro Nacional de Memoria Histórica (CNMH) (2015) Una nación desplazada: informe nacional del desplazamiento forzado en Colombia. CNMH-UARIV, Bogotá, Colombia.

Centro Nacional de Memoria Histórica (CNMH), Culma, E. and Guerra, J. (2015) Petróleo, coca, despojo territorial y organización social en Putumayo. CNMH, Bogotá, Colombia.

Centro Nacional de Memoria Histórica (CNMH), Londoño Botero, R., Castro, C., Delgado, A. and Landínez, J. (2016) Tierras y conflictos rurales. Historia, políticas agrarias y protagonistas. CNMH, Bogotá, Colombia.

Contraloría General de la República (2012) Informe del estado de los recursos naturales y del ambiente 2011–2012. Contraloría General de la República, Bogotá, Colombia.

Comisión Nacional de Reparación y reconciliación (CNRR) (2009) El despojo de tierras y territorios. Aproximación conceptual. CNRR, IEPRI, Universidad Nacional de Colombia, Bogotá, Colombia. Available at: www.centrodememoriahistorica.gov.co/descargas/informes2010/tierra_conflicto/despojo_tierras_baja.pdf (accessed 19 March 2018).

Comisión Nacional de Reparación y Reconciliación CNRR – Grupo de Memoria Histórica GMH, Machado, A. and Meertens, D. (2010) La tierra en disputa, Memorias del despojo y resistencias campesinas en la costa caribe 1960–2010. CNRR, CMH, Fundación Semana, Taurus, Bogotá, Colombia.

Culma Vargas, E. (2010) La presencia de las AUC en Putumayo a partir de la entrada en vigencia del Plan Colombia. BA Thesis. National University of Colombia, Bogotá, Colombia.

De Moor, N. and Cliquet, A. (n.d.) Environmental displacement: a new challenge for European migration policy. Refugee Studies Centre, University of Oxford, Oxford, UK.

Defensoría del Pueblo (2002) Resolución Defensorial Nacional No. 026. Derechos Humanos y Derecho Internacional Humanitario en el marco del Conflicto Armado y de las fumigaciones de los cultivos de coca en el Departamento del Putumayo. Bogotá, Colombia, 9 October 2002.

Duncan, G. (2006) *Los señores de la guerra*. Planeta, Bogotá, Colombia.

Fajardo, D. (2002) *Tierra, poder político y reformas agraria y rural*. Vol. 1. Instituto Latinoamericano de Servicios Legales Alternativos, Bogotá, Colombia.

García, H. and Calderón, L. (2012) Evaluación de la política de biocombustibles en Colombia. Centro de investigación económica y social, Fedesarrollo, Bogotá, Colombia. Available at: www.repository.fedesarrollo.org.co/bitstream/handle/11445/338/Repor_Octubre_2012_Garcia_y_Calderon.pdf?sequence=3&isAllowed=y (accessed 19 March 2019).

Gemenne, F., Brücker, P. and Ionesco, D. (eds) (2013) The state of environmental migration 2013 – a review of 2012. IDDRI-SciencesPo, Paris, France and IOM, Geneva, Switzerland.

Goebertus, J. (2008, January–June) Palma de aceite y desplazamiento forzado en zona bananera. *Colombia Internacional, Revista del departamento de Ciencia política*, Number 67. Department of Social Sciences, Universidad de los Andes, Bogotá, Colombia, pp. 152–175.

Gutiérrez Sanín, F. (2014) Propiedad, seguridad y despojo: el caso paramilitar. *Revista Estudios Socio-Jurídicos* 16(1), 43–74.

Human Rights Network of San Miguel, Putumayo (2013, 9 November) Denuncia pública. *La Dorada*.

Inter-Church Commission (2007) Agribusiness of palm and banana in Bajo Atrato. Environmental and socioeconomic impacts. Working paper. Available at: www.semillas.org.co/apc-aa-files/5d99b14191c59782eab3da99d8f95126/palma_y_bio_1.pdf. (accessed 19 March 2018).

Kalmanovitz, S., and López, E. (2006) *La agricultura colombiana en el siglo XX*. FCE, Bogotá, Colombia.

López Duque, A. (2010) Estimación de conflictos de uso de la tierra por dinámica de cultivos de palma africana usando sensores remotos, Caso: Departamento del Cesar. Universidad Nacional de Colombia, Medellín, Colombia.

Maughan, M. (2011, April) Land grab and oil palm in Colombia. Presented at the Global Land Grabbing, University of Sussex.

Organización de las Naciones Unidas (ONU) (1998) Principios rectores de los desplazamientos internos. Principios Deng. Oficina de Coordinación de Asuntos Humanitarios de las Naciones Unidas (OCHA), New York.

Restrepo, J.C. and Bernal Morales, A. (2014) *La cuestión agraria*. DEBATE.

Reyes, A. (2009) *Guerreros y campesinos. El despojo de la tierra en Colombia*. Norma – Fescol, Bogotá, Colombia.

Robinson, C. (2003) Risks and rights: the causes, consequences, and challenges of development-induced displacement. In: *The Brookings Institution-SAIS Project on Internal Displacement*. Available at: www.brookings.edu/fp/projects/idp/articles/didreport.pdf (accessed 19 March 2018).

SIDIH-OCHA Colombia (2009) Perfil departamental. Sistema Integrado de Información Humanitaria para Colombia. Available at: http://colombiassh.org/info (accessed 19 March 2018).

Soler, J.P. and León, D. (2009) Impactos ambientales de la expansión de palma aceitera en el Magdalena Medio, hablan los pobladores, Estudio de caso Las Pavas, municipio El Peñón, Bolívar, Programa de desarrollo y paz del Magdalena Medio PCPMM. Program of Development and Peace of Magdalena Medio PDPMM, Strategy of Defense of the Right to Land and Territory, under the coordination Elizabeth Ruiz Thorrens. Available at: http://prensarural.org/spip/IMG/pdf/Informe_de_Impactos_Ambientales_Version_FINAL._1_.pdf (accessed 19 March 2018).

Terminski, B. (2012, 23 March) Environmentally-induced displacement, theoretical frameworks and current challenges. Research paper. Centre d'Etudes de l'Ethnicite et des Migrations, Université de Liège. Available at: www.cedem.ulg.ac.be (accessed 19 March 2018).

Twomey, H. (2014, July) Displacement and dispossession through land grabbing in Mozambique. In: *The Limits of International and National Legal Instruments*. Refugee Studies Centre, University of Oxford, Oxford, UK.

Van Hear, N. (1998) *New Diasporas: The Mass Exodus, Dispersal and Regrouping of Migrant Communities*. University of Washington Press, Seattle, Washington.

Verdad abierta (2013, 31 October) El Copey: tierra de desapariciones, despojo y muerte. Available at: www.verdadabierta.com/masacres-seccion/5001-el-copey-tierra-de-desapariciones-despojo-y-muerte (accessed 19 March 2018).

World Bank (2017) *Forcibly Displaced: Toward a Development Approach Supporting Refugees, the Internally Displaced, and Their Hosts*. World Bank, Washington, DC.

Appendix 1. Leaders assassinated in 2016

Leader	Social struggle	Place
Erley Monroy	Leader belonging to the ASCAL-G Environmental Small Farmer Association of Losada Guayabero. Monroy defended national parks (Tinigua, Picachos and La Sierra de la Macarena), and recently protested mining and hydrocarbon exploitation in order to protect these areas. He also demanded the establishment of a small farming reserve between the Losada and Guayabero Rivers	San Vicente del Caguan – Caquetá
Joel Meneses and two other people from the same social movement (Nereo Meneses and Ariel Sotero)	Leader of the Committee for the Integration of the Colombian Massif (CIMA) and the Committee of Small Farmer and Indigenous Process of Almaguer (Procaminas). Prior to his assassination, he had been threatened because he defended water sources from illegal mining	Almaguer – Cauca
Nestor Iván Martínez	Leader of the community council of black communities in the middle of the department of Cesar. He had publicly denounced the expansion of coal mining by the multinational corporation, Drummond	Chiriguaná – Cesar
Wiliam García Cartagena	Member of the Committee for Human Rights of Segovia, municipality of the department of Antioquia. Defender of small informal miners	Segovia – Antioquia
Guillermo Veldaño	Community leader of Puerto Asís (Putumayo). He was part of the Small Farmers Union and fought to defend water sources and a dignified quality of life in his region	Puerto Asís – Putumayo
Aldemar Parra García	President of the Beekeeping Association of El Hatillo, in the department of Cesar. He received threats because of denouncements he had made of environmental contamination due to mining by the Pribbenow, Drummond and Colombian Natural Resources (CNR) corporations, of changes in agricultural land, and of the proposal to resettle the community due to the contamination	El Paso, Cesar
Hernan Agames	Leader of the Small Farmer Association of Southern Córdoba and the Committee of Coca Cultivators. He was an opponent of mining	Montelíbano – Córdoba

18 Prior Consultation and the Defence of Indigenous Lands in Latin America

Marcela Torres Wong*
Latin American Faculty of Social Science, Ciudad de México, Mexico

Introduction

For many indigenous rights supporters, the right to prior consultation recognized through the International Labour Organization (ILO) Convention 169 (1989) was a suitable mechanism to keep indigenous territories free of industrialized resource extraction. This study evidences, however, that the implementation of prior consultation procedures in Latin America does not deter the advancement of extractive industries. Instead, these procedures generally result in indigenous acceptance of extractive projects. In turn, our findings indicate that extractive industries are only blocked when prior consultation does not take place.

International non-governmental organizations (NGOs) advocating for the implementation of prior consultation regimes in Latin America have relied on two assumptions. First, and typical of industrialized societies, is that institutions are born strong (Levitsky and Murillo, 2013). Second, is that indigenous peoples are both environmentalists and communitarian (Eisenstadt, 2011).[1] Yet, we find that the role of the extractive industry sector in sustaining Latin America's economic growth influences prior consultation outcomes in spite of national political differences. Past research shows that Latin America's weak institutions enable governments to non-enforce legislation according to the interests they have over the matters at stake (O'Donnell, 1993; Levitsky and Murillo, 2013). In addition, previous studies have shown that most indigenous groups are actually in favour of extractive projects as they represent opportunities to access resources (Arellano-Yanguas, 2011; Arce, 2014). For these reasons, groups who are consulted by their governments are in favour of extraction and end up approving extractive projects.

This chapter is organized as follows: I first situate extractive industries in the Latin American context. Then I distinguish between indigenous groups seeking to profit from the extractive industry, referred to as 'pro-extractivist', from those seeking to prohibit extraction on their lands, referred to as 'anti-extractivist'. In this section, we find that viable agricultural economies influence anti-extractivist attitudes in opposition to subsistence agricultural economies, which generally lead to pro-extractivist stances. In the second section, I explore the development of the right to prior consultation in the Latin American region. I analyse the implementation of prior consultation in the cases of Bolivia, Peru and Mexico. These three countries exhibit wide variation in the political influence of indigenous movements, legal design of prior consultation rights and government ideology – three variables that political scientists associate with better outcomes regarding the

---

* marcela.torres@flacso.edu.mx

advancement of indigenous rights (Yashar, 2005; Van Cott, 2008; Lucero, 2009; Albó, 2012; Jaskosky, 2013; Belle-Antoine, 2015; Due Process of Law Foundation, 2015). This section shows that national variation does not seem to account for different prior consultation results. As of the time of writing of this chapter, all 66 prior consultation procedures conducted in Bolivia, Peru and Mexico over hydrocarbon and mining projects resulted in indigenous approval. Some of these procedures resulted in economic benefits for indigenous groups, whereas others served as a formality to advance extraction. Finally, the chapter shows that those groups with viable agricultural economies in place, and who do not participate in prior consultation, are able to ban extraction within their lands.

Extractive Industries and Prior Consultation in Latin America

In most Latin American countries, the expansion of extractive industries has been the direct consequence of neoliberal reforms implemented in the 1990s (Slack, 2009; Arce, 2014). Neoliberal policies sought to limit the scope of the state in order to grant greater autonomy to private investors and achieve a dramatic increase in export revenues. Such policies included the privatization of strategic mining and hydrocarbon industries. Latin American economic growth since the turn of the century has relied primarily upon the export of commodities, fuelled in part by skyrocketing Chinese demand.[2]

As of 2016, 27% of global investment in mining and 52% of the worldwide investment in gold and copper was concentrated in Latin America. Foreign investment in mining projects was followed by the introduction of the controversial open pit method and the use of cyanide in extractive operations. The use of sophisticated technology allows the removal of sizeable expanses of land to extract as much mineral as possible, negatively impacting the agricultural economies that are often found near mining projects (Bury, 2004; Bebbington *et al.*, 2008; Himley, 2010; Perreault, 2014). The need for great amounts of water for extractive operations and the discharge of waste water into natural waterways often impact the quantity and quality of water sources needed by adjacent agricultural communities (Bridge, 2004). In addition, the modern mining industry is capital-intensive and does not require a large labour force.[3]

The lack of job opportunities for locals and ecological degradation to agricultural lands together are cause of strong opposition by surrounding indigenous communities. The mining industry is responsible for at least 200 socio-environmental conflicts taking place in the Latin American region (Observatory of Mining Conflicts in Latin America, n.d.; 'Mapa de conflictos mineros, proyectos y empresas mineras en América Latina'). Unfortunately, conflicts over mining projects often engender high levels of violence as they tend to radicalize. State repression frequently takes place and confrontations generally result in deadly outcomes (Observatory of Mining Conflicts in Latin America, n.d.).

The hydrocarbons industry is also central to the Latin American economy. The region holds 20% of proven global oil reserves, and 4% of global gas reserves (BP Statistical Review of World Energy, 2016). Whereas the majority of large-scale mining projects is controlled by transnational corporations, the majority of hydrocarbon production is carried out by state-owned companies (Bebbington and Bury, 2013, p. 54). State control of this sector has spurred a sort of 'resource nationalism' in countries such as Bolivia and Mexico, preventing anti-oil mobilization from eventuating (De la Fuente, 2013; Perreault, 2013). In comparison with the mining industry, social conflicts over this industry have been less numerous (De la Fuente, 2013; Perreault, 2013; The Dialogue, 2015; Eisenstadt and West, 2017).

Still, environmental contamination is associated with this industry. In Latin America, hydrocarbon extraction is generally advanced in isolated areas where there is little oversight of the extractive process (Bebbington and Scurrah, 2013, p. 173). Numerous ethnic groups living in places such as the ecologically fragile Amazon basin have to suffer the contamination of their rivers and fauna due to oil spills, pipeline ruptures and the deficient closure of oil wells (Environmental Health News, 2014; Eisenstadt and West, 2017).[4] Moreover, falling oil prices since 2014 have diminished funds available for governments in Venezuela, Mexico, Argentina, Brazil, Bolivia, Colombia and Ecuador (BBC Mundo, 2005). Currently, these countries struggle to maintain incomes generated from the exports of oil and gas while attempting to fulfil their domestic energy demands. Exploration activities are encouraged in areas that formerly prohibited extractive industries. National parks and ecological reserves in the Amazon – shared by Brazil, Peru, Bolivia, Ecuador, Colombia, Venezuela, Surinam – and El Chaco desert – shared by Bolivia, Paraguay and Argentina – are increasingly used to search for new energy sources.

What Do Indigenous Peoples Fight for?

Indigenous movements in the developing world are typically framed as struggles against the dispossession of their natural resources (Harvey, 2003). According to this framework, indigenous communities mobilize against the extraction of minerals and hydrocarbons to avoid losing control of their agricultural lands and water resources (Bebbington et al., 2008; Bebbington and Bury, 2009). Yet, some scholars argue that no real 'ecological communities' would exist if the extractivists offered those communities legitimate opportunities to escape poverty (Devlin and Yap, 2008, p. 20). Following this argument, research on conflicts surrounding extractive industries demonstrates that most of anti-mining protests are undertaken with the objective of obtaining a share in the mining revenues (Arellano-Yanguas, 2011; Arce, 2014).

Research also shows, however, that barring economic depravity as the root cause for mobilization, there are communities that are sceptical of the environmental sustainability of extractive projects. Moreover, such communities do not believe that these industries are capable of improving their current living conditions (Zavaleta, 2014). Confirming these findings, research on conflicts over extractive industries in Peru shows that a small portion of resource-based protestors actually seek to block extraction to protect their natural environment (Arellano-Yanguas, 2011; Arce, 2014). Yet, distinguishing anti-extractivist groups from those seeking to obtain economic resources is not an easy task. This is because the main strategy of pro-extractivist groups is pretending to be radical ecologists defending their lands. It is likely that this strategy allows them to gain the support of a broader audience and to better position the movement to put pressure on the state (Arellano-Yanguas, 2011, p. 154).

This chapter takes the position that most indigenous groups are pro-extractivists but there are also indigenous communities mobilizing against extractive projects to prevent the degradation of their lands. This duality within indigenous communities corresponds to the different conditions existing in indigenous territories. Social movement theorists conceptualize social struggles as responses to disruptive changes that either pose a new threat to some segment of the population or grant new opportunities/leverage to potential challengers (Tilly, 1978; Tarrow, 1998; McAdam, 1999; Goldstone and Tilly, 2001). Whether extractive industries pose a threat or represent an opportunity depends on the socioeconomic reality the group faces at the time disruptive changes takes place.

Pro-Extractivist Indigenous vs Anti-Extractivist Indigenous

Economic structures of indigenous communities influence how they perceive the extractive industry sector. Viability of agricultural economies, market integration and access to information about the impacts of the extractive industry contribute to the formation of indigenous stances. Indigenous communities living on subsistence agriculture, fishing and hunting usually lack sufficient resources to cover their basic needs. Subsistence economies are usually found in geographically isolated areas that do not have access to broader markets, curtailing these communities' economic mobility. Extractive companies generally expand local economies. Employees of such companies require provision of lodging, food and entertainment. Increasing demands for services create conditions for the emergence of new businesses near the extraction sites (Wise and Shtylla, 2007, p. 9). The possibility of greater incomes prompts disadvantaged indigenous groups to accept an extractive project.

This is the case of most Amazonian tribes in Peru and Bolivia complying with hydrocarbon extraction in their lands. Geographic disconnection and insufficient subsistence economies push these groups to pursue resource extraction, with the expectation of greater access to jobs and services. Pro-extractivist indigenous are also found in places where extractive companies operate for long periods. An example of this are Guaraní indigenous communities in Bolivia, co-existing with the hydrocarbon industry for long periods, and using prior consultation to negotiate extractive resources. Over time, the presence of extractive companies generates local economic dependence and discourages the emergence of alternative sources of income (MMSD, 2002, p. 276). For these types of community, resource extraction does not represent a major threat to their living, as they have little to lose and potentially much to gain from relinquishing their land.

Previous case studies show that anti-extractivist indigenous communities constitute a minority of cases of indigenous mobilization. Most of these studies depict ecologically oriented communities as the most economically fragile in the developing world, fighting to preserve their means of subsistence (Goebel, 2010, p. 130).[5] Generally, the assumption that indigenous ecological protestors are extremely poor is based on misleading or outdated statistical information on incomes of rural households in Latin America (Grosskoff, 1998).[6] In turn, previous research on extractive conflicts in Latin America suggests that ecological protestors have viable economic systems in place that are threatened by extractive companies (Arellano-Yanguas, 2011; Arce, 2014). Indigenous peoples engaging in anti-extractivist mobilization are convinced that the damages caused by resource extraction would be greater than the benefits they could reap from it (Zavaleta, 2014, p. 6).

For instance, Arce finds that anti-mining struggles are likely to emerge and succeed in cases in which well-established, agricultural-based social organizations exist, as mineral extraction jeopardizes local control of agricultural lands (Arce, 2014, p. 22). Following this argument, radical opposition to 'big mining' projects (particularly to open pit mining) originates because opponents perceive that these projects threaten existing agricultural economies. 'New mining' uses highly sophisticated technology that makes local labour redundant, but it still requires extensive amounts of water and removal of enormous pieces of land to extract the largest quantity of minerals (O'Huallachain and Matthews, 1996).

We build upon these findings yet we conclude that agriculture must represent a viable economic system for the indigenous (in opposition to subsistence agriculture) in order to create conditions for anti-extractivist mobilization. This requires that agriculturally based economies exhibit specific characteristics. Mexican anthropologist Aracely Burguete argues that indigenous communities with access to external revenue sources are most likely to be defensive of their lands and natural resources. According to Burguete, communities with 'mixed economies' have more to lose from damage to their environment than communities that only have subsistence economies like agriculture, fishing or hunting, to cover their basic needs (Aracely Burguete, personal communication, 11 March 2015).

Connections to major cities – crucial for indigenous populations to trade their products and obtain cash to buy products lacking in their own communities – bolster indigenous economies. When connectivity exists, migration rates are generally low as people do not have to leave their hometown in search of jobs.[7] In addition, connectivity allows indigenous populations to have access to circulating narratives about the negative ecological impacts of extractive companies. This argument is illustrated by case studies of powerful anti-mining mobilizations emerging within the municipalities of Challapata in Bolivia and San Esteban de Chetilla in Peru (see the cases below). Both municipalities, catalogued as extremely poor by official statistics, exhibit well-established 'mixed economies' based on both agriculture and informal commerce. Connections to capital cities allow informal indigenous traders to sell their products abroad without having to migrate to other cities searching for jobs, and also grant them access to information about environmental impacts associated with open pit mining.

The emergence of non-traditional economic sectors among the indigenous also create new conditions for anti-extractivist mobilization to manifest. Ecological tourism, for instance, is emerging within some indigenous municipalities as an additional source of income. The purpose of ecotourism is the exchange of resources between indigenous populations in need of money, and foreigners seeking untouched environments and sustainable use of natural resources (Valcuende del Río et al., 2012). The implementation of extractive projects represents a threat to ecotourism, as it damages the image of indigenous communities as ecologically harmonious (see the case below).

Hence, conditions for anti-extractivist indigenous mobilization are found within indigenous communities exhibiting agricultural economies that provide them access to cash incomes. Connections to capital cities, access to broader markets and low migration rates are typically associated with this type of economy. When extractive industries threaten this type of economy, anti-extractivist movements are likely to emerge. Table 18.1 illustrates this argument.

Development of the Right to Prior Consultation in Latin America

Whereas the United States and Canada have refused to date to sign ILO Convention 169, 14 out of the 22 states that ratified this international law in the 1990s are Latin American. Domestic implementation of the right to prior consultation, however, did not start until almost two decades later in this part of the world. Violence between indigenous communities and extractive companies over the control of resource-rich lands skyrocketed during the 2000s commodity boom (Arellano-Yanguas, 2011, p. 22). Latin American governments were then forced to pass legislation in order to carry out formerly ignored prior consultation procedures. Currently, Latin America is the region with the greatest legal development of prior consultation regimes (Due Process of Law Foundation, 2015).

Latin American countries did not use the same legislative formula to address the right to prior consultation. The most progressive legislation, associated with leftist governments with pro-indigenous rights agendas, incorporated this right into constitutions (Bolivia and Ecuador); more conservative governments, which are usually associated with neoliberal ideologies, passed a framework law to regulate prior consultation procedures (Peru, Chile and Colombia), or only recognized this right through sectoral legislation (Mexico), whereas still others did not implement prior consultation at all. Within the set of countries adopting legal regulations of the right to prior consultation, the legislative spectrum of these initiatives is illustrated by Fig. 18.1.

Prior Consultation in Bolivia, Peru and Mexico

These countries were selected for comparison, in part due to their wide variation across the three variables that arguably influence prior consultation outcomes: the national influence of indigenous movements, the legal design used to address the right to prior consultation and the ideology of the government in office. The evidence

Table 18.1. Economic structures and indigenous attitudes towards the extractive industry.

Type of local economy	Characteristics	Indigenous attitudes towards the extractive industry
Subsistence agriculture (farming, fishing, hunting)	Isolation No access to cash incomes	Pro-extractivist
Subsistence agriculture coexists with the extractive industry	The population entirely depends or supplements their subsistence activities with jobs in the extractive industry	Pro-extractivist
Viable local economy (flourishing agriculture, commerce, ecotourism)	Access to cities and broader markets Low migration rates	Anti-extractivist

Progressive ◀───────────────────────────▶ Conservative

Constitution   Framework law   Sectoral legislation
(Bolivia and Ecuador) (Peru, Chile, Colombia) (Mexico)

Fig. 18.1. Legislative spectrum

presented in this chapter demonstrates, however, that in spite of national differences and legislative models, in none of these countries do prior consultation procedures prevent extractive projects from being implemented.

Indigenous movements in Bolivia have had a much greater influence on national politics than respective movements in Mexico or Peru. Such movements led to the development of one of the most progressive prior consultation systems in all of Latin America, as well as the 2006 election of a leftist indigenous president, Evo Morales. The hydrocarbon industry was nationalized and indigenous communities living near hydrocarbon reserves were granted veto power over gas and oil extraction. Unprecedented constitutional incorporation of indigenous rights took place, including the right to prior consultation. Such changes led many scholars to argue that Bolivia had become an experimental laboratory for the realization of long-standing indigenous economic and political demands (Eisenstadt, 2011).

Peru differs from Bolivia across the three theoretical variables outlined above. Indigenous mobilization remains weak at the national level, yet conflicts between indigenous communities and extractive companies have increased since the democratic transition in 2001 (Yashar, 2005; Paredes, 2015). Prior consultation was enacted in 2011, but did not grant veto power to consulted indigenous populations. Nor was prior consultation added to the Constitution. Neoliberal policies of the 1990s remain in force to this day. Therefore, extractive industries are still in hands of foreign companies.

Mexico also differs from Bolivia and Peru across the variables detailed above. Indigenous movements have not reached the same level of political salience as in Bolivia, yet ethnic organizing at the local level is stronger than it is in Peru (Yashar, 2005; Silva, 2009). Ethnic insurgency made it so that indigenous rights were partially regulated through electoral indigenous autonomy. However, prior consultation regimes remain almost completely unimplemented. Unlike the cases of Bolivia and Peru, there is no constitutional mandate or framework law to regulate prior consultation procedures in Mexico despite increasing political violence over control of resource-rich lands. Sectoral legislation exists, yet it does not specify whether the right to prior consultation grants indigenous groups veto power. Neoliberal reforms in Mexico have been in force since the 1980s. Despite never having been as radical as those in Peru, such reforms have expanded under the current Peña Nieto government.

Pro-Extractivist Indigenous and Prior Consultation in Bolivia, Peru and Mexico

Bolivia

Although indigenous veto was legally granted in this country and, in theory, indigenous communities could say 'no' to extraction, all 52 prior consultation procedures resulted in indigenous acceptance of the projects (Falleti and Riofrancos, 2014; CEDIB, 2015). Within the hydrocarbon industry, the majority of these procedures were carried out with the Guaraní indigenous group – a sizeable, politically skilled and highly mobilized population living in the Chaco region. The Guaraní are the third largest indigenous group in Bolivia, after the Quechua and the Aymara.[8] They are settled agriculturalists, and rely on semi-subsistence-based farming, hunting, fishing and the gathering of wild plants and fruits (Perreault, 2008). Extraction of hydrocarbons is not new for the Guaraní as such activity dates back to the period of dominance under the estate system (Marco Gandarillas, personal communication, 16 April 2015). As Guaraní communities started claiming lands enriched with sizeable gas reserves, violent confrontation between these communities and extractive operators began. In these contentious relationships, indigenous leaders used environmental discourses to negotiate economic compensation with oil operators. For the Guaraní, resources from oil companies are used to supplement their incomes. With the legalization of prior consultation in 2005, it is no surprise that the Guaraní use these procedures as platforms to negotiate better economic terms with oil operators.

Whereas Guaraní territories have benefited from prior consultation, not all indigenous groups benefit in the same way. This is the case of prior consultation with the Amazonian Mosetén tribes over the Lliquimuni oil block in 2008. Such consultation resulted in acceptance of the project, yet politically weak indigenous groups were not able to obtain economic compensation in exchange (Diaz-Vidaurre, 2010). Unlike the Guaraní, hydrocarbon extraction was new for the Mosetén and indigenous leaders lacked the capacity to negotiate extractive resources.

Prior consultation, although on the books in Bolivia by 2009, was not implemented over mining projects until June 2015, when it was used by the indigenous community of Huacuyo (Municipality of Antequera, Department of Oruro).[9] As part of fieldwork conducted by the author, this community was visited the week after prior consultation was scheduled to take place. There, interviews were carried out with several municipal officials, who stated that they had not yet been informed of any consultation process in the area (Municipal employees, personal communication, 8 July 2015). Once in Huacuyo, it was verified that small-scale extraction of minerals was already taking place despite the fact that no prior consultation with the local population had been undertaken. It was also verified that members of the mining cooperative Minera Monserrat Ltda, the company requiring permission to operate in that jurisdiction, were also members of Huacuyo. Such members were extracting the minerals yet were also the indigenous members to be consulted about mineral extraction in their lands. Back in Oruro city, the installations of the State Office in Charge of Mining Affairs (Autoridad Jurisdiccional Administrativa Minera or AJAM) were visited, to find out why the prior consultation procedure did not occur as was announced in the newspapers. State officials said that, due to administrative delays, they had not been able to conduct the prior consultation yet, but they also said that 'prior' consultation was going to be carried out 'later' that month. On 19 July 2015, the Bolivian government announced that the first prior consultation over mining had been successfully completed in the indigenous community of Huacuyo, and that mining operations had been approved by the community (La Razón, 2015).

Huacuyo represents a large proportion of indigenous communities in the Bolivian highlands. For these communities, prior consultation is just a formality to advance small-scale mineral extraction. Indigenous actors are main stakeholders in resource extraction. Thus, it is unlikely that they will use prior consultation to oppose extractive operations.

Peru

Until June 2016, prior consultation regarding hydrocarbon projects has been used a total of 11 times in Peru. Hydrocarbon projects were accepted by consulted indigenous groups in all 11 cases (Perupetro Database on Prior Consultations, 2016). Most of these consultations were conducted in the Amazon basin. However, while Bolivian Guaraní were able to derive some benefits from prior consultation, Amazonian groups from Peru have not been as fortunate. Only one of Peru's 11 prior consultations over hydrocarbon projects resulted in state distribution of resources in favour of consulted indigenous communities. Even though indigenous consent is not required in extant Peruvian legislation, the Achuar, Quechua and Kiwchua led several waves of protests in parallel with prior consultation, finally forcing the state to comply with their demands.[10]

Most Amazonian groups are primarily dependent upon subsistence activities. Due to extreme geographic conditions the Amazonian indigenous population is disconnected from broader economic markets. In this context, the case of the Achuar, Quechua and Kiwchua living close to oil block 192 is exceptional. Similar to the Guaraní in Bolivia, the hydrocarbon industry was not new for these three groups. Oil block 192 (formerly Block 1-AB), the biggest oil producer in Peru, has been active since 1970 and indigenous political organizing in this area was shaped by interaction with oil companies (Bebbington et al., 2012). By the time these Amazonian groups engaged in prior consultation with the state, they already had plenty of experience bringing forth demands for economic compensation to make up for ecological damage caused by industrial activity (Scurrah and Chaparro, 2011).

In the mining sector, Minister of Mining Rosa Maria Ortiz decided to undertake the first ever prior consultation in the highlands of the country in September 2015. This decision was made 2 months after a group of journalists broadcasted that two former ministers had secretly granted 25 mining concessions within lands belonging to Quechua communities. These communities were legally entitled to prior consultation, however, the government failed to uphold such rights, completely skipping over the consultation process (Ojo Público, 2015). Amid these reports, prior consultation with the indigenous community of Parobamba (Province of Calca, Department of Cusco), began on 8 September 2015. Prior consultation included the legal authorization to Canadian mining company, Focus, to build the 'Aurora' mine within Quechua communities. A month after the initiation of the procedure, community members of Parobamba accepted the mining project. Questions over the validity of the procedure were immediately posed by several NGOs, which argued that there was not much left to consult about with indigenous members of Parobamba by the time consultation began, as environmental impact studies had already been approved and legal authorization for the use of communal lands had already been granted by the government (Cooperaccion, 2015). Three more prior consultation procedures over the mining projects 'Misha', 'Toropunto' and 'La Merced' followed Parobamba's consultation delivering similar results. In none of these cases was extraction forbidden, neither were the consulted groups granted economic compensation for accepting mining operations within their lands.

Like in Bolivia, Peru's four prior consultations over mining projects resulted in approval of extraction. Some NGOs argue that these consultations are 'window-dressing' to advance aggressive extractivism into new regions. Studies also show that consulted indigenous groups lacked capacitation and sufficient information about the implications of consultation procedures (Ocampo and Urrutia, 2016). Until now in Peru, there is no evidence of how these procedures would result if they were applied within regions where salient anti-mining movements exist and the population is politically skilled.

Mexico

As of the autumn of 2014, the Mexican government has completed one prior consultation procedure in the hydrocarbon industry. The consulted project was a gas pipeline to be built across lands belonging to the Yaqui tribes located in the state of Sonora who ended up complying with the consulted project. Yet, unlike Guaraní or Amazonian tribes, Yaqui members do not live off subsistence agriculture. Most Yaqui communities currently obtain cash incomes from renting their lands to farmers in Ciudad Obregon. This ethnic group has a long-standing reputation of political mobilization. The Yaqui resisted the influence of Spaniard colonizers and the Catholic Church during the colonial period, as well as attempted land dispossession by the Porfirio Díaz government at the end of the 19th century (Memoria Política de México, n.d.). The political impact of Yaqui struggles led President Lázaro Cárdenas to legally recognize Yaqui ownership of their lands and political self-government as early as 1940 (De la Maza, 2004).

Prior consultation with the Yaqui lasted 14 months and was formally concluded with an economic agreement. Divisions and conflict among Yaqui members followed prior consultation, however. A group of Yaqui communities did not agree with the terms agreed in negotiations. Violent confrontations between factions left one dead and several injured in 2016 (La Jornada, 2016). Whereas the Yaqui still fight over the allocation of state resources, there is no evidence that those resources are being used to address potential environmental impacts of the gas pipeline on Yaqui lands.

Within the mining industry, Mexico has not carried out any prior consultation until the writing of this chapter. Legislation regulating prior consultations over mining projects does not exist, and the Mexican government continues to avoid its international commitments.

Table 18.2 provides a summary of cross-national variation in the three variables discussed in this chapter – influence of indigenous movements in national politics, legal design of the right to prior consultation and political ideology of the government – as well as prior consultation outcomes in the three countries.

Who Can Benefit from Prior Consultation?

Whereas indigenous communities who live off subsistence agriculture are likely to have pro-extractivist stances, not all of them are equipped to benefit from these industries. Pre-existing

Table 18.2. Summary of cross-national variation and prior consultation implementation.

	Influence of indigenous movements	Indigenous consent	Government ideology	Implementation outcomes
Bolivia	Far-reaching	Required	Statist	All consultations resulted in indigenous approval of extractive projects
Peru	Incipient	Not required	Neoliberal	All consultations resulted in indigenous approval of extractive projects
Mexico	Moderate	Not defined	Neoliberal	All consultations resulted in indigenous approval of extractive projects

negotiation capacities enable indigenous groups to obtain better economic terms in cases of resource extraction. Cases such as the Guaraní in Bolivia, the Achuar, Kiwchua and Quechua in Peru, and the Yaqui people in Mexico detailed above, exemplify this argument. Usually, negotiation skills are found in places where the population has experience with extractive industries.

In turn, politically weak indigenous population are generally unable to obtain favourable economic terms from prior consultation. Recall that 10 out of the 11 prior consultations carried out over hydrocarbon projects in Peru left the indigenous in the same situation they were in before accepting the extractive project (or worse due to environmental contamination likely to result from hydrocarbon extraction). Likewise in Bolivia, the Mosetén did not obtain extractive resources from prior consultation in the same way the Guaraní did. This is likely to happen in cases where indigenous communities lack sufficient negotiation skills and experience with extractive companies.

Anti-Extractivist Movements and Prior Consultation

Previous studies demonstrate that prior consultation procedures produce divisions among indigenous communities increasing internal tensions within indigenous organizations (Diaz-Vidaurre, 2010; Rodríguez-Garavito, 2011). Accordingly, anti-extractivist indigenous communities – aware of the negotiation inherent to these procedures and the negative impacts that they might have on their organizational structures – refuse to be consulted. Instead, they choose to rely on their native decision-making mechanisms. Through their Community Assemblies, ecologically oriented groups 'consult among themselves,' make decisions and form strategic alliances with other groups to build political power. Internal consultations have been used by approximately 700,000 people in Colombia, Peru, Bolivia, Ecuador, Mexico, Argentina, Guatemala and Canada to organize politically and fight mining (Mining Watch, 2012). Three cases from Bolivia, Peru and Mexico were selected that exemplify this argument.

Bolivia

In the indigenous municipality of Challapata (Oruro), the population opposed mining operations within their lands, contrasting with most pro-mining communities that define the Oruro region. Challapata has 27,046 inhabitants, the majority of whom belong to the Quechua indigenous group. Challapata is politically organized into a municipal council, headed by a mayor. At the same time, the population is organized into the 'ayllu' system, an ancestral model of political organization still in force in some parts of the Andes. In 1961, the Bolivian government carried out the building of the Tacagua dam, intended to improve water provision and develop the cattle and agriculture industry in Challapata. Following this, a third type of political organization was created. Several ayllu members organized into an Irrigators Association with the aim of administering the use of the dam and assign water ships to indigenous users (Madrid, 2014).

Over time, the effective management of the Tacagua Dam gave the Irrigators Association significant credibility vis-à-vis indigenous communities. Challapata's optimum soil conditions for agriculture and its proximity to the capital city of Oruro (only 2 h away) and to the department of Potosí (3 h away using the Potosí–Oruro highway) boosted the cattle raising and agriculture industries. The economy of the municipality flourished and the population is no longer dependent on subsistence activities. Basic services are mostly covered, communication services such as internet and radio are present and agrarian careers are available for the youth.

Since the 1990s, the Bolivian government has made several attempts to implement mining projects in Challapata. In 2011, through a General Assembly, indigenous leaders decided to prohibit all forms of mineral extraction. A resolution was approved stating that the native indigenous people of Bolivia prohibited the mining industry with the ultimate goal of protecting Mother Earth. Subnational authorities of the Oruro department, the National Council of Ayllus and Markas of the Qullasuyu (or CONAMAQ),[11] and environmental NGOs supported the indigenous anti-mining stance. The resolution of the General Assembly was later handed to President Evo Morales in a symbolic ceremony during his visit to the municipality (Madrid, 2014). After a decade of social mobilization, the government was finally persuaded to cancel the mining project. The right to prior consultation was not offered by the government. Neither was such a right demanded by Challapatan organizations. In the words of the anthropologist Emilio Madrid: 'The ayllus claimed that they had already agreed in their General Assembly that they did not want the mine, therefore there was nothing left for the government to consult' (E. Madrid, 2015, personal communication). Implicitly, prior consultation was perceived as a means of negotiation and indigenous leaders were not willing to trade their agricultural economy for uncertain benefits from the mining industry.

Peru

In the indigenous municipality of San Estaban de Chetilla (Cajamaraca), indigenous communities have resisted mining projects for more than a decade. Through Community Assemblies and the 'Rondas Campesinas' local system of justice, Chetilla's authorities have discouraged the entrance of mining employees to their lands. The population of Chetilla totals 4294, the majority of whom speak the Quechua language. Yet Chetillans do not identify with ethnically centred identities and instead they self-identify as peasants. Chetilla is politically organized into a municipal council headed by a mayor. Simultaneously, Chetilla is a peasant community ruled by a traditional type of political authority, elected through customary norms. The peasant political organization comprises a Community Assembly – in charge of communal lands – and a Rondas Campesinas justice system in charge of punishing local crime. The Community Assembly and the Rondas Campesinas are strongly rooted in Chetilla due to the effectiveness of their practices for the benefit of the people. For this reason, municipal authorities generally coordinate their decisions with traditional community leaders (A. Ramírez, 22 June 2014, personal communication).

Chetillans depend on agriculture and cattle raising for the most part, thus access to water is critical for the population. The proximity to Cajamarca city (only 2 h away), where a powerful anti-mining movement exists, allowed Chetillans contact with the discourse over the negative impacts that the mining industry has on water sources. Cajamarca has the largest reserve of gold in South America, which is controlled by the American mining company Yanacocha. Environmental disasters attributed to Yanacocha, as well as the numerous conflicts between the company and adjacent communities, have influenced radical anti-mining attitudes in Chetilla (H. Alcántara, 25 June 2014, personal communication).

In 2003, the government granted the Canadian Shield Company a mining concession over the Colpayoc Mountain located within Chetilla's lands. Exploration activities were initiated without consulting with the people. In a Community Assembly, traditional authorities decided to send Rondas Campesinas members to capture the engineers working in the Colpayoc. In an interview with Aníbal Ramírez, then-member of the municipal council, he recounts that hundreds of community members caught mining engineers. The leaders burned the engineers' mining equipment, took away their shoes and some of their clothing, and expelled them from the municipality. Again in 2009, the same company under the name of Estrella Gold Peru sent employees to Chetilla to resume exploration activities yet they received the same hostile treatment. Estrella Gold Peru withdrew from the project in 2013 arguing that 'social costs' were too high and that local conditions had to change in order to advance mining operations. In 2014, the company sold its mining concession to the Australian mining company Wild Acre, however, this company was also forced to postpone operations (Enlace Mineria, 2014).

Today, Chetilla's mineral reserves remain unexploited. While the government is reticent to apply prior consultations in Cajamarca, community demands for these procedures are not salient either. Only by relying on their customary mechanisms of decision-making, the political leadership in Chetilla managed to prohibit extraction more than once.

Mexico

In Capulálpam de Méndez (Oaxaca), indigenous municipal authorities engaged in several years of socio-legal mobilization finally forcing the federal government to suspend mining projects. The population of this municipality totals 3000 and ascribes to the Zapoteca group although most people speak Spanish and only a few still speak the Zapoteca language. The population is politically organized into a municipal council headed by a president, however, in the mid-1990s residents of Capulálpam de Méndez decided to convert to the Usos y Costumbres municipal system. Because of this, the population no longer abides by state rules to elect municipal authorities. Instead, such authorities are elected through customary norms and are accountable to a Community Assembly (Development Plan of the Municipality of Capulálpam de Méndez, 2009).

The quality of life is good in Capulálpam and main basic needs such as functioning education and health care systems are available. Sustainable use and sale of forest resources and ecotourism are the main economic activities (Development Plan of the Municipality of Capulálpam de Méndez, 2009). In 2002, the government granted the Canadian company Continuum Resources rights for exploration and exploitation of silver and gold in lands that were part of Capulálpam de Méndez. In 2005, municipal authorities filed a complaint against the government's decision to authorize mineral extraction within their lands. The argument used to question the legality of the mining project was that mining activities were causing damage to the environment. The lack of state response prompted indigenous communities to adopt forceful measures, which included seizing mining company offices, cutting off mine access to employees and state officials and blockading the highway connecting Capulálpam de Méndez with Oaxaca City (Aquino, 2011). The Federal Environmental Office PROFEPA was finally persuaded to inspect mining company operations. The agency verified that ecological damages had been caused by mining activities and suspended the mining project (La Jornada, 2013). Prior consultation was never conducted.

The cases of Challapata (Bolivia), Chetilla (Peru) and Capulálpam de Méndez (Mexico), illustrate the anti-extractivist type of indigenous group. The indigenous people in these municipalities have economic models in place based on the sustainability of environmental resources. Agricultural economies engender successful anti-extractivist movements to the extent that they are integrated to the market and provide indigenous population access to cash incomes. In all three cases, access to information over the impacts of mining on water sources fuelled anti-extractivism among local authorities who were able to mobilize support from the majority of community members. The population perceived that the benefits of mining would be less than the negative impacts these activities would have on their existing economic models. Negotiation was not an option for this type of population and prior consultation was not demanded. Instead, the communities decided to ban mining using their customary norms and projects were prohibited.

Conclusion

This chapter finds that prior consultation is not useful to prevent the expansion of extractive industries into indigenous agricultural lands in Latin America. On the contrary, the evidence suggests that prior consultation needs to be absent if the indigenous objective is to prohibit extraction.

In recent years, much attention has been given to the legal design of prior consultation. Indigenous rights activists and lawyers have pushed for more substantial recognition of indigenous voices in state policies. For many, this will only result by giving veto power to consulted indigenous groups. However, the policy implications of this study do not indicate that such changes will bring about different results. Not even in cases such as Bolivia, where indigenous consent was legally required by the 2005 hydrocarbon law, have prior consultation procedures been used to deter aggressive extractive policies.

The importance of the extractive industry conditions not only the way Latin American states address indigenous prior consultation rights, but also how indigenous groups engage with extractive projects. This chapter demonstrates that the driver of indigenous struggles is economic. Pro-extractivist groups mobilize to profit from extraction. For this type of actor, extractive projects come to supplement local income and even replace insufficient subsistence agricultural models. In these cases, mining and oil companies are seen as sources of jobs, services and infrastructure long denied by a failing state. For anti-extractivist groups, on the other hand, environmental protection is the driver of conflicts only to the extent that environmental resources offer venues to indigenous peoples for escaping poverty. As shown by the cases presented, market-based agricultural development is what principally allows indigenous groups to successfully push back against extractive activities.

The results of indigenous movements in Bolivia, Peru and Mexico demonstrate that as long as Latin American states and/or NGOs remain incapable of providing tools for economic viability to impoverished indigenous communities, extractivism will be advanced even in ecologically fragile areas. The implementation of prior consultation procedures is embedded in these conditions. In this vein, those groups concerned with the protection of indigenous agricultural systems should concentrate their efforts in making local agricultural models economically viable for indigenous communities.

Notes

[1] In this regard, the ILO Convention 169 reads: '(...) calling attention to the distinctive contributions of indigenous and tribal peoples to the cultural diversity and social and ecological harmony of humankind'.
[2] From 1990–2009, the percentage of extractive industries exports in the share of total exports of some Latin American countries doubled (Rights and Resources Initiative, 2013).
[3] Mining accounted for 0.6% of the workforce in Latin America, on average, from 2000–2010 (Bebbington and Bury, 2013, p. 52).
[4] One of the bloodiest conflicts over hydrocarbons in the Amazon was the Baguazo massacre in Peru in 2009, where 34 people were killed including indigenous protestors and police forces.
[5] The 'ecology of the poor' refutes the thesis proposed by Ronald Inglehart about poor societies being incapable of having post-materialist values such as ecology. This chapter confirms that indigenous communities can have ecological attitudes but only when environmental resources are the base of viable economies.
[6] In Latin America, household incomes are usually underreported as incomes generated from independent work (including informal sources) are the most difficult to measure. Fieldwork was carried out in each research site to account for more accurate figures regarding the economic situation of indigenous protestors.
[7] Bates finds an inverse relationship between rural prosperity and migration. Because market penetration in rural areas is limited, the capacity to develop productive agricultural models is lower when the distance to urban centres is larger. In this case, migration is the best option to escape rural poverty (Bates, 1976).
[8] The total Guaraní population reached nearly 60,000 people according to the 2012 Bolivian Census.
[9] Unlike the 2005 hydrocarbon law, the 2014 mining law does not grant veto power to indigenous communities. Differences across extractive industries correspond to the different political contexts in which reforms were passed. As of this writing, the Bolivian government is less enthusiastic about prior consultation than it was in early 2006, when Evo Morales had recently taken office.
[10] Conflicts over oil block 192 did not disappear as a consequence of prior consultation. Oil spills continue to take place, and indigenous groups continue to mobilize as they claim that the government has failed to fulfil its commitments to indigenous communities.
[11] CONAMAQ was created in 1997 with NGO support, with the objective of reconstituting indigenous forms of political organization within Bolivia's highlands.

References

Albó, X. (2012) Hacia el poder indígena en Ecuador, Perú y Bolivia. In: Betancur, A.C. (ed.) *Movimientos indígenas en América Latina. Resistencia y nuevos modelos de integración*. IWGIA, Copenhagen, pp. 133–166.

Aquino, S. (2011) La lucha por el control del territorio en Capulálpam. Diferentes maneras acerca de la comprensión del subsuelo, el oro, la plata, la ley y el capital. Available at: www.encuentroredtoschiapas.jkopkutik.org/BIBLIOGRAFIA/MOVIMIENTOS_POLITICA_CULTURA_Y_PODER/La_lucha_por_el_control_territorio.pdf (accessed 26 April 2016).

Arce, M. (2014) *Resource Extraction and Protest in Peru*. University of Pittsburgh Press, Pittsburgh, Pennsylvania.

Arellano-Yanguas, J. (2011) Minería sin fronteras? Conflicto y desarrollo en regiones mineras del Perú. IEP, Lima.

Bates, R. (1976) *Rural Responses to Industrialization: A Study of Village Zambia*. Yale University Press, New Haven, Connecticut.

BBC Mundo. (2005) Hidrocarburos en América Latina. Available at: http://news.bbc.co.uk/hi/spanish/specials/newsid_4562000/4562409.stm (accessed 30 May 2016).

Bebbington, A. and Bury, J. (2009) Confronting the institutional challenge for mining and sustainability: the case of Peru. *Proceedings of the National Academy of Sciences* 106(41), 17296–17301.

Bebbington, A. and Bury, J. (2013) *Subterranean Struggles. New Dynamics of Mining, Oil and Gas in Latin America*. University of Texas Press, Austin, Texas.

Bebbington, A. and Scurrah, M. (2013) Hydrocarbon conflicts and indigenous people in the Peruvian Amazon: mobilization and negotiation along the Rio Corrientes. In: Bebbington, A. and Bury, J. (eds) *Subterranean Struggles. New Dynamics of Mining, Oil and Gas in Latin America*. University of Texas Press, Austin, Texas.

Bebbington, A., Dani, A., de Haan, A. and Walton, M. (2008) (eds) *Institutional Pathways to Equity: Addressing Inequality Traps*. World Bank, Washington, DC.

Bebbington, A., Scurrah, M. and Bielich, C. (2012) Los movimientos sociales y la política de la pobreza en el Perú. Instituto de Estudios Peruanos. CEPES, Lima.

Belle-Antoine, R.M. (2015) Interview on prior consultation. Available at: https://www.youtube.com/watch?v=7E7mcKjjDBg (accessed 19 March 2018).

BP Statistical Review of World Energy (2016) BP statistical review of world energy June 2016. Available at: https://www.bp.com/content/dam/bp/pdf/energy-economics/statistical-review-2016/bp-statistical-review-of-world-energy-2016-full-report.pdf (accessed 31 May 2017).

Bridge, G. (2004) Contested terrain: mining and the environment. *Annual Review of Environmental Resources* 29, 205–259.

Bury, J. (2004) Livelihoods in transition: transnational gold mining operations and local change in Cajamarca, Peru. *Geographical Journal* 170(1), 78–91.

CEDIB (2015) Consulta Previa. Available at: www.cedib.org/tag/consulta-previa (accessed 19 March 2018).

Cooperaccion (2015) Primera consulta en minería? Available at: http://cooperaccion.org.pe/main/advanced-stuff/cooperaccion-informa/440-primera-consulta-previa-en-mineria (accessed 15 April 2017).

De la Fuente Lopez, A. (2013) La explotación de los hidrocarburos y los minerales en México: un Análisis Comparativo. Heinrich Boll Stieftung, Mexico city, Mexico.

De la Maza, F. (2004) Gobierno indígena y política social. Los yaquis de Sonora, México (1989–2003). Available at: https://www.aacademica.org/v.congreso.chileno.de.antropologia/148.pdf (accessed 19 March 2018).

Development Plan of the Municipality of Capulálpam de Méndez (2009) Plan de desarrollo municipal Capulálpam de Méndez. Available at: https://www.finanzasoaxaca.gob.mx/pdf/inversion_publica/pmds/08_10/247.pdf (accessed 31 May 2017).

Devlin, J. and Yap, N. (2008) Contentious politics in environmental assessment: blocked projects and winning coalitions. *Impact Assessment and Project Appraisal* 26(1), 17–27. DOI: 10.3152/146155108X279939.

Diaz-Vidaurre, M. (2010) Proceso de consulta realizado al pueblo indígena Mosetén en Alto Beni, proyecto exploración sísmica 2D, Bloque Lliquimun. *Lecciones Aprendidas sobre la Consulta Previa*. CEJIS, La Paz, Bolivia.

Due Process of Law Foundation (2015) Derecho a la consulta y al consentimiento previo, libre e informado en América Latina. DPLF, Washington, DC.

Eisenstadt, T. (2011) *Politics, Identity, and Mexico's Indigenous Rights Movements*. Cambridge University Press, Cambridge, UK.

Eisenstadt, T. and West, K. (2017) Where the debate between development and environmentalism get personal: public opinion, vulnerability, and living with extraction on Ecuador's oil frontier. *Comparative Politics* 49(2), 231–251.

Enlace Mineria (2014) Wild Acre posterga proyecto Peruano Colpayoc. Available at: http://enlacemineria.blogspot.com/2014/12/wild-acre-posterga-proyecto-peruano.htm (accessed 19 March 2018).

Environmental Health News (2014) Oil spill in the Amazon sickens villagers, kills fish. Available at: www.scientificamerican.com/section/environmental-health-news/?page=5 (accessed 19 March 2018).

Falleti, T. and Riofrancos, T. (2014) Participatory democracy in Latin America: the collective right to prior consultation in Ecuador and Bolivia. Paper presented at Lasa Conference, Chicago, 27–30 May.

Goldstone, J.A. and Tilly, C. (2001) Threat (and opportunity): popular action and state response in the dynamics of contentious action. In: Aminzade, R.R., Goldstone, J.A., McAdam, D., Perry, E.J., Sewell, W.H.J., Tarrow, S. and Tilly, C. (eds) *Silence and Voice in the Study of Contentious Politics*. Cambridge University Press, Cambridge, UK.

Goebel, A. (2010) Ecology of the poor and social marginality: vehicles of complementary and bridges of a dialogue. *Reflexiones* 89(1), 127–142.

Grosskoff, R. (1998) Comparación de las estadísticas de ingresos provenientes de encuestas de hogares con estimaciones externas. Paper prepared for the Taller Regional de Medición del Ingreso en las Encuestas de Hogares. Buenos. Available at: www.cepal.org/deype/mecovi/docs/taller2/26.pdf (accessed 19 March 2018).

Harvey, D. (2003) *The New Imperialism*. Oxford University Press, Oxford.

Himley, M. (2010) Frontiers of capital: mining, mobilization, and resource governance in Andean Peru. Doctoral Dissertation Syracuse University.

International Labour Organization (ILO) (1989) ILO Convention 169. Available at: www.ilo.org/dyn/normlex/en/f?p=NORMLEXPUB:12100:0::NO::P12100_INSTRUMENT_ID:312314 (accessed 19 March 2018).

Jaskosky, M. (2013) The local politics of project approvals in the Peruvian mining and Bolivian gas sectors. Paper presented at APSA Conference, Chicago, Illinois, 1 September.

La Jornada (2013) Capulálpam de Méndez contra la explotación minera. Available at: www.jornada.unam.mx/2011/05/05/opinion/024a1pol (accessed 30 May 2016).

La Jornada (2016) Se enfrentan Yaquis por gasoducto en Sonora; un muerto y ocho heridos. Available at: www.jornada.unam.mx/2016/10/22/estados/023n1est (accessed 23 April 2017).

La Razón (2015) AJAM hace inédita consulta previa minera.. Available at: www.la-razon.com/suplementos/financiero/AJAM-inedita-consulta-previa-minera-financiero_0_2309169173.html (accessed 30 June 2016).

Levitsky, S. and Murillo, M.V. (2013) Building institutions on weak foundations: lessons from Latin America. *Journal of Democracy* 24(2), 93–107.

Lucero, J.A. (2009) Decades lost and won: indigenous movements and multicultural neoliberalism in the Andes. In: Burdick, J., Oxhorn, P. and Roberts, K. (eds) *Beyond Neoliberalism in Latin America: Societies and Politics at the Crossroads*. Palgrave Macmillan, New York.

Madrid, E. (2014) Challapata: resistencia communal a la desposesion de la mineria. In: Perreault, T. (ed.) *Mineria, Agua y Justicia Social en los Andes*. PIEB, La Paz, Bolivia.

McAdam, D. (1999) *Political Process and the Development of Black Insurgency, 1930–1970*. University of Chicago Press, Chicago, Illinois.

Memoria Política de México (n.d.) Sublevación de los yaquis; las compañías deslindadoras extranjeras los despojan, valiéndose de la ley de deslinde de terrenos baldíos. Available at: www.memoriapoliticademexico.org/Efemerides/7/31071899.html (accessed 8 November 2016).

Mining Watch Canada (2012) Local votes and mining in the Americas. Available at: http://miningwatch.ca/blog/2012/5/14/local-votes-and-mining-americas (accessed 15 November 2016).

MMSD (2002) *Abriendo Brecha. Minería, Minerales y Desarrollo Sustentable*. IIED and World Business Council for Sustainable Development, London.

Observatory of Mining Conflicts in Latin America (n.d.) Mapa de conflictos mineros, proyectos y empresas mineras en América Latina. Available at: http://mapa.conflictosmineros.net/ocmal_db (accessed 30 May 2017).

Ocampo, D. and Urrutia, I. (2016) La implementación de la consulta en el sector minero: una mirada a los primeros procesos. In: Implementacion del Derecho de Consulta. Available at www.consulta-previa.org.pe/publicaciones/Consulta_Previa_paginas.pdf (accessed 15 April 2017).

O'Donnell, G. (1993) On the state, democratization and some conceptual problems: A Latin American view with glances at some postcommunist countries. *World Development* 21(8), 1355–1369.

O'Huallachain, B. and Matthews, R. (1996) Restructuring of primary industries: technology, labor and corporate strategy and control in the Arizona cooper industry. *Economic Geography* 72(2), 196–215.

Ojo Público (2015) Los secretos mineros detrás de la lista de comunidades indígenas del Peru. Available at: http://ojo-publico.com/77/los-secretos-detras-de-la-lista-de-comunidades-indigenas-del-peru (accessed 19 May 2016).

Paredes, M. (2015) transnational networks acting from below: indigenous prior consultation and the Peruvian paradox. Paper presented at Lasa Conference. San Juan, 27–30 May.

Perreault, T. (2008) Natural gas, indigenous mobilization and the Bolivian state: identities, conflict and cohesion. Programme Paper 12. United Nations Research Institute for Social Development, Geneva.

Perreault, T. (2013) Nature and nation: the territorial logics of hydrocarbon governance in Bolivia. In: Bebbington, A. and Bury, J. (eds) *Subterranean Struggles. New Dynamics of Mining, Oil and Gas in Latin America*. University of Texas Press, Austin, Texas, pp. 67–90.

Perreault, T (ed.) (2014) Mineria, agua y justicia social en los andes: experiencias comparativas de Perú y Bolivia. Justicia Hídrica, Centro de Ecología and Pueblos Andinos Fundación PIEB, La Paz, Bolivia.

Perupetro Database (2016) Prior consultation. Available at: https://www.perupetro.com.pe/wps/wcm/connect/perupetro/site/consulta%20previa/Ley%20de%20Consulta%20Previa (accessed 9 June 2016).

Rights and Resources Initiative (2013) Impacto de las industrias extractivas en los derechos colectivos de sobre territorios y bosques de los pueblos y comunidades. Available at: http://theredddesk.org/sites/default/files/resources/pdf/2013/rrifull.pdf (accessed 31 May 2017).

Rodríguez-Garavito, C. (2011) Ethnicity.gov: global governance, indigenous peoples, and the right to prior consultation in social minefields. *Indiana Journal of Global Legal Studies* 18(1), 263–305.

Scurrah, M. and Chaparro, A. (2011) Estrategias indígenas, gobernanza territorial e industrias extractivas en la Amazonia peruana. Presented in the Congress 'Desarrollo territorial y extractivismo: luchas y alternativas en la región andina' at the Bartolomé de las Casas Center, Cusco, 7 November.

Silva, E. (2009) *Challenging Neoliberalism in Latin America*. Cambridge University Press, New York.

Slack, K. (2009) Digging out from neoliberalism: responses to environmental (mis) governance of the mining sector in Latin America. In: Burdick, J., Oxhorn, P. and Roberts, K. (eds) *Beyond Neoliberalism in Latin America: Societies and Politics at the Crossroads*. Palgrave Macmillan, New York, pp. 117–134.

Tarrow, S. (1998) *Power in Movement: Social Movements and Contentious Politics*. Cambridge University Press, Cambridge.

The Dialogue (2015) Local conflicts and natural resources. Available at: www.thedialogue.org/wp-content/uploads/2015/05/Local-Conflicts-and-NaturalResources-FINAL.pdf (accessed 29 May 2016).

Tilly, C. (1978) *From Mobilization to Revolution*. Addison-Wesley, Reading, Massachusetts.

Valcuende del Río, J.M., Murtagh, C. and Rummenhoeller, K. (2012) Turismo y poblaciones indígenas: espacio tiempo y recursos. Available at: www.ub.edu/geocrit/sn/sn-410.htm (accessed 7 May 2016).

Van Cott, D.L. (2008) *Radical Democracy in the Andes*. Cambridge University Press, New York.

Wise, H. and Shtylla, S. (2007) The role of the extractive sector in expanding economic opportunities. The Fellows of Harvard College, Boston, Massachusetts. Available at: https://www.hks.harvard.edu/m-rcbg/CSRI/publications/report_18_EO%20Extractives%20Final.pdf (accessed 19 March 2018).

Yashar, D. (2005) *Contesting Citizenship in Latin America. The Rise of Indigenous Movements and the Postliberal Challenge*. Cambridge University Press, New York.

Zavaleta, M. (2014) *La Batalla por los Recursos Naturales*. Pontificia Universidad Catolica del Peru, Lima.

19 The Political Mediation of Indigenous Land Conflicts in Argentina

Matthias vom Hau*

Institut Barcelona d'Estudis Internacionals (IBEI), Barcelona, Spain

---

The past decades have witnessed a striking transformation in Latin America. Historically among the poorest and most marginalized sectors of the population (Hall and Patrinos, 2005; Mahoney, 2010), indigenous people have recently become a formidable political force in their own right, something unthinkable even a generation ago. Their core demands are the restitution of communal land rights and control over the territories that indigenous communities have historically used and occupied.

The rise of indigenous political mobilization marks a substantial juncture when compared to previous periods, in which social class identities patterned the interest mediation between states and their citizens (Collier and Collier, 1991; Yashar, 2005; Silva, 2009). Its causes are multifaceted and linked to major changes in global and national *opportunity structures* – most prominently the emergence of a global human rights regime, new international legal resources such as the International Labour Organization's (ILO) Convention No. 169, and constitutional multiculturalism (Brysk, 2000; Van Cott, 2000). The main *motivation* for local activists to make use of these new laws and norms, however, is recent economic changes linked to trade liberalization and the (re)emergence of development models focused on raw material extraction (Bebbington *et al.*, 2008; Burchardt and Dietz, 2014).

As smallholders, rural indigenous peoples have been affected by the current pressures on land in a particularly severe manner. The expansion of export-oriented agriculture and extractive industries have increased land sales but also the practice of land grabbing (e.g. Zoomers, 2000; Svampa and Viale, 2015). The consequences have been new forms of agrarian conflict, especially because rural indigenous communities often lack tenure rights and do not hold a title, and therefore are vulnerable to (the threat of) eviction from the lands they live on and use.

Argentina is illustrative of this broader regional trend. Even though dominant discourses about national identity continue to depict Argentina as a 'white nation' of European immigrants (Bastia and vom Hau, 2014), more than 955,000 Argentines, or about 2.4% of the total population, self-identify as belonging to one of 35 different indigenous groups (INDEC, 2010). In fact, the country has witnessed a dramatic increase in indigenous political mobilization over the past two decades (Gordillo and Hirsch, 2010).

In this chapter I focus on indigenous land struggles in northwestern Argentina. The three neighbouring provinces of Tucumán, Salta and Catamarca are surprisingly similar in their geography, political economy and their relative levels of human development. At the same time, the provinces vary dramatically in the implementation

---

* Email: mvomhau@ibei.org

of a nationwide survey that gauges the land claims made by indigenous peoples. The survey constitutes a true watershed in state–indigenous community relations: It indicates that – for the first time in the country's post-colonial history – the national state actually acknowledges and engages with indigenous land claims as a political reality.

Drawing on insights from social movement theory and historical institutionalism, I argue that indigenous land survey implementation needs to be understood as crucially driven by variations in indigenous mobilization, but also mediated by local power configurations and institutional structures. In a *quiescent path*, as illustrated by Catamarca, a weak indigenous movement did not achieve its goals. In a *contentious path*, as exemplified by Salta, strong indigenous land rights activism confronted powerful opposition from within the provincial state, leading to recurrent and often intense conflict around the implementation of the land survey. Finally, in a *transformative path*, as shown by Tucumán, well-organized indigenous activists did not confront institutional blockage and could rely on support from within the state, leading to a relatively smooth survey implementation. Seen in this light, the chapter suggests that, even though economic pressures are often at the root of why vulnerable rural groups mobilize for rights to (customary) lands, a focus on economic structures and commercial interests alone cannot account for distinct consequences of land conflict. Rather, it is the organizational strength – their networks, organization and leadership – of subordinate actors and specific patterns of state formation that shape how land conflicts unfold.

In developing this argument, the chapter builds on yet also moves beyond a variety of broader theoretical and substantive debates in the study of agrarian transformations. First, much of the existing scholarship on land tenure regimes is primarily concerned with the conflicts and tensions that result from the adoption of private land rights (e.g. in the form of individual or family-based titling) in contexts previously characterized by the prevalence of customary land tenure (e.g. Ubink *et al.*, 2009; Boone, 2014). The case of Argentina, by contrast, provides an opportunity to explore a distinct and less studied process, namely the introduction of communal land rights into a wider context where private property rights constitute the dominant land tenure system. Second, the chapter also engages discussions around the consequences of indigenous land struggles. While not at the centre of this analysis, my interviews with indigenous activists indicate that a communal land title provides indigenous communities with an important legal resource to remain on and maintain access to their lands in the face of the expanding frontier of extraction (see vom Hau and Wilde, 2010; vom Hau, 2016). A formal land title also enables indigenous communities to demand the implementation of other constitutionally granted collective rights such as bilingual schools or medical services. These findings echo recent works on the empowering consequences of communal land titles in subsequent struggles for social justice and redistribution (Saffon, 2015). Third and finally, this research is not just of scholarly interest. Exploring the consequences of indigenous land rights movements also provides activists and policy makers with a stronger basis to make informed decisions about universalist and multicultural policy options, and to assess the viability of particular mobilization tactics.

The chapter employs a qualitative methodology that combines a variety of sources. I draw on more than 120 interviews with indigenous activists, non-activists, state officials and economic elites in the three provinces that I, together with a team of researchers, conducted during several research trips between November 2008 and December 2014. The interviews were complemented by consultations with scholars from various universities and research institutions in San Miguel de Tucumán, the city of Salta, San Fernando del Valle de Catamarca and Buenos Aires.

The reminder of the chapter is organized as follows: the next two sections provide the context of indigenous land struggles in Argentina and develop the empirical puzzle to be explained – the dramatic subnational variations in the implementation of a mandatory survey of indigenous land claims. The third section evaluates alternative accounts concerned with the demographic composition of the population, economic structures, and the cohesion and balance of power among economic elites, state interests, subnational political regimes and historical levels of state capacity, but ultimately finds them to be limited in their explanatory power. The fourth and fifth sections develop the explanatory argument focused on social movement characteristics and state structures in more detail and apply it to the cases at hand. The final section concludes.

National Context: Left Wing Populism, Commodity Boom and Indigenous Land Struggles

During the 1990s most countries in Latin America introduced new multicultural rights in their respective constitutions, which usually included the recognition of indigenous communal lands (Van Cott, 2000). Argentina was no exception. The 1994 Constitution recognized Argentina as a pluriethnic nation and acknowledged the ethnic and cultural pre-existence of indigenous peoples.

The constitutional recognition of communal property certainly increased indigenous activism around land and territorial rights. Yet, what ultimately instigated local activists to draw on this multicultural legal framework was a new political opening, combined with the intensification of threats to their livelihoods deriving from a major economic transformation. In the aftermath of the 2001 economic and financial crisis the left-populist governments of Nestor Kirchner (2003–2007) and Cristina Fernandez de Kirchner (2007–2015) directly engaged with various social movements and their demands – as long as protests did not stand in the way of these governments' broader agenda of economic recovery, equitable growth and social provision. In fact, encouraged by booming global demand, the Kirchners embraced the large-scale exportation of primary goods as the basis from which to generate revenues and finance more generous and inclusive public services. Seen in this light, the increased pressures on established land use patterns faced by local indigenous communities were to an important extent the consequence of the commodity-driven development model pursued by a left-populist national government in Buenos Aires (Briones, 2015; vom Hau and Higuera, 2018).

The Puzzle: Subnational Variations in the Governance of Indigenous Land Claims

The economic and political transformations at the national level had distinct implications in different parts of Argentina. For a long time, the economies of Tucumán and Salta have been primarily oriented towards agriculture. Sugarcane, tobacco and citrus fruits cultivation dominated in the subtropical lowlands, while in the Andean valleys vegetable farming and cattle herding constituted the main sources of subsistence. In Catamarca the mining of gold and copper already had greater importance for much of the 20th century, while agriculture (e.g. livestock and sugarcane) was relatively less relevant for the provincial economy.

Historically, indigenous communities in the three provinces engaged in a mixture of pastoralism, small-scale farming and seasonal work in the sugarcane harvest. Over the past 20 years many changes transformed local subsistence strategies. In the lowlands the large-scale mechanization of sugarcane production reduced the demand for seasonal workers, while the dramatic expansion of soy production increased pressures on land holdings, led to the massive sale even of public lands and accelerated deforestation. Meanwhile, in the Andean valleys the growing importance of export-oriented wine production and tourism intensified land sales to speculators. This often resulted in the fencing of historically open pastures used by local communities or even their eviction from now more valuable lands for which they have no formal title. Moreover, mining activities in Catamarca led to increased water scarcity and reduced access to arable lands.

In Tucumán, Salta and Catamarca there are currently about 105,500 individuals who self-identify as indigenous and belong to at least ten different ethnic groups, the most populous among them being the Diaguita, the Kolla and the Wichi (INDEC, 2010). In light of growing tenure insecurity many local communities have started to emphasize their indigenous heritage and claim communal land rights. When asked about their reasons to mobilize for land rights, activists stress economic security. Not having a title exposes communities to the risk of possible eviction, even from lands they have lived on for generations. Land rights also provide the basis for claiming social benefits such as public housing and infrastructural investments, resources that cannot be accessed in the absence of a title.

In order to justify their land claims indigenous activists portray themselves as the original inhabitants who occupied the land long before the onset of Spanish colonialism. Land claims usually emphasize a continuity between precolonial and contemporary patterns of indigenous land use. To strengthen their claims to historical precedence indigenous leaders often prepare maps that document spatial memories and show the settlements, ceremonial centres and pasture areas used by them. As one activist from Salta stresses, 'we have proofs, archaeological traces, that we existed before'. Other forms of evidence include legal documents, kinship trees and academic works – especially scholarship in ethnohistory and archaeology.

At first glance, indigenous land struggles are far from achieving their main goal, the titling of communal lands. In Tucumán, Salta and Catamarca, local indigenous communities have (so far) only in exceptional circumstances obtained a formal title. Yet, a second look reveals a different story. After witnessing mounting land conflicts between indigenous communities, investors and/or current legal title holders, Nestor Kirchner signed a new law in 2006, the Law 26.160. The law mandated a nationwide land survey (or *relevamiento* in Spanish), with the aim of documenting and registering all indigenous land claims by the end of 2016. It also prohibited any further evictions of indigenous communities, an increasingly common tactic used by the current legal owners to preempt possible claims on their titles, until the survey is completed.

The survey has been funded, logistically supported and ultimately approved at the national level, by the national state. Yet, the responsibility to conduct the survey is at the subnational level. Argentina is a federal state and the governance of resources such as land and water falls under the jurisdiction of the 23 provinces in the country. It is thus provincial state authorities that hire and supervise the survey teams composed of lawyers, anthropologists and geographers to collect relevant information on indigenous land use for the assessment of current and future land claims.

As summarized in Table 19.1, there are dramatic subnational variations in the implementations of the land survey. In Tucumán, as of December 2016, the relevamiento has been completed, and the process has been relatively speedy and smooth. Responsibility for the survey changed once, in 2010, from the University of Tucumán to the Ombudsman Office, because indigenous communities complained about a too technical approach and lack of local knowledge among the university-based team. A reshuffled survey team then enjoyed widespread support and cooperation. Current debates among indigenous leaders in Tucumán focus primarily on what to do with the results: whether to pursue legal expropriation of current landowners, or alternatively, by drawing on Bolivia as a model, push for the constitutional recognition of indigenous territories.

In Salta, by contrast, the survey of indigenous land claims has been very slow and contested. The relevamiento was initially in the hands of the Instituto Provincial de Pueblo Indígenas de *Salta* (IPPIS), the provincial state agency in charge of indigenous affairs. The organization rejected any involvement by non-indigenous technical staff. In 2011, after most of the federal funds were gone, but only six communities had been surveyed, the national state formally intervened, took away the implementation responsibility from IPPIS and threatened to sue the province for compensation. From that point onwards the survey team reported directly to Buenos Aires. As of early 2016, a new reconciliatory attempt to put together a team that is backed both by the provincial state and national state is

Table 19.1. Variations in land survey implementation.

	Tucumán	Salta	Catamarca
Implementation	Almost completed; results accepted	Not completed; results contested	Not started
Responsibility	Province	Shift between province and national government	
Survey team characteristics	High technical competence, high embeddedness	Province: low technical competence, high embeddedness; federal government: reverse	

underway. What remains unclear at this stage is whether the surveys that have been done by the second nationally backed team are legally valid in the eyes of the province. Finally, in Catamarca, the survey has not even started, and there is no evidence that it will any time soon.

The differences are also striking when it comes to the composition and training of the survey teams. In Tucumán the land survey team – about 30 anthropologists, lawyers, engineers and geographers – had the necessary funds and expertise, and, after it was run from the Ombudsman Office, one of the most respected state agencies in the province, also the necessary credentials among community groups. The team was put together by a respected public figure and led by an anthropologist with long working experience in the area. Many team members had previously worked on indigenous affairs, and a significant part (but not all) were self-identified Diaguita.

In Salta the first survey team put together by IPPIS, considered one of the most corrupt and ineffective state agencies in the province, was based on clientelist networks. The second team, put together by Instituto Nacional de Asuntos Indígenas (INAI) in Buenos Aires, had the necessary technical expertise but did not have a designated office. Meetings took place in different coffee shops, and team members often had no previous work experience in indigenous communities and lacked local ties. In Catamarca, there is currently no provincial state infrastructure that deals with indigenous policy, either from the provincial or national state. The INAI does not have an office in the province, and the provincial state does not have a specialized agency in place. The reasons for this are multifaceted, but what stands out is the power of the mining sector, together with a long history of political tensions between the national government and the province: Catamarca has been a stronghold of the Radical Party, which opposed the dominance of Peronism at the national level.

Alternative Explanations and their Limitations

Why do the provincial states of Tucumán, Salta and Catamarca vary in indigenous land survey implementation? What accounts for these striking subnational differences? This section reviews a number of plausible explanations related to ethnic demographics, economic structure, state interests, political regime type and state capacity, but ultimately finds each of them wanting.

Ethnic fractionalization

A large and influential body of work has argued that ethnic fractionalization inhibits collective action. This is primarily because of heterogeneous preferences. Citizens belonging to different ethnic groups are less likely to take each other's welfare into account and to agree on what should be done about their situation, and this prevents them from effectively pressuring the state for the implementation of desired policies, such as the surveying of indigenous land claims (Habyarimana et al., 2009, p. 8).

At first glance, there is some merit to such an explanation. As shown in Table 19.2, Tucumán, Salta and Catamarca differ both in the relative size and the composition of their indigenous population. Even when taking into account that available census data from 2010 might be conservative estimates and do not capture the recently observed 'boom' in indigenous self-identification, they establish distinct regional patterns. In Salta, a comparatively larger part of the overall population, 6.5%, identifies as indigenous, when compared to Tucumán and Catamarca, where 1.3% and 1.9% of the population consider themselves as Argentines of indigenous origin, respectively. Yet, the indigenous population itself is far more fragmented in Salta than in the other two provinces. In Salta there exist solidly established ethnolinguistic and ethnocultural distinctions between at least nine different indigenous groups, the most numerous among them being the Diaguita, Kolla, Guaraní and Wichi. By contrast, in Tucumán and Catamarca the indigenous population almost exclusively identifies as Diaguita. A fractionalization-based explanation thus suggests that the more ethnically heterogeneous indigenous population in Salta should face greater difficulties in agreeing on shared goals and coordinating their lobbying efforts.

While this might seem plausible for Salta, such an account has difficulties explaining the contrast in land survey implementation between

Table 19.2. Ethnic demographics. (From: INDEC (2010).)

	Tucumán	Salta	Catamarca
Total population (2010)	1,448,000	1,214,000	368,000
Indigenous population (2010)	19,300 (1.3% of total population)	79,200 (6.5% of total population)	6900 (1.9% of total population)
Composition of indigenous population	Homogeneous (predominantly Diaguita)	Diverse (several groups of similar size: Diaguita, Kolla, Wichi)	Homogeneous (predominantly Diaguita)

Tucumán and Catamarca. From a demographic perspective, those two provinces are strikingly similar both in the size and composition of the indigenous population. In addition, a more careful look at Salta reveals that there is only limited empirical support for the proposed causal mechanism – preference heterogeneity – through which fractionalization is expected to impede land survey implementation. Interviews with indigenous leaders in Salta reveal that – regardless of their specific ethnic background – these activists agree that a mandatory land survey is an important first step towards establishing indigenous land rights and natural resource control.

Economic structures and elite interests

Another explanatory approach shifts the focus from ethnic demographics to the structure of the economy, and related to that, the interests and power resources of economic elites. A growing interdisciplinary scholarship emphasizes elite politics, and in particular levels of elite cohesion and the configurations of elite coalitions, as major drivers to account for state provision of public goods and services (Waldner, 1999; Amsden et al., 2012). This literature has also established that the upper classes of the dominant ethnic group tend to be particularly concerned about 'ethnic threats' from minorities to property relations and accumulation regimes (Lieberman, 2003; Slater, 2010). Seen through this analytical lens, non-indigenous economic elites are expected to oppose indigenous land claims, and they are more likely to effectively do so if they act in a unified manner and are endowed with political and organizational power.

This argument is complemented by a regionally specific literature on the recent rise of (neo)extractivism in Latin America – an economic strategy organized around the exploitation of raw materials through practices such as mining, gas production or agro-industrial monocultures (e.g. soy). Since the 2000s a substantial increase in global demand has led to consistently high prices for raw materials, propelling a tendency towards 'reprimarization' and the formation of economies of extraction. Within this changing economic context large multinational mining corporations and agro-businesses have emerged as particularly powerful actors (Bebbington et al., 2008), often simply because of the scale of their operations vis-à-vis their local competitors or even other economic sectors. Thus, where multinational mining companies or agro-exporters dominate local economies, they are more likely to effectively oppose indigenous land claims, whether through lobbying the government or paying off indigenous leaders.

An initial look at the empirical record supports such a focus on economic structure and the power resources of economic elites. As shown in Table 19.3, mining dominates the economy of Catamarca. This economic sector generates 20.8% of the provincial gross domestic product (GDP). Even more strikingly, this sector is almost entirely controlled by one large gold and copper mining corporation, La Alumbrera. And indeed, interviews with indigenous activists and scholars specializing on Catamarca reveal that La Alumbrera is in opposition to indigenous land rights and actively combats initiatives such as the survey of indigenous land claims, especially when those claims intersect with areas designated for future mining activity. Most prominently, the corporation marshals significant influence over provincial politicians, who often turn from major critics to outspoken supporters of mining as soon as they have been elected as provincial governors or ministers in Catamarca. Thus,

Table 19.3. Economic structure. (From: INDEC (2010), CEPAL (2000).)

	Tucumán	Salta	Catamarca	Argentina (average)
2008 GDP per capita (US$)	4200	3900	6000	8300
2010 income poverty (% of total population)	27.7	29.7	24.3	17.8
Agriculture (% of 2008 GDP)	6.0	8.1	3.6	5.0
Mining (% of 2008 GDP)	0.1	7.6	20.8	2.7
Manufacturing (% of 2008 GDP)	12.6	10.4	12.1	16.3
Service/construction (% of 2008 GDP)	81.3	73.9	63.5	76.0

there are good reasons to believe that in the province of Catamarca a large mining corporation has the political and organizational power necessary to effectively oppose the implementation of the indigenous land survey.

A more careful look complicates this line of argument, however. For one thing, a focus solely on the cohesion and power resources of economic elites cannot account for the variation in indigenous land survey implementation between Tucumán and Salta. In the two provinces economic elites are more fragmented across and within distinct economic sectors. In both Tucumán and Salta a number of smaller corporations engage in more diversified forms of mining, while large agro-exporters are further divided among those who primarily engage in wine, soy, citrus fruits or sugarcane production. Not surprisingly, in the two provinces economic elites are also politically more divided into distinct interest groups and business associations.

Yet, there is widespread elite unity with respect to indigenous land rights. Whether in Tucumán and Salta, and whether producing wine, soy, citrus fruits or sugarcane, agricultural elites share an equal disdain for indigenous land claims. Indigenous activists from both provinces report fierce opposition by large landowners against the land survey. In fact, the organizational power of the mining industry is more strongly felt in Tucumán than in Salta. It is indigenous communities in the former that are also at the receiving end of La Alumbrera's corporate social responsibility programmes. Similar to their counterparts in Catamarca, Diaguita leaders report that the mining company supports local infrastructure projects. Some interviewees even assert that a number of prominent Diaguita leaders in Tucumán are receiving pay-offs from the mining company in exchange for a softer stance on indigenous land rights (which might affect, for example, the transit routes of mining trucks through territories claimed by indigenous communities in Tucumán). Seen in this light, then, an explanatory approach solely focused on economic structures and corporate power cannot account for the puzzle why the survey of indigenous lands was implemented more quickly in Tucumán than in Salta.

Democratic quality and levels of state capacity

Understanding states as (potentially) autonomous organizations and institutional configurations provides a distinct analytical perspective. One line of research emphasizes the centrality of political regimes – that is, institutions involved in regulating access to political decision-making power – in shaping policy implementation. Specifically, this literature suggests that the presence of a democratic as opposed to an authoritarian regime greatly facilitates the adoption of policies valued by non-elite forces in society (e.g. Haggard and Kaufman, 2008; Huber and Stephens, 2012). Everything being equal, subordinate sectors that lack economic power and political influence are more likely to organize collectively and make their demands heard in democracies than in authoritarian settings. Moreover, in democratic regimes politicians have on average greater incentives to be responsive to subordinate demands because of electoral incentives to win re-election (Brown and Hunter, 1999).

The main problem with such a focus on regime characteristics is that an empirical association between access to power and the implementation of the indigenous land survey in Tucumán, Salta, and Catamarca does not exist. At the national level Argentina is broadly classified as a democracy, but there are significant subnational variations in this federal state (Giraudy, 2015). More than 30 years after the end of the most recent military dictatorship some Argentine provinces continue to exhibit outright authoritarian regimes. As shown in Fig. 19.1, those subnational differences, however, do not map onto the variation to be explained. Tucumán and Salta are more or less similar with respect to indicators of subnational democracy. In fact, a focus on subnational political regimes would predict the land survey implementation to be speedier and less complicated in Salta than in Tucumán, largely because in the former province politicians had greater incentives to cater to the demands of a comparatively larger indigenous population.

Another critical aspect of state institutions concerns the exercise of power – or the capabilities of states to implement policy decisions. A substantial literature on state capacity highlights the difference between bureaucratic and patrimonial states. Even when ruled autocratically, the former are more likely to have the administrative competencies (e.g. meritocratic recruitment, an esprit de corps among state officials, proactivity in relations to society) in place for effective policy implementation (Kurtz, 2013; Soifer, 2015). Scholars also emphasize the historical path dependency of state capacity. The organizational structures that sustain state capacity only form slowly, over the course of decades or even generations (Ertman, 1997). Seen in this perspective, then, contrasting capacities to implement a survey of indigenous lands is likely to be endogenous to broader levels of state capacity found in a particular Argentine province.

At first glance, there is some truth to this explanatory approach. Even though most research converges in the assessment that the national state in Argentina is marked by institutional weakness and relatively limited state capacity (e.g. Levitsky and Murillo, 2005), there are again substantial subnational variations. Most prominently, recent research singles out the province of Tucumán as having the organizational competence to effectively enforce new environmental regulation, even though two major polluting agro-industries, sugar and citrus processing, dominate the provincial economy (Amengual, 2015). Yet, more systematic comparative evidence on variations in bureaucratic competence across Argentine provinces does not find any significant differences between Tucumán and Salta. As shown in Fig. 19.1, the two provinces are quite similar in their levels of bureaucratization, which is measured by the analysis of horizontal accountability (i.e. the independence of the judiciary), the extent of patronage (i.e. the size of the provincial public administration) and the ruler's fiscal discretion (i.e. rules of fiscal allocation) (Giraudy, 2015, p. 48). In sum, then, a focus on neither ethnic demographics, economic structures, political regime characteristics, nor state capacity can fully explain subnational differences in indigenous land survey implementation between Tucumán, Salta and Catamarca.

Theoretical Framework: Social Movements, State Structures and Land Governance

How to account for the variation then? In this section I draw on insights derived from social movement theory and historical institutionalism to develop my explanatory argument. While scholarship on the consequences of social movements is comparatively small, existing works provide some important clues, both for identifying the specific mechanisms of *how* indigenous movements might influence public policy implementation, and the specific conditions *when* indigenous movements are more likely to do so.

I start with the causal mechanisms. While social movements are usually not part of the state bureaucracy, there is substantial evidence that they carefully follow the implementation of valued policies, and, if necessary, engage in the naming and shaming of state officials (Brysk, 1994; Tilly, 1999). Moreover, movements often

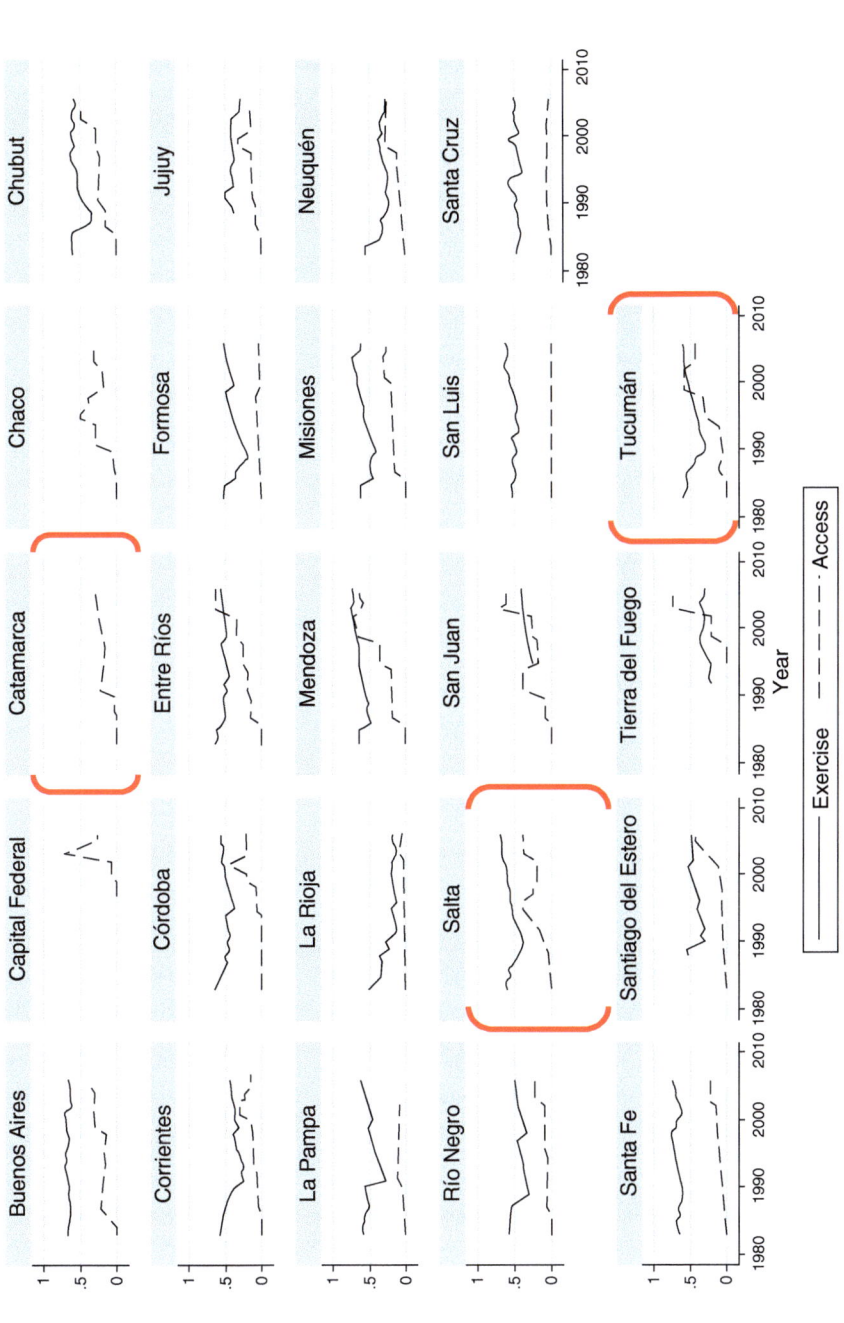

Fig. 19.1. Subnational Regime and State Characteristics (Giraudy 2015).

Access and exercise in Argentina (1983–2006). Higher values of access indicate higher levels of subnational democracy; 0 and near 0 scores denote subnational undemocratic regimes. Higher values of exercise indicate higher levels of subnational bureaucratization.'

foster new linkages between state organizations and society, and these 'bridging ties' provide public officials with more accurate knowledge about local needs and wants, thereby facilitating policy implementation as well (Tendler, 1997; Putnam, 2000).

Exploring movement impact by focusing on these causal mechanisms raises the question of why some indigenous movements are better able to monitor state bureaucracies, establish information linkages and – if necessary – name and shame state officials. Existing scholarship on social movements, especially the resource mobilization approach, emphasizes that movement organization is key. Movement success in bringing about policy change crucially depends on the networks, resources and leadership of a movement (Cress and Snow, 2000; Andrews, 2004). Movements with a well-established organizational infrastructure are more likely to facilitate the implementation of policies valued by them. In other words, those movements that can rely on skilled leaders, formal organization and wider civil society ties are more likely to achieve their aims.

Yet, as illustrated by numerous present and past examples (McAdam et al., 2001), even if endowed with significant organizational strength, movements might not achieve their goals. Another strand in social movement theory, the political mediation approach, therefore puts greater emphasis on the interaction between movement organization and the wider institutional context (Amenta, 2006; Amenta et al., 2010). In this view mobilization is not sufficient, but interacts with the prevailing state structures to shape movement consequences: The basic idea here is that the likelihood of movement impact decreases when movements confront state institutions (e.g. agencies charged with indigenous affairs) with formally defined decision-making powers to act as veto players (Tsebelis, 2002). If these state institutions regard the implementation of a movement-supported policy as a threat to their political survival, they are likely to use their position to prevent the effective monitoring of officials and accurate information flows.

The historical–institutional dimension enters when we look at the second-order causes of why state agencies act as veto players. Obviously the territorial organization of states plays a major role. Federal states are more likely to have those kinds of veto players. Yet, perhaps even more importantly, prior conflicts between states and social movements, and how states dealt with those contentious episodes, inscribe themselves into the current state structure. Specifically, I argue that historical 'divide and rule' responses likely led to the formation of state agencies with formal veto power and capacity to 'parochialize' contentious demands, whereas 'repress and ignore' responses were less likely to create these kinds of state agencies.

Taken together, then, the analytical framework developed in this section highlights two factors that together explain distinct capacities to mediate land conflicts and survey indigenous land claims. I argue that the organizational infrastructure of movements and the state structures in which they operate shape the implementation of desired policies. In other words, it is the networks, organization and leadership of indigenous movements, and the state institutions they encounter, that shape distinct patterns of land conflict, such as differences in indigenous land survey implementation across Argentina.

The Argument: Explaining Variations in Indigenous Land Survey Implementation

Equipped with this analytical framework, let's now return to Argentina. The main empirical argument is summarized in Table 19.4.

In Catamarca, indigenous mobilization is incipient and very fragmented and localized, and has not scaled up into a sustained collective challenge. Contentious practices and a sense of common purpose only occasionally move beyond the level of individual indigenous settlements or communities.

In Tucumán and Salta, by contrast, indigenous activism is quite intense and based on collective action and a shared sense of mission that cross-cuts distinct local communities and their concerns, and even distinct ethnic groups. A clear indication for the organizational strength of indigenous activism in Tucumán and Salta is that local indigenous communities collectively stopped paying pasture rents to the current legal title holders. Ending these payments brought

Table 19.4. The argument for the three empirical cases.

	Movement organizational strength	Provincial response to prior mobilization	Land survey implementation
Tucumán (transformative path)	High	'Ignore and repress'	Complete
Salta (contentious path)	High	'Divide and rule'	Contested
Catamarca (quiescent path)	Low	–	Blocked

some modest economic improvements, yet the most important consequences were symbolic. By refusing to be charged for the use of the land they consider as traditionally theirs, indigenous communities put their land claims into practice.

In both provinces indigenous land rights activism is coordinated by major umbrella organizations. The Unión de los Pueblos de la Nación Diaguita en Tucumán (UPNDT) and its counterpart in Salta, the Unión de los Pueblos de la Nación Diaguita en Salta (UPNDS), connect local Diaguita communities and their leaders with each other, while also maintaining close working relationships with state agencies, national and international non-governmental organizations (NGOs), and other civil society organizations. A similar organizational infrastructure can be found for other ethnic groups in Salta, such as the Kolla or Guaraní (Hirsch, 2003; Schwittay, 2003). Moreover, local indigenous activists from distinct ethnic groups are also linked to each other through cross-cutting networks facilitated by the provincial chapters of the Encuentro Nacional de Organizaciones Territoriales de Pueblos Originarios (ENOTPO), a nationwide alliance of indigenous advocacy networks.

Yet, Tucumán and Salta are very different with respect to the institutional context that indigenous land rights activism confronts. During the 1980s Salteño state officials dealt with a prior round of indigenous mobilization, and popular contention more generally, through a 'divide and rule' strategy. This entailed the controlled cooptation of indigenous leaders and the establishment of provincial state institutions to directly deal with indigenous affairs. Accordingly, the IPPIS was formed in 1986 with the explicit aim of controlling and containing the indigenous protest that emerged in the aftermath of Argentina's transition to democracy. IPPIS constitutes the official representative organ of indigenous communities vis-à-vis the provincial state. Accordingly, it is in charge of granting indigenous communities legal status in the province. The IPPIS budget is primarily used for patronage-oriented spending on discretionary benefits for particular individuals or communities. Interviews both with indigenous activists and IPPIS representatives reveal that the organization sought to resist the nationally mandated land survey, largely because the organization feared that this would strengthen the national state at the expense of its own mandate in Salta.

The provincial state in Tucumán chose a different strategy. During the 1980s and 1990s it ignored and repressed popular movements, and did not build any institutional structures comparable to IPPIS. As a result, the dominant interlocutor between Diaguita activists on the one hand, and the province and the national state on the other, is the provincial office of the INAI, the main national state agency responsible for dealing with indigenous communities. Diaguita leaders report INAI tactics of political fragmentation. In the Valley of Tafí, for example, INAI officials encouraged the formation of several small, individual communities, apparently with the intent to prevent the formation of a larger community representing the whole valley. At the same time, though, the INAI often works together with representatives of the UPNDT to support local land struggles, for instance, by covering the legal costs of lawsuits with current title holders. Moreover, INAI staff members also work in close coordination with other subnational state agencies to facilitate the implementation of the nationally mandated land survey.

Conclusion

Based on a comparison of the Argentine provinces of Tucumán, Salta and Catamarca, the chapter has sought to explain contrasting pathways in the implementation of a survey of indigenous lands (*relevamiento*), which constitutes an important first step in the activation of multicultural rights and the recognition of indigenous land claims as a political reality. Combining insights from social movements theory and historical institutionalism, the chapter has put the analytical focus on both the characteristics of indigenous mobilization and provincial state structures to account for those subnational differences. Specifically, the case studies have shown that in Tucumán the *organizational strength* of indigenous activism in the province, together with an *institutional context* characterized by the absence of formal veto players, greatly contributed to the relatively smooth and speedy implementation of the survey. In Salta and Catamarca by contrast, the lack of organizational strength and/or the presence of major institutional blockage prevented a similarly effective land survey implementation.

In developing this argument, the chapter cautions against interpreting the survey of indigenous land claims as a mere manifestation of James Scott's (1998) 'legibility thesis'. It is certainly true that states, including the Argentine state, conduct surveys and censuses to make complex local practices 'legible' and thus better controllable to them, and to signal modern nationhood to the wider international community. But the survey is not just about state control. While indigenous activists are acutely aware of the downsides of the survey – most importantly its privileging of communal land claims made by local indigenous communities rather than larger ethnic groups (e.g. Mapuche, Diaguita) – they nonetheless converge in seeing the survey as an opportunity to stake a claim over the territories they consider historically theirs, and gain official recognition for it. Moreover, initial evidence from research on the consequences of the land survey (vom Hau and Higuera, 2018) shows that surveyed indigenous communities are more likely to obtain a communal land title than communities that have not been surveyed.

Moreover, the chapter has pushed away and forward from the existing scholarship on land conflict. Even though economic pressures are often at the root of why vulnerable groups demand rights to (customary) lands, a focus on economic structures and commercial interests alone cannot account for distinct patterns of land governance. Rather, states play a pivotal role in their own right, and such perspective opens up new possibilities to bring history and politics into the equation. More precisely, locally specific patterns of state formation, and the distinct institutional structures resulting from it, shape how contemporary state officials respond to demands for (customary) land rights by vulnerable groups. Moreover, close attention to collective mobilization is warranted. The room for manoeuvre of subordinate actors is not a direct function of their economic conditions. Rather, it is their organizational strength that matters to achieve their goals.

Future research would benefit from further unpacking the internal hierarchies and conflicts of indigenous communities, and their role in shaping land struggles. Most prominently, gender differences are crucial to consider. In some indigenous groups (e.g. the Diaguita in Tucumán, Salta and Catamarca) women actively participate in land rights protests and frequently emerge as leaders, and sometimes their activism for land rights also becomes a springboard for these indigenous women to become politically active in other fields, too. Yet, indigenous women's activism is obviously highly context-dependent and crucially rests on relative gender equality in terms of educational attainment and the prevalence of female labour migration. Where those conditions are not met (e.g. among the Mbya in Misiones), indigenous activism around land rights remains confined to men, while women are largely excluded from politics (vom Hau, 2015).

The findings of the chapter also have major actionable implications, both for policy makers and indigenous activists. The close link between communal land rights and indigenous well-being emphasized by activists makes it critical for policy makers to reassess prevailing strategies to combat indigenous destitution. If we take the claims and analysis of indigenous movements seriously, then the improvement of indigenous well-being cannot be achieved solely through the instigation of economic growth or demand-side social programmes such as conditional cash transfers. Gains from these currently dominant strategies will remain shallow unless they are coupled with more profound changes of national development models focused on raw material extraction, which are at the roots of indigenous land struggles to begin with.

Acknowledgements

A research grant received from the Spanish Ministry of Economy and Competitiveness (MINECO Grant CSO2015-67558-P) supported the field research for this chapter. I would also like to thank Cyrus Afshar and Diego Higuera for excellent research assistance.

References

Amengual, M. (2015) *Politicized Enforcement in Argentina: Labor and Environmental Regulation*. Cambridge University Press, New York.

Amenta, E. (2006) *When Movements Matter: The Townsend Plan and the Rise of Social Security*. Princeton University Press, Princeton, New Jersey.

Amenta, E., Caren, N., Chiarello, E. and Su, Y. (2010) The political consequences of social movements. *Annual Review of Sociology* 36, 287–307.

Amsden, A.H., DiCaprio, A. and Robinson, J.A. (2012) *The role of Elites in Economic Development*. Oxford University Press, Oxford, UK.

Andrews, K.T. (2004) *Freedom is a Constant Struggle: The Mississippi Civil Rights Movement and Its Legacy*. University of Chicago Press, Chicago, Illinois.

Bastia, T. and vom Hau, M. (2014) Migration, race and nationhood in Argentina. *Journal of Ethnic and Migration Studies* 40, 475–492.

Bebbington, A., Humphreys Bebbington, D., Buryc, J., Lingand, J., Muñoz, J.O. and Scurrah, M. (2008) Mining and social movements: struggles over livelihood and rural territorial development in the Andes. *World Development* 36, 2888–2905.

Boone, C. (2014) *Property and Political Order in Africa: Land Rights and the Structure of Politics*. Cambridge University Press, New York.

Briones, C. (2015) Políticas indigenistas en Argentina: entre la hegemonía neoliberal de los años noventa y la 'nacional y popular' de la última década. *Antípoda*, 21–48.

Brown, D.S. and Hunter, W. (1999) Democracy and social spending in Latin America, 1980–92. *American Political Science Review* 93, 779–790.

Brysk, A. (1994) *The Politics of Human Rights in Argentina: Protest, Change, and Democratization*. Stanford University Press, Stanford, California.

Brysk, A. (2000) *From Tribal Village to Global Village: Indian Rights and International Relations in Latin America*. Stanford University Press, Stanford, California.

Burchardt, H.J. and Dietz, K. (2014) (Neo-)extractivism – a new challenge for development theory from Latin America. *Third World Quarterly* 35, 468–486.

Collier, R.B. and Collier, D. (1991) *Shaping the Political Arena*. University of Notre Dame Press, Notre Dame, Indiana.

Comisión Económica para América Latina (CEPAL) (2000) Anuario estadístico de América Latina y el Caribe. Available at: https://en.wikipedia.org/wiki/List_of_Argentine_provinces_by_gross_domestic_product (accessed 24 May 2017).

Cress, D.M. and Snow, D.A. (2000) The outcomes of homeless mobilization: the influence of organization, disruption, political mediation, and framing. *American Journal of Sociology* 105, 1063–1104.

Ertman, T. (1997) *Birth of the Leviathan: Building States and Regimes in Medieval and Early Modern Europe*. Cambridge University Press, New York.

Giraudy, A. (2015) *Democrats and Autocrats: Pathways of Subnational Undemocratic Regime Continuity within Democratic Countries*. Oxford University Press, New York.

Gordillo, G.R. and Hirsch, S.M. (2010) *Movilizaciones Indígenas e Identidades en Disputa en la Argentina*. La Crujía, Buenos Aires, Argentina.

Habyarimana, J., Humphreys, M., Posner, D.N. and Weinstein, J.M. (2009) *Coethnicity: Diversity and the Dilemmas of Collective Action*. Russell Sage Foundation, New York.

Haggard, S. and Kaufman, R.R. (2008) *Development, Democracy, and Welfare States: Latin America, East Asia, and Eastern Europe*. Princeton University Press, Princeton, New Jersey.

Hall, G. and Patrinos, H.A. (2005) *Pueblos Indígenas, Pobreza y Desarrollo Humano en América Latina: 1994– 2004*. World Bank, Washington, DC.

vom Hau, M. (2015) Gendered mobilization: women and the politics of indigenous land claims in Argentina. In: Archambault, C. and Zoomers, A. (eds) *Shifting Grounds: Gender Impacts of Global Trends in Land Tenure Reform*. Earthscan, London, pp. 93–109.

vom Hau, M. (2016) Advancing while losing: indigenous land claims and development in Argentina. In: Blumberg, R. and Cohn, S. (eds) *Development in Crisis: Threats to Human Well Being in the Global South and the Global North*. Routledge, London, pp. 173–191.

vom Hau, M. and Higuera, D. (2018) *Implementing Multiculturalism: The Governance of Indigenous Land Claims in Argentina*. Manuscript in progress. Institut Barcelona d'Estudis Internacionals, Barcelona, Spain.

vom Hau, M. and Wilde, G. (2010) 'We have always lived here': indigenous movements, citizenship, and poverty in Argentina. *Journal of Development Studies* 46, 1283–1303.

Hirsch, S.M. (2003) Bilingualism, pan-Indianism and politics in northern Argentina: the Guaraní's struggle for identity and recognition. *Journal of Latin American Anthropology* 8, 84–103.

Huber, E. and Stephens, J.D. (2012) *Democracy and the Left: Social Policy and Inequality in Latin America*. University of Chicago Press, Chicago, Illinois.

Instituto Nacional de Estadísticas y Censos de la República Argentina (INDEC) (2010) Encuesta Complementaria de Pueblos Indígenas. Available at: www.indec.gob.ar/nivel4_default.asp?id_tema_1=2&id_tema_2=21&id_tema_3=99 (accessed 24 May 2017).

Kurtz, M.J. (2013) *Latin American State Building in Comparative Perspective: Social Foundations of Institutional Order*. Cambridge University Press, New York.

Levitsky, S. and Murillo, M.V. (2005) *Argentine Democracy: The Politics of Institutional Weakness*. Penn State Press, College Station, Pennsylvania.

Lieberman, E.S. (2003) *Race and Regionalism in the Politics of Taxation in Brazil and South Africa*. Cambridge University Press, New York.

Mahoney, J. (2010) *Colonialism and Postcolonial Development: Spanish America in Comparative Perspective*. Cambridge University Press, New York.

McAdam, D., Tarrow, S. and Tilly, C. (2001) *Dynamics of Contention*. Cambridge University Press, New York.

Putnam, R. (2000) *Bowling Alone: The Collapse and Revival of American Community*. Simon & Schuster, New York.

Saffon, M.P. (2015) When theft becomes grievance dispossessions as a cause of redistributive land claims in 20th century Latin America. PhD thesis. Columbia University, New York.

Schwittay, A.F. (2003) From peasant favors to indigenous rights: the articulation of an indigenous identity and land struggle in northwestern Argentina. *Journal of Latin American Anthropology* 8, 127–154.

Scott, J.C. (1998) *Seeing Like a State: How Certain Schemes to Improve the Human Condition have Failed*. Yale University Press, New Haven, Connecticut.

Silva, E. (2009) *Challenging Neoliberalism in Latin America*. Cambridge University Press, New York.

Slater, D. (2010) *Ordering Power: Contentious Politics and Authoritarian Leviathans in Southeast Asia*. Cambridge University Press, New York.

Soifer, H.D. (2015) *State Building in Latin America*. Cambridge University Press, New York.

Svampa, M. and Viale, E. (2015) *Maldesarrollo: la Argentina del Extractivismo y el Despojo*. Katz editores, Buenos Aires.

Tendler, J. (1997) *Good Government in the Tropics*. Johns Hopkins University Press, Baltimore, Maryland.

Tilly, C. (1999) From interactions to outcomes in social movements. In: Giugni, M., McAdam, D. and Tilly, C. *How Social movements Matter*. University of Minnesota Press, Minneapolis, Minnesota, pp. 253–270.

Tsebelis, G. (2002) *Veto Players: How Political Institutions Work*. Princeton University Press, Princeton, New Jersey.

Ubink, J.M., Hoekema, A.J. and Assies, W.J. (2009) *Legalizing Land Rights: Local Practices, State Responses and Tenure Security in Africa, Asia and Latin America*. Leiden University Press, Leiden, The Netherlands.

Van Cott, D.L. (2000) *The Friendly Liquidation of the Past: The Politics of Diversity in Latin America*. University of Pittsburgh Press, Pittsburgh, Pennsylvania.

Waldner, D. (1999) *State Building and Late Development*. Cornell University Press, Ithaca, New York.

Yashar, D. (2005) *Contesting Citizenship in Latin America: The Rise of Indigenous Movements and the Postliberal Challenge*. Cambridge University Press, New York.

Zoomers, A. (2000) *Current Land Policy in Latin America: Regulating Land Tenure under Neoliberalism*. Royal Tropical Institute, Amsterdam, The Netherlands.

20 The Role of Land Reform in Rural Development: Promoting Productivity or Democracy?

Matthew Hoffman*

RURALIS – Institute for Rural and Regional Research, Trondheim, Norway

Introduction

This chapter uses the recent case of land reform in Scotland – in which large private estates are being transferred to community ownership – to reconsider the role of land reform in rural development and, more broadly, the role of property rights. In contrast to reforms that redistribute land (or those that enforce collectivization) in order to improve agricultural productivity, the Scottish land reform is unusual in that it neither changes the size of land holdings, nor does it change the rights of individual landholders, and in fact has very little to do with the practice of agriculture at all. Nevertheless, smallholder communities in the Highlands and islands are looking to land reform as a driver of rural community development. Drawing on the rationale for land reform in Scotland, this chapter argues that, whereas land reform has classically been regarded as a way of increasing agricultural productivity, it can also be viewed as a way of increasing democracy – and that increasing democracy can be a better path to community development than focusing primarily on increasing farm productivity.

Private Property Rights, Land Reform and Rural Development

The term land reform is generally applied to any sweeping change in the distribution or nature of rights to land. Sometimes it is applied to situations where landholders' rights are altered or clarified, with no land changing hands, but usually it refers to the breaking up of large, private estates in order to distribute the land to peasant farmers, either in numerous small private parcels or through the creation of collectives.

The dominant philosophy of land reform in academic and policy circles has long been one that advocates the creation of widely held individual private ownership as a means of enhancing productivity (Deininger and Binswanger, 1999). Smaller landholdings are characterized as being more productive (Binswanger *et al.*, 1995; Prosterman and Hanstad, 2003) and improved tenure security is expected to encourage long-term investment and provide greater access to credit. Facilitating productivity in this way is considered important by those who claim that 'low agricultural productivity is the root cause of rural poverty' (Kiresur *et al.*, 2010, p. 29) and that 'increasing the productivity of

---

* Email: mdh32@cornell.edu

smallholder agriculture holds the key to poverty reduction' (Abro et al., 2014, p. 461).

Economists argue that private property rights do two things to promote the efficient use of resources: first, investment and conservation are encouraged when the rewards and costs of decisions are brought home to the decision-making owners; and second, the right to transfer ownership enables trade, as well as the use of property to secure credit. It is believed that these two functions of private property – known as the internalizing and allocative functions – will cause resources to gravitate to the most efficient users, resulting in greater wealth creation overall (The World Bank Group, 2001; Anderson and McChesney, 2003; O'Driscoll and Hoskins, 2003).

This market-based philosophy of land reform places great emphasis on the importance of property rights that are not only clear and secure, but which are also private and represent a sphere of unfettered control. In a private property system that 'gives individuals the exclusive right to use their resources as they see fit', O'Driscoll and Hoskins (2003, p. 8) tell us, 'dominion over what is theirs leads property owners to take full account of all the benefits and costs', a process that they claim produces the most efficient outcomes and 'translates into higher living standards for all'. It is argued that owners should hold complete bundles of property rights because complete control allows for greater entrepreneurial activity and people who hold 'thin bundles' are less likely to invest in property or to take good care of it (Boudreaux, 2005; cf. Heller, 1998).

The ongoing land reform in Scotland presents itself as an interesting case in contrast to this market-based philosophy of land reform. While it is squarely focused on economic development – the first consultation paper of the Land Reform Policy Group reported that the current concentration of land ownership in Scotland inhibits enterprise (LRPG, 1998) – it neither redistributes land nor does it redefine the rights of individual landholders regarding their own land. Instead, it responds to highly concentrated private ownership by facilitating the transfer of large private estates to community ownership. In doing so, the Scottish land reform advances what MacGregor (1993) terms the 'social rights perspective' over the 'individual rights perspective' and seems to align with Bryden and Hart's (2000) claim that the fundamental goal of land reform should be 'the extension of democracy and democratic practice'.

The next two sections examine the rationale for community ownership as a development strategy in the Scottish context, based on the perspectives of people living in communities where such transfers have taken place. In the discussion that comes after, I use the Scottish case to make sense of why the extension of democracy and democratic practice should be a central priority for land reform, in contrast to a primary focus on productivity and private wealth creation.

Background to the Scottish Case

Private land ownership in Scotland is said to be the most heavily concentrated in the developed world, with more than half of all privately owned land being in the hands of just 432 owners (Hunter et al., 2014) – a pattern of ownership that has long been associated with an acute lack of rural development and has been the source of persistent calls for land reform.

The process by which land came to be so concentrated in so few private hands is the story of how land was transformed over time from being the common heritage of the people who lived there to being the private property of their leader, and later, their landlord – a slow transition in the nature of property ownership under the influence of both medieval and early modern forces, culminating in mass displacement of the rural population in the late 18th and early 19th centuries.

The effect of feudalism on early clan society was to strengthen the hereditary connection between the clan chief and the territory controlled by clan, while also redefining the chief's power as coming from above, rather than from below. The full significance of these changes remained latent until the 18th century, when clan society came to an end. Clan chiefs had already begun to regard their land and their people as a source of personal income, and when they were stripped of all their authority in 1746 after the failed Jacobite rising – except, significantly, for the right to collect rent – their transformation from leaders to landlords was complete. The relationship between tenant and landlord was further attenuated as large estates came increasingly to be bought and sold and landlords began to employ professional estate managers.

With rising wool prices in the late 18th century, southern sheep farmers were eager to rent land in the Highlands, and proprietors soon realized that a great deal more rent could be collected with a great deal less trouble from a small number of sheep ranches than from a large number of subsistence farmers. In order to do this, landlords needed to get rid of the Highland people, and this was achieved through mass evictions that came to be known as the Clearances. Valleys that had once held several townships and numerous small farms would come to be occupied by a pair of shepherds, their dogs and thousands of sheep.

At first, people were settled along the rocky coast on plots of land that were intentionally too small to support a family. This was done in order to compel them to take up work in the nascent kelp and fishing industries (Bangor-Jones, 2001). When the price of kelp fell in the early 19th century, forced emigration began (Prebble, 1963). The brutality with which evictions were carried out lingers in popular memory and has been well documented by historians (MacKenzie, 1883; Prebble, 1963; Richards, 1999; Hunter, 2015).

Falling wool prices in the late 19th century led to a decline in sheep farming, but the shooting of grouse and deer was becoming extremely popular sport among the upper classes and Highland estates were increasingly let, sold and managed specifically for these activities. By 1893 sporting rents contributed, on average, 44% of the rental income on Highland estates (Jarvie and Jackson, 1998). Thus, the Highlands were transformed from being a site of agricultural production to being an upper class recreational area where the imagery of clan society added romantic flavour to a landscape viewed as scenic wilderness.

It is clear on the face of it that, first by putting the Highlands under sheep, and later by making them the occasional hunting grounds of a small number of people, private property rights and market forces were not allocating land to the most productive uses or users. Neither could they be said to have improved the general welfare. The lives of most people living in the Highlands and islands continued to be ones of poverty and hardship.

In response to increasing incidences of rural unrest and a growing public awareness of persistent poverty in the region, the Napier Commission was appointed in 1883 to conduct a public inquiry into the conditions of crofters and cottars[1] in the Highlands and islands. Public testimony was taken at hearings throughout the region and, in 1886, the Crofters Act was passed, guaranteeing crofters a fair rent, protection from eviction and ownership of the improvements on their crofts. An amendment to the Act in 1891 formalized crofters' rights to common grazing land. Modern crofting tenure can thus be seen as a result, first, of settlement patterns created by the Clearances and, second, of subsequent legislation for the protection of crofters.

What the Crofters Act did not do was make more land available – a failure that was especially unsatisfying to cottars, who had no formal access to land. This shortage of access to land, combined with a declining fishing industry and poor economic conditions, led to continued unrest throughout the region, including incidences of land-raiding, rent strikes and clashes with police. Discontent was exacerbated by the frustrated expectations of World War I veterans, who had been given reason to believe that their service would entitle them to land (Leneman, 1989); and in some cases, crofters and cottars occupied land from which their parents or grandparents had been evicted (Hunter, 2000; Willis, 2001). Although new crofts were created by public agencies throughout the early 20th century, progress was slow; poverty remained widespread; and highly concentrated patterns of landownership continued to be associated with a lack of rural development, thus keeping the cry for land reform alive at the end of the 20th century (Bryden, 1999; Hunter, 2000).

In 1997 a Land Reform Policy Group was established to examine land ownership in Scotland; to assess possibilities for land reform; and to make recommendations for action on this issue to the new Scottish Parliament, which in 1999 convened for the first time in almost 300 years. During its first session the Land Reform (Scotland) Act 2003 was passed. The most remarkable feature of this act was the Community Right to Buy, which allowed communities of fewer than 10,000 persons to register an interest in land and thereby to have a preemptive right to buy that land should it come up for sale

(Land Reform Act 2003: Part 2). Crofting communities were given the right to force a sale without waiting for the land to come onto the market (Land Reform Act 2003: Part 3; Donnelley, 2009). The Land Reform (Scotland) Act 2016 expands on the 2003 Act by creating a process whereby non-crofting communities can also apply to force a sale of land if it is in the interest of furthering sustainable community development (Land Reform Act 2016: Part 5).

In order to register an interest in land and to exercise a preemptive right to buy or to force a sale, a community needs to meet a number of requirements, including the formation of a community organization that is a company limited by guarantee,[2] with a board of directors elected by the company's members. All local residents have the right to be voting members and a majority of the board must be local residents.

The community organization has the right to own land and other assets, as well as to make contracts, but the company's property and income do not belong to its members and cannot be distributed (except in the form of reasonable payment for goods and services). The company is constrained to manage its assets for the long-term benefit of the local community and in accordance with its articles of incorporation and the will of its members. These companies can vary somewhat in how they are set up, but all must meet these and other statutory requirements. A more detailed understanding can be gained from looking at the model articles of association (Scottish Government, 2006) and guidance for communities on the websites of the Community Ownership Support Service (DTAS, 2017) and Highlands and Islands Enterprise (HIE, 2017), both government-funded entities tasked with helping communities to take ownership of local assets.

When an estate is transferred to community ownership, the relationship between tenants and the community organization, in its capacity as landowner, is technically the same relationship that existed between the tenants and their previous landlord. The significant difference, in principle, is that the new landlord is a democratic organization of which they are voting members, and one that is required to manage the land for local public benefit.

It is important to note that several community buy-outs and other occurrences resulting in community ownership happened prior to 2003 and outside the provisions of the Land Reform Act, and that it was emphasized to me by a number of interviewees that the community land movement does not centre entirely on this legislation. Money for community buy-outs has come from a variety of sources, including public funding from the lottery and, in most cases, extensive fundraising efforts on the part of the communities.

By March of 2014, 175,418 ha of land in the Highlands and islands were under the ownership of 52 community organizations (HIE, 2014). In only one instance has land been acquired by a forced sale, that being the case of the Pairc Estate (Scottish Government, 2017) on the Isle of Lewis in December of 2015 – an area of 10,840 ha that is now owned by the Pairc Trust on behalf of local residents (Pairc Trust, 2017).

Why Community Ownership? The Perspective from Scottish Communities

The reasons for land reform in Scotland are typical enough: highly concentrated private ownership of rural land; a history of injustice at the hands of landlords; and an acute need for economic development. The strategy that land reformers are pursuing, however – transferring estates to community ownership – stands in stark contrast to a more mainstream emphasis on diffuse private ownership, fairly complete property rights bundles and markets for land. It also touches very little on agriculture.

In order to understand why Scottish land reformers have picked such an unusual path and how they expect community ownership to help people in the Highlands and islands, I conducted loosely structured in-depth interviews in nine communities where land has been transferred to community ownership (Hoffman, 2013).Interviewees in these communities were all either involved in the land transfer at the time it happened or were working for the community organization at the time of the interview in either 2007 or 2008. Interviews in communities were supplemented by interviews with staff members at the government agency responsible for helping communities to acquire land, and my understanding of local circumstances was greatly increased by numerous informal discussions with crofters and other community members.

Although the circumstances leading to community ownership varied greatly between the communities, a number of clear trends emerged regarding the rationale. One of these was that people did not generally view land reform as something that would enable them to do things with their own land that they could not do otherwise. Rather, they saw land reform as something that would enable them to carry out community projects and promote local development.

The protections granted by the Crofters Act of 1886 already provide crofters with many of the benefits of private ownership and, although they have long had the right to acquire full ownership of their crofts at well below market price (Crofting Reform Act, 1976), very few of them have taken advantage of this option. What then do they need land reform for?

Communities in the Highlands and islands perceive that they face a double challenge: generating local economic development and capturing the benefits of that development. Many people expressed the concern that whereas private entrepreneurship might generate wealth, it will not necessarily benefit the community – an observation that was frequently coupled with the long-standing complaint about private landlords not reinvesting rents in their estates.

It was repeatedly pointed out that development is most fundamentally about making sure that people's needs are met. When a local grocery store, gas station or post office closes in a rural community, people may have to drive long distances to access these basic services in some other town. Once they do, they are likely to run all of their errands on the same trip. Thus, the closure of one essential rural business can trigger a cascading loss of other businesses as well, sending the community into a downward spiral as it becomes harder to attract and retain residents in a place with few amenities or jobs, and harder to attract businesses to a place with few customers.

In 2009, more than 400 village shops closed in the UK (Davies, 2009) and pubs were closing at a rate of 39 per week (Brignall, 2009). These businesses were not necessarily unprofitable, but may have been less profitable than other investment opportunities elsewhere, and often closed their doors when the owners retired. The problem was not limited to the private sector: during the previous 2 years, 2500 post office branches closed (Post Office®, 2010a), bringing the number of post office branches from 25,000 in 1960 (Shankleman, 2007) to about 12,000 in 2010 (Post Office®, 2010b). The response to this has been a wave of community ownership across Britain (Brignall, 2009), with the establishment of more than 260 community-owned shops, including 18 in Scotland (Plunkett Foundation, 2012).

Encouraging the development of assets that may be ignored by mobile capital goes hand in hand with the challenge of making sure that these assets are developed in a way that meets local needs. In one community, for example, the fishermen needed a new wharf and were capable of paying fees just sufficient to cover the cost of construction and maintenance. A private investor seeking to employ his or her capital profitably anywhere in the world would probably ignore this property or develop it in a way that is inconsistent with the fishermen's needs. The community-as-landlord, however, is in a position to either develop the property themselves or to offer incentives for private investment that does meet their needs. In this case, the community was planning to tie renovation of the fishing wharf to the lease of a hotel in the same location.

Thus, communities seek not only to stimulate development, but also to guide it and to capture a portion of the value created. An important insight, frequently mentioned in this regard, is that the value of any given property is strongly affected – for better or for worse – by how neighbouring properties are used. When the community is the landlord of not just one property, but all the neighbouring properties as well, they have a strong incentive to protect and enhance the value of each of these properties, making sure that all of them are used in a complementary fashion, such that none of them is used in a way that adversely affects the use or value of the others. This is also, of course, one of the basic functions of public land use planning – a point that will be taken up below.

A related issue is the desire to ensure that the benefits of development are evenly spread. Whereas some assets, such as wildlife and commercialized hunting opportunities (still a major source of income on estates) can only be managed at a large scale and would be greatly impaired by landscape fragmentation, other assets, such as wind power, can be developed in just one place. Without community ownership, however, the benefits of such development would be concentrated, while the adverse impacts would be diffuse. It was also frequently mentioned that in addition to taking an integrated landscape perspective, communities want to expand the timeframe of their development thinking, taking the needs of future generations into account.

In sum, five things were most commonly mentioned across all interviews. People in the Highlands and islands are pursuing community ownership in order to make sure that:

- Wealth generated from the land remains in the community.
- The benefits of development are evenly spread.
- Needed services are provided.
- The population is maintained.
- Resources are managed for the long-term benefit of the community.

One thing that is conspicuously absent from the land reform agenda in Scotland is agriculture. Although there is a strong association in most people's minds between land reform and crofting, many crofters are only minimally engaged in farming activities and none of the communities in which I conducted interviews had any concrete plans to either expand the practice of farming or its productivity. It was explained to me, both by community members and government agency staff, that there is no expectation that crofting should be an adequate source of livelihood. The focus instead is on generating additional sources of employment. This helps to explain why the Scottish land reform leaves both the size of crofts and the conditions of tenure virtually unchanged – focusing instead on replacing the private landlord with a democratic community organization intended, for the reasons given above, to generate local community development.

Further Reflections on Property and Development

There is a tension at the heart of our local and global conversations about development, a tension between whether the central problem is one of productivity and wealth creation, or one of fairness in distribution. The distribution question is further complicated by whether we are talking about the distribution of wealth after it is created or the distribution of resources prior to wealth creation.

Discussions about land reform often blend the issues of productivity and fairness by suggesting that the distribution of property rights has an effect on productivity. In some cases the reasons for this seem straightforward, as they did in 19th-century Ireland, where 'scenes of wretchedness were surrounded by wide-ranging pastures from which the villagers or their fathers had been evicted'.

> To look over the fence of the famine-stricken village and see the rich green solitudes which might yield full and plenty spread out at the very doorsteps of the ragged and hungry peasants was to fill a stranger with a sacred rage and make it an unshirkable duty to strive towards undoing the unnatural divorce between the people and the land.
>
> (O'Brien, 1910, pp. 85–86)

In many cases, however, the redistribution of rights to land is intended not simply to give peasants access to a means of subsistence, but to stimulate an intensification of production (Prosterman and Hanstad, 2003).

The problem comes when intensified production and markets for land lead to a more concentrated pattern of ownership, further separating people from access to the increased productivity. This is the point that a sympathetic Scottish landlord was making in the early 19th century when he commented that 'a population may be in a wretched condition although their country is very well farmed' (Laing, 1837, p. 37). The truth of this has been borne out by the persistence of hunger alongside the surpluses of the Green Revolution (Lappé et al., 1998), as well as by the numerous famines that have occurred in countries that were exporting substantial quantities of food at the time (Ponting, 2007), including Ireland in the mid-19th century.

Because the pursuit of greater agricultural productivity not only can be divorced from rural well-being, but it is further 'possible for agricultural growth to be associated with worsening of the distribution of income', and 'even cause parts of the rural population to become poorer in absolute terms' (Alexandratos, 1995, p. 224), it is clear that this is an insufficient path to development and thus an insufficient goal for land reform if, as El-Ghonemy (1999, p. iii) describes land tenure policy after 1980, it places the 'emphasis on resource use efficiency and output growth, irrespective of distributional consequences'.

In his classic mid-20th-century study of California agriculture, Walter Goldschmidt (1978) described the effect of ownership structure on community well-being when productivity is held as a constant, claiming that diffuse ownership – that is, having many small farms instead of a few large ones – results in higher levels of community well-being. The reasons for this have been theorized by Mills and Ulmer (1970) and by Lyson et al. (2001) to be not strictly related to income distribution, but rather to the higher levels of civic engagement that are associated with diffuse ownership. This suggests that improving local democracy can improve community well-being, even when productivity is held constant – and that improved democracy is an expected result of more diffuse ownership.

So long as productivity and diffuse ownership appear to go hand in hand, it is easy to let the two mingle in policy discourse. When they appear to diverge, however, a clarification of goals becomes necessary. It has been suggested at various times that because crofts in the Highlands and islands are by design too small to support a family, they ought to be amalgamated into appropriately sized units for the adoption of more efficient farming (Gillanders, 1968). On average, 70% of crofting household income is from non-agricultural activities (Shucksmith, 2008) and the expectation expressed by the Committee of Inquiry on Crofting in their vision for the future is that this will continue to be the case (Committee of Inquiry, 2008). Government assistance to crofters in the latter half of the 20th century, coming as it did in the form of agricultural improvement grants, betrayed 'a lack of insight into this central reality of crofting life' (Hunter, 2000, p. 284).

Crofts are places where people live, participate in a way of life and enjoy a certain baseline of security. The point therefore is not to increase the productivity of crofts, but to increase the viability of crofting communities through expanded employment opportunities. Agricultural modernization and the consolidation that goes with it would run counter to this need, as any measures that reduce the number of people on the land would exacerbate the problem of dwindling population and the consequent loss of services. This, then, explains why the community land movement today touches very little upon agriculture and leaves the tenure of individual crofters virtually unchanged, and why its primary emphasis is on generating economic development that will supplement rather than enhance farm income. To borrow a distinction from Shucksmith and Rønningen (2011), it is an expression of rural policy rather than agricultural policy.

The land tenure history of Scotland and the rationale for community ownership as expressed by the people I interviewed cast doubt on the classic arguments for private property based on its supposed internalizing and allocative functions. The community's need to make sure that neighbouring land uses are complementary has long been recognized in the planning literature, where the interdependence of property values and the impossibility of internalizing the effects of land use decisions on small parcels has led many writers to liken the situation of private landowners to a prisoner's dilemma game (Davis and Whinston, 1961; Mandelker, 1965; McMillan, 1974; Glück, 2000). Loehr's (2012, p. 837) claim that the individualization of property rights via land reform 'leads to a decoupling of the benefits and costs of land use', turns the internalizing argument for private property on its head. The community-as-landlord, on the other hand, because they capture a portion of all property values via rent, has a strong incentive to protect neighbouring property values and to seek the best mix of compatible uses. This is beneficial both to individual tenants and to the public at large, offering a model by which communities can act on Bryden's (1996) warning that 'we have paid too much attention to the growth of individual enterprises and too little to the development of the public goods on which the development of enterprises and communities both depend'.

It might well be asked, however, why community ownership is necessary in order to carry out what would seem to be the basic land use planning functions of local government, with rent in this case acting as a substitute for property taxes. The answer to this, to quote the commission on local government and the Scottish Parliament, is that 'Scotland today simply does not have a system of local government in the sense in which many other countries still do' (McIntosh, 1999, chapter 6, Line 155). When people speak of local government in Scotland, they are referring to unitary councils such as that which serves the entire Highland area. Community councils exist only to communicate local opinion to higher levels of government and possess neither the resources nor the authority to govern in their own right (McIntosh, 1999; ONS, 2004). Thus, it may not be wholly off the mark to suggest that the Scottish land reform, consciously or not, is fundamentally an attempt to reinvent local, democratic governance in a place where it does not otherwise exist.

In this regard, land reform in Scotland is following a similar path philosophically to that described by Rosset (2013, p. 727) for La Via Campesina as they have moved from demanding land for agriculture to demanding territory 'for (re)constructing and defending community'. Why democracy is important, specifically in relation to the local governance of land, has been clearly articulated by people living in Scottish communities where community ownership has been pursued: local development depends on the community being able to encourage and guide the development of local assets in ways that meet the community's needs, and also on being able to capture a portion of publicly created value.

Whether this is done via community ownership or through the strengthening or reinvention of local government in some other fashion, land reform, as Bryden and Hart (2000) have put it, must fundamentally be 'a change in the balance of power between individual property owners, communities and the state', the goal of which should be 'the extension of democracy and democratic practice'.

In light of the all-too-frequent divorce between agricultural productivity and rural well-being, the idea of Scottish land reformers – that changes in land rights can contribute to rural development by advancing democratization as much as by advancing productivity – deserves our increased attention in the years to come.

Notes

[1] A croft is a form of agricultural tenancy unique to the Highlands and islands of Scotland. Crofts are usually small and situated on land of marginal agricultural value. The holder of a croft, a crofter, owns the buildings and other improvements but pays rent to the landlord on whose land the croft is situated. Crofting tenure is statutory and regulated by the Crofting Commission (see crofting.scotland.gov.uk). A cottage, by contrast, was a small house with no agricultural land attached to it and no formal access to common grazing land. Such tenancies were often located on agricultural properties where the cottar was employed.
[2] A company limited by guarantee is a common form of incorporation for not-for-profit organizations in the UK. It does not have shareholders and cannot distribute profits.

References

Abro, Z.A., Alemu, B.A. and Hanjra, M.A. (2014) Policies for Agricultural Productivity Growth and Poverty Reduction in Rural Ethiopia. *World Development* 59, 461–474.
Alexandratos, N. (ed.) (1995) *World Agriculture: Towards 2010. An FAO Study*. Food and Agriculture Organization of the United Nations and John Wiley and Sons, Chichester, UK.
Anderson, T.L. and McChesney, F.S. (2003) *Property Rights: Cooperation, Conflict, and Law*. Princeton University Press, Princeton, New Jersey.
Bangor-Jones, M. (2001) *The Assynt Clearances*. The Assynt Press, Dundee, UK.
Binswanger, H.P., Deininger, K. and Feder, G. (1995) Power, distortions, revolt and reform in agricultural land relations. In: Behrman, J. and Srinivasan, T.N. (eds) *Handbook of Development Economics*. Elsevier Science, Amsterdam, pp. 2659–2728.
Boudreaux, K. (2005) *The Role of Property Rights as an Institution: Implications for Development Policy*. Mercatus Center, George Mason University, Arlington, Virginia.

Brignall, M. (2009) Last orders? Locals fight back. *The Guardian*, 21 March 2009. Available at: www.guardian.co.uk/money/2009/mar/21/rural-communities-buyout (accessed 21 March 2018).

Bryden, J. (1996) Land tenure and rural development. 3rd Annual John McEwen Lecture. Available at: www.caledonia.org.uk/land/bryden.htm (accessed 21 March 2018).

Bryden, J. (1999) Scottish land reform: the great debate. Paper presented at Kings College Conference Centre, University of Aberdeen, Aberdeen.

Bryden, J. and Hart, K. (2000) Land reform, planning and people: an issue of stewardship? Paper given at the RSE/SNH Millennium Conference 'The Future for the Environment in Scotland: Resetting the Agenda?' in Edinburgh, March 2000.

Committee of Inquiry (2008) *Towards the Future of Crofting: Vision Statement of the Committee of Inquiry on Crofting*. Edinburgh: Rural Directorate, Scottish Government, Edinburgh, UK.

Crofting Reform (Scotland) Act 1976, Chapter 21. Available at: www.legislation.gov.uk/ukpga/1976/21/pdfs/ukpga_19760021_en.pdf (accessed 21 March 2018).

Davies, C. (2009) The Archers helps to save village shops: radio soap spreads word about community-owned stores. *The Guardian*, 30 December 2009. Available at: www.guardian.co.uk/uk/2009/dec/30/archers-village-shops-community-owned (accessed 21 March 2018).

Davis, O.A. and Whinston, A.B. (1961) The economics of urban renewal. *Law and Contemporary Problems* 26(1), 105–117.

Deininger, K. and Binswanger, H. (1999) The evolution of the world bank's land policy: principles, experience, and future challenges. *The World Bank Research Observer* 14(2), 247–276.

Development Trusts Association Scotland (DTAS) (2017) Community Ownership Support Service, administered by the Development Trusts Association Scotland. Available at: www.dtascommunityownership.org.uk/resources (accessed 21 May 2017).

Donnelley, R.R. (2009) *Part 3 of the Land Reform (Scotland) Act 2003, Crofting Community Right to Buy: Introduction*. The Scottish Government, Edinburgh, UK. Available at: www.gov.scot/Resource/Doc/274633/0082188.pdf (accessed 21 March 2018).

El-Ghonemy, M.R. (1999) *The Political Economy of Market-Based Land Reform*. DP 104. United Nations Research Unit for Social Development, Geneva.

Gillanders, F. (1968) The economic life of Gaelic Scotland today. In: Thomson, D.S. and Grimble, I. (eds) *The Future of the Highlands*. Routledge, London, pp. 95–150.

Glück, P. (2000) Policy means for ensuring the full value of forests to society. *Land Use Policy* 17(3), 177–185.

Goldschmidt, W. (1978) *As You Sow*. Allanheld, Osmun, Montclair, New Jersey.

Heller, M.A. (1998) The tragedy of the anticommons: property in transition from Marx to markets. *Harvard Law Review* 111(3), 621–688.

Highlands and Islands Enterprise (HIE) (2014) Community land ownership in the highlands and islands (map). Available at: www.hie.co.uk/community-support/community-assets/assets-and-buyout-map.html (accessed 18 April 2017).

Highlands and Islands Enterprise (HIE) (2017) Highlands and Islands Enterprise, guidance documents and other resources. Available at: www.hie.co.uk/community-support/community-assets/resources.html (accessed 31 May 2017).

Hoffman, M. (2013) Why community ownership? Understanding land reform in Scotland. *Land Use Policy* 31, 289–297.

Hunter, J. (2000) *The Making of the Crofting Community*. John Donald, Edinburgh.

Hunter, J. (2015) *Set Adrift Upon the World: The Sutherland Clearances*. Birlinn, Edinburgh.

Hunter, J., Peacock, P., Wightman, A. and Foxley, M. (2014) 432:50 – Towards a comprehensive land reform agenda for Scotland. A Briefing Paper for the House of Commons Scottish Affairs Committee. Available at: https://www.parliament.uk/documents/commons-committees/scottish-affairs/432-Land-Reform-Paper.pdf (accessed 21 March 2018).

Jarvie, G. and Jackson, L. (1998) Deer forests, sporting estates and the aristocracy. *The Sports Historian* 18(1), 24–54.

Kiresur, V.R., Melinamani, V.P., Kulkarni, V.S., Bharati, P. and Yadav, V.S. (2010) Agricultural productivity, rural poverty and nutritional security: a micro evidence of inter-linkages from Karnataka state. *Agricultural Economics Research Review* 23, 29–40.

Laing, S. (1837) *Journal of a Residence in Norway during the Years 1834, 1835, and 1836; with a View to Enquire into the Moral and Political Economy of that Country, and the Condition of its Inhabitants*. Longman, Orme, Brown, Green, and Longmans, London.

Land Reform (Scotland) Act 2003, asp 2, Part 2: *The Community Right to Buy*. Available at: www.legislation.gov.uk/asp/2003/2/pdfs/asp_20030002_en.pdf (accessed 21 March 2018).

Land Reform (Scotland) Act 2003, asp 2, Part 3: *The Crofting Community Right to Buy*. Available at: www.legislation.gov.uk/asp/2003/2/pdfs/asp_20030002_en.pdf (accessed 21 March 2018).

Land Reform (Scotland) Act 2016, asp 18, Part 5: *Right to Buy Land to Further Sustainable Development*. Available at: www.legislation.gov.uk/asp/2016/18/pdfs/asp_20160018_en.pdf (accessed 21 March 2018).

Lappé, F.M., Collin, J. and Rosset, P. (1998) *World Hunger: 12 Myths*. Grove Press, New York.

Leneman, L. (1989) *Fit for Heroes?: Land Settlement in Scotland After World War I*. Aberdeen University Press, Aberdeen, UK.

Loehr, D. (2012) Capitalization by formalization? – Challenging the current paradigm of land reforms. *Land Use Policy* 29, 837–845.

Land Reform Policy Group (LRPG) (1998) Land Reform Policy Group: identifying the problems. The Scottish Office. Available at: www.scotland.gov.uk/library/documents1/lrpg00.htm (accessed 21 March 2018).

Lyson, T., Torres, R.J. and Welsh, R. (2001) Scale of agricultural production, civic engagement, and community welfare. *Social Forces* 80(1), 311–327.

MacGregor, B. (1993) Land tenure in Scotland. 1st Annual John McEwen Lecture. Available at: www.caledonia.org.uk/land/macgrego.htm (accessed 21 March 2018).

MacKenzie, A. (1883) *History of the Highland Clearances*. A. & W. MacKenzie, Inverness, UK.

Mandelker, D.R. (1965) The role of law in the planning process. *Law and Contemporary Problems* 30(1), 26–37.

McIntosh, N. (1999) The Report of The Commission on Local Government and The Scottish Parliament. The Scottish Office, Edinburgh, UK. Available at: www.scotland.gov.uk/deleted/library/documents-w10/clg-00.htm (accessed 21 March 2018).

McMillan, M. (1974) Open space preservation in developing areas: an alternative policy. *Land Economics* 50(4), 410–418.

Mills, C.W. and Ulmer, M. (1970) Small business and civic welfare. In: Aiken, M. and Mott, P. (ed.) *The Structure of Community Power*. Random House, New York, pp. 124–154.

O'Brien, W. (1910) *An Olive Branch in Ireland and Its History*. MacMillan and Company, London.

O'Driscoll Jr., G.P. and Hoskins, L. (2003) Property rights: the key to economic development. Cato Institute Policy Analysis No. 482. Available at: https://www.cato.org/publications/policy-analysis/property-rights-key-economic-development (accessed 21 March 2018).

Office for National Statistics (ONS) (2004) Parishes/communities (under Scotland, under administrative geography). ONS, London. Available at: www.statistics.gov.uk/geography/parishes.asp#sc (accessed 21 March 2018).

Pairc Trust (2017) Urras na Pàirce. Available at: www.pairctrust.co.uk (accessed 18 April 2017).

Plunkett Foundation (2012) What we do. Available at: www.plunkett.co.uk/whatwedo/rcs/ruralcommunityshops.cfm (accessed 21 March 2018).

Ponting, C. (2007) *A Green History of the World: The Environment and the Collapse of Great Civilizations*. Penguin, London.

Post Office® (2010a) See the network change programme under the Post Office® network. Available at: www.postoffice.co.uk/portal/po/content2?catId=57600693&mediaId=57600697 (accessed 21 March 2018).

Post Office® (2010b) See About us > About the Post Office®. Available at: www.postoffice.co.uk/portal/po/content2;jsessionid=MHVOXYIOCY3MYFB2IGDEOSQUHRAYUQ2K?catId=20000192&mediaId=103100763 (accessed 21 March 2018).

Prebble, J. (1963) *The Highland Clearances*. Penguin, London.

Prosterman, R.L. and Hanstad, T. (2003) *Land Reform in the 21st Century: New Challenges, New Responses*. Rural Development Institute, Seattle, Washington.

Richards, E. (1999) *Patrick Sellar and the Highland Clearances: Homicide, Eviction and the Price of Progress*. Polygon (Edinburgh University Press), Edinburgh, UK.

Rosset, P. (2013) Re-thinking agrarian reform, land and territory in La Via Campesins. *Journal of Peasant Studies* 40(4), 721–775.

Scottish Government (2006) Model company articles of association. Available at: www.gov.scot/Topics/People/engage/AssetTransfer/Resources/CompanyLtdByGuaranteeArticlesandGuidance?refresh=0.8613091444252661 (accessed 21 March 2018).

Scottish Government (2017) Crofting community right to buy. Available at: www.gov.scot/Topics/farmingrural/Rural/rural-land/right-to-buy/crofting (accessed 18 April 2017).

Shankleman, M. (2007) Why are post offices at-risk? BBC News, 2 October 2007. Available at: http://news.bbc.co.uk/2/hi/uk_news/7024969.stm (accessed 18 April 2017).

Shucksmith, M. (2008) Committee of inquiry on crofting: final report. RR Donnelley, Edinburgh, UK.

Shucksmith, M. and Rønningen, K. (2011) The uplands after neoliberalism? – the role of the small farm in rural sustainability. *Journal of Rural Studies* 27, 275–287.

The World Bank Group (2001) Question and answer on land issues at the World Bank. Available at: http://lnweb18.worldbank.org/ESSD/ardext.nsf/24ByDocName/QuestionAnsweronLandIssuesattheWorldBank (accessed 21 March 2018).

Willis, D. (2001) *Crofting*. John Donald, Edinburgh, UK.

Index

active paramilitary dispossession 234
Adivasis, India
    agrarian question
        common property resources 206
        contemporary Maoism 205
        intent of policy 206
        Maoist mobilization 206
        National Environment Policy 206
        penetration of foreign capital 206
        politics 206
        post-Nehruvian Green Revolution era 206
        Public Trust Doctrine 207
        state capitalism, critique of 206
    identity 205
    integration of 203
    landlessness 207, 208
    Naxalbari 204
    proletarianization 203–204
African drylands
    Mali and the Sahelian belt 165–167, 169–170
    political-institutional landscape, reconfiguration 171
    Somalia and the Horn of Africa 162–165
    see also pastoralism
agrarian proletariat 109
agrarian question 53, 190
    in the 21st century 16–17
    in Adivasi, India 205–209
    agrarian and agrarian transition 13–14
    Class forces variant 15
    classical 14
    and conflict 59–61
    contemporary 14–16
    corporate food regime, variant 15
    decoupled variant 15
    ecological 15–16, 190, 200
    gendered 15
    of the south 16
Agricultural Co-operative Bank (ACB) 109
agricultural production
    estimating 82
    quantifying changes 73, 78–80
agriculture
    in fragile and conflict affected states 4–6
    importance 4
    last frontier of 35
agro-pastoral systems, Ferghana Valley 186
anti-colonialism 133
anti-extractivist 246
    movements and prior consultation
        Bolivia 254–255
        Mexico 256
        Peru 255–256
    vs. pro-extractivist indigenous 248–250
anti-modernity 150
Arab Spring
    agriculture and ancien regimes
        Egypt 92–94
        Tunisia 94–95
    farmers and resistance
        Egypt 98–99
        Tunisia 100
Argentina
    explanations and limitations
        democratic quality and state capacity levels 267–269
        economic structures and elite interests 266–267
        ethnic fractionalization 265–266
        indigenous land survey implementation, variations 270–271

Argentina (*continued*)
    indigenous political mobilization 261
    motivation 261
    national context 263
    opportunity structures 261
    puzzle 263–265
    qualitative methodology 262
    quiescent path 262
    social movement theory 262
    theoretical framework
        bridging ties 270
        'divide and rule' responses 270
        land conflicts 270
        'repress and ignore' responses 270
        social movements 268
        state structures 270
    transformative path 262
    white nation 261
aridity line 158
Armed Conflict Location and Event Dataset (ACLED) 42
ayllu 254, 255

Ba'athist government 105
    radical policies 108–109
bilateral water conflict 57–58
biofuel policy (2008) 236, 238
biomass estimation 80–81
blue water 54
Bolivia
    prior consultation 250–251
        anti-extractivist movements and 254–255
        pro-extractivist indigenous and 251–252
    resource nationalism 247

Cambodia and Vietnam
    agrarian transition and contained conflict
        on highlands 221–224
        in lowlands 219–221
    analysis 224–226
        agrarian crisis and conflict, dialectics of 224
        failed development 225
        French colonial system 224
        Hun Sen regime 226
        peasants' mobilization 225
        rubber plantations 226
        Scott's 'everyday resistance' 226
        values and norms 225
    analytical framework and methodology 218–219
    crisis, revolts and revolution under the French (1884–1954) 216–217
    hidden resistance and killing fields 217–218
    places of conflict 220
    poor-on-poor conflict 224
    relief and main cities 215
    revolutionary potential 215
    stagnation and peasants' agitation 216
capitalism 16
    biophysical contradictions 190
    robbing of the soil 31
Chashma Canal Right Bank Project (CRBIP) 192
climate change 11, 36, 40–49, 58, 59, 60, 61, 192, 200, 236
    climate variability and conflict
        conceptualizing conflict 42
        quantitative research and gaps 42–43
        theoretical mechanisms 43–44
        weather 41
    consequences for society 40–41
    institutions 45–47
    migration 47–48
    and water scarcity 61
climate-smart agriculture technologies 114
Cold War 92
    political weapon 92
    and rise of opium 65–67
    SAP 161
    and trade liberalization 92
collateral effect 234
Colombia
    causes of displacement 234–236
    conflict and forced displacement 232–234
    department of Putumayo 236–238
    displacement and dispossession 240
    palm oil crops 238–239
Committee of Inquiry on Crofting 281
communities of sufferers 43
conflict and agriculture
    agrarian dynamics in 21st-century 17–19
    agrarian question
        in the 21st century 16–17
        agrarian and agrarian transition 13–14
        classical 14
        contemporary 14–16
    bidirectional relationship of 12–13
    cause of 9–11
    extent and evolution 7–8
    impact of 11–12
    land grab phenomenon 10
    land issues 9
    linkages of 10
    theoretical models 8–9
conflict over water 53
    levels of 55
    national/bilateral water conflict 57–58
    resources 54
    sub-national/local level 58–59
    water allocation 55
    water conflict on global level 56–57

Conflict Shoreline (book) 158
corporate food regime 56–57, 60, 61
Corrective Movement of November 109
Crofters Act 277, 279
crop production 12
cropland 5
cropland changes, Iraqi Kurdistan
    in Duhok (1984–2014) 122–123
    in Mangesh and Semel 123–124
    satellite data, land dynamics 122

Dandakaranya:AdivasiKisanMazdoorSangathana (DAKMS) 210
de facto tolerance 69
degraded capital formation in Syria 111–113
Đổi Mới era 220, 223
drug-trafficking, fight against 236
drugs and war 70–71
Duhok province 119
    cropland changes 122–123

ecological agrarian question 15–16, 190, 200
economic displacement 235
Economic Land Concessions (ELC)
    policy 221–222
economic wilderness 29
EcoPark 221
Egypt
    agriculture and ancien regimes
        agricultural investment scheme 94
        food insecurity 93
        neoliberal market reform 93
        political economy 93
    farmers and resistance 98–99
    rebranding of 96
    unequal farmer access 92
emergency life-saving interventions 114
enclosure food regime 15
Enhanced Vegetation Index (EVI) 74
environmental displacement 236
environmental refugees 236

Farm Bill (1933) 60
farm subsidies 33
farmers and resistance
    Egypt 98–99
    health insurance 95
    pluriactivity 91
    Tunisia 100
Ferghana Valley 176
    agro-pastoral activities 176
    conflict in transborder areas 178
    elevation and political map 177
    Kyrgyz–Tajik border 177

long-lasting agricultural crisis 186
methodology and focal area 178–179
pene-enclaves 177
post-Soviet period in 181–183
pre-Soviet histories 176
Soviet period in 179–181
Uzbek–Kyrgyz border 177
Vorukh and Ak-Say
    map of 179
    natural resource management policy for 184–186
first food regime 28, 30–31, 33, 36, 56
flash-flood irrigation 132
Food for Peace programme 28, 33
food for war programme 33
food production 40
food regime 11, 56, 60, 93
food security 92
    macro-economic policy 93
    monitoring schemes 73
    Syria 112–113
    trade-based 92, 94
    version of 54
food self-sufficiency, Syria 108–109
food sovereignty 95
food supplies 28
    urbanization and 29–30
food trade, politicization 33–35
forced displacement
    categories 235
    conflict and 232–234
    development projects 235
    public policies 236
foreign grain supplies 30
Forest Rights Act 209, 211
fragile and conflict-affected situations (FCS)
    countries 4
    cross-country comparisons 6
    economic activity and livelihoods 6
    Gini coefficients 5–7
    global economy 5
    sub-national data 5
French colonial system 224

gendered agrarian question 15
genocide 16
geo-spatial analysis software 78
geographic information systems (GIS) *see* GIS and remote sensing
geopolitics
    first food regime 30–31
    global food crisis (2008) 35–36
    second food regime 33–35
    urbanization and food supplies 29–30
    World Wars I and II 31–33
Gini coefficients 5–7

GIS and remote sensing 73
   agricultural production
      changes in 78–80
      estimating 82
   controlling for rainfall 81–82
   deriving biomass estimates 80–81
   materials and methods 74–75
   of satellite imagery 74–77
   software 79
   vegetation detection 75, 77–78
   wheat and barley production (Syria) 82–85
global food crisis (2008) 35–36
global level water conflict 56–57
global positioning systems (GPS) 73
global reserve army of labour variant 15
global warming 41
   see also climate change
grain–livestock complexes 28
great grain robbery 34
Green Revolution 191
gross domestic product (GDP) 107
   decomposition in Syria 107
   global agriculture 5
   Palestinian 145–146, 148
   Tajikistan and Kyrgyzstan 182
   Yemen 135
gross fixed capital formation (GFCF) 112
gross national product (GNP) 133
Guano Islands Act (1856) 31

hill tribe economy 68
Hirak movement 140
hotspots 238
household coping strategies 12, 140
hunger blockade 32
Hungerplan 32
hydraulic civilizations 29
hydro-agrarian question 54, 61
hydro-hegemony framework 54
hydropolitics 54

illegal drug plant cultivation
   Cold War and rise of opium 65–67
   drugs and war 70–71
   war and drug production 67–70
India
   Adivasis *see* Adivasis, India
   agrarian capitalism and conflicts 203–204
   agrarian question in Adivasi 205–209
   Implementation of the Compensatory
      Afforestation scheme 209
   Maoist operations
      agrarian relations and change,
         insurgency 210–212
      contemporary history of 209–210

      Naxalite movement, rural roots and origins
         of 204–205
indigenous 145, 190, 191, 194, 198, 203, 219,
   221, 223, 224, 225, 232, 236, 238, 242,
   245, 246–257, 261–272
indigenous land claims, Argentina
   commodity boom and 263
   subnational variations in governance 263–265
institutions, conflict-resolving
   civil war setting 45
   customary or informal 45
   defined 45
   formal 45
   in-migration 46–47
   pre-colonial hierarchization 46
   regulating resource conflicts 46
Instituto Provincial de Pueblo Indígenas de *Salta*
   (IPPIS) 264
integrated water resource management (IWRM) 183
International Fund for Agricultural Development
   (IFAD) 34
international grain trade 28
International Labour Organization (ILO)
   Convention 169 246
investment 112–113
Iraqi Kurdistan *see* Kurdistan Region of Iraq
Israeli occupation, Palestinian agriculture
   Bedouin communities 149
   Gaza Strip 148
   Jewish National Fund's plan 149
   policies of domination 147
   political and economic rights 147
   UNCTAD 148

Johnston Plan 57
Jordan basin water plan 58

ketmen wars 184
Khadafy regime 166
Khmer-Rouge collectivization project 218
KrantikariMahila Adivasi Sangathana (KMAS) 210
Kriegsbrot 31
Kurdistan Region of Iraq
   cropland changes
      in Duhok (1984–2014) 122–123
      in Mangesh and Semel 123–124
      satellite data, land dynamics 122
   historical background 118–122
      Anfal campaign 120
      food sufficiency and security 121
      March Manifesto 119
      Oil-for-Food programme 120
      political-economic development 121
      satellite-based data 121
   qualitative approach 125–126

La tierra en disputa 233
labour-intense techniques 139
Land Centre for Human Rights (LCHR) 98–99
land grab phenomenon 10
land ownership 132
land reform 10, 18, 98, 99, 105, 106, 108, 211, 217, 275–282
Land Reform (Scotland) Act (2016) 278
land reform in Scotland
    Clearances 277
    community ownership 278–280
    Crofters Act 277, 279
    feudalism 276
    Highlands 277, 278, 280
    individual rights perspective 276
    land tenure history 281
    market-based philosophy 276
    Napier Commission 277
    philosophy of 275
    private land ownership 276
    private property rights 275–276
    productivity 275
    reflections on property and development 280–282
    social rights perspective 276
Land Reform Policy Group 277
last frontier of agriculture 35
Latin America
    anti-extractivist movements and prior consultation
        Bolivia 254–255
        Mexico 256
        Peru 255–256
    indigenous movements 248
    prior consultation
        benefit from 253–254
        in Bolivia, Peru and Mexico 250–251
        cross-national variation and 254
        development of right to 250
        extractive industries and 247–248
    pro-extractivist indigenous and prior consultation
        Bolivia 251–252
        Mexico 253
        Peru 252–253
    pro-extractivist vs. anti-extractivist indigenous 248–250
left wing populism, Argentina 263
legibility thesis 272
livestock export, Somali 164
local level water conflict 58–59
long-distance grain trade 29–30
    first food regime and 30–31

Mahatma Gandhi National Rural Employment Guarantee Act (MgNREGA) 211
maladaptation 12

Mali and the Sahelian belt
    civil war 166
    commercial movements and trade 165
    droughts 166
    Fulani population 166–167
    hybrid threat 169
    Khadafy regime 166
    nomad pastoralists, diffusion 166
    pastoralism 165
    poverty rate, decline 169
    reinventing pastoral routes 169
    smuggling 167
    Touareg population 167
    transnational trafficking 167
Malthusian mechanism 47
Mangesh, cropland changes 123–124
Maoism 203
Maoists 203
    agrarian relations and change, insurgency 210–212
    cover organizations 210
    DAKMS and KAMS 210
    liberated area 210
    operations contemporary history 209–210
    PWG 209
    voluntary labour 210
March Manifesto 119
MCD12 land cover product 81
metabolic rift 30
Mexico
    neoliberal reforms 251
    prior consultation 250–251
        anti-extractivist movements and 256
        pro-extractivist indigenous and 253
    resource nationalism 247
migration
    conditions for migrants 47
    displacement link 47
    environmental change on 41
    ethical issues 48
    Malthusian mechanism 47
    natural disasters to drive 41
    population datasets 48
    state-to-state 47
mining and energy policy (2010) 236
Mujammaʿāt 120
Multani 194

narcotics 65
national/bilateral water conflict 57–58
national question 190, 200
National Water Law 138
natural disasters 236
Naxalbari movement 204, 211
Naxalite movement, India 204–205
neoliberal market reform 93

neoliberalism, protectionism 109–111
    'Corrective Movement' of November 109
    liberalization 110
    modernisation 110
    privatization of state farms 110
    resilience of farming system 111
    social market economy 110
net primary productivity (NPP) 80
Nile Water Agreement (1959) 57
Normalized Difference Vegetation Index (NDVI) 74, 78, 122
North West Frontier Province (NWFP), Pakistan 68

oil discovery, Yemen 137–139
Oil for Food programme 34
oil hothouse, Yemen 133–135
opium production, illegal
    Cold War and 65–67
    opium revenue 65
    opium trade 64
    opium wars 67
opportunity cost (OC) mechanism 43
out-migration 133

Pakistan
    hydraulic mission 191
    industrialization 191
    iron triangle 192
    see also Southern Punjab
Palestinian agriculture sector
    average daily wages 146
    colonization process 144
    cultivated areas 146
    de-development 145
    deterioration 147
    food self-reliance 146
    GDP 145–146
    holdings 146
    impact of Israeli occupation
        Bedouin communities 149
        Gaza Strip 148
        Jewish National Fund's plan 149
        policies of domination 147
        political and economic rights 147
        UNCTAD 148
    neoliberalism 144
    olive oil production 146
    PA and agriculture 150–151
    use of water 147
Palestinian Authority (PA) 144
    agro-industrial zones 151
    anti-modernity 150
    consequences 150
    empowering farmers 150
    high risk investment 151

neoliberal policies 151
policy makers 150
Parti Communiste Indochinois (PCI) 217
pastoralism 58, 157
    crisis-and conflict-prone 158
    geo-political system 158
    pastoralists 157
    pastoral settings 158
    trends in pastoral areas
        Conflict Shoreline 159
        crisis and conflict 162
        hybrid governance 160
        limited statehood areas 160
        reconfiguring margins 160–161
        'safe haven myth' 161
        socio-political contract 160
        strategies of change and transformation 161–162
path-dependent variant 15
Peace Research Institute Oslo (PRIO) 42
People's War Group (PWG) 209
Peru
    extractive industries 248
    prior consultation 250–251
        anti-extractivist movements and 255–256
        pro-extractivist indigenous and 252–253
petroleum-intense farming 134
Plan of Action for Syria 2016–2017 114
pluriactivity, farmers 91
policy of plenty 30
politics of denial 70
The Politics of Heroin (book) 64
poor-on-poor conflicts 224
post-Soviet period, Ferghana Valley
    agrarian reforms 181
    collective *dekhan* farms 182
    IWRM 183
    Kyrgyzstan and Tajikistan 182
    labour migration 182–183
    natural resources 182
    Pasture User Unions (PUUs) 183
    Water Code 183
    Water Users Associations (WUAs) 183
Pravda, Vorukh 181, 184
primitive accumulation 28
pro-extractivist 246
    indigenous and prior consultation
        Bolivia 251–252
        Mexico 253
        Peru 252–253
    vs. anti-extractivist indigenous 248–250
Public Law 480 56
Putumayo, department of
    impacts on human health 237

oil exploration and exploitation 237–238
oil industry and glyphosate spraying 236–237

rainfall, controlling 81–82
re-engineering of institutions 97
Reagan Doctrine on agricultural trade 34
Red Corridor 203
reflectance, vegetation detection 75
relative deprivation (RD) mechanism 43–44
remote sensing 73
    see also GIS and remote sensing
reprimarization 266
risk-proof infrastructure and plantation 114

Sahelian belt *see* Mali and the Sahelian belt
satellite imagery
    changes in agricultural production 75
    sensors 76–77
    spatial resolution 75
    vegetation detection 75
Saths 194, 197
Scotland, land reform in *see* land reform in Scotland
second agricultural revolution 30
second food regime 28, 33–35, 36, 58, 92
Semel, cropland changes 123–124
Service de documentation extérieure et de contre-espionnage (SDECE) 66
sharecropping 132
smallholder agriculture 4
social conflict 91
social market economy 110
socio-environmental conflicts 235
socio-political contract 160
soil organic matter 114
Somali ecosystem 162, 163
Somali pastoral economy 164, 165
Somalia and the Horn of Africa
    agro-ecological conditions 162
    Al-Shabab 165
    berkaad establishment 164
    corridors 162
    import–export dynamics 164
    livelihood strategies 162
    livestock export 164
    pastoral economy 164
    Siad Barre's regime, failure 163
    social innovation 165
    Somali ecosystem 162, 163
Southern Punjab
    consequences of floods 197–199
        cynicism 197
        environmentalism of poor 198
        Paris Agreement (2015) 199
        political impact 198
        tribunal in Mauza Spira 199

flood causality and politics of blame
    agrarian question 196
    criticism of *waderas* 196
    feudalism 196
    irrigation 197
    Muzaffargarh 195
    'pondage' areas 195
floods in context (2010) 191–193
irrigation policy 191
methodology and data 193–195
Soviet period, Ferghana Valley
    agricultural production 179
    agro-pastoral practices 179
    Basmachi movement 180
    collective farms 180–181
    de facto border 181
    semi-nomadic livestock 180
    in Vorukh and Ak-Say 180
state-set pricing policy 109
state subsidized credit 138
structural adjustment program (SAP) 94, 160
sub-national/local level water conflict 58–59
Suez Canal crisis 58
sustainable irrigation, Yemen
    development projects 137
    egalitarian techniques 137
    GDP 135
    land devoted to commodity crops 136
    marketing channels 136
    qat production 136
    unsustainable techniques 135
symptomatic silence 91
synergy 64–65
Syria
    agricultural sector 107–108
    Assad government 109–111
    Ba'athist radical policies 108–109
    degraded capital formation 111–113
    food industries 108
    GDP decomposition 107
    irrigation techniques 106
    land reform 105
    liberalization 110
    neoliberal reforms 106

theoretical models, conflict
    armed conflict 9
    economic demand and opportunity 8
    natural resources 8
    political structures 8–9
    urban bias 9
third food regime 36, 92
Tierras y conflictos rurales (book) 238
Tunisia
    agriculture and ancien regimes

Tunisia (*continued*)
    Fifth Five Year Plan 94
    food security 94
    structural adjustment program (SAP) 94
    smallscale family farming 94
    water-intensive agriculture 94
  farmers and resistance 100
  unequal farmer access 92

United States Agency for International Development (USAID) 93
Uppsala Conflict Data Program (UCDP) 42
urbanization and food supplies 29–30

value-chain development 114
vegetation detection 75, 77–78
Victims and Land Restitution Law 233
Vietnam *see* Cambodia and Vietnam
virtual water 54
Volta basin 58

war
  on drugs 66, 70–71
  Houthis and 139–140
  and illegal agricultural drug production
    cannabis cultivation 70
    cash crops 67
    hill tribe economy 68
    North West Frontier Province (NWFP), Pakistan 68
    opium wars 67
  water 54
water challenge 54–55
water conflict *see* conflict over water
water-saving irrigation methods 55
water scarcity 61
water wars 54
wheat and barley production (Syria) 82–85
wheat detection index (WDI) 81
World Food Programme (WFP) 34, 140

Yemen
  agricultural system 131
  cereal production 134
  exacerbating factors 131
  GNP of North Yemen and South Yemen 133
  Houthis and war 139–140
  before oil 132–133
  oil discovery 137–139
  oil hothouse 133–135
  reconstruction 140–141
  remittances 133
  sorghum and millet production 132
  sustainable irrigation *see* sustainable irrigation, Yemen